Protective Relaying

ELECTRICAL ENGINEERING AND ELECTRONICS

A Series of Reference Books and Textbooks

1. Rational Fault Analysis, *edited by Richard Saeks and S. R. Liberty*
2. Nonparametric Methods in Communications, *edited by P. Papantoni-Kazakos and Dimitri Kazakos*
3. Interactive Pattern Recognition, *Yi-tzuu Chien*
4. Solid-State Electronics, *Lawrence E. Murr*
5. Electronic, Magnetic, and Thermal Properties of Solid Materials, *Klaus Schröder*
6. Magnetic-Bubble Memory Technology, *Hsu Chang*
7. Transformer and Inductor Design Handbook, *Colonel Wm. T. McLyman*
8. Electromagnetics: Classical and Modern Theory and Applications, *Samuel Seely and Alexander D. Poularikas*
9. One-Dimensional Digital Signal Processing, *Chi-Tsong Chen*
10. Interconnected Dynamical Systems, *Raymond A. DeCarlo and Richard Saeks*
11. Modern Digital Control Systems, *Raymond G. Jacquot*
12. Hybrid Circuit Design and Manufacture, *Roydn D. Jones*
13. Magnetic Core Selection for Transformers and Inductors: A User's Guide to Practice and Specification, *Colonel Wm. T. McLyman*
14. Static and Rotating Electromagnetic Devices, *Richard H. Engelmann*
15. Energy-Efficient Electric Motors: Selection and Application, *John C. Andreas*
16. Electromagnetic Compossibility, *Heinz M. Schlicke*
17. Electronics: Models, Analysis, and Systems, *James G. Gottling*

18. Digital Filter Design Handbook, *Fred J. Taylor*

19. Multivariable Control: An Introduction, *P. K. Sinha*

20. Flexible Circuits: Design and Applications, *Steve Gurley, with contributions by Carl A. Edstrom, Jr., Ray D. Greenway, and William P. Kelly*

21. Circuit Interruption: Theory and Techniques, *Thomas E. Browne, Jr.*

22. Switch Mode Power Conversion: Basic Theory and Design, *K. Kit Sum*

23. Pattern Recognition: Applications to Large Data-Set Problems, *Sing-Tze Bow*

24. Custom-Specific Integrated Circuits: Design and Fabrication, *Stanley L. Hurst*

25. Digital Circuits: Logic and Design, *Ronald C. Emery*

26. Large-Scale Control Systems: Theories and Techniques, *Magdi S. Mahmoud, Mohamed F. Hassan, and Mohamed G. Darwish*

27. Microprocessor Software Project Management, *Eli T. Fathi and Cedric V. W. Armstrong (Sponsored by Ontario Centre for Microelectronics)*

28. Low Frequency Electromagnetic Design, *Michael P. Perry*

29. Multidimensional Systems: Techniques and Applications, *edited by Spyros G. Tzafestas*

30. AC Motors for High-Performance Applications: Analysis and Control, *Sakae Yamamura*

31. Ceramic Materials for Electronics: Processing, Properties, and Applications, *edited by Relva C. Buchanan*

32. Microcomputer Bus Structures and Bus Interface Design, *Arthur L. Dexter*

33. End User's Guide to Innovative Flexible Circuit Packaging, *Jay J. Miniet*

34. Reliability Engineering for Electronic Design, *Norman B. Fuqua*

35. Design Fundamentals for Low-Voltage Distribution and Control, *Frank W. Kussy and Jack L. Warren*

36. Encapsulation of Electronic Devices and Components, *Edward R. Salmon*

37. Protective Relaying: Principles and Applications, *J. Lewis Blackburn*

Additional Volumes in Preparation

Electrical Engineering-Electronics Software

1. Transformer and Inductor Design Software for the IBM PC, *Colonel Wm. T. McLyman*
2. Transformer and Inductor Design Software for the Macintosh, *Colonel Wm. T. McLyman*
3. Digital Filter Design Software for the IBM PC, *Fred J. Taylor and Thanos Stouraitis*

Protective Relaying
PRINCIPLES AND APPLICATIONS

J. Lewis Blackburn
Bothell, Washington

MARCEL DEKKER, INC. New York and Basel

Library of Congress Cataloging-in-Publication Data

Blackburn, J. Lewis.
Protective relaying.

(Electrical engineering and electronics ; 37)
Includes bibliographies and index.
1. Protective relays. I. Title. II. Series.
TK2861.B59 1987 621.31'7 86-32869
ISBN 0-8247-7445-0

MARCEL DEKKER, INC.
270 Madison Avenue, New York, New York 10016

Current printing (last digit):
10 9 8 7

PRINTED IN THE UNITED STATES OF AMERICA

To my wife Peggy
whose patience and understanding
made this possible

Preface

Protective relaying is a vital part of any electric power system; quite unnecessary during normal operation, but very important during trouble, faults, and abnormal disturbances. Properly applied, protective relaying initiates the disconnection of the trouble area while operation and service in the rest of the system continue.

This book presents the fundamentals and basic technology of application of protective relays in electric power systems and documents the protection practices in common use. The objective is to provide a useful reference for practicing engineers and technicians as well as a practical book for college-level courses in the power field. Applications with examples are included for both utility and industrial-commercial systems operating generally above 480 volts.

Protective relay designs vary with different manufacturers and are constantly changing, especially as solid state technology impacts this area. However, protective relaying applications remain basically the same: relatively independent of design and their trends. As a result design aspects are not emphasized in this book. This area is best covered by individual manufacturer's information.

Protective relaying can be considered a vertical specialty with a horizontal vantage point, thus, while specialized, it is involved with and requires knowledge of all of the equipment required in the generation, transmission, distribution, and utilization of electrical power. In addition it requires an understanding of how the system performs normally as well as during faults and abnormal conditions. As a result this subject provides an excellent background for specialized study in any of the individual areas, and is especially important for system planning, operation, and management.

Fifty years of friends, associates and students within Westinghouse, the IEEE, CIGRE, many utilities, and industrial companies around the world have directly or indirectly contributed to this book. Their contributions and support are gratefully acknowledged.

Special acknowledgement and thanks are extended to Rich Duncan for his enthusiastic encouragement and support during the preparation of this manuscript, and to W. A. Elmore, T. D. Estes, C. H. Griffin, R. E. Hart, C. J. Heffernan and H. J. Li for photographic and additional technical help.

And my gratitude to Dr. Eileen Gardiner of Marcel Dekker, Inc., who most patiently encouraged and supported this effort.

J. Lewis Blackburn

Contents

Protective Relaying

1

Introduction and General Philosophies

1.1 INTRODUCTION AND DEFINITIONS

What is a relay; more specifically, what is a protective relay? The Institute of Electrical and Electronic Engineers (IEEE) defines a *relay* as "an electric device that is designed to interpret input conditions in a prescribed manner and after specified conditions are met to respond to cause contact operation or similar abrupt change in associated electric control circuits." A note amplifies: "Inputs are usually electric, but may be mechanical, thermal, or other quantities. Limit switches and similar simple devices are not relays" (IEEE 100-1984).

Relays are utilized in all aspects of activity; the home, communication, transportation, commerce, and industry, to name a few. Wherever electricity is utilized, there is a high probability that relays are involved. They are used in heating, air conditioning, stoves, dishwashers, clothes washers and dryers, elevators, telephone networks, traffic controls, transportation vehicles, automatic process systems, robotics, space activities, and many other applications.

In this book we focus on one of the more interesting and sophisticated applications of relays, the protection of electric power systems. The IEEE defines a protective relay as "a relay whose function is to detect defective lines or apparatus or other power system conditions of an abnormal or dangerous nature and to initiate appropriate control circuit action" (IEEE 100-1984).

Fuses are also used in protection. IEEE defines a *fuse* as "an overcurrent protective device with a circuit-opening fusible part that is heated and severed by the passage of the overcurrent through it" (IEEE 100-1984).

Thus protective relays and their associated systems (often abbreviated simply as relays and relay systems) are compact units of analog, discrete component, and/or digital networks connected through a power system for the purpose of sensing problems. They are utilized in all parts of electric power systems, together with fuses for the detection of intolerable conditions.

A primary objective of all power systems is to maintain a very high level of continuity of service, and when intolerable conditions occur, to minimize the outage times. Loss of power, voltage dips, and overvoltages will occur, however, because it is impossible as well as impractical to avoid the consequences of natural events, physical accidents, equipment failure, or misoperation due to human error. Many of these result in faults: inadvertent, accidental connections or "flashovers" between the phase wires or from the phase wire(s) to ground.

Natural events that can cause short circuits (faults) are lightning (induced voltage or direct strikes), wind, ice, earthquake, fire, explosions, falling trees, flying objects, physical contact by animals, and contamination. Accidents include faults resulting from vehicles hitting poles or contacting live equipment, unfortunately people contacting live equipment, digging into underground cables, and so on. Considerable effort is made to minimize damage possibilities, but the elimination of all such problems is not yet achievable.

A dramatic illustration of the need and importance of power system protection is shown in Figure 1.1. This spectacular lightning stroke occured over Seattle during a storm on July 31, 1984 and in a region where lightning is infrequent. The isokeraunic charts for this area of the Pacific Northwest indicate that the probability of storm days when thunder is heard is five or less per year. (*Electrical Transmission and Distribution Reference Book*, 4th ed. Westinghouse Electric Corp., East Pittsburg, PA, 1964). While some 12,000 homes lost power during this storm, no major damage nor prolonged outages were experienced by the local utilities. Fortunately, lightning protection and many relays operated to minimize the problems.

The large majority of faults in an electrical utility system with a network of overhead lines are one-phase-to-ground faults resulting primarily from lightning-induced transient high voltage, and falling trees and tree limbs. In the overhead distribution systems, momentary tree contact caused by wind is another major cause of faults. Ice, freezing snow, and wind during severe storms can cause many faults and damage.

In many cases the flashover caused by such events does not result in permanent damage if the circuit is interrupted quickly. A common practice is to open the faulted circuit, permit the arc to extinguish naturally, and then reclose the circuit. In many cases, this enhances the continuity of service by causing only a momentary outage and voltage dip. Typical outage times are in the order of 1/2 to 1 or 2 minutes rather than many minutes and hours.

System faults usually, but not always, provide significant changes in the system quantities, which can be used to distinguish between tolerable and intolerable system conditions. These changing quantities include overcurrent, over-or undervoltage, power, power factor or phase angle, power or current direction, impedance, frequency, temperature, physical movements, pressure, and contamination of the insulating

Fig. 1.1 Lightning over Seattle — a vivid illustration of the importance of power system protection (Greg Gilbert/Seattle Times photo).

quantities. The most common fault indicator is a sudden and generally significant increase in the current, so overcurrent protection is widely used.

Protection is the science, skill, and art of applying and setting relays and/or fuses to provide maximum sensitivity to faults and undesirable conditions, but to avoid their operation on all permissible or tolerable conditions. The basic approach throughout this book is to define the tolerable and intolerable conditions that may exist and to look for definite differences ("handles") that the relays or fuses can sense.

It is important to recognize that the "time window" of decision in power system protection is very narrow, and when faults have occurred, a recheck for verification or a decision-making procedure that involves additional time is not desirable. It is vital (1) that a correct decision be made by the protective device as to whether the trouble is intolerable and thus demands quick action, or whether it is a tolerable or transient situation that the system can absorb; and (2) that the protective device operate if necessary to isolate the trouble area quickly and with a minimum of system disturbance. This trouble time may be and often is associated with high extraneous "noise," which must not "fool" the device or cause incorrect operation.

Both failure to operate and incorrect operation can result in major system upsets involving increased equipment damage, increased personnel hazards, and possible long interruption of service. These stringent requirements with high potential consequences tend to make protection engineers somewhat conservative. They often wish to continue using protection equipment that has a long history of reliability.

Problems can and do occur in protective equipment; nothing is perfect. To minimize the potential catastrophic problems that can result in the power system from a protection failure, the practice is to use several relays or relay systems operating in parallel. These can be at the same location (primary backup), at the same station (local backup), and/or at various remote stations (remote backup). All three are used together in many applications. In higher-voltage power systems this concept is extended by providing separate current and/or voltage measuring devices, separate trip coils on the circuit breakers, and separate tripping battery sources.

The various protective devices must be properly coordinated so that the primary relays assigned to operate at the first sign of trouble in their assigned protective zone operate first. Should they fail, various backup systems must be available and able to operate to clear the trouble. An adequate, high-protection redundancy capability is very important.

1.2 TYPICAL PROTECTIVE RELAYS AND RELAY SYSTEMS

A typical logic representation of relay is shown in Fig. 1.2. The components can be electromechanical, solid state, or both. The logic functions are quite general, so that in any particular unit they may be combined or in some cases not required.

Specific designs and features vary widely with application requirements, different manufacturers, and the time period of the particular design. Originally, all protective relays were of the electromechanical type, and these are still in wide use, but solid-state designs are becoming more common. Although this trend will continue, it may be a long time before electromechanical devices are completely superseded. Solid-state devices are particularly advantageous in terms of accurate timing, frequency sensing, and for multiple logic decision systems, such as "pilot" protection. Solid-state devices are used in low-voltage protection systems where the relay and circuit breaker are a common unit. Here, electromechanical relays were generally relatively inaccurate, sometimes insensitive, and difficult to test and maintain. It is in these areas where solid-state relays are most commonly used today.

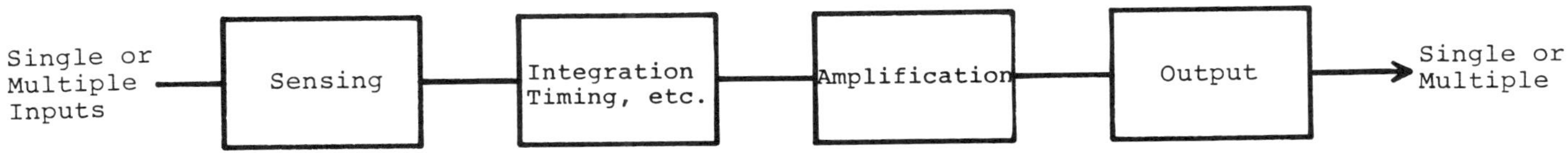

Fig. 1.2 Logic representation of electric relays.

The fundamental characteristics of relay designs necessary to understand the applications are outlined in Chapter 6 and augmented as required in subsequent chapters. For those not familiar with protective relays, a representative collection of electromechanical types in current use is shown in Fig. 1.3. These relays have been removed from their mounting cases. The cases are normally permanently mounted on switchboard panels such as that shown in Fig. 1.4. The relays on this panel provide backup protection for two 500-kV transmission lines. The relay units or chassis, similar to those shown in Fig. 1.3 have been inserted in their respective cases. Connections to the external circuits are made by a connecting plug or knife switches and then a protective cover installed. The relay second from the top right does not have its cover in place.

Three sections of a control-protection switchboard for 500-kV circuits are shown in Figure 1.5. This illustrates the combination of electromechanical and solid-state relays. Solid-state relays commonly are rack mounted, as shown. Doors provide access to the removable printed-circuit boards and test-adjustment facilities.

Fig. 1.3 Typical representative protective relay units used for power system protection. The chassis are shown without switchboard mounting cases. (Courtesy of Puget Sound Power and Light.)

Fig. 1.4 Typical relay protection for the backup protection of two 500-kV transmission lines utilizing electromechanical protective relays. (Courtesy of Georgia Power Company.)

Fig. 1.5 Typical relay protection, control, and instrumentation switchboards in a modern power station. Both electromechanical and solid-state relays are shown. (Courtesy of Georgia Power Company.)

Two panels of solid-state relays during factory test are illustrated in Fig. 1-6. The front covers have been removed. The right panel has one of the relay units pulled out.

1.3 TYPICAL POWER CIRCUIT BREAKERS

Protective relays provide the "brains" to sense trouble, but as low-energy devices they are not able to open and isolate the problem area of the power system. Circuit breakers and various types of circuit interrupters, including motor contactors and motor controllers, are used for this and provide the "muscle" for fault isolation. Thus protective relays and circuit breaker-interrupting devices are a "team," both necessary for the prompt isolation of a trouble area or damaged equipment. A protective relay without a circuit breaker has no basic value except possibly for alarm. Similarly, a circuit breaker without relays has minimum value, that being for manually energizing or de-energizing a circuit or equipment.

Typical circuit breakers used to isolate a faulted or damaged area are shown in Figs. 1.7 and 1.8. Figure 1.7 shows a long row of three-phase 115-kV oil circuit breakers with pneumatic controls in an outdoor substation. These are known as dead-tank breakers; the tank or breaker housing is at ground potential. Toroidal wound-bushing current transformers are mounted in the pods just under the fluted porcelain insulators at the top of the tank. This general type is in wide use in many different designs and variations. The media employed for circuit interruption include air, air blast, compressed air, gas, and vacuum in addition to oil.

Figure 1.8 shows a 500-kV live-tank circuit breaker. Here the interrupting mechanisms and housing is at the high-voltage level and insulated from ground through the porcelain columns. The current transformers are mounted on separate porcelain columns as shown at the left of each phase breaker.

Dead-tank breakers such as that illustrated in Fig. 1.7 usually have a single trip coil that initiates opening all three-phase breaker poles simultaneously. The live-tank types generally have a trip coil and mechanism for operating each pole or phase independently. This is evident in Fig. 1.8. For these types the relays must energize all three trip coils to open the three-phase power circuit. It is possible to connect the three trip coils in parallel or in series for tripping all three poles. Three trip coils in series is preferred. This arrangement permits easier monitoring of circuit continuity and requires less trip current.

In the United States the practice for many years has been to open all three phases for all types of faults even though one or two of the phases may not be involved in the fault. Connecting the three

Fig. 1.6 Typical solid-state relays for the protection of transmission lines during factory testing. (Uniflex Systems, courtesy of Westinghouse Electric Corporation.)

Fig. 1.7. Row of typical three-phase 115-kV oil circuit breakers. The pneumatic operating mechanism is visible within the open cabinet. (Courtesy of Puget Sound Power and Light Company.)

independent trip coils in series for this purpose is known as independent pole tripping.

Any failure of the opening mechanism on single-trip coil breakers would require the backup protection to isolate the trouble area by opening all other breakers feeding the fault. With independent pole tripping there is a low probability that all three poles would fail to open by protective relay action; hopefully, at least one or two of the poles would open correctly. Should the fault be three-phase, which is the most severe fault for a power system, the opening of at least one or two poles reduces this type of fault to a less severe one from a system stability standpoint, either double phase or single phase.

Since most transmission-line faults are transient single-line-to-ground type, opening only the faulted phase would clear it. With a transient fault such as that resulting from lightning-induced overvoltage, immediate reclosing of the open faulted phase would restore three-phase service. Known as single-pole tripping, this tends to reduce the shock on the power system. It is discussed further in Chapters 13 and 14.

Fig. 1.8 Typical 500-kV live-tank circuit breaker. Two interrupting mechanisms operated in series are mounted on insulating columns, with the associated current transformer on its column at the left. These are single-phase units; three are shown for the three-phase system. (Courtesy of Georgia Power Company.)

As indicated above, at the lower voltages the circuit breaker (interrupter) and relays frequently are combined into a single operating unit. The circuit-breaker switches commonly installed in the service entrance cabinet in modern residential homes and commercial buildings are typical examples. In general, this type of arrangement is used up through 480 to 600 V. Primarily, the protection is overcurrent, although overvoltage may be included. The low accuracy and difficulties of calibration and testing have resulted in a wider application of solid state in these designs. Since the relay units and breaker are together physically, and the voltage exposure level is low, the problems of extraneous voltages and noise that affect solid-state designs are minimized.

1.4 NOMENCLATURE AND DEVICE NUMBERS

The nomenclature and abbreviations used generally follow common practice in the United States. The functions of various relays and equipment are identified by the ANSI/IEEE standardized device function numbers (IEEEC37.2-1979). A brief review is in order.

The phases of the three-phase system are designated as A, B, C or a, b, c rather than 1, 2, 3, also used in the United States, or r, s, t, used in Europe. 1, 2, 3 is avoided because 1 also designates positive sequence, and 2, negative sequence. Letters avoid possible confusion. Capital letters are used on one side of wye-delta transformer banks, with lowercase letters on the other side. Although normally not followed in practice, this emphasizes that there is a phase shift and voltage difference across the transformer bank.

Device numbers with suffix letter(s) provide convenient identification of the basic functions of electrical equipment, such as circuit breakers, relays, switches, and so on. Where several units of the same type of device are used in a circuit or system, a number preceding the device number is used to differentiate between them. Letters following the device number provide additional information on the application, use, or actuating quantities. Unfortunately, the same letter may be used with quite different connotations or meanings. Normally, this will be clear from the use.

Letters and abbreviations frequently used include the following:

A	Alarm
AC or ac	Alternating current
B	Bus, battery, blower
BP	Bypass
BT	Bus tie
C	Current, close, control, capacitor
CC	Closing coil, coupling capacitor
CS	Control switch, contactor switch
CT	Current transformer
CCVT	Coupling capacitor voltage device
D	Down, direct, discharge
DC or dc	Direct current
E	Exciter, excitation
F	Field, feeder, fan
G*	Ground, generator
GND	Ground
H	Heater, housing
L	Line, lower, level, liquid
M	Motor, metering
MOC	Mechanism-operated contact
MOS	Motor-operated switch
N*	Neutral, network
NC	Normally closed
NO	Normally open
O	Open
P	Power, pressure

PB	Pushbutton
PF	Power factor
R	Raise, reactor
S	Speed
T	Transformer, trip
TC	Trip coil
U	Up
V	Voltage, vacuum
VAR	Reactive power
VT	Voltage transformer
W	Watts, water
X,Y,Z	Auxiliary relays

Device numbers frequently used are listed below. A complete list and definitions are given in Standard IEEE C37.2-1979.

1. Master element: normally used for hand-operated devices. A common use is the spring-return-to-center control switch for circuit breakers, where the switch contacts are 101T (trip), 101c (close), 101SC (closed when turned to close and remains closed when released, opens when turned to trip and remains open when released). Where several breakers are involved they are identified by 101, 201, 301, and so on.
2. Time-delay starting or closing relay: except device functions 48, 62, and 79
3. Checking or interlocking relay
4. Master contactor
5. Stopping device
6. Starting circuit breaker
7. Anode circuit breaker
8. Control power disconnecting device
9. Reversing device
10. Unit sequence switch
12. Overspeed device
13. Synchronous-speed device
14. Underspeed device
15. Speed- or frequency-matching device
17. Shunting or discharge switch
18. Accelerating or decelerating device
19. Starting-to-running transition contactor

---

* N and G (or n and g) are used in circuits involving ground. A common convention but not standardized is the use of G when a relay is connected to a CT in the grounded neutral circuit, and N when connected in the neutral of three wye-connected CTs. Similar usage is applied to voltage.

20. Electrically operated valve
21. Distance relay
22. Equalizer circuit breaker
23. Temperature control device
24. Volts/hertz relay
25. Synchronizing or synchronism-check device
26. Apparatus thermal device
27. Undervoltage relay
28. Flame detector
29. Isolating contactor
30. Annunciator relay
31. Separate excitation device
32. Directional power relay
33. Position switch
34. Master sequence device
36. Polarity or polarizing voltage device
37. Undercurrent or underpower relay
40. Field relay
41. Field circuit breaker
42. Running circuit breaker
43. Manual transfer or selector device
46. Reverse-phase or phase-balance relay
47. Phase-sequence voltage relay
48. Incomplete-sequence relay
49. Machine or transformer thermal relay
50. Instantaneous overcurrent or rate-of-rise relay
51. Ac time overcurrent relay
52. Ac circuit breaker: mechanism-operated contacts are:
    a. 52a, 52aa: open when breaker contacts open, closed when breaker contacts closed
    b. 52b, 52bb: closed when breaker contacts open, open when breaker contacts closed

    52aa and 52bb operate just as mechanism motion starts; known as high-speed contacts
53. Exciter or dc generator relay
55. Power factor relay
56. Field application relay
57. Short-circuiting or grounding device
59. Overvoltage relay
60. Voltage or current balance relay
61. current, phase balance
62. Time-delay stopping or opening relay
63. Pressure switch
64. Ground detector relay
65. Governor
66. Notching or jogging device
67. AC directional overcurrent relay
68. Blocking relay

69. Permissive control device
70. Rheostat
71. Level switch
72. Dc circuit breaker
74. Alarm relay
76. Dc overcurrent relay
77. Pulse transmitter
78. Phase-angle measuring or out-of-step protective relay
79. Ac reclosing relay
80. Flow switch
81. Frequency relay
82. Dc reclosing relay
83. Automatic selective control or transfer relay
84. Operating mechanism
85. Carrier or pilot-wire receiver relay
86. Lockout relay
87. Differential protective relay
89. Line switch
90. Regulating device
91. Voltage directional relay
92. Voltage and power directional relay
93. Field-changing contactor
94. Tripping or trip-free relay

1.5 TYPICAL RELAY AND CIRCUIT BREAKER CONNECTIONS

Protective relays utilizing electrical quantities are connected to the power system through current (CT) and/or voltage (VT) transformers. These input devices or instrument transformers provide insulation from the high-power system voltages, and reduce the magnitudes to practical secondary levels for the relays. An important element of the protection system, these units are discussed in Chapter 5. In circuit schematics and diagrams they are represented as shown in Fig. 1-9. This diagram shows a typical "one-line" ac schematic and a dc trip circuit schematic.

The protective relay system is connected to the ac power system through the current transformers commonly associated with the circuit breaker, and if necessary to the voltage transformers. These are shown connected to the station ac bus, but often at the higher voltages the voltage devices are connected to the transmission line. The circuit breaker is designated as device 52 following the ANSI/IEEE device number system (IEEE C37.2-1979).

In the dc schematic the contacts are always shown in their de-energized position. Thus when the circuit breaker is closed and in service, its 52a contact is closed. When a system fault operates the

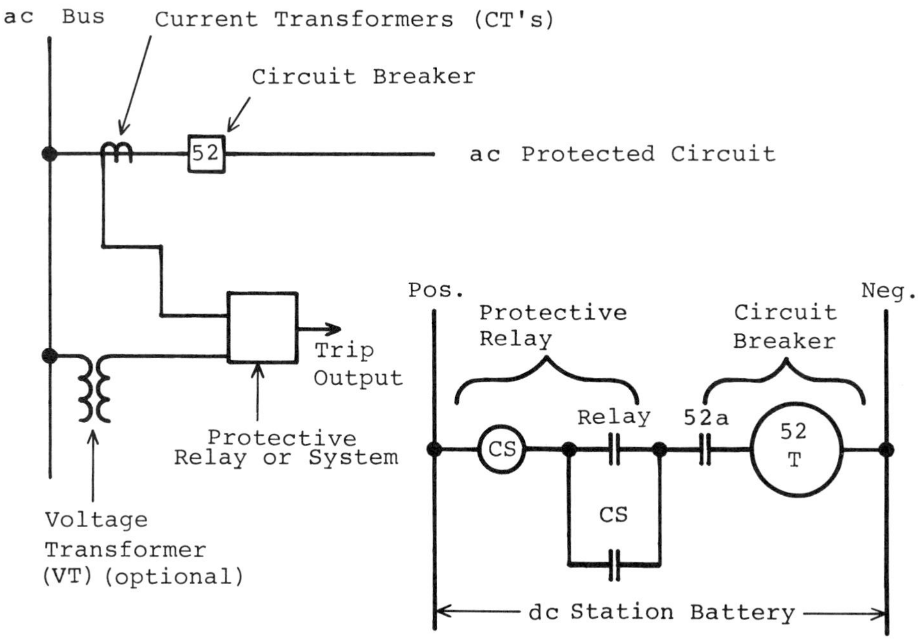

Fig. 1.9 Typical single-line ac connections of a protective relay with its dc trip schematic.

protective relay, its output contact closes to energize the circuit breaker trip coil 52T, which functions to open the breaker main contacts and deenergize the connected power circuit.

The relay contacts basically are not designed to interrupt the circuit breaker trip coil current, so a common practice is to use an auxiliary dc-operated relay designated CS (contactor switch) to "seal in" or bypass the protective relay contacts as shown. When the circuit breaker opens, the 52a switch will open to deenergize the trip coil 52T. The interruption of the fault by the circuit breaker will have opened the protective relay contacts before 52a contact opens.

The various power-interrupting devices are operated either by the overcurrent that passes through them during a fault, or by a dc-operated trip coil such as that shown in Fig. 1.9. The first types are designated as series trip, direct acting, direct release, indirect release, and overcurrent release. Usually, these have built-in overcurrent relay units which determine the level of the ac current at and above which their contacts will open. All of these types are used at the lower power system voltage level.

At the higher power system voltages each station where circuit breakers are installed has a station battery to supply direct current to the breaker trip coils, the control and protective relay circuits as required, emergency alarms and lighting, and so on. In the United States this is generally 125 V dc; 250 V dc is used in some large power stations, and 48 V dc is often used for electronic and solid-state devices. This dc supply is another vital part of the protection system and requires careful attention and maintenance for high system and protection reliability.

Many protective relays are packaged as individual phase and ground units, so for complete phase- and ground-fault protection, four units are commonly used. Typical three-phase ac connections for a set of relays and their associated current and voltage transformers are shown in Fig. 1.10.

1.6 BASIC OBJECTIVES OF SYSTEM PROTECTION

The fundamental and basic objective of system protection is to provide isolation of a problem area in the system quickly so that as much as possible of the rest of the power system is left to continue service. Within this context, there are five basic facets of protective relay application.

Before discussing these it should be noted that the use of the term "protection" does not indicate or imply that the protection equipment can prevent trouble, such as faults and equipment failures. It cannot anticipate trouble. The protective relays act only after an abnormal or intolerable condition has occurred, with sufficient indication to permit their operation. Thus "protection" does not mean "prevention" but rather minimizing the duration of the trouble and limiting the damage, outage time, and related problems that may otherwise result.

The five basic facets are:

1. *Reliability:* assurance that the protection will perform correctly
2. *Selectivity:* maximum continuity of service with minimum system disconnection
3. *Speed of operation:* minimum fault duration and consequent equipment damage
4. *Simplicity:* minimum protective equipment and associated circuitry to achieve the protection objectives
5. *Economics:* maximum protection at minimum total cost

As these are the underlying foundation of all protection, further discussion is in order.

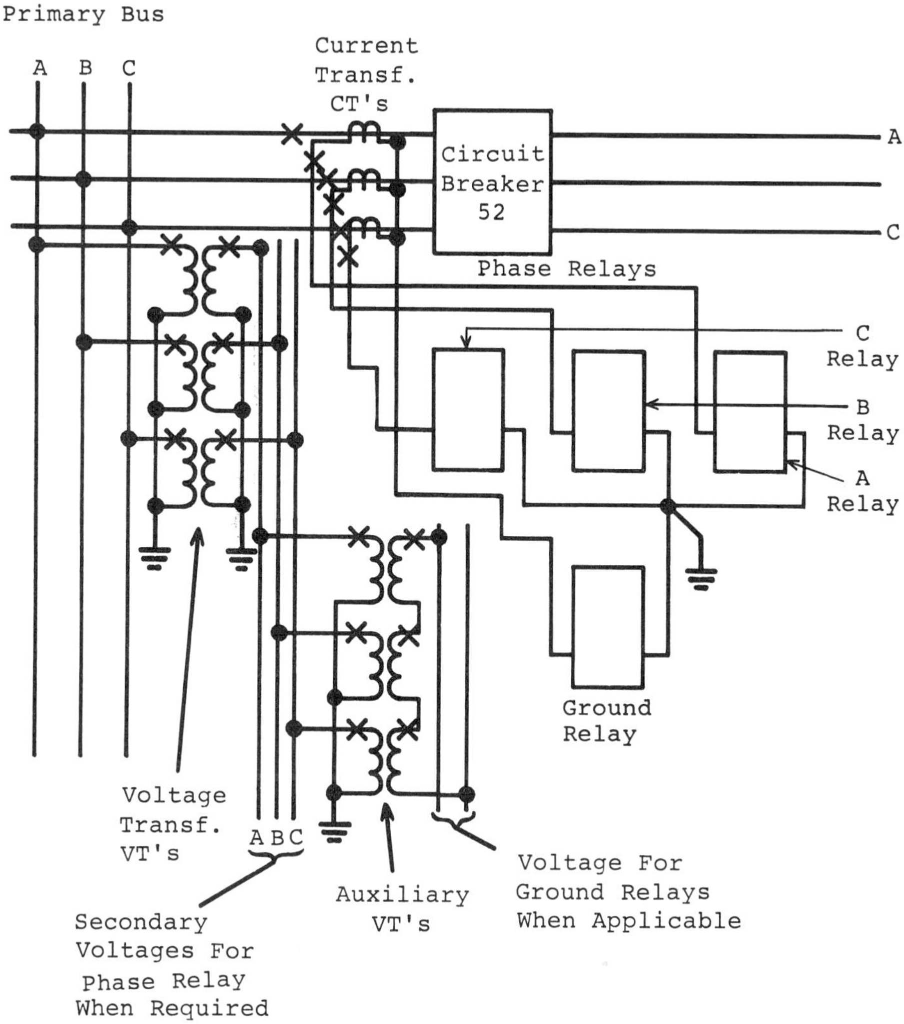

Fig. 1.10 Typical three-phase ac connections of a set of phase and ground relays for the protection of an ac power system.

1.6.1 Reliability

Reliability has two aspects: *dependability* and *security*. Dependability is defined as "the degree of certainty that a relay or relay system will operate correctly" (IEEE C37.2-1979). Security "relates to the degree of certainty that a relay or relay system will not operate incorrectly" (IEEE C37.2-1979). In other words, dependability indicates the ability of the protection system to perform correctly when required, while security is its ability to avoid unnecessary operation during normal day-after-day operation, and for faults and problems outside the designated zone of operation. There is often a very fine line between the tolerable transients which the power system can operate through successfully, and those, such as light faults, that may develop and result in a major problem if not quickly isolated. Thus the protection must be secure (not operate on tolerable transients), yet dependable (operate on intolerable transients and permanent faults). It is these somewhat conflicting requirements, together with the speculation of what trouble may occur when and where, that help make power system protection a most interesting technical science and art.

Dependability is easy to ascertain by testing the protection system to assure that it will operate as intended when the operating thresholds are exceeded. Security is more difficult to ascertain. There can be almost an infinite variety of transients that might upset the protective system, and predetermination of all these possibilities is difficult or impossible.

Manufacturers often use elaborate power system simulations, computers, and sometimes staged fault tests on energized power systems to check both dependability and security. The practical and best answer to both security and dependability is the background experience of the designers, confirmed by field experience. Thus actual in-service installations provide the best and final laboratory. This should only confirm the reliability, not be used basically for the development.

As a generality, enhancing security tends to decrease the dependability, and vice versa. As an illustration, the single relay trip contact shown in Fig. 1.9 represents high dependability, but it has the potential of being accidentally closed by unanticipated transients or human error and thus result in an undesired operation. To minimize this potential problem, a second relay, such as a fault detector, can be used with its operating contact in series in the dc trip circuit. Now both contacts must close to trip the circuit breaker, which should occur for intolerable conditions or faults. This has increased security since it is less likely that extraneous transients or problems would cause both relays to operate simultaneously. However, the dependability has been decreased, as it now requires two relays to operate correctly. This arrangement is used, as the dependability is still quite high, together with improved security.

Security is thus very important (as is dependability), as relays are connected for their lifetime to the power system as "silent sentinels," "waiting" for intolerable conditions and experiencing all the transients and external faults not in their operating zone. The hope always is that there will be no faults or intolerable conditions and hence no occasion for the relays to operate. Fortunately, there are revlatively few faults, on average, in a power system. It is estimated that, in general, the cumulative operating time (the times the relay is sensing and operating for an internal fault) during a relay's lifetime probably averages on the order of seconds to a few minutes, depending on the speed of the particular relay type. This contrasts dramatically with a life of over 30 years for many electromechanical relays. Thus relays basically do not wear out from operations — indeed, more wear, as such, will occur from maintenance testing and similar use.

Similar experience will undoubtedly result with solid-state relays, except that because of the still rapidly changing technology, the lifetime of many of these relays will probably be much shorter.

In general, experiences in power systems, both large and small, utilities and industrials, indicate that their protective relay systems have greater than 99% reliability — a commendable tribute to the industry.

1.6.2 Selectivity

Relays have an assigned area known as the primary protection zone but may properly operate in response to conditions outside this zone. In these cases they provide backup protection for the area outside their primary zone. This is designated as the backup or overreached zone.

Selectivity (also known as relay coordination) is the process of applying and setting the protective relays which overreach other relays so that they operate as fast as possible within their primary zone but have delayed operation in their backup zone. This is necessary to permit the primary relays assigned to this backup or overreached area time to operate. Otherwise, both sets of relays may operate for faults in this overreached area; the assigned primary relays for the area and the backup relays. Operation of the backup protection is incorrect and undesirable unless the primary protection of that area fails to clear the fault. Thus selectivity or relay coordination is important to assure maximum service continuity with minimum system disconnection. This process is discussed in more detail in later chapters.

1.6.3 Speed

Obviously, it is desirable that the protection isolate a trouble zone as rapidly as possible. In some applications this is not difficult, but in others, particularly where selectivity is involved, faster operation can be accomplished by more complex and generally higher-cost protection. Zero-time or very high speed protection, although inherently desirable, may result in an increased number of undesired operations. As a broad generality, the faster the speed of operation, the higher the probability of incorrect operation. Time, generally a very small amount, remains one of the best means of distinguishing between tolerable and untolerable transients.

A high-speed relay is one that operates in less than 50 ms (three cycles on a 60-Hz basis) (IEEE 100-1984). The term "instantaneous" is defined to indicate that no (time) delay is purposely introduced in the action of the device (IEEE 100-1984). In practice, "instantaneous" and "high-speed" are used interchangeably to describe protective relays that operate in 50 ms or less.

Modern high-speed circuit breakers operate in the range of 17 to 50 ms (one to three cycles at 60 Hz); others operate at less than 83 ms (five cycles at 60 Hz). Thus the total clearing time (relays plus breaker) typically ranges from approximately 35 to 130 ms (two to eight cycles at 60 Hz).

In the lower-voltage systems, where time coordination is required between protective relays, relay operating times generally will be slower; typically on the order of 0.2 to 1.5 sec. for the primary zone. Primary-zone relay times greater than 1.5 to 2.0 sec. are unusual for faults in this zone, but they are possible and do exist. Thus speed is important, but it is not always absolutely required nor always practical to obtain high speed without additional cost and complexity, which may not be justified.

1.6.4 Simplicity

A protective relay system should be kept as simple and straightforward as possible while still accomplishing its intended goals. Each added unit or component which may offer enhancement of the protection but which is not necessarily basic to the protection requirements should be considered very carefully. Each addition provides a potential source of trouble and added maintenance. As has been emphasized, incorrect operation or unavailability of the protection can result in catastrophic problems in a power system. Problems in the protective system can have a great impact on the system — in general, probably more so than for any other power system component.

The increasing use of solid-state and digital technologies in protective relaying provides many convenient possibilities for increased sophistication. Some will enhance the protection, others add things that are nice to have. All adjuncts should be evaluated carefully to assure that they really, and significantly, contribute to improved system protection.

1.6.5 Economics

It is fundamental to obtain the maximum protection for the minimum cost, and cost is always a major factor. The lowest-priced initial cost-protective system may not be the most reliable system; further, it may involve greater difficulties in installation and in operation, as well as higher maintenance costs. Protection "costs" are considered "high" when considered alone, but they should be evaluated in the light of the much higher cost of the equipment they are protecting, and the cost of an outage and/or loss of the protected equipment through improper protection. Saving to reduce the first costs can result in spending many more times this saving to repair or replace equipment damaged or lost because of inadequate or improper protection.

1.6.6 General Summary

It would indeed be Utopia if all five basic objectives could be achieved to their maximum level. Real-life practical considerations require common sense and compromise. Thus the protection engineer must maximize these as a group for the protection problem at hand and for the requirements of the system. This is an exciting challenge that will produce many different approaches and answers.

1.7 FACTORS AFFECTING THE PROTECTION SYSTEM

There are four major factors that influence protective relaying: (1) economics, (2) the personality of the relay engineer and the character of the power system, (3) the location of disconnection and isolating devices (circuit breakers and switches) and input devices (current and/or voltage transformers), and (4) available measures of faults.

1.7.1 Economics

Economics has been discussed in Section 1.6.5 and is always very important. Protection does not produce revenue and is not required for normal system operation. Fortunately, faults and troubles are relatively infrequent, so it is easy to decide not to spend money on protection since there have not been any problems. Certainly, the protection engineer hopes that the protection will never be called on to operate, but when trouble does occur, protection is vital for the life of the system. A single fault where the protection promptly and correctly isolates the trouble zone, thereby minimizing the outage time and reducing equipment damage, can more than pay for the protection required.

1.7.2 The Personality Factor

What, when, and where an intolerable condition will occur in the power system is unpredictable. Almost an infinity of possibilities exist. Thus the engineer must design the protective system for the most probable events based on past experiences, anticipated possibilities that seem most likely to occur, and the equipment manufacturer's recommendations, all well seasoned by good practical judgment. This tends to make protection an art as well as a technical science. Since the personalities of protection engineers as well as that of the power system as reflected by the management and operating considerations are different, so is the protection that results. Although there is much common technology, protection systems and practices are far from standardized. Thus protection reflects the "personality" of the engineers and the system, again making the art and practice of system protection most interesting.

1.7.3 Location of Disconnecting and Input Devices

Protection can be applied only where there are circuit breakers or similar devices to enable isolation of the trouble area and where current and voltage transformers, where required, are available to provide information about faults and trouble in the power system.

1.7.4 Available Fault Indicators

The troubles, faults, and intolerable conditions must provide a distinguishable difference from the normal operating or tolerable condition. Some signal or change in the quantities — "handle" — is necessary to cause relay operation or detection of the problem. To repeat, common handles available are current, voltage, impedance, reactance, power, power factor, power or current direction, frequency, temperature, and pressure. Any significant change in these may provide a means to detect abnormal conditions and so be utilized for relay operation.

The key to the selection and application of protection is first to determine what measures (handles) exist to distinguish between tolerable and intolerable conditions. From this information a relay or relay system can be found or designed if necessary to operate on the detectable difference(s).

If a significant difference does not exist between the normal and abnormal conditions, protection is limited at best or not possible at all, An example of this exists in distribution systems, where accidents or storms may result in an energized line being near or on the ground. This is totally intolerable, but the fault current can be very small or zero and all other system parameters, such as voltage, power, and frequency, may remain within normal limits. Thus in these situations no handle exists for any type of relay to detect and isolate the intolerable condition.

1.8 CLASSIFICATION OF RELAYS

Relays may be classified in several different ways, such as by function, input, performance characteristics, or operating principles. Classification by function is most common. There are five basic function types: (1) protective; (2) regulating; (3) reclosing, synchronism check, synchronizing; (4) monitoring; and (5) auxiliary.

1.8.1 Protective Relays

Protective relays and associated systems (and fuses) operate on the intolerable power system conditions and are the main thrust of this book. They are applied to *all* parts of the power system; generators, buses, transformers, transmission lines, distribution lines and feeders, motors and utilization loads, capacitor banks, and reactors. For the most part, the relays discussed are separate devices that are connected to the power system through current and voltage transformers from the highest system voltage (765 kV at present) down to service levels of 480 V. In general, distribution equipment below 480 V is protected by fuses or protection devices integral with the equipment. Such devices are not discussed in depth here.

1.8.2 Regulating Relays

Regulating relays are associated with tap changers on transformers and on governors of generating equipment to control the voltage levels with varying loads. Regulating relays are used during normal system operation and do not respond to system faults unless the faults are left on the system far too long. This is not normal. This type relay is not discussed in this book.

1.8.3 Reclosing, Synchronism Check, Synchronizing Relays

Reclosing, synchronism check, and synchronizing relays were formerly classed as "programming," but since this term is now widely used in a different context as related to computers, a name change has been made. Relays of this type are used in energizing or restoring lines to service after an outage and in interconnecting preenergized parts of systems.

1.8.4 Monitoring Relays

Monitoring relays are used to verify conditions in the power system or in the protective system. Examples in power systems are fault detectors, voltage check, or directional-sensing units which confirm power system conditions but do not directly sense the fault or trouble. In a protection system they are used to monitor the continuity of circuits, such as pilot wires and trip circuits. In general, alarm units serve as monitoring functions.

1.8.5 Auxiliary Relays

Auxiliary units are used throughout a protective system for a variety of purposes. Generally there are two categories: contact multiplication and circuit isolation. In relaying and control systems there are frequent requirements for (1) more outputs for multiple tripping, alarms, and operating other equipment, such as recording and data acquisition, lockout, and so on, (2) contacts that will handle higher currents or voltages in the secondary systems; and (3) electrical and magnetic isolation of several secondary circuits.

The seal-in (CS) relay of Fig. 1.9 is an auxiliary relay application. The trip and closing relays used with circuit breakers are auxiliary relays.

1.8.6 Other Relay Classifications

Protective relays classified by *input* are known as current, voltage, power, frequency, and temperature relays. Those classified by *operating* principle include electromechanical, solid-state, digital, percentage differential, multirestraint, and product units. Those classified by *performance characteristics* are known as distance, reactance, directional overcurrent, inverse time, phase, ground, definite, high speed, slow speed, phase comparison, overcurrent, undervoltage, and overvoltage, to name a few.

1.9 PROTECTIVE RELAY PERFORMANCE

It is difficult to specify or evaluate protective relay performance by statistical techniques. As indicated, relays are connected to the power system and energized, but they are basically inactive (i.e., in a quiescent mode) until an intolerable situation occurs within their operating zones. Thus it is impossible to know from the experience of normal operation whether the relays will respond and operate correctly at the onset of an emergency. This is one of the primary reasons for regular maintenance; with relatively few faults the question arises: will the protection operate correctly when a fault occurs? If a failure has occurred or a defective condition exists in the relay system, it may not be evident until the next fault, unless it happens to be a type that can cause misoperation on its occurrence. Hence maintenance for relays tends primarily to be for "assurance" more than for problem correction, with little wear occurring from their normal infrequent operation. As discussed previously, there are many relays still on the job after 30 to 40 or more years in service.

When a fault occurs in a power system the primary relays for the faulted zone are expected to operate and initiate isolation. However, many relays surrounding the trouble area will be alerted by the fault and begin to operate. They do not initiate tripping if the primary relays operate correctly. Information on these backup relay operations is not available when the relays do not trip; nevertheless, they are an important part of the system protection — the backup in case of a primary protection failure.

Many protection systems employ several primary relays that can respond to some faults, but frequently only one of the relays actually initates tripping of the circuit breaker. Again, the other relays are in a correct operating mode, although there may be no direct evidence of this.

Relay performance is documented by those relays that provide direct or specific evidence of operation, and can be classified as: (1) correct operation, (a) wanted or (b) unwanted, (2)Incorrect operation, either failure to trip or false tripping, (a) unwanted or (b) acceptable for the particular situation, and (3) no conclusion.

1.9.1 Correct Operation

Correct operation indicates that (1) at least one of the primary relays operated correctly, (2) none of the backup relays operated to trip for the fault, and (3) the trouble area was properly isolated in the time expected. As indicated, which and how many other relays were alerted and involved cannot easily be determined and so is seldom known.

Around 99% or all relay operations are corrected and wanted, i.e., operation is as planned and programmed. The correct but un wanted operation describes the few cases where all relays and associated equipment performed correctly. However, their operation was not desired or anticipated. An example of this was the 1965

Northeast power blackout. This was initiated by an unanticipated operating condition during the unusual system emergency at that time. This resulted in "correct but unwanted relay operations" which cascaded in many more of the same types of relay operations over a very wide area. The relays performed correctly since the system conditions of the emergency provided currents and voltages that were within the relay operating zones.

1.9.2 Incorrect Operation

Incorrect operations result from a failure, a malfunction, or an unanticipated or unplanned operation of the protective system. This can cause either incorrect isolation of a no-trouble area, or a failure to isolate a trouble area. The reasons for incorrect operation can be one or a combination of (1) misapplication of relays, (2) incorrect setting, (3) personnel errors, and (4) equipment problems or failures (relays, breakers, CTs, VTs, station battery, wiring, pilot channel, auxiliaries, and so on).

As indicated earlier, it is practically impossible to anticipate and provide protection for the "infinity" of possible power system problems. With the best of planning and design there will always be a potential situation that may not be "protected," or an error not detected. Occasionally these are "covered" by an incorrect operation that can be classified as "acceptable for the particular situation." While these are very few, they have saved power systems and minimized embarrassments.

1.9.3 No Conclusion

"No conclusion" refers to cases where one or more relays have or appear to have operated, such as the circuit breaker tripping, but no cause can be found. No evidence of a power system fault or trouble, nor apparent failure of the equipment, causes an extremely frustrating situation. It is fundamental and very important for protection engineers as well as operating personnel to understand all system operations. Thus no-conclusion cases involve considerable concern accompanied by many hours of postmortem investigation. It is suspected that many of these cases are the result of personnel involvement that is not reported and from intermittent troubles that do not become apparent during testing and investigation. Fortunately, modern oscillographs and data recording equipment are used in many power systems. These often provide direct evidence or clues as to the problem, as well as indicating possibilities that could not have occurred.

1.10 PRINCIPLES OF RELAY APPLICATION

The power system is divided into protection zones defined by the equipment and the available circuit breakers. Six categories of protection zones are possible in each power system: (1) generators and generator-transformer units, (2) transformers, (3) buses, (4) lines (transmission, subtransmission, distribution), (5) utilization equipment (motors, static loads, etc.), and (6) capacitor and/or reactor banks (when separately protected).

Most of these zones are illustrated in Figure 1.11. Although the fundamentals of protection are quite similar, each of these six categories has protective relays specifically designed for the primary protection which are based on the characteristics of the equipment being protected. The protection of each zone normally includes relays that can provide backup for the relays protecting the adjacent equipment.

The protection in each zone should overlap that in the adjacent zone; otherwise, a primary protection void would occur between the protection zones. This overlap is accomplished by the location of the current transformers, the key sources of power system information for the relays. This is shown in Fig. 1.11, and more specifically in Fig. 1.12. Faults between the two current transformers (Fig. 1.12) result in both zone X and zone Y relays operating and both tripping the associated circuit breaker.

For case (a) this fault probably involves the circuit breaker itself and so may not be cleared until the remote breakers at either end are opened. For case (b) zone Y relays alone opening the circuit breaker would clear faults between the two current transformers from the left fault source. The relays at the remote right source must also be opened for these faults. The operation of the zone X relays are not required but cannot be prevented. Fortunately, the area of exposure is quite small and the possibility of faults low. Without this overlap, primary protection for the area between the current transformers would not exist, so this overlap is standard practice in all applications.

1.11 INFORMATION FOR APPLICATION

One of the most difficult aspects of applying protection is often an accurate statement of the protection requirements or problem. This is valuable as an aid to a practical efficient solution, and is particularly important when assistance is desired from others who might be involved or might assist in the solution, such as consultants, manufac-

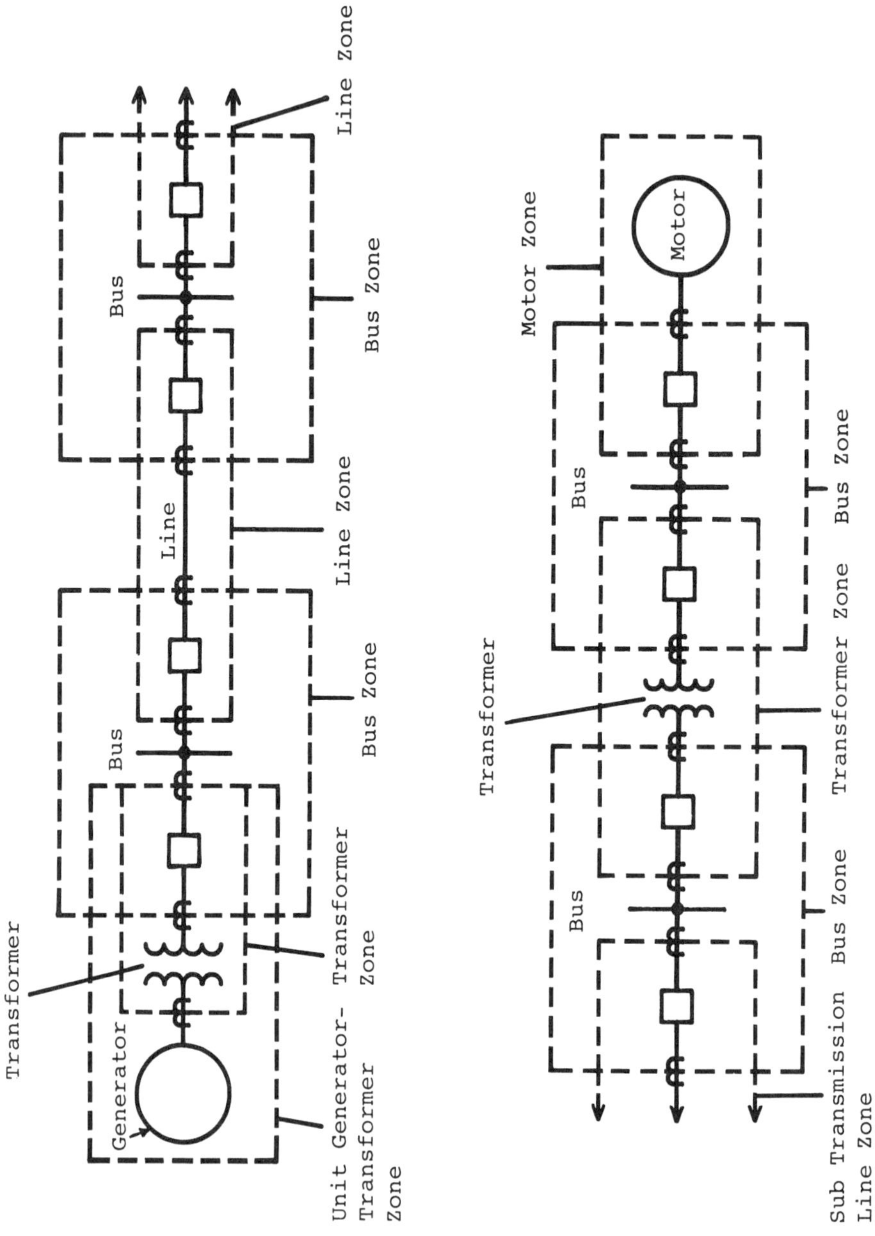

Fig. 1.11 Typical relay primary protection zones in a power system.

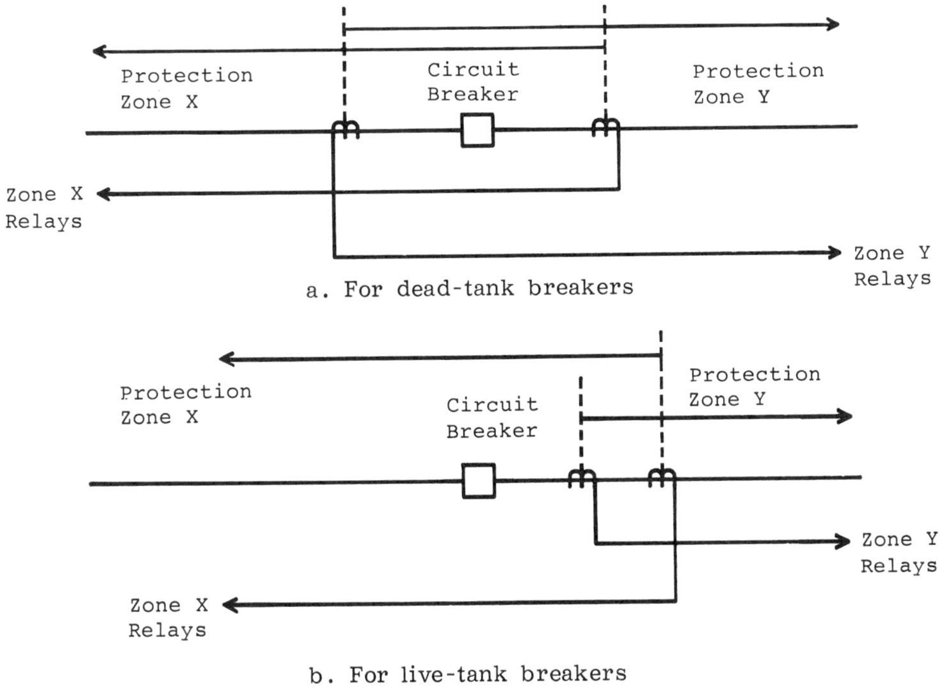

Fig. 1.12 Overlapping protection zones with their associated current transformers.

turers, and other engineers. The following checklist of information required has been developed from many years of assisting relay engineers with their protection problems. It outlines the information required, some of which is invariably overlooked in the first definition of the problem.

1.11.1 System Configuration

A single-line diagram for application considerations or a three-line diagram for complete connections and panel wiring drawings documenting the area to be studied and of the surrounding system. The location of the circuit breakers, current and voltage transformers, generators, buses, and taps on lines should be shown. The transformer bank connections and system grounding are necessary when considering ground-fault protection.

1.11.2 Impedance and Connection of the Power Equipment, System Frequency, System Voltage, and System Phase Sequence

Most of this information is usually included on the single-line diagram, but often omitted are the connections and grounding of the power transformer banks, and the circuit impedances. Phase sequence is necessary when a three-line connection diagram is required.

1.11.3 Existing Protection and Problems

If it is a new installation, this does not apply, but should be indicated. Otherwise, information about the existing protection and any problems may permit its updating or integration with the changes desired.

1.11.4 Operating Procedures and Practices

Additions or changes should conform to the existing practices, procedures, and desires. Where these affect the protection, they should be indicated. Often, this can be accomplished by indicating that certain types of equipment or practices are not acceptable.

1.11.5 Importance of the System Equipment Being Protected

This is often apparent by the system voltage level and size. For example, high-voltage transmission lines are usually protected by high-speed pilot protection and low-voltage systems by time-overcurrent protection. However, this should be clarified as to the desires of the protection engineers or the requirements of the system. In general, the more important the equipment to be protected is to the power system and its ability to maintain service, the more important it becomes to provide full and adequate high-speed protection.

1.11.6 System Fault Study

A fault study is important for most protection applications. For phase-fault protection a three-phase fault study is required, while for ground-fault protection a single-line-to-ground fault study is required. These latter should include the zero-sequence voltages and the negative-sequence currents and voltages, which can be useful where directional sensing of ground faults is involved.

On lines, information regarding a fault on the line side at an open breaker (known as a "line-end" fault) is important in many cases. The currents recorded should be those which will flow through the relays or fuses rather than the total fault current.

The fault study should indicate the units (i.e., in volts or amperes at a specified voltage base, or in per unit with the base clearly specified. Experience has shown that quite often the base of the quantities is not shown or clearly indicated.

1.11.7 Maximum Loads and System Swing Limits

The maximum load that will be permitted to pass through the equipment during short-time or emergency operation for which the protection must not operate should be specified. Where known, the maximum system swing from which the power system can recover after a transient disturbance is important in some applications and should be specified.

1.11.8 Current and Voltage Transformer Locations, Connections, and Ratios

This information is often shown on the one-line drawing, but often the data are incomplete or not clear. Where multiratio devices exist, the specific tap or ratio in use should be specified. The grounding of the voltage transformer or voltage devices should be clear.

1.11.9 Future Expansion

The system growth or changes that are likely to occur within a reasonable time and are known or planned should be indicated.

Not all of the items above necessarily apply to a specific problem or system requirement, but this checklist should assist in providing a better understanding of the protection problems and requirements. Usually, the fault study, together with related information, will provide information on the measurable quantities (handles) to which the protective relays can respond. Where this is not apparent, the first priority for any application is to search for the handles that can be utilized to distinguish between tolerable and intolerable conditions.

BIBLIOGRAPHY

The following provides references for this chapter plus general references applicable to all chapters.

The IEEE Power System Relaying committee of the Power Engineering Society has documented a Bibliography of Relay Literature from 1927. In recent years this is issued as a committee report every two years. These are in the AIEE and IEEE Transactions as follows:

Period Covered	Transactions reference		
1927-1939,	Vol. 60,	1941;	pp. 1435—1447
1940-1943,	Vol. 63,	1944;	pp. 705—709
1944-1964,	Vol. 67, pt. I,	1948;	pp. 24—27
1947-1949,	Vol. 70, pt. I,	1951;	pp. 247—250
1950-1952,	Vol. 74, pt. III,	1955;	pp. 45—48
1953-1954,	Vol. 76, pt. III,	1957;	pp. 126—129
1955-1956,	Vol. 78, pt. III,	1959;	pp. 78—81
1957-1958,	Vol. 79, pt. III,	1960;	pp. 39—42
1959-1960,	Vol. 81, pt. III,	1962;	pp. 109—112
1961-1964,	Vol. PAS-85, No. 10;	1966;	pp. 1044—1053

1965-1966,	Vol. PAS-88,	No. 3;	1969;	pp.	244— 250
1967-1969,	Vol. PAS-90,	No. 5;	1971;	pp.	1982—1988
1970-1971,	Vol. PAS-92,	No. 3;	1973;	pp.	1132—1140
1972-1973,	Vol. PAS-94,	No. 6;	1975;	pp.	2033—3041
1974-1975,	Vol. PAS-97,	No. 3;	1978;	pp.	789— 801
1976-1977,	Vol. PAS-99,	No. 1;	1980;	pp.	99— 107
1978-1979,	Vol. PAS100,	No. 5;	1981;	pp.	2407—2415
1980-1981,	Vol. PAS102,	No. 4;	1983;	pp.	1014—1024
1982-1983,	Vol. PAS104,	No. 5;	1985;	pp.	1189—1197

IEEE Standard Dictionary of Electrical and Electronics Terms, Centennial (3rd) Edition, ANSI/IEEE Standard 100-1984, IEEE Service Center, 445 Hoes Lane, Piscataway, NJ 08854. This dictionary is referred to throughout the book as IEEE 100-1984.

Standard Electrical Power System Device Function Numbers, ANSI/IEEE Standard C37.2-1979, IEEE Service Center.

Definitions for Power Switchgear, ANSI/IEEE Standard C37.100-1981, IEEE Service Center.

IEEE Standard Letter Symbols for Units of Measurement, ANSI/IEEE Standard 260-1978, IEEE Service Center.

IEEE Standard Letter Symbols for Quantities Used in Electrical Science and Electrical Engineering, ANSI/IEEE Standard 280-1985, IEEE Service Center.

IEEE Recommended Practice for Preferred Metric Units for Use in Electrical and Electronics Science and Technology, ANSI/IEEE Standard 945-1984, IEEE Service Center.

IEEE Recommended Practice for Protection and Coordination of Industrial and Commercial Power Systems, IEEE Buff Book, Standard 242-1975, IEEE Service Center.

Application Guide for AC High-Voltage Circuit Breakers Rated on a Symmetrical Current Basis, ANSI/IEEE Standard C37.010-1979, IEEE Service Center.

Applied Protective Relaying, Westinghouse Electric Corp., Coral Springs, FL, 1982.

Beeman, D., *Industrial Power Systems Handbook*, McGraw-Hill Book Company, New York, 1955.

Electrical Transmission and Distribution Reference Book, 4th ed. Westinghouse Electric Corp., East Pittsburgh, PA, 1964.

Electric Utility Engineering Reference Book, Vol. 3: *Distribution Systems*, Westinghouse Electric Corp., East Pittsburg, PA, 1965.

Fink, D. G., and Beaty, H. W., *Standard Handbook for Electrical Engineers*, McGraw-Hill Book Company, New York, 1968.

Horowitz, S. H., *Protective Relaying for Power systems*, IEEE Press, 1980, IEEE Service Center.

Mason, C. R., *The Art and Science of Protective Relaying*, John Wiley and Sons, Inc., New York, 1956.

Protective Relays Application Guide, 2nd ed., GEC Measurements, Stafford, England, 1975.

Recommended Practice for Electrical Power Distribution for Industrial Plants, IEEE Red Book, Standard 141-1976, IEEE Service Center.

Recommended Practice for Industrial and Commercial Power system Analysis, IEEE Brown Book, Standard 399-1980, IEEE Service Center.

Relays and Relay Systems Associated with Electric Power Apparatus, ANSI/IEEE Standard C37.90-1978, IEEE Service Center.

Review of Recent Practices and Trends in Protective Relaying, IEEE Power System Relaying Committee Report, *IEEE Trans. Power Appar. Syst.*, PAS 100, No. 8, 1981, 4054-4064.

Van C. Warrington, A. R., *Protective Relays, Their Theory and Practice:* Vol. I, John Wiley & Sons, Inc., New York, 1962; Vol. II, Chapman & Hall, Ltd., London, 1974.

2

Fundamental Units: Per Unit and Percent Values

2.1 INTRODUCTION

Power systems operate at voltages where kilovolt (kV) is the most convenient unit for expressing voltage. Also, these systems transmit large amounts of power, so that kilovolt-ampere (kVA) and megavolt-ampere (MVA) are used to express the total (general or apparent) three-phase power. These quantities, together with kilowatts, kilovars, amperes, ohms, flux, and so on, are usually expressed as a per unit or percent of a reference or base value. The per unit and percent nomenclatures are widely used because they simplify specification and computations, especially where different voltage levels and equipment sizes are involved.

This discussion is for three-phase electric systems which are assumed to be balanced or symmetrical up to a point or area of unbalance. This means that the source voltages are equal in magnitude and are 120° displaced in phase relations, and that the impedances of the three-phase circuits are of equal magnitude and phase angle. From this as a beginning, various shunt and series unbalances can be analyzed, principally by the method of symmetrical components. This method is reviewed in Chapter 4.

2.2 PER UNIT AND PERCENT DEFINITIONS

Percent is 100 times per unit. Both are used as a matter of convenience or of personal choice and it is important to designate either percent (%) or per unit (pu).

The per unit value of any quantity is the ratio of that quantity to its base value, the ratio expressed as a nondimensional decimal number. Thus actual quantities, such as voltage (V), current (I), power (P), reactive power (Q), volt-amperes (VA), resistance (R), reactance (X), and impedance (Z), can be expressed in per unit or percent as follows:

$$\text{Quantity in per unit} = \frac{\text{actual quantity}}{\text{base value of quantity}} \tag{2.1}$$

$$\text{Quantity in percent} = (\text{quantity in per unit}) \times 100 \tag{2.2}$$

where "actual quanity" is the scalar or complex value of a quantity expressed in its proper units, such as volts, amperes, ohms, or watts. "Base value of quantity" refers to an arbitrary or convenient reference of the same quantity chosen and designated as the base. Thus per unit and percent are dimensionless ratios which may be either scalar or complex numbers.

As an example for a chosen base of 115 kV, voltages of 92, 115, and 161 kV become 0.80, 1.00, and 1.40 pu or 80%, 100%, and 140%, respectively.

2.3 ADVANTAGES OF PER UNIT AND PERCENT

Some of the advantages of using per unit (or percent) are:

1. Its representation results in more meaningful data where the relative magnitudes of all similar circuit quantities can be compared directly.
2. The per unit equivalent impedance of any transformer is the same when referred to either the primary or the secondary side.
3. The per unit impedance of a transformer in a three-phase system is the same regardless of the type of winding connections (wye-delta, delta-wye, wye-wye, or delta-delta).
4. The per unit method is independent of voltage changes and phase shifts through transformers where the base voltages in the windings are proportional to the number of turns in the windings.
5. Manufacturers usually specify the impedance of equipment in per unit or percent on the base of its nameplate rating of power (usually kVA) and voltage (V or kV). Thus the rated impedance can be used directly if the bases chosen are the same as the nameplate ratings.
6. The per unit impedance values of various ratings of equipment lie in a narrow range, while the actual ohmic values may vary widely. Therefore, where actual values are not known, a good approximate value can be used. Typical values for various types of equipment are available from many sources and reference

books. Also, the correctness of a specified unit can be checked knowing the typical values.

7. There is less chance of confusion between single-phase power and three-phase power, or between line-to-line voltage and line-to neutral voltage.
8. The per unit method is very useful for simulating the steady-state and transient behavior of power systems on computers.
9. The driving or source voltage usually can be assumed to be 1.0 pu for fault and voltage calculations.
10. With per unit the product of two quantities expressed in per unit is expressed in per unit itself. However, the product of two quantities expressed in percent must be divided by 100 to obtain the result in percent. For this reason it is desirable to use per unit rather than percent in computations.

2.4 GENERAL RELATIONSHIPS BETWEEN CIRCUIT QUANTITIES

Before continuing the discussion of the per unit method, a review of some general relationships between circuit quantities applicable to all three-phase power systems is in order. This will focus on the two basic

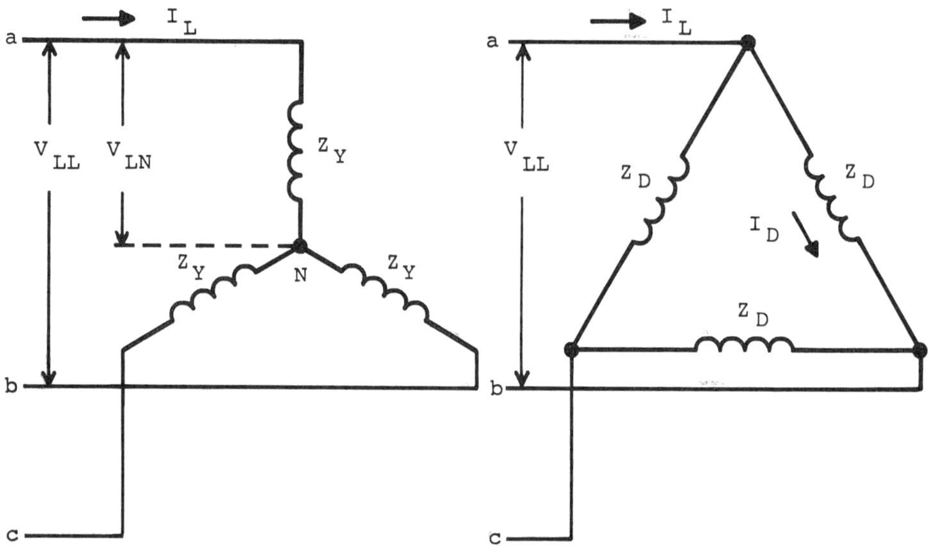

a. Wye connected impedances b. Delta connected impedances

Fig. 2.1 Impedances in three-phase wye (left) and delta (right) circuits.

types of connections, wye and delta, as shown in Fig. 2.1. For either of these the following basic equations apply*:

$$S_{3\phi} = \sqrt{3}\, V_{LL} I_L \tag{2.3}$$

$$V_{LL} = \sqrt{3}\, V_{LN} \angle{+30^\circ} \tag{2.4}$$

$$I_L = \frac{S_{3\phi}}{\sqrt{3}\, V_{LL}} \tag{2.5}$$

From these three equations the value of the impedances and the delta current can be determined.

1. *Wye-Connected Impedances (Fig. 2.1a)*

$$Z_Y = \frac{V_{LN}}{I_L} = \frac{V_{LL} \angle{-30^\circ}}{\sqrt{3}} \times \frac{\sqrt{3}\, V_{LL}}{S_{3\phi}}$$

$$= \frac{V_{LL}^2 \angle{-30^\circ}}{S_{3\phi}} \tag{2.6}$$

2. *Delta-Connected Impedances (Fig. 2.1b)*

$$I_D = \frac{I_L \angle{+30^\circ}}{\sqrt{3}} \tag{2.7}$$

$$Z_D = \frac{V_{LL}}{I_D} = \frac{\sqrt{3}\, V_{LL} \angle{-30^\circ}}{I_L} = \frac{\sqrt{3}\, V_{LL} \angle{-30^\circ}}{} \times \frac{\sqrt{3}\, V_{LL}}{S_{3\phi}}$$

$$= \frac{3V^2_{LL} \angle{-30^\circ}}{S_{3\phi}} \tag{2.8}$$

$$I_D = \frac{V_{LL}}{Z_D} = \frac{S_{3\phi} \angle{+30^\circ}}{3V_{LL}} \tag{2.9}$$

---

*S is the apparent or complex power in volt-amperes (VA, kVA, MVA), P is the active power in watts (W, kW, MW), and Q is the reactive power in vars (var, kvar, Mvar). Thus $S = P + jQ$.

These equations show that the circuit quantities S, V, I, and Z are so related that the selection of any two of them determines the values of the remaining two quantities. Usually, the wye connection is assumed, so Eq. (2.3) through (2.6) are most commonly used for power system calculations. A great deal of confusion can be avoided by clearly remembering that wye connections are assumed and not delta connections, or vice versa. If a delta connection is given, it can be converted into an equivalent wye connection for calculation purposes. Alternatively, Eq. (2.8) and (2.9) can be used directly, if the need arises, to express the impedance and current in terms of delta circuit quantities.

2.5 BASE QUANTITIES

In the following chapters it is more convenient to use the notation kVa or MVA instead of S, and kV instead of V. The base quantities are scalar quantities, so that phasor notation is not required for the base equations. Thus equations for the base values can be expressed from Eq. (2.3), (2.5), and (2.6) with the subscript B to indicate a base quantity as follows:

For base power: $$kVA_B = \sqrt{3}\, kV_B I_B \qquad (2.10)$$

For base current: $$I_B = \frac{kVA_B}{\sqrt{3}\, kV_B} \qquad (2.11)$$

For base impedance: $$Z_B = \frac{kV_B^2 \times 1000}{kVA_B} \qquad (2.12)$$

and since 1000 × the value of MVA = kVA (2.13)

the base impedance can also be expressed as

$$Z_B = \frac{kV_B^2}{MVA_B} \qquad (2.14)$$

In three-phase electric power systems the common practice is to use the standard or nominal system voltage as the voltage base, and a convenient MVA or kVA quantity as the power base. 100 MVA is a widely used power base. The system voltage commonly specified is the voltage between the three phases (i.e., the line-to-line voltage). This is the voltage used as a base in Eq. (2.10) through (2.14). As a shortcut and for convenience, the line-to-line subscript designation (LL) is omitted. With this practice it is always understood that the voltage is the line-to-line value unless indicated otherwise. The major exception is in the method of symmetrical components, where line-to-neutral phase voltage is used. This should always be specified carefully, but there is sometimes a tendency to overlook this step. Similarly, current is always the phase or line-to-neutral current unless otherwise specified.

Power is always understood to be three-phase power unless otherwise indicated. General power, also known as complex or apparent power, is designated by MVA or kVA, as indicated above. Three-phase power is designated by MW or kW. Three-phase reactive power is designated by RMVA or RkVA.

2.6 PER UNIT AND PERCENT IMPEDANCE RELATIONSHIPS

Per unit impedance is specified in ohms (Z_Ω) from Eq. (2.1) by substituting Eq. (2.14):

$$Z_{pu} = \frac{Z_\Omega}{Z_B} = \frac{MVA_B Z_\Omega}{kV_B^2} \text{ or } \frac{kVA_B Z_\Omega}{1000\ kV_B^2} \tag{2.15}$$

or, in percent notation,

$$\%Z = \frac{100 MVA_B Z_\Omega}{kV_B^2} \text{ or } \frac{kVA_B Z_\Omega}{10 kV_B^2} \tag{2.16}$$

Where the ohm values are desired from per unit, percent values, the equations are

$$Z_\Omega = \frac{kV_B^2 Z_{pu}}{MVA_B} \text{ or } \frac{1000 kV_B^2 Z_{pu}}{kVA_B} \tag{2.17}$$

$$Z_\Omega = \frac{kV_B^2 (\%Z)}{100\ MVA_B} \text{ or } \frac{10 kV_B^2 (\%Z)}{kVA_B} \tag{2.18}$$

The impedance values may be either scalars or phasors. The equations are also applicable for resistance and for reactance calculations.

Per unit is recommended for calculations involving division, as it is less likely to result in a decimal-point error. However, the choice of per unit or percent is personal. It is often convenient to use both, but care should be used.

Careful and overredundant labeling of all answers is recommended very strongly. This is valuable in identifying a value or answer, particularly later, when you or others refer to the work. Too often, answers such as 106.8, for example, are indicated without any label. To others, or later when memory is not fresh, questions can arise, such as: "What is this? amperes? volts? per unit what?" Initially, the proper units were obvious, but to others, or later, they may not be. A little extra effort and the development of the good habit of labeling leaves no frustrating questions, doubts, or tedious rediscovery later.

Currents in amperes and impedances in ohms should be referred to a specific voltage base or to primary or secondary windings of transformers. Voltages in volts should be clear as to whether they are primary, secondary, high, low, and so on, quantities.

When per unit or percent values are specified for impedances, resistance, or reactance, *two bases* must be indicated. These are the MVA (or kVA) and the kV bases using Eq. (2.15) through (2.18). Without the two bases the per unit or percent values are meaningless. For electrical equipment these two bases are the rated values cited on the equipment nameplate or on the manufacturer's drawings or other data supplied. Where several ratings are specified, generally it is correct to assume that the normal rated values were used to determine the per unit or percent values specified. Fundamentally, the manufacturer should specifically indicate the bases where several ratings exist.

System drawings should clearly indicate the MVA (or kVA) base with the base voltages indicated for the various voltage levels shown, where all the impedance components have been reduced to one common base value. Otherwise, the per unit or percent impedances with their two bases must be indicated for every piece of equipment or circuit on the drawing.

For per unit or percent voltages, only the voltage base is required. Thus a 90% voltage on a 138 kV system would be 124.2 kV. For per unit or percent currents, one or two bases are required. If the base current is specified, that is sufficient. A 0.90 pu current with a 1000-A base specifies that the current is 900 A. Where the more common MVA (or kVA) and kV bases are given, Eq. (2.11), with Eq. (2.13), provides the base current required. Thus with 100 MVA 138-kV bases, the base current is

$$I_B = \frac{1000 \times 100}{\sqrt{3} \times 138} = 418.37 \text{ A at } 138 \text{ kV} \qquad (2.19)$$

Thus 418.37 A is 1 pu or 100% current in the 138 kV system.

2.7 PER UNIT AND PERCENT IMPEDANCES OF TRANSFORMER UNITS

As indicated in Section 2.3, a major advantage of the per unit (percent) system is its independence of voltage and phase shifts through tranformer banks, *where the base voltages on the different terminals of the transformer are proportional to the turns in the corresponding windings.*

This can be demonstrated by the following analysis. From basic fundamentals, the impedance on one side of a transformer is reflected through the transformer by the square of the turns ratio, or where the voltages are proportional to the turns by the square of the voltage ratio. Thus for one phase of a transformer as shown in Fig. 2.2, the impedance Z_y on the N_y turns winding appears as Z_x on the N_x turns winding side, as

$$Z_x = \left(\frac{N_x}{N_y}\right)^2 Z_y = \left(\frac{V_x}{V_y}\right)^2 Z_y \tag{2.20}$$

The impedance bases on the two sides of the transformer are, from Eq. (2.14),

$$Z_{xB} = \frac{kV_x^2}{MVA_B} \quad \text{where } kV_x \text{ is the x-side base} \tag{2.21}$$

$$Z_{yB} = \frac{kV_y^2}{MVA_B} \quad \text{where } kV_y \text{ is the y-side base} \tag{2.22}$$

Taking the ratio of Z_{xB} and Z_{yB} yields

$$\frac{Z_{xB}}{Z_{yB}} = \frac{kV_x^2}{kV_y^2} = \left(\frac{N_x}{N_y}\right)^2 \tag{2.23}$$

where the turns are proportional to the voltages.

The per unit impedances are, from Eq. (2.1), (2.20), and (2.24).

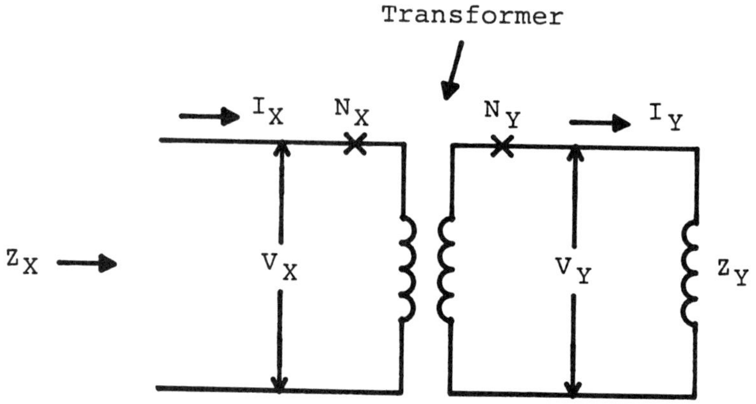

Fig. 2.2 Impedances through one phase of a three-phase transformer.

$$Z_X pu = \frac{Z_x \text{ ohms}}{Z_{xB}} = \left(\frac{N_x}{N_y}\right)^2 \left(\frac{N_y}{N_x}\right)^2 \frac{Z_y \text{ ohms}}{Z_{yB}}$$

$$= \frac{Z_y \text{ ohms}}{Z_{yB}} = Z_y pu \tag{2.24}$$

Thus the per unit impedance is the same on either side of the bank.

2.7.1 Transformer Bank Example

Consider a transformer bank rated 50 MVA with 34.5 kV and 161 kV windings connected to a 34.5 and 161 kV power system. The bank reactance is 10%. Now when looking at the bank from the 34.5 kV system, its reactance is

10% on a 50-MVA 34.5 kV base (2.25)

and when looking at the bank from the 161 kV system its reactance is

10% on a 50-MVA 161 kV base (2.26)

This equal impedance in percent or per unit on either side of the bank is independent of the bank connections: wye-delta, delta-wye, wye-wye, or delta-delta.

This means that the per unit (percent) impedance values throughout a network can be combined independently of the voltage levels as long as all the impedances are on a common MVA (kVA) base and the transformer windings ratings are compatible with the system voltages. This is a great convenience.

The actual transformer impedances in ohms are quite different on the two sides of a transformer with different voltage levels. This can be illustrated for the example. Applying Eq. (2.18), we have

$$jX = \frac{34.5^2 \times 10}{100 \times 50} = 2.38\ \Omega \text{ at } 34.5\text{ kV} \tag{2.27}$$

$$= \frac{161^2 \times 10}{100 \times 50} = 51.84\ \Omega \text{ at } 161\text{ kV} \tag{2.28}$$

This can be checked by Eq. (2.20), where for the example x is the 34.5 kV winding side, and y is the 161 kV winding side, Then,

$$2.38 = \frac{34.5^2}{161^2} \times 51.84 = 2.38 \tag{2.29}$$

2.8 CHANGING PER UNIT (PERCENT) QUANTITIES TO DIFFERENT BASES

Normally, the per unit or percent impedances of equipment is specified on the equipment base, which generally will be different from the power system base. Since all impedances in the system must be expressed on the same base for per unit or percent calculations, it is necessary to convert all values to the common base selected. This conversion can be derived by expressing the same impedance in ohms on two different per unit bases. From Eq. (2.15) for a MVA_1, kV_1 base and a MVA_2, kV_2 base,

$$Z_{1pu} = \frac{MVA_1 Z_\Omega}{kV_1^2} \tag{2.30}$$

$$Z_{2pu} = \frac{MVA_2 Z_\Omega}{kV_2^2} \tag{2.31}$$

By ratioing these two equations and solving for one per unit value, the general equation for changing bases is

$$\frac{Z_{2pu}}{Z_{1pu}} = \frac{MVA_2}{kV_2^2} \times \frac{kV_1^2}{MVA_1} \tag{2.32}$$

$$Z_{2pu} = Z_{1pu} \frac{MVA_2}{MVA_1} \times \frac{kV_1^2}{kV_2^2} \tag{2.33}$$

Equation (2.33) is the general equation for changing from one base to another base. In most cases the turns ratio of the transformer is equivalent to the different system voltages, and the equipment-rated voltages are the same as the system voltages, so that the voltage-squared ratio is unity. Then Eq. (2.33) reduces to

$$Z_{2pu} = Z_{1pu} \frac{MVA_2}{MVA_1} \tag{2.34}$$

It is very important to emphasize that the voltage-square factor of Eq. (2.33) is used *only* in the same voltage level and where slightly different voltage bases exist. It is *never* used where the base voltages are proportional to the transformer bank turns, such as going from the high to the low side across a bank. In other words, Eq. (2.33) has nothing to do with transferring the ohmic impedance value from one side of a transformer to the other side.

Several examples will illustrate the applications of Eq. (2.33) and (2.34) in changing per unit and percent impedances from one base to another.

2.8.1 Example: Base Conversion with Eq. (2.34)

The 50-MVA 34.5:161 kV transformer with 10% reactance is connected to a power system where all the other impedance values are on a 100-MVA 34.5 kV or 161 kV base. To change the base of the transformer, Eq. (2.34) is used since the transformer and system base voltages are the same. This is because if the fundamental Eq. (2.33) were used,

$$\frac{kV_1^2}{kV_2^2} = \left(\frac{34.5}{34.5}\right)^2 \text{ or } \left(\frac{161}{161}\right)^2 = 1.0 \tag{2.35}$$

so Eq. (2.34) results, and the transformer reactance becomes

$$jX = 10\% \times \frac{100}{50} = 20\% \text{ or } 0.20 \text{ pu} \tag{2.36}$$

on a 100 MVA 34.5 kV base from the 34.5 kV side or on a 100 MVA 161 kV base from the 161 kV side.

2.8.2 Example: Base Conversion Requiring Eq. (2.33)

A generator and transformer, shown in Fig. 2.3, are to be combined into a single equivalent reactance on a 100 MVA 110 kV base. With the transformer bank operating on its 3.9 kV tap, the low-side base voltage corresponding to the 110 kV high-side base is

$$\frac{kV_{LV}}{110} = \frac{3.9}{115} \quad \text{or} \quad kV_{LV} = 3.73 \text{ kV} \tag{2.37}$$

Since this 3.73-kV base is different from the specified base of the generator subtransient reactance, Eq. (2.33) must be used:

$$jX_d'' = 25\% \times \frac{100 \times 4^2}{25 \times 3.73^2} = 115\% \text{ or } 1.15 \text{ pu} \text{ on 100 MVA 3.73 kV base or on 100 MVA 110 kV base} \tag{2.38}$$

Similarly, the transformer reactance on the new base is

$$jX_T = 10\% \times \frac{100 \text{ X } 3.9^2}{30 \text{ X } 3.73^2} = 10\% \times \frac{100 \text{ X } 115^2}{30 \text{ X } 110^2}$$

$$= 36.43\% \text{ or } 0.364 \text{ pu} \quad \text{on 100 MVA 3.73 kV base or on 100 MVA 110 kV base} \tag{2.39}$$

Now the generator and transformer reactances can be combined into one equivalent source value by adding:

$$115\% + 36.43\% = 151.43\%$$

or

$$1.15 \text{ pu} + .3643 \text{ pu} = 1.514 \text{ pu, both on a 100 MVA 110 kV base} \tag{2.40}$$

The previous warning bears repeating and emphasizing. Never, *never*, NEVER use Eq. (2.33) with voltages on the opposite sides of

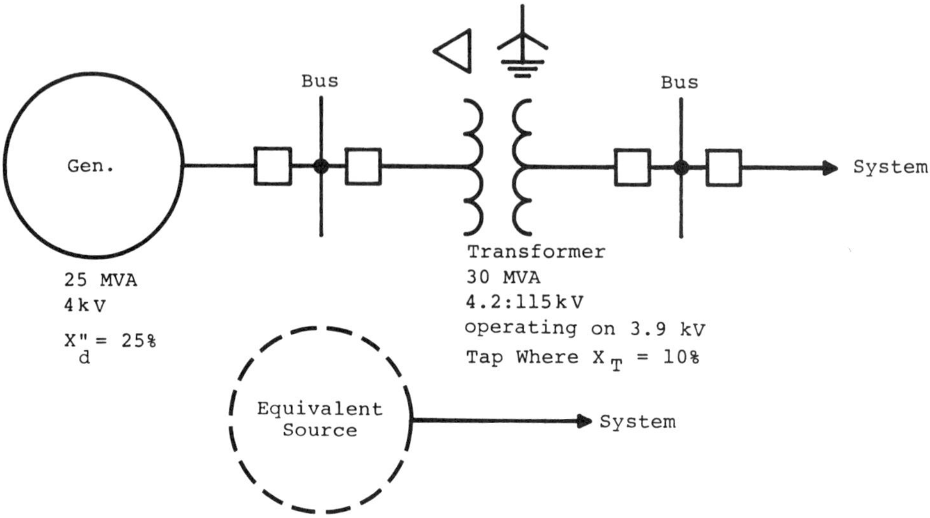

Xeg = 151.43% = 1.154 pu on 100 MVA, 110kV Base

Fig. 2.3 Typical example for combining a generator and a transformer into an equivalent source.

transformers. Thus the factors $(115/3.9)^2$ and $(110/3.73)^2$ in Eq. (2.33) are *incorrect*.
See the bibliography at the end of Chapter 1 for additional information.

BIBLIOGRAPHY

Fitzgerald, A. E., Kingsley, C., and Umans, S. D., *Electric Machinery*, McGraw-Hill Book Company, New York, 1983.

Seidman, A. H., Mahrous, H., and Hicks, T. G., *Handbook of Electric Power Calculations*, McGraw-Hill Book Company, New York, 1983.

Stevenson, W. D., Jr., *Elements of Power Systems Analysis*, 3rd Ed., McGraw-Hill Book Company, New York, 1975.

Weedy, B. M., *Electric Power Systems*, 3rd Ed., John Wiley & Sons, Inc., New York, 1979.

3

Phasors and Polarity

3.1 INTRODUCTION

Phasors and polarity are two important and useful tools in power system protection. They aid in the understanding and analysis of the connections, operation, and testing of relays and relay systems. Also, these concepts are essential in understanding power system performance during both normal and abnormal operation. Thus a sound theoretical and practical knowledge of phasors and polarity is a fundamental and valuable resource.

3.2 PHASORS

The IEEE Dictionary (IEEE 100-1984) defines a phasor as "a complex number. Unless otherwise specified, it is used only within the context of steady state alternating linear systems." It continues: "the absolute value (modulus) of the complex number corresponds to either the peak amplitude or root-mean-square (rms) value of the quantity, and the phase (argument) to the phase angle at zero time. By extension, the term 'phasor' can also be applied to impedance, and related complex quantities that are not time dependent."

In this book, phasors will be used to document various ac voltages, currents, fluxes, impedances, and power. For many years phasors were referred to as "vectors," but this use is deprecated to avoid confusion with space vectors. However, the former use lingers on, so occasionally a lapse to "vectors" may occur.

3.2.1 Phasor Representation

The common pictorial form for representing electrical and magnetic phasor quantities uses the cartesian coordinates with x (the abscissa) as the axis of real quantities and y (the ordinate) as the axis of imaginary quantities. This is illustrated in Fig. 3.1. Thus a point c on the complex plane x-y can be represented as shown in this figure, and mathematically documented by the several alternative forms given in Eq. (3.1).

Phasor form	Rectangular form	Complex form	Exponential form	Polar form
c =	$x + jy$ =	$\lvert c\rvert(\cos\phi + j\sin\phi)$ =	$\lvert c\rvert e^{j\phi}$ =	$\lvert c\rvert \angle{+\phi}$

$$c = x + jy = |c|(\cos\phi + j\sin\phi) = |c|\,e^{j\phi} = |c|\angle{+\phi} \tag{3.1}$$

Sometimes useful is the conjugate form:

$$c^* = = x - jy = |c|(\cos\phi - j\sin\phi) = |c|\,e^{-j\phi} = |c|\angle{-\phi} \tag{3.2}$$

where c = the phasor

c^* = its conjugate

x = the real value (alternate: Re c or c')

y = the imaginary value (alternate: Im c or c")

$|c|$ = the modulus (magnitude or absolute value)

ϕ = the phase angle (argument or amplitude) (alternate: arg c)

The modulus (magnitude or absolute value) of the phasor is

$$|c| = \sqrt{x^2 + y^2} \tag{3.3}$$

From Eq. (3.1) and (3.3),

$$x = \frac{1}{2}(c + c^*) \tag{3.4}$$

$$y = \frac{1}{2j}(c - c^*) \tag{3.5}$$

3.2.2 Phasor Diagrams for Sinusoidal Quantities

In applying the notation above to sinusoidal (ac) voltages, currents, and fluxes, the axes are assumed fixed, with the phasor quantities rotating at constant angular velocity. The international standard is that *phasors always rotate in the counterclockwise direction*. However, as a convenience, on the diagrams the phasor is always shown "fixed" for the given condition. The magnitude of the phasor (c) can be either the maximum peak value or the rms value of the corresponding sinusoidal quantity. In normal practice it represents the rms maximum value

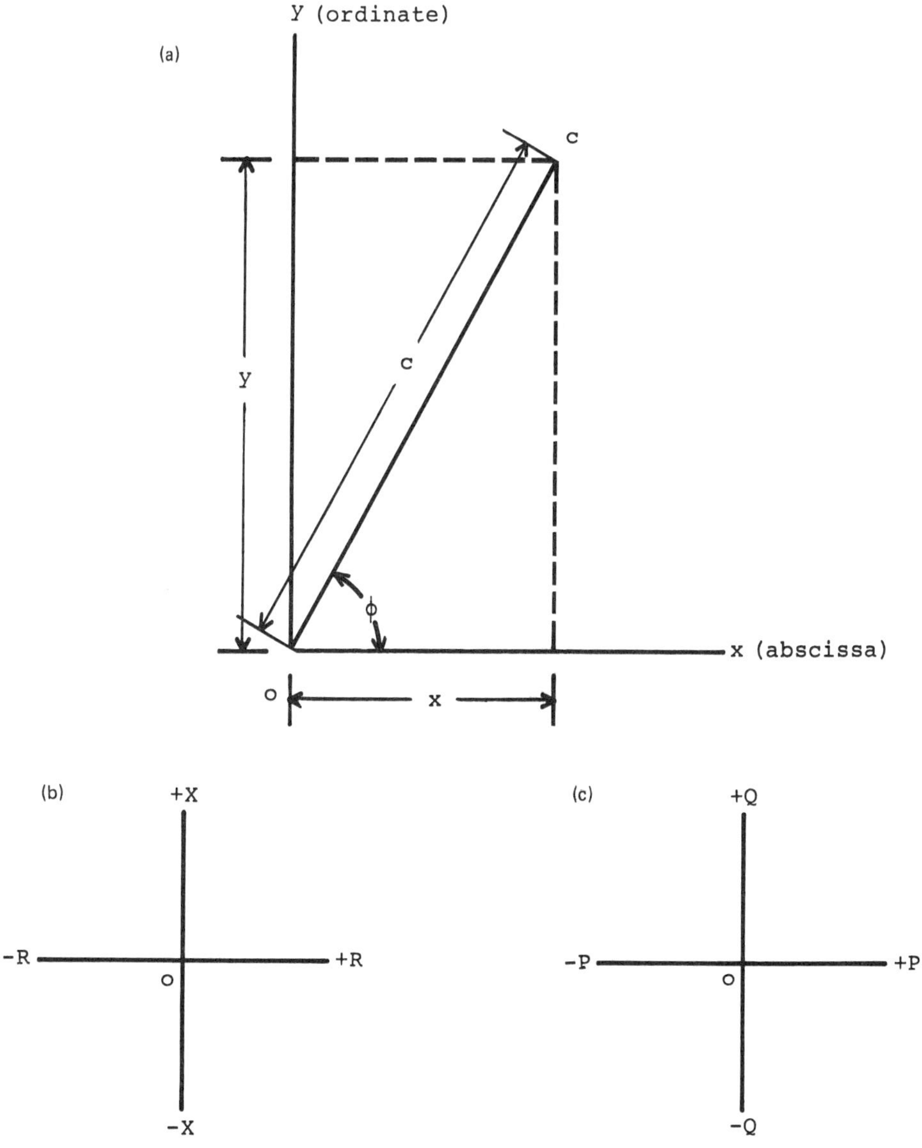

Fig. 3.1 Reference axes for phasor quantities: (a) cartesian x - y coordinates; (b) impedance phasor axes; (c) power phasor axes.

of the positive half cycle of the sinusoid unless otherwise specifically stated.

Thus a phasor diagram shows the respective voltages, currents, fluxes, and so on, existing in the electrical circuit. It should document *only* the magnitude and relative phase-angle relations between these various quantities. Thus all phasor diagrams require a scale or complete indications of the physical magnitudes of the quantities shown. The phase-angle reference usually is between the quantities shown, so that the zero (or reference angle) may vary with convenience. As an example, in *fault calculations* using reactance (X) only, it is convenient to use the voltage (V) reference at +90°. Then I = jV/jX and the j value cancels, so the fault current does not involve the j factor. On the other hand, in *load calculations* it is preferable to use the voltage (V) at 0° or along the x axis so that the angle of the current (I) represents its actual lag or lead value.

Other reference axes that are in common use are shown in Fig. 3.1b and c. For plotting impedance, resistance, and reactance, the R-X axis of Fig. 3-1b is used. Inductive reactance is +X and capacitive reactance is −X.

For plotting power phasors, Fig. 3.1c is used. P is the real power (W, kW, MW) and Q is the reactive power (var, kvar, Mvar). These impedance and power diagrams are discussed in later chapters. While represented as phasors, the impedance and power "phasors" do not rotate at system frequency.

3.2.3 Combining Phasors

The various laws for combining phasors are present for general reference:

Multiplication

The magnitudes are multiplied and the angles added:

$$VI = |V||I| \underline{/\phi_V + \phi_I} \tag{3.6}$$

$$VI^* = |V||I| \underline{/\phi_V - \phi_I} \tag{3.7}$$

$$II^* = |I|^2 \tag{3.8}$$

Division

The magnitudes are divided and the angles subtracted:

$$\frac{V}{I} = \frac{|V|}{|I|} \underline{/\phi_V - \phi_I} \tag{3.9}$$

Powers

$$(I)^n = \left(|I|e^{j\phi}\right)^n = |I|^n e^{j\phi n} \tag{3.10}$$

$$\sqrt[n]{I} = \sqrt[n]{|I|e^{j\frac{\phi}{n}}} \tag{3.11}$$

3.2.4 Phasor Diagrams Require a Circuit Diagram

The phasor diagram defined above has a indeterminate or vague meaning unless it is accompanied by a circuit diagram. The circuit diagram identifies the specific circuit involved with the location and assumed direction for the currents and the location and assumed polarity for the voltages to be documented in the phasor diagram. The assumed directions and polarities are not critical, as the phasor diagram will confirm if the assumptions were correct, and provide the correct magnitudes and phase relations. These two complementary diagrams (circuit and phasor) are preferably kept separate to avoid confusion and errors in interpretation. This is discussed further in Section 3.3.

3.2.5 Nomenclature for Current and Voltage

Unfortunately, there is no standard nomenclature for current and voltage, so confusion can exist among various authors and publications. The nomenclature used throughout this book has proven to be flexible and practical over many years of use, and it is compatible with power system equipment polarities.

Current and Flux

In the circuit diagrams, current or flux is shown by either (1) a letter designation, such as I or θ, with an arrow indicator for the assumed direction of flow; or (2) a letter designation with double subscripts, the order of the subscripts indicating the assumed direction. The direction is that assumed to be the flow during the positive half cycle of the sine wave. This convention is illustrated in Fig. 3.2a. Thus in the positive half cycle the current in the circuit is assumed to be flowing from left to right, as indicated by the direction of the arrow used with I_S, or denoted by subscripts, as with I_{ab}, I_{bc}, and I_{cd}. The single subscript such as I_S, is a convenience to designate currents in various parts of a circuit and has no directional indication, so an arrow for the direction must be associated with these. Arrows are not required with I_{ab}, I_{bc}, or I_{cd} but are often used for added clarity and convenience.

It is very important to appreciate that in these circuit designations, the arrows do *not* indicate phasors. They are only assumed directional and location indicators.

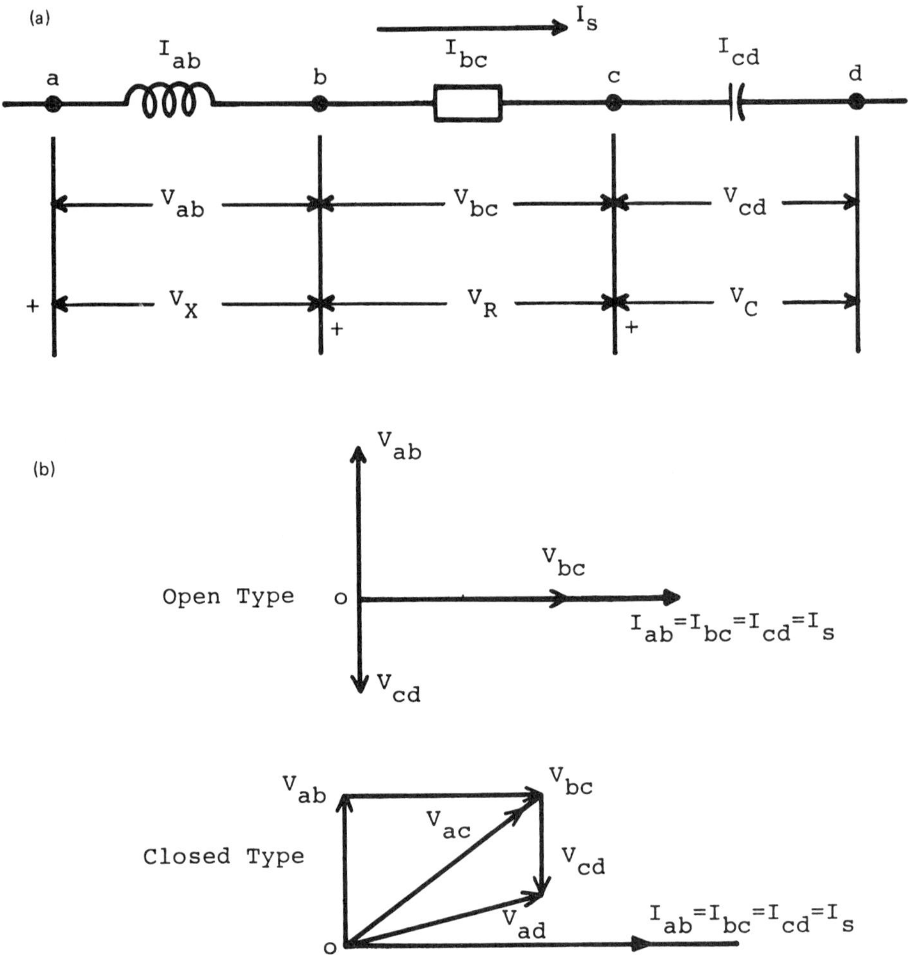

Fig. 3.2 Phasor diagram for the basic circuit elements: (a) circuit diagram showing location and assumed directions of current and of voltage drops; (b) phasor diagrams showing current and voltage magnitudes and phase relations.

Voltage

Voltages can be either drops or rises. Much confusion can result by not clearly indicating which is intended or by mixing the two practices in circuit diagrams. This can be avoided by standardizing on one, and *only one* practice. As voltage drops are far more common throughout the power system, *all* voltages are shown and *always* considered to be *drops* from a higher voltage to a lower voltage. This convention is independent of whether the letter V or E is used for the voltage.

The consistent adoption of *only drops* throughout need cause no difficulties. A generator or source voltage becomes a minus drop since current flows from a lower voltage to a higher voltage. This practice does not conflict with the polarity of equipment, such as transformers, and it is consistent with fault calculations using symmetrical components.

Voltages (always drops) are indicated by either (1) a letter designation with double subscripts, or (2) a small plus (+) indicator shown at the point assumed to be at a relatively high potential. Thus during the positive half cycle of the sine wave, the voltage drop is indicated by the order of the two subscripts when used, or from the "+" indicator to the opposite end of the potential difference. This is illustrated in Fig. 3.2a, where both methods are shown. It is preferable to show arrows at both ends of the voltage-drop designations, to avoid possible confusion. Again, it is most important to recognize that both of these designations, especially where arrows are used, in the circuit diagrams are only location and direction indicators, *not phasors*.

It may be helpful to consider current as a "through" quantity and voltage as an "across" quantity. In this sense, in the representative Fig. 3.2a, the same current flows *through* all the elements in series, so that $I_{ab} = I_{bc} = I_{cd} = I_S$. By contrast, voltage V_{ab} applies only *across* nodes a and b, voltage V_{bc} *across* nodes b and c, and V_{cd} *across* nodes c and d.

3.2.6 The Phasor Diagram

With the proper identification and assumed directions established in the circuit diagram, the corresponding phasor diagram can be drawn from calculated or test data. For the circuit diagram of Fig. 3.2a, two types of phasor diagrams are shown in Fig. 3.2b. The top diagram is referred to as an "open-type" diagram, where all the phasors originate from a common origin. The bottom diagram is referred to as a "closed-type," where the voltage phasors are summed together from left to right for the same series circuit. Both types are useful, but the open type is preferred to avoid the confusion that may occur with the closed type. This is amplified in Section 3.3.

3.3 CIRCUIT AND PHASOR DIAGRAMS FOR A BALANCED THREE-PHASE POWER SYSTEM

A typical section of a three-phase power system is shown in Fig. 3.3a. Optional grounding impedances (Z_{Gn}) and (Z_{Hn}) are omitted with solid grounding. This topic is covered in Chapter 7. (R_{sg}) and (R_{ssg})

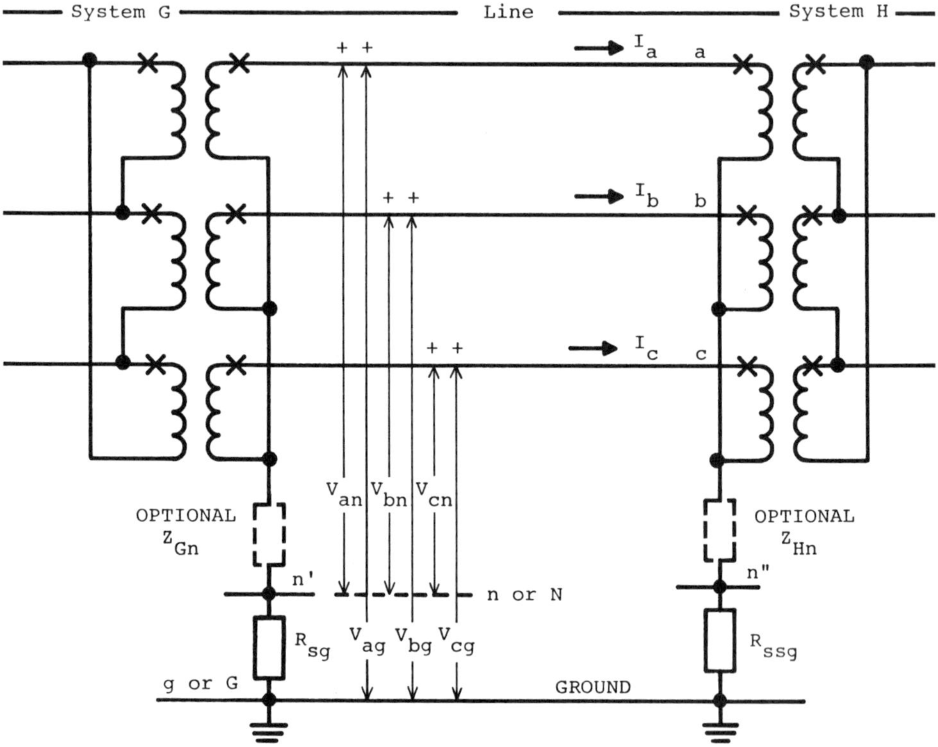

(a)

Fig. 3.3 Phasor diagram for a typical three-phase circuit operating with balanced or symmetrical quantities: (a) circuit diagram showing location and assumed directions of current and voltage drops; (b) phasor diagrams showing current and voltage magnitudes and phase relations.

represent the ground-mat resistance in the station or substation. Ground g or G represents the potential of the true earth, remote ground plane, and so on. The system neutrals n', n or N, and n" are not necessarily the same unless a fourth wire is used, as in a four-wire three-phase system. Upper- or lowercase N and n are used interchangeably as convenient for the neutral designation.

The various line currents are assumed to flow through this series section as shown, and the voltages are indicated for a specific point on the line section. These follow the nomenclature discussed previously. To simplify the discussion at this point, symmetrical or balanced operation of the three-phase power system is assumed. Therefore, no

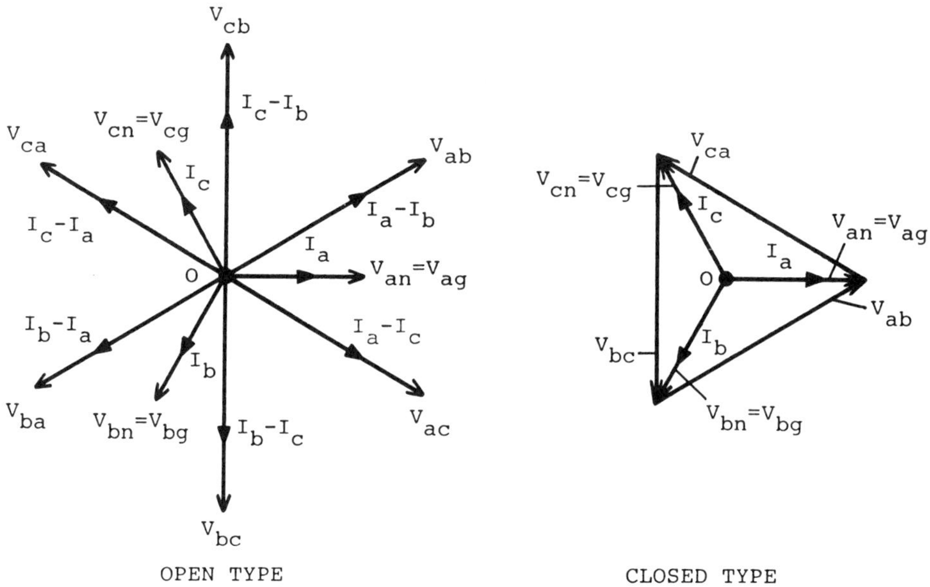

(b)

Fig. 3.3 (Continued)

current can flow in the neutrals of the two transformer banks, so that with this simplification there is no difference of voltage between n', n or N, n'' and the ground plane g or G. As a result, $V_{an} = V_{ag}$, $V_{bn} = V_{bg}$, and $V_{cn} = V_{cg}$. Again, this is true only for a balanced or symmetrical system. With this the respective currents and voltages are equal in magnitude and 120° apart in phase, as shown in the phasor diagram (Fig. 3.3b), in both the open and closed types. The phasors for various unbalanced and fault conditions are discussed in Chapter 4.

The open-type phasor diagram permits easy documentation of all possible currents and voltages, some of which are not convenient in the closed-type phasor diagram. The delta voltage V_{ab}, representing the voltage (drop) from phase a to phase b, is the same as $V_{an} - V_{bn}$. Similarly, $V_{bc} = V_{bn} - V_{cn}$ and $V_{ca} = V_{cn} - V_{an}$.

As indicated, the closed-type phasor diagram can lead to difficulties. As seen in Fig. 3.3b, its shape lends itself mentally to an assumption that the three vertices of the triangle represent the a, b, and c phases of the power system, and that the origin 0 represents n = g. Questions arise with this closed-type phasor diagram as to why $V_{an} = V_{ag}$ has its phasor arrow as shown since the voltage drop is from phase a to neutral; similarly for the other two phases. Also why V_{ab}, V_{bc}, and V_{ca} are pointing as shown, as they are drops from phase a to phase b, phase b to phase c, and phase c to a, respectively. It would appear that they should be pointing in the opposite direction.

The phasors shown on this closed phasor diagram (Fig. 3.3b) are absolutely *correct* and must not be changed. The difficulty is in the combining of a circuit diagram with the phasor diagram by the mental association of "a," "b," and "c" to the closed triangle. The open type avoids this difficulty. This also emphasizes the desirability of having *two separate* diagrams, a circuit diagram and a phasor diagram. Each serves particular but quite different functions.

3.4 PHASOR AND PHASE ROTATION

"Phasor" and "phase rotation" are two entirely different terms, although they almost look alike. *Ac phasors* are always rotating counterclockwise at the system frequency. The fixed diagrams, plotted such as in Fig. 3.3b, represent what would be seen if a stroboscopic light of system frequency were imposed on the system phasors. The phasors would appear fixed in space as plotted.

In contrast, *phase rotation* or *phase sequence* refers to the order in which the phasors occur as they rotate counterclockwise. The standard sequence today is: a, b, c; A, B, C; 1, 2, 3; or in some areas r, s, t. In Fig. 3.3b the sequence is a, b, c. The IEEE Dictionary (IEEE 100-1984) only defines phase sequence, so this is preferred. However, phase rotation has been used over many years and is still used in practice.

Not all power systems operate with phase sequence a, b, c or its equivalent. There are several large electric utilities in the United States that operate with a, c, b phase sequence. In some cases this sequence is used throughout the system; in others, one voltage level may be a, b, c, and another voltage level, a, c, b. The specific phase sequence is only a name designation that was established arbitrarily early in the history of a company, and it is difficult to change after many years of operation.

Knowing the existing phase sequence is very important in three-phase connections of relays and other equipment, so should be clearly indicated on the drawings and information documents. This is especially true if it is not a, b, c. The connections from a, b, c to a, c, b, or vice versa, can generally be made by completely interchanging phases b and c for both the equipment and the connections.

3.5 POLARITY

Polarity is important in transformers and in protection equipment. A clear understanding of polarity is useful and essential for the chapters that follow.

3.5.1 Transformer Polarity

The polarity indications for transformers are well established by standards that apply to all types of transformers. There are two varieties of polarity: substractive and additive. Both follow the same rules. Power and instrument transformers are subtractive, while some distribution transformers are additive. The polarity marking can be a dot, a square, or an X, or it can be indicated by the standardized transformer terminal markings, the practices having varied over the years. It is convenient to designate polarity by an X in this book.

The two fundamental rules of transformer polarity are illustrated in Fig. 3.4 and apply to both varieties. These are:

1. Current flowing in at the polarity mark of one winding flows out of the polarity mark of the other winding. Both currents are substantially in phase.
2. The voltage drop from polarity to nonpolarity across one winding is essentially in phase with the voltage drop from polarity to nonpolarity across the other winding(s).

The currents through, and the voltages across, the transformers are substantially in phase since on the magnetizing current and the impedance drop through the transformers is very small and can be considered negligible. This is normal and practical for these definitions.

The current transformer polarity markings are shown in Fig. 3.5. It will be noted that the direction of the secondary current is the same independently of whether the polarity marks are together on one side or on the other.

For current transformers (CTs) associated with circuit breakers and transformer banks, it is the common practice for the polarity marks to be located on the side away from the associated equipment.

The voltage drop rule is often omitted in the definition of transformer polarity, but it is an extremely useful tool to check the phase relations through wye-delta transformer banks, or in connecting up a transformer bank for a specific phase shift required by the power system. The ANSI/IEEE standard for transformers states that the high voltage should lead the low voltage by 30° with wye-delta or delta-wye banks. Thus different connections are required if the high side is wye than if the high side is delta. The connections for these two cases are shown in Fig. 3.6. The diagrams below the three-phase transformer connection illustrate the use of the voltage-drop rule to provide or check the connections. Arrows on these voltage drops have been omitted (preferably not used), as they are not necessary and can cause confusion.

In Fig. 3.6a the check is made by noting that a to n from polarity to non polarity on the left-side winding is in phase with A to B from polarity to nonpolarity on the right-side winding. Similarly, b to n

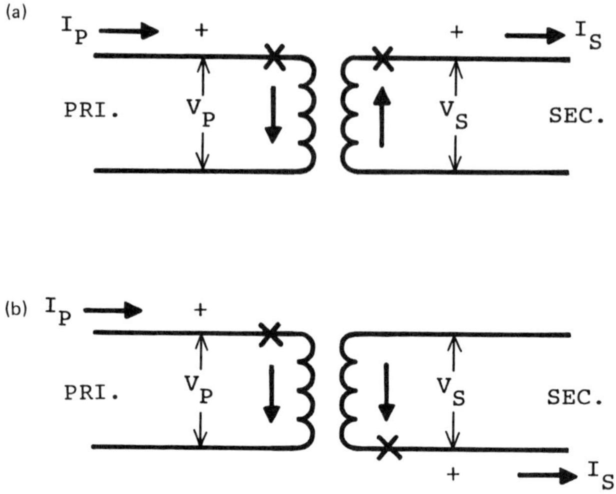

Fig. 3.4 Polarity definitions for transformers: (a) subtractive polarity; (b) additive polarity.

(polarity to nonpolarity) is in phase with B to C (polarity to nonpolarity) across the middle transformer, and c to n (polarity to nonpolarity) is in phase with C to A (polarity to nonpolarity) across the lower transformer. From this, by comparing the line-to-neutral voltages on the two sides, it is observed that phase-a-to-n voltage leads phase-A-to-neutral voltage. Thus the wye side would be the high-voltage side if this is a ANSI/IEEE standard transformer.

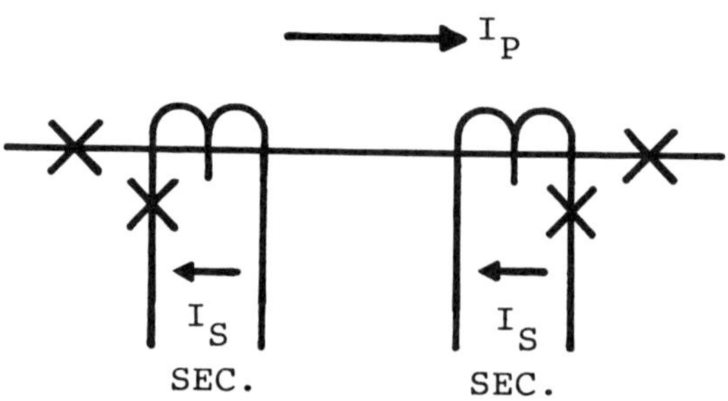

Fig. 3.5 Polarity markings for current transformers.

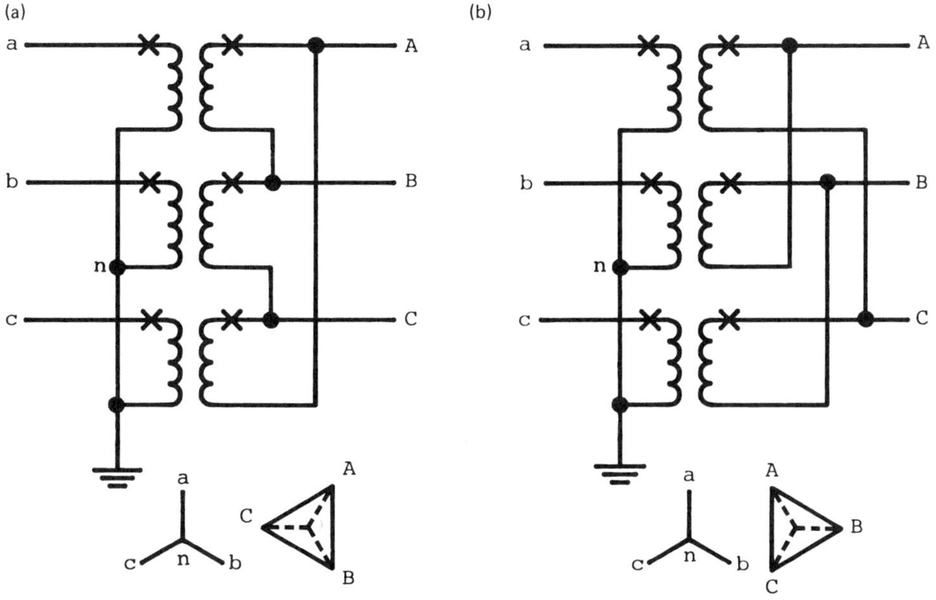

Fig. 3.6 Voltage-drop polarity rule useful in checking or connecting wye-delta transformer banks: (a) wye-connected side leads, delta-connected side 30°; (b) delta-connected side leads, wye-connected side 30°.

This same technique of applying voltage drops to Fig. 3.6b shows that for this three-phase bank connection the voltage drop polarity to nonpolarity or phase a to n is in phase with the voltage drop polarity to nonpolarity or phase A to phase C. Similarly, voltage drop phase b to n is in phase with voltage drop phase B to phase A, and voltage drop phase c to n is in phase with voltage drop phase C to phase B. By comparing similar voltages on the two sides of the transformer, phase-A-to-neutral voltage drop leads the phase-a-to-n voltage drop by 30°, so the delta winding would be the high-voltage side if this is an ANSI/IEEE standard transformer bank. This technique is very useful to make the proper three-phase transformer connections from a desired or known voltage diagram or phase-shift requirement. It is a very powerful tool, simple and straightforward to use.

Since the ANSI/IEEE standards have been in existence for a number of years, most transformer banks in service today follow this standard except where it is not possible because of preexisting system conditions. Many years ago, in the absence of a standard, a great many different connections were used. Some of the older references and textbooks reflect this.

3.5.2 Relay Polarity

Relays involving interaction between two input quantities from the power system may have the polarity marking that is necessary for their correct operation. There are no standards in this area, so if the polarity of the relay connections is important, the relay manufacturer must both specify the polarity markings and clearly document their meaning. Relays that sense the *direction of current* (or *power*) *flow* at a specific location, and thus indicate the direction of the fault, provide a good practical example of relay polarity. Directional units are usually not applied alone but rather, in combinations with other units, such as fault sensors or detectors. A common practice is to use the output of the directional-sensing unit to control the operation of the fault sensors, which often is an instantaneous or an inverse-time-overcurrent unit, or both units together. Thus if the current flow is in the desired operating direction (trip direction) and its magnitude is greater than the fault sensor's minimum operating current (pickup), the relay can operate. If the current is in the opposite direction (nontrip or nonoperate direction or zone), no operation can occur even though the magnitude of the current is above the pickup threshold current.

A directional-sensing unit requires a *reference quantity* that is reasonably constant against which the current in the protected circuit can be compared. For relays intended to provide operation for phase-type faults, one of the system voltages of Fig. 3.3b can be used as a reference. For all practical purposes most system voltages do not change their phase positions significantly during a fault. In contrast, line currents can shift around 180° (essentially reverse their direction or flow) for faults on one side of the circuit CTs relative to a fault on the other side of the CTs.

Typical polarity indications for three commonly used directional-sensing units are shown in Fig. 3.7. This uses the custom of showing several loops for voltage coils and a single loop for current coils, of placing the reference circuit or voltage circuit above the current circuit, and of placing the polarity markings diagonally, all as shown on the relay schematics in Fig. 3.7.

The reference quantity is commonly called the "polarizing" quantity, especially for ground-fault relaying, where either or both current and/or voltage polarizing is used. The polarity marks (Fig. 3.7) are small plus symbols (+) placed, as illustrated, above one end of each coil, diagonally as shown, or on the opposite diagonal. As shown in Fig. 3.5, relay operation is not affected whether the polarity marks are on one diagonal or the other.

The meaning of the polarity for a specific relay must be stated clearly in words or by a diagram such as the one shown in Fig. 3.7. These show the basic design characteristics of an individual relay independent of any connection or association to the power system.

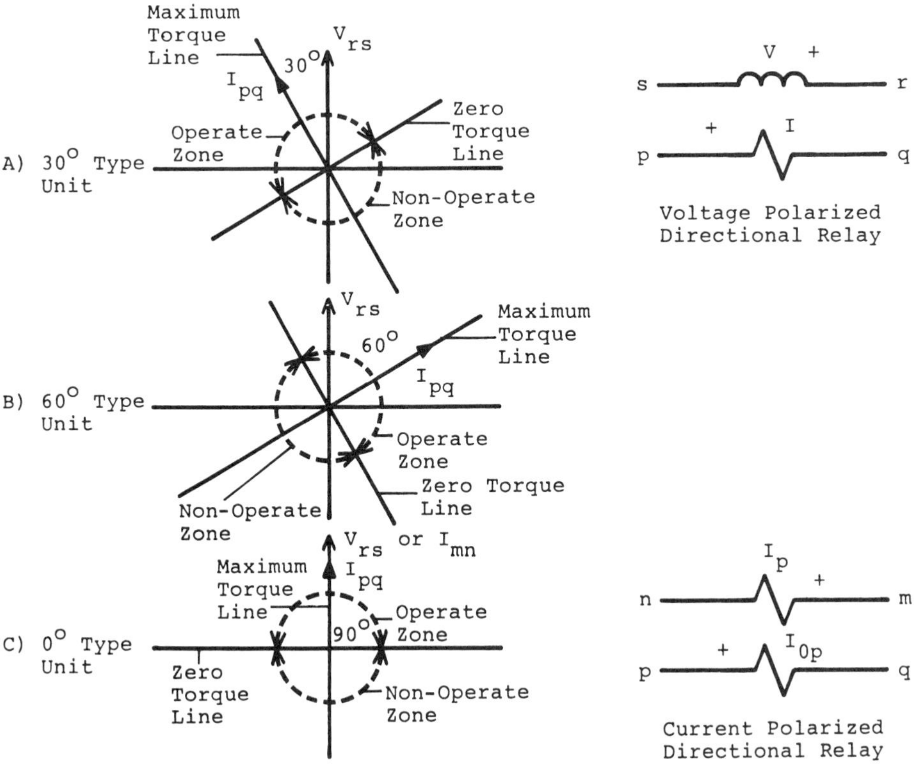

Fig. 3.7 Typical directional relay characteristics.

The terms "maximum torque line" and "zero torque line" come from the electromechanical designs long used and still common in the industry. With recent solid-state designs these would be the operating lines or thresholds, but the well-established terminology no doubt will continue for many years for all types of designs.

The interpretation of relay polarity is illustrated in Fig. 3.7 for three typical units. In Fig. 3.7A the maximum operating energy occurs when the current flow from polarity to nonpolarity (I_{pq}) leads by 30° the voltage drop from polarity to nonpolarity (V_{rs}). The minimum pickup of the directional unit is specified at this maximum torque or operating condition. As seen, the unit will operate for currents from almost 60° lagging the reference voltage V_{rs} to almost 120° leading. The operate (trip, contact close) zone or area is represented by the half plane, bordered on one side by the zero-torque (nonoperating) line and extending in the direction that contains both the reference (polarizing) and operating quantities. Higher current values will be required if I_{pq} deviates from the maximum-torque line. The operating torque is a function of the cosine of the angle between the current (I_{pq}) and the maximum-torque line, as well as the magnitudes of the operating quantities.

The 60° unit and zero (or watt) unit are illustrated in Fig. 3.7B and C, respectively. These are used in ground-fault protection and are discussed at the end of the next section.

A similar type of directional-sensing unit to that shown in Fig. 3.7A has its maximum-torque line at 45° leading instead of 30° leading. Both of these units are in wide use for phase-fault protection.

3.6 APPLICATION OF POLARITY FOR PHASE-FAULT DIRECTIONAL SENSING

A number of phase voltages, as shown in Fig. 3.3b, exist within the power system and are available for consideration as the reference quantity for directional relaying. Five different connections for phase-fault directional sensing have been used over the years. These are outlined in Table 3.1. In recent years connections 4 and 5 have been used almost exclusively, so these will be discussed. The other three, in very limited applications, are outlined in Table 3.1 for reference only.

Connections 4 and 5 are fundamentally the same and they are known as the "90° connection." The only difference between them is the angle that the system current lags the system voltage for maximum operating torque or energy. Either 60° or 45° is the typical angle of the fault current for maximum energy or torque. The difference is of no significance because the cos (60° − 45°) = 0.97 and typical pickup of these types of directional units is about 2 to 4 VA or less. With the normal 120 V available to the relay, this represents a current sensitivity on the order of 0.02 to 0.04 A. As a result, the normal power load in the operating zone will operate the phase directional units, but the relay will not operate unless a fault has occurred to increase the current above the fault-sensing unit's pickup.

3.6.1 The 90°–60° Connection for Phase-Fault Protection

The 90° connection (4 and 5 of Table 3.1) applies a power system voltage that lags the power system *unity power factor* current by 90°. These voltages and currents are obtained from the power system through voltage and current transformers. Typical three-phase connections are shown in Fig. 3.8. Three separate units are used, one for each of the three phases of the power system. Only the directional sensing units are illustrated, with the fault sensors or detectors omitted for

Table 3.1 Connection Chart for Phase-Fault Directional Sensing

		Phase A		Phase B		Phase C		Maximum torque
Connection	Unit type	I	V	I	V	I	V	occurs when
1 30°	Fig. 3.7C	I_a	V_{ac}	I_b	V_{ba}	I_c	V_{cb}	I lags 30°
2 60° delta	Fig. 3.7C	$I_a - I_b$	V_{ac}	$I_b - I_c$	V_{ba}	$I_c - I_a$	V_{cb}	I lags 60°
3 60° wye	Fig. 3.7C	I_a	$-V_c$	I_b	$-V_a$	I_c	$-V_b$	I lags 60°
4 90°–45°	Fig. 3.7A but maximum torque at 45°	I_a	V_{bc}	I_b	V_{ca}	I_c	V_{ab}	I lags 45°
5 90°–60°	Fig. 3.7A	I_a	V_{bc}	I_b	V_{ca}	I_c	V_{ab}	I lags 60°

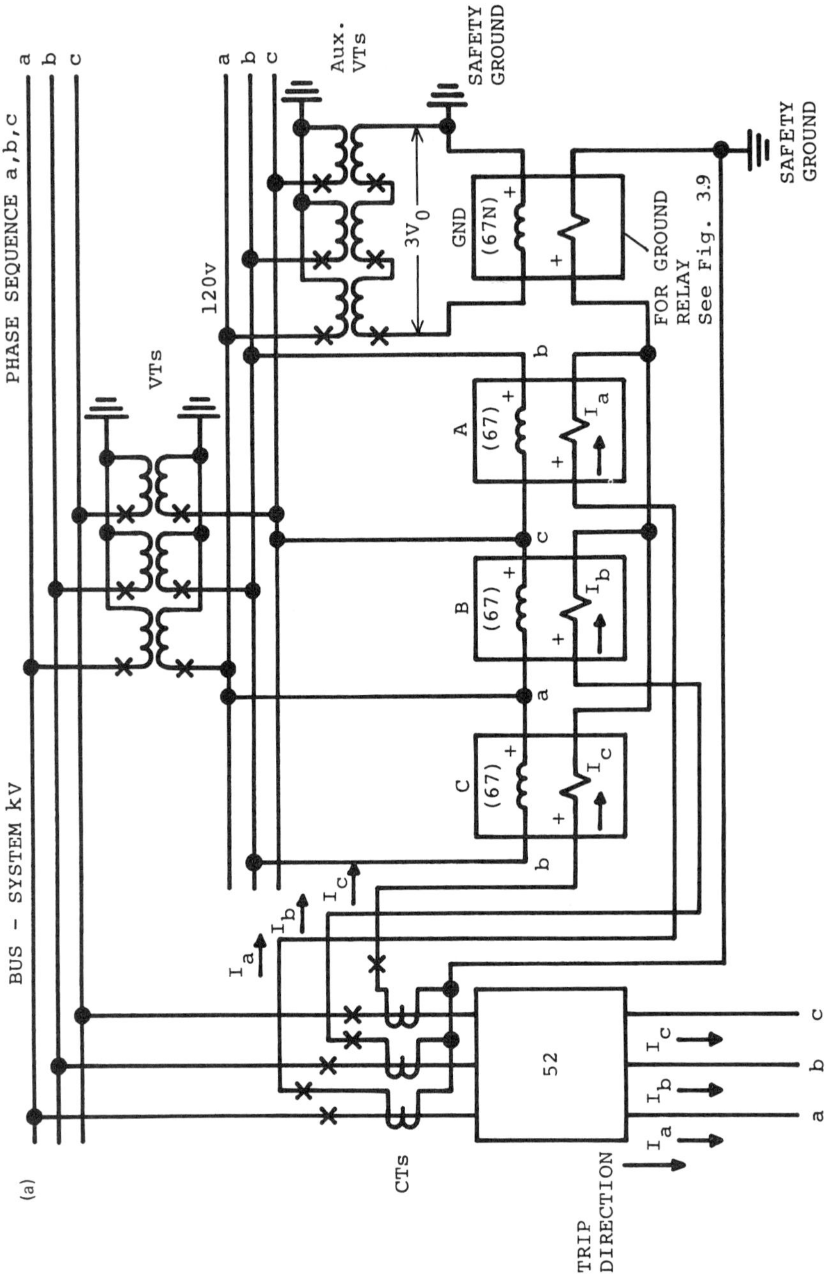

(a)
BUS - SYSTEM kV
PHASE SEQUENCE a,b,c
a
b
c
VTs
120v
Aux. VTs
SAFETY GROUND
3V0
GND
(67N)
FOR GROUND RELAY See Fig. 3.9
SAFETY GROUND
A
(67)
B
(67)
C
(67)
Ia
Ib
Ic
CTs
52
TRIP DIRECTION

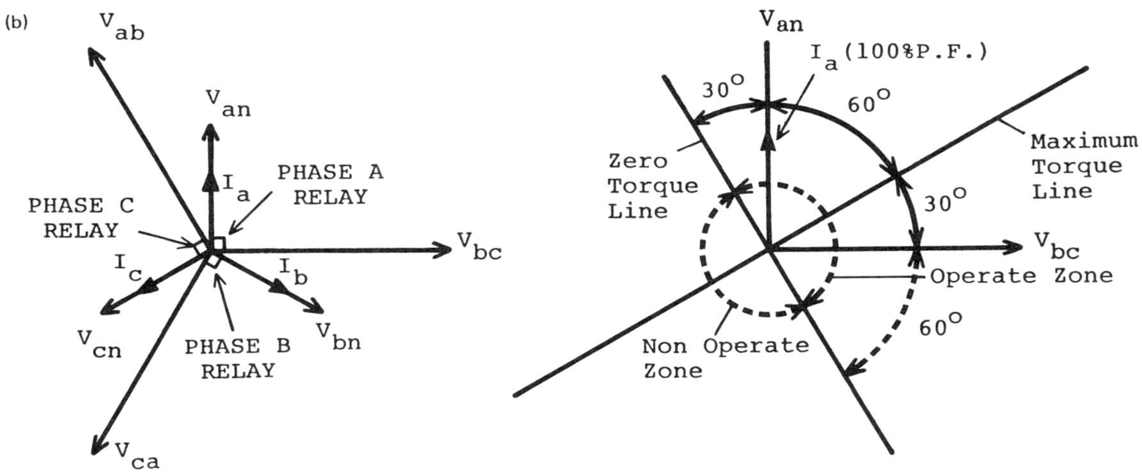

Fig. 3.8 Typical three-line connections for phase-fault directional sensing using the 30° unit of Fig. 3.7a, and for ground-fault directional sensing using the 60° unit of Fig. 3.7b.

this discussion. They are shown packaged by phase, but other packaging combinations are possible.

The phase A directional unit receives I_a and from the system phasors of Fig. 3.3b, the 90° lagging voltage is V_{bc}. The phase B directional unit receives I_b where the 90° lagging voltage is V_{ca}, and the phase C directional unit receives I_c where the 90° lagging voltage is V_{ab}. These are also shown in Table 3.1 for connections 4 and 5 and in Fig. 3.8.

In Fig. 3.8 the currents are connected so that when I_a, I_b, and I_c are flowing in the trip direction indicated by the "trip direction" arrow, the secondary currents flow through the directional units from polarity to nonpolarity. The polarity of the CTs does not have to go to the polarity of the relay, although often that is convenient, as in this example.

With the trip direction of the currents established in the directional unit current coils, the voltages V_{bc} on unit A, V_{ca} on unit B, and V_{ab} on unit C must be connected from polarity to nonpolarity on the directional unit voltage coils as shown. The right-hand phasor diagram of Fig. 3.8b applies the directional unit characteristic of Fig. 3-7a to the power system phasors. The maximum-torque line leads the voltage by 30°, so with V_{bc} polarity to nonpolarity on the relay voltage winding, the maximum-torque line is drawn 30° leading, as illustrated in the lower right-hand phasor diagram. This is 60° lagging the unity power factor position of the current phasor I_a. Thus whenever the phase a current in the power system lags by 60°, the directional unit will operate at maximum torque with the lowest pickup value and the highest sensitivity. Since most system faults provide relatively large currents, the range of possible operation is for power system currents from almost 30° leading to 150° lagging in the "trip direction." This is the operating zone shown in Fig. 3.8. Similar relations exist for the other two-phase units using I_b and I_c phase currents.

Thus the discussion above describes the 90°–60° connection, where a 90° lagging voltage is used and maximum operation occurs when the phase current lags in the system by 60°. The 90°–45° connection is identical except that the relay design provides maximum torque, leading the reference voltage by 45° rather than by 30° used for illustration.

Wye-wye-connected voltage transformers (VTs) are shown in the typical connections of Fig. 3.8. Open-delta connections using only two VTs to provide three-phase voltage can be utilized as an alternate. This is applicable only for phase-fault protection, not for ground protection.

3.7 DIRECTIONAL SENSING FOR GROUND FAULTS: VOLTAGE POLARIZATION

The connections of a directional sensing unit for ground-fault protection utilizing a voltage reference (voltage polarization) are shown in Fig. 3.8 and in greater detail in Fig. 3.9. While the phase relays of Section 3.6 were connected and analyzed using balanced three-phase voltages and currents, it is necessary to assume a fault involving ground for the ground relay. Thus a phase-a-to-ground fault is assumed in the trip direction as shown in Fig. 3.9. The characteristics of this type of fault generally are a collapse of the faulted phase voltage (V_{ag}) with an increase and lag of the faulted phase current (I_a), such as typically illustrated in the left phasor diagram. In many cases the unfaulted (b and c) phase currents are small and negligible practically, so that their phase-to-ground voltages are essentially uncollapsed. The assumption here is that $I_b = I_c = 0$, so that $I_a = 3I_0$. This, together with V_0 or $3V_0$, is a zero-sequence quantity, reviewed in Chapter 4.

In Fig. 3.9, voltage polarizing (voltage reference) is shown which uses the $3V_0$ zero-sequence voltage derived from a "broken-delta" connection of the VTs (in this example the auxiliary VTs). This voltage is the sum of the three-phase-to-ground voltages as shown in the connections and in middle phasor diagram. For balanced conditions this $3V_0$ voltage is zero.

For ground-fault protection a directional 60° unit (Fig. 3.7B) is used. The connections show that with I_a fault current flowing in the tripping direction and to the fault, the secondary current flows in the ground relay from its polarity-to-nonpolarity direction. To provide proper operation it becomes necessary to apply $-3V_0$ to the ground relay voltage coil from polarity to nonpolarity, as shown in the connections and the right phasor diagram, where the operation is as indicated. With $-3V_0$ connected to the ground relay from polarity to nonpolarity, the relay characteristic of Fig. 3.7B indicate that maximum torque will occur when the current polarity to nonpolarity lags in the power system by 60°. Thus the maximum-torque line is drawn as shown in the lower right phasor diagram of Fig. 3.9. As long as the magnitudes of $-3V_0$ and $3I_0$ are above the pickup of the directional unit, it will operate for currents from nearly 30° leading to 150° lagging.

An alternative check of these connections may be made by assuming that the VTs are a "ground source" with current flowing from the ground through the primary of the VT to the fault. Tracing this through the VT windings, this assumed current, together with the fault current from the system, will flow from polarity to nonpolarity in both of the ground relay windings.

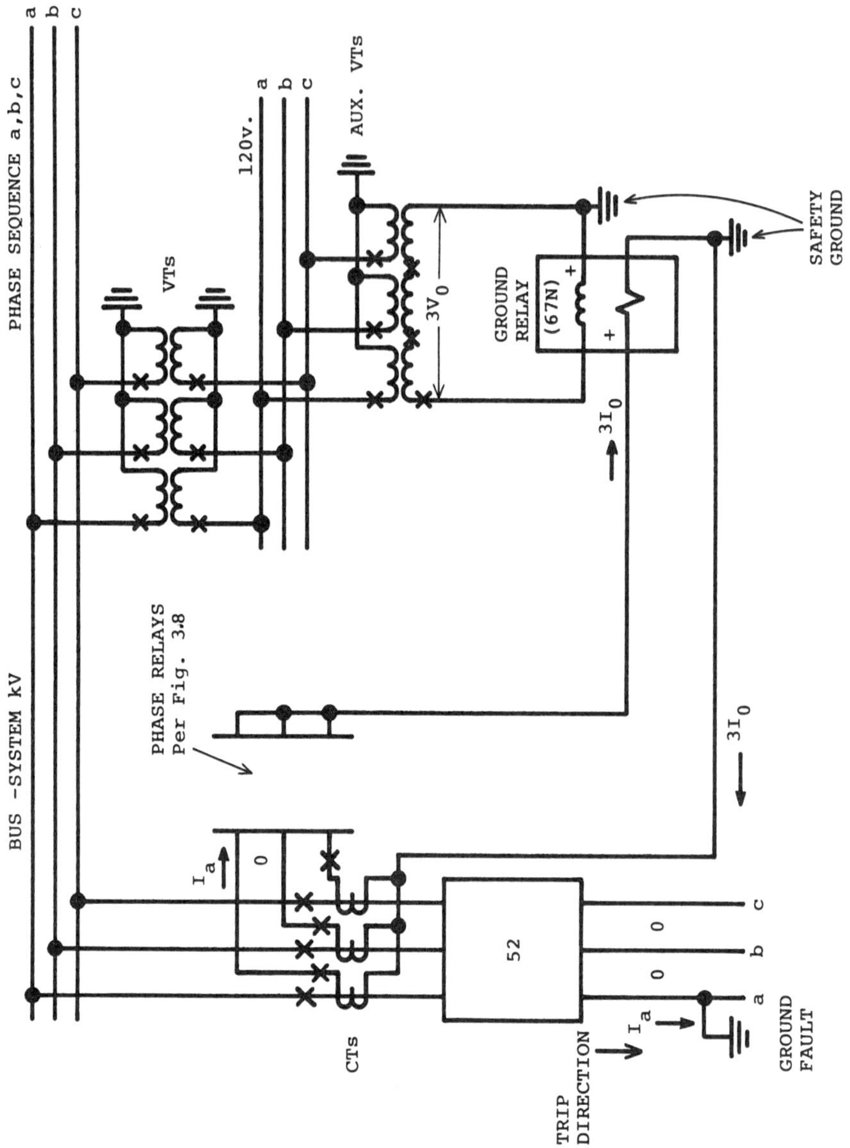
PHASE SEQUENCE a,b,c
BUS -SYSTEM kV
a
b
c
VTs
120v.
AUX. VTs
3V0
GROUND RELAY
(67N)
SAFETY GROUND
3I0
PHASE RELAYS Per Fig. 3.8
Ia
CTs
52
TRIP DIRECTION
GROUND FAULT

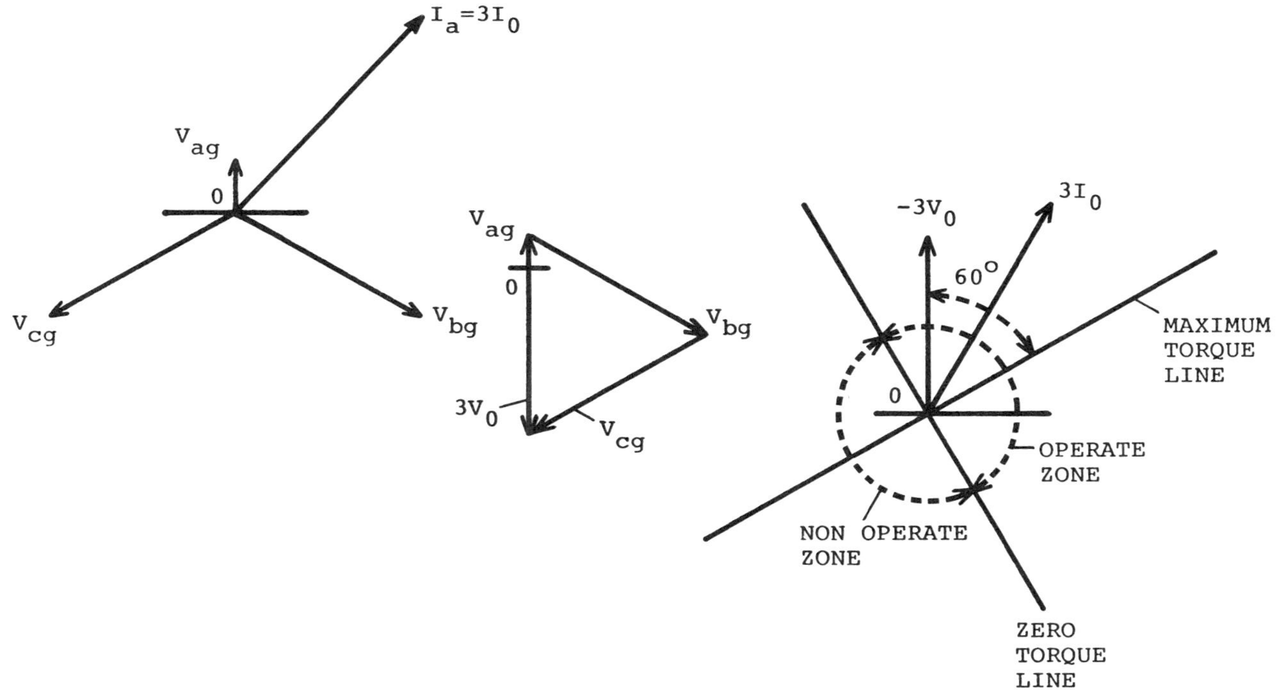

Fig. 3.9 Typical three-line connections for ground-fault directional sensing with voltage polarization using the 60° unit of Fig. 3.7B.

3.8 DIRECTIONAL SENSING FOR GROUND FAULTS: CURRENT POLARIZATION

The current that flows in the grounded neutral of a wye-delta power or distribution transformer bank can be used as a reference or polarizing quantity for ground-fault protection. Typical connections are shown in Fig. 3.10. Again no current can flow to the ground relay either from the fault or from the transformer bank neutral if the power system is balanced. Thus a phase-a-to-ground fault is shown on phase a in the trip direction. For simplification, I_b and I_c phase currents are assumed to be zero. For all practical purposes the current flowing into the fault will be essentially in phase with the current flowing up the transformer bank neutral, so the 0° type ground relay with characteristics as shown in Fig. 3.7C is applicable and is the one used in Fig. 3.10 connections.

To illustrate and to emphasize that the polarity marks on the current transformers do not have to be connected to the polarity-marked terminal of the relay, the fault I_a current from the CTs on the line have been connected arbitrarily so that I_a flows from nonpolarity to polarity on the relay coil. Therefore, the polarizing secondary current I_a must be connected from nonpolarity to polarity when the primary fault I_n flows up the neutral of the transformer.

With the currents I_a and I_n in phase, maximum operating torque will occur as per Fig. 3.7C. Operation is still possible, as one current leads or lags by almost 90° from the other as long as the magnitudes are above the required pickup values for the directional unit. It should be obvious that the connections of Fig. 3.10 are also correct (also for Fig. 3.9) if the operating quantity I_a ($3I_0$) and the polarizing quantity are *both* reversed at the directional relay.

3.9 OTHER DIRECTIONAL-SENSING CONNECTIONS

Various other directional-sensing, connections can be derived to measure different power factor areas by connecting the different (or other) units of Fig. 3.7 to various combinations of currents or current and voltage. One type of connection is used to operate on power, either watts or vars, flowing in the power system. These relays (designated device 32) are available with various operating power levels.

For watt measurement, the 0° unit of Fig. 3.7C can be used with star (or delta) current or voltage. For example, this unit, used with I_a phase current and V_{an} voltage, will give maximum torque when these two quantities are in phase. Also, $I_a - I_b$ with V_{ab} could be used. For var measurement, I_a with V_{bc} provides maximum torque when I_a lags 90° and zero torque when I_n is at unity power factor and flowing in either direction.

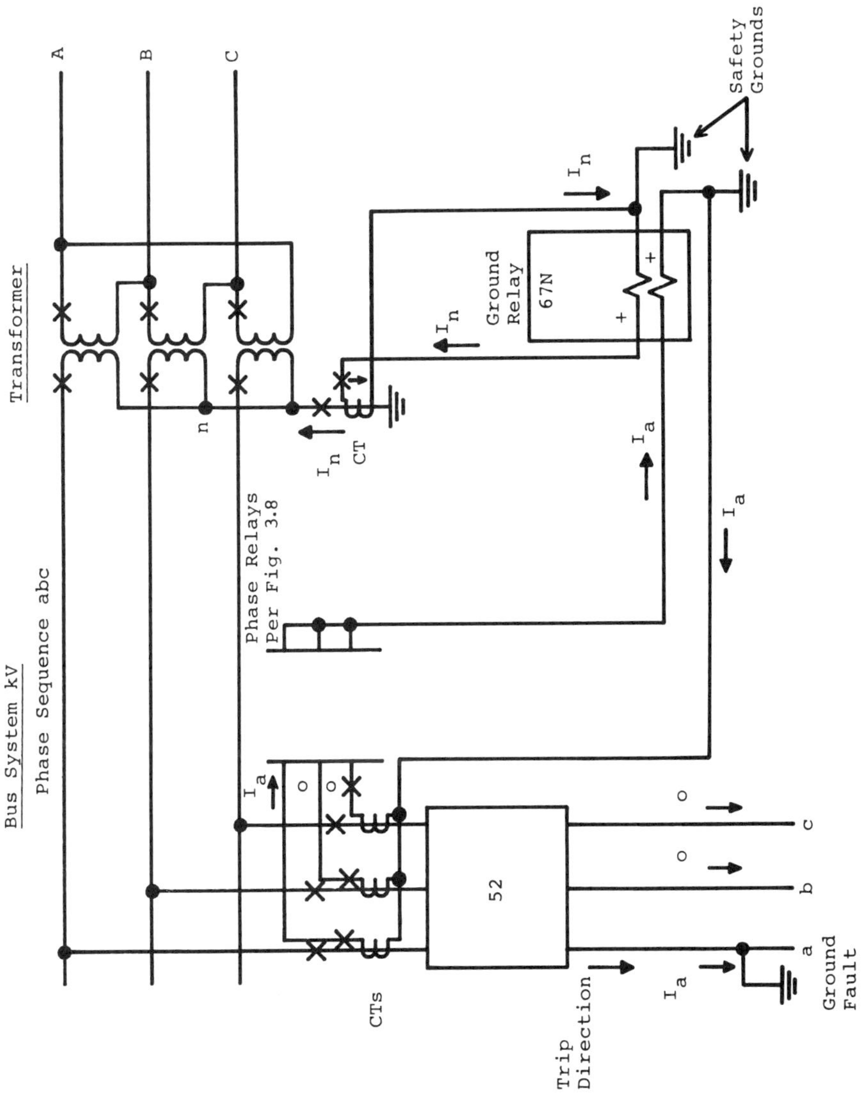

Fig. 3.10 Typical three-line connections for ground-fault directional sensing with current polarization using the 0° unit of Fig. 3.7c.

A watt-power relay (32) can also be obtained by using the 30° unit of Fig. 3.7A with I_a and V_{ac}. This places the maximum-torque line in phase with the unit power factor position of I_a. Similarly, a var-type relay can be obtained using this 30° unit with I_a and V_{bn}. This provides maximum torque when I_a lags by 90°.

3.10 SUMMARY

The fundamental methodology of phasors and polarity reviewed in this chapter will be employed throughout the rest of the book. As stressed previously, these concepts are essential as useful aids in the selection, connection, operation, performance, and testing of the protection for all power systems.

4

Symmetrical Components: A Review

4.1 INTRODUCTION AND BACKGROUND

The method of symmetrical components provides a practical technology for understanding and analyzing power system operation during unbalanced conditions such as those caused by faults between phases and/or ground, open phases, unbalanced impedances, and so on. Also, many protective relays operate from the symmetrical component quantities. Thus a good understanding of this subject is of great value and another very important tool in protection.

In a sense "symmetrical components" can be called the "language" of the relay engineer or technician. Its value is both in thinking or visualizing unbalances, as well as a means of detailed analysis of them from the system parameters. In this simile it is like a language, in that it requires experience and practice for easy access and application. Faults and unbalances occur infrequently, and many do not require detailed analysis, so it becomes difficult to "practice the language." This has increased with the ready availability of fault studies via computers. These provide rapid access to voluminous data, often with little user understanding of the background or method that provides the data. Hence this review of the method is intended to provide the fundamentals, basic circuits and calculations, and an overview directed at clear understanding and visualization.

The method of symmetrical components was discovered by Charles L. Fortescue, who was investigating mathematically the operation of induction motors under unbalanced conditions late in 1913. At the 34th Annual Convention of the AIEE on June 28, 1918, in Atlantic city, he presented an 89-page paper entitled "Method of Symmetrical Co-ordinates Applied to the Solution of Polyphase Networks." The six discussants, including Steinmetz, added 25 pages. Practical application for system fault analysis was developed by C. F. Wagner and R. D. Evans in the later part of 1920s and early 1930s, with

W. A. Lewis adding valuable simplifications in 1933. E. L. Harder provided tables of fault and unbalance connections in 1937. Edith Clarke was also developing notes and lecturing in this area at this time, but formal publication of her work did not occur until 1943.

Only symmetrical components for three-phase systems will be reviewed. For these systems there are three distinct sets of components: positive, negative, and zero for both current and voltage. Throughout this discussion, the sequence quantities are *always line to neutral* or *line to ground*, as appropriate. This is an exception for voltage, where in the power system *line to line* voltages commonly are indicated, but in symmetrical components they are always *line to neutral* (or possibly line to ground).

4.2 POSITIVE-SEQUENCE SET

This set consists of the balanced three-phase currents and line-to-neutral voltages supplied by the system generators. Thus they are *always* equal in magnitude and phase displaced by 120°. With the power system phase sequence of a, b, c, Fig. 4.1 shows a positive-sequence set of phase currents. A voltage set is similar except for line-to-neutral voltage of the three phases, again all equal in magnitude and displaced 120°. These are phasors rotating counterclockwise at the system frequency.

To document the angle displacement, it is convenient to use a unit phasor with an angle displacement of 120°. This is designated as "a" so that,

$$\begin{aligned} a &= 1\angle 120^\circ = -0.5 + j0.866 \\ a^2 &= 1\angle 240^\circ = -0.5 - j0.866 \\ a^3 &= 1\angle 360^\circ = 1\angle 0^\circ = 1.0 + j0 \end{aligned} \tag{4.1}$$

Thus the positive-sequence set can be designated as

$$\begin{aligned} I_{a1} &= I_1 & V_{a1} &= V_1 \\ I_{b1} &= a^2 I_{a1} = a^2 I_1 = I_1\angle 240^\circ & V_{b1} &= a^2 V_1 = V_1\angle 240^\circ \\ I_{c1} &= aI_{a1} = aI_1 = I_1\angle 120^\circ & V_{c1} &= aV_1 = V_1\angle 120^\circ \end{aligned} \tag{4.2}$$

It is most important to emphasize that the set of sequence currents or sequence voltages *always* exists as defined. I_{a1} or I_{b1} or I_{c1}

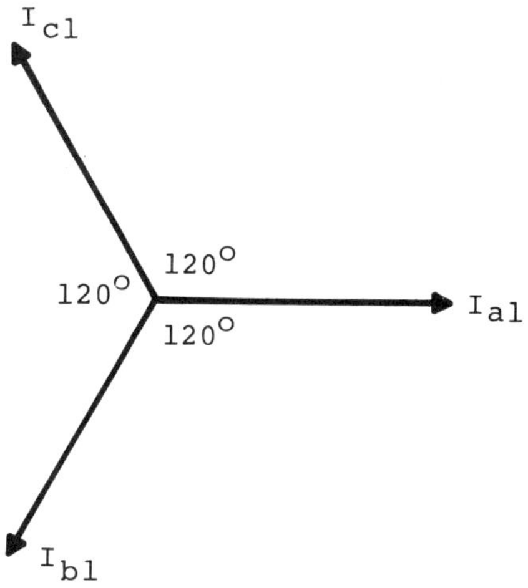

Fig. 4.1 Positive-sequence current phasors.

can *never* exist alone or in pairs, *always* all three. Thus it is necessary to define only one of the phasors (any one), from which the other two will be as documented in Eq. (4.2).

4.3 NOMENCLATURE CONVENIENCE

It will be noted that the designation subscript for phase a was dropped in the second expression for the currents and voltages in Eq. (4.2) (and also in the following equations). This is a common shorthand notation used for convenience. Whenever the phase subscript does not appear, it can be assumed that the reference is to phase a. If phase b or phase c quantities are intended, the phase subscript must be correctly designated; otherwise, it is assumed to be phase a. This shortcut will be used throughout the book and is common in practice.

4.4 NEGATIVE-SEQUENCE SET

This set is also balanced with three equal magnitude quantities at 120° apart, but with the phase rotation or sequence reversed as illustrated in Fig. 4.2. Thus if positive sequence is a, b, c, negative will be a, c, b. Where positive sequence is a, c, b, as in some power systems, negative sequence is a, b, c.

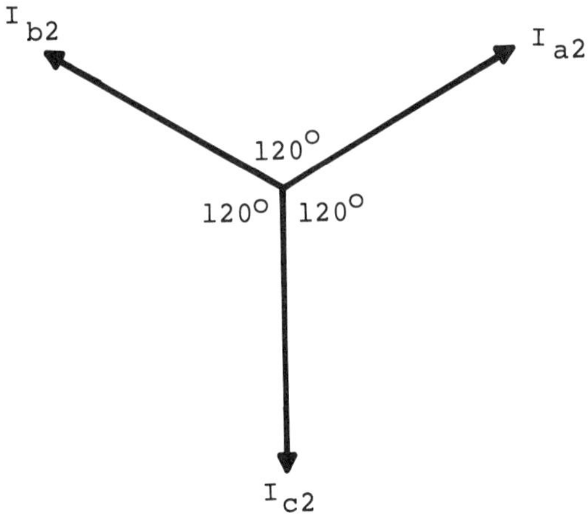

Fig. 4.2 Negative-sequence current phasors.

The negative-sequence set can be designated as

$$I_{a2} = I_2 \qquad V_{a2} = V_2$$

$$I_{b2} = aI_{a2} = aI_2 = I_2\angle 120° \qquad V_{b2} = aV_2 = V_2\angle 120°$$

$$I_{c2} = a^2I_{a2} = a^2I_2 = I_2\angle 240° \qquad V_{c2} = a^2V_2 = V_2\angle 240° \tag{4.3}$$

Again, negative sequence always exists as a set of current or voltage as defined above or in Fig. 4.2. I_{a2} or I_{b2} or I_{c2} can *never* exist alone. When one current or voltage phasor is known, the other two of the set can be defined as above.

4.5 ZERO-SEQUENCE SET

The members of this set of rotating phasors are always equal in magnitude and always in phase (Fig. 4.3).

$$I_{ao} = I_{bo} = I_{co} = I_o \qquad V_{ao} = V_{bo} = V_{co} = V_o \tag{4.4}$$

Again, I_o or V_o, if it exists, exists equally in all three phases, *never* alone in one phase.

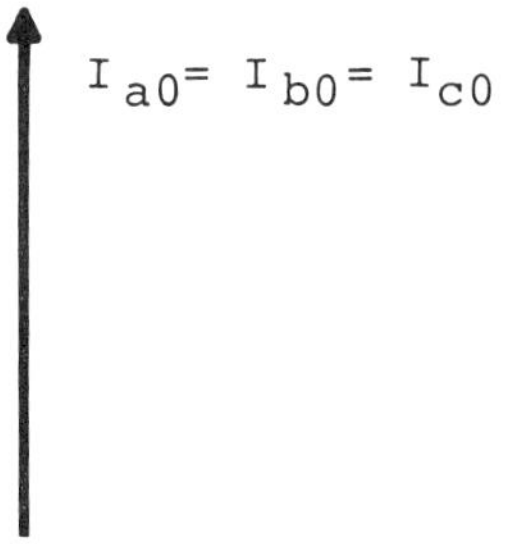

Fig. 4.3 Zero-sequence current phasors.

4.6 GENERAL EQUATIONS

Any unbalanced current or voltage can be determined from the sequence components from the following fundamental equations:

$$I_a = I_1 + I_2 + I_o \qquad V_a = V_1 + V_2 + V_o \tag{4.5}$$

$$I_b = a^2 I_1 + aI_2 + I_o \qquad V_b = a^2 V_1 + aV_2 + V_o \tag{4.6}$$

$$I_c = aI_1 + a^2 I_2 + I_o \qquad V_c = aV_1 + a^2 V_2 + V_o \tag{4.7}$$

where I_a, I_b, and I_c or V_a, V_b, and V_c are general unbalanced line-to-neutral phasors.

From these, equations defining the sequence quantities from a three-phase unbalanced set can be determined:

$$I_o = \frac{1}{3}(I_a + I_b + I_c) \qquad V_o = \frac{1}{3}(V_a + V_b + V_c) \tag{4.8}$$

$$I_1 = \frac{1}{3}(I_a + aI_b + a^2 I_c) \qquad V_1 = \frac{1}{3}(V_a + aV_b + a^2 V_c) \tag{4.9}$$

$$I_2 = \frac{1}{3}(I_a + a^2 I_b + aI_c) \qquad V_2 = \frac{1}{3}(V_a + a^2 V_b + aV_c) \tag{4.10}$$

These last three fundamental equations are the basis for determining if the sequence quantities exist in any given set of unbalanced three-phase currents or voltages. They are used for protective relaying operation from the sequence quantities. For example,

Fig. 4.4 shows the physical application of current transformers (CTs) and voltage transformers (VTs) to measure zero sequence as required by Eq. (4.8), and used in ground-fault relaying.

Networks operating from CTs or VTs are used to provide an output proportional to I_2 or V_2 and are based on physical solutions of (Eq. 4.10). This can be accomplished with resistors, transformers, reactors, or with operational amplifier components (see Westinghouse Electric Corp., *Applied Protective Relaying*).

4.7 SEQUENCE INDEPENDENCE

The key that makes dividing unbalanced three-phase quantities into the sequence components practical is the independence of the components in a balanced system network. For all practical purposes electric power systems are balanced or symmetrical from the generators to the point of single-phase loading except in an area of a fault or unbalance such as an open conductor. In this essentially balanced area, the following conditions exist:

1. Positive-sequence currents flowing in the symmetrical or balanced network produce *only* positive-sequence voltage drops; *no* negative- or zero-sequence drops.
2. Negative-sequence currents flowing in the balanced network produce *only* negative-sequence voltage drops; *no* positive- or zero-sequence voltage drops.
3. Zero-sequence currents flowing in the balanced network produce *only* zero-sequence voltage drops; *no* positive- or negative-sequence voltage drops.

This is not true for any unbalanced or nonsymmetrical point or area such as an unsymmetrical fault, open phase, and so on. In these:

4. Positive-sequence current flowing in an unbalanced system produces positive- and negative- and possibly zero-sequence voltage drops.
5. Negative-sequence currents flowing in an unbalanced system produces positive-, negative-, and possibly zero-sequence voltage drops.
6. Zero-sequence current flowing in an unbalanced system produces all three: positive-, negative-, and zero-sequence voltage drops.

This important fundamental permits setting up three independent networks, one for each of the three sequences, which can be interconnected only at the point or area of unbalance.

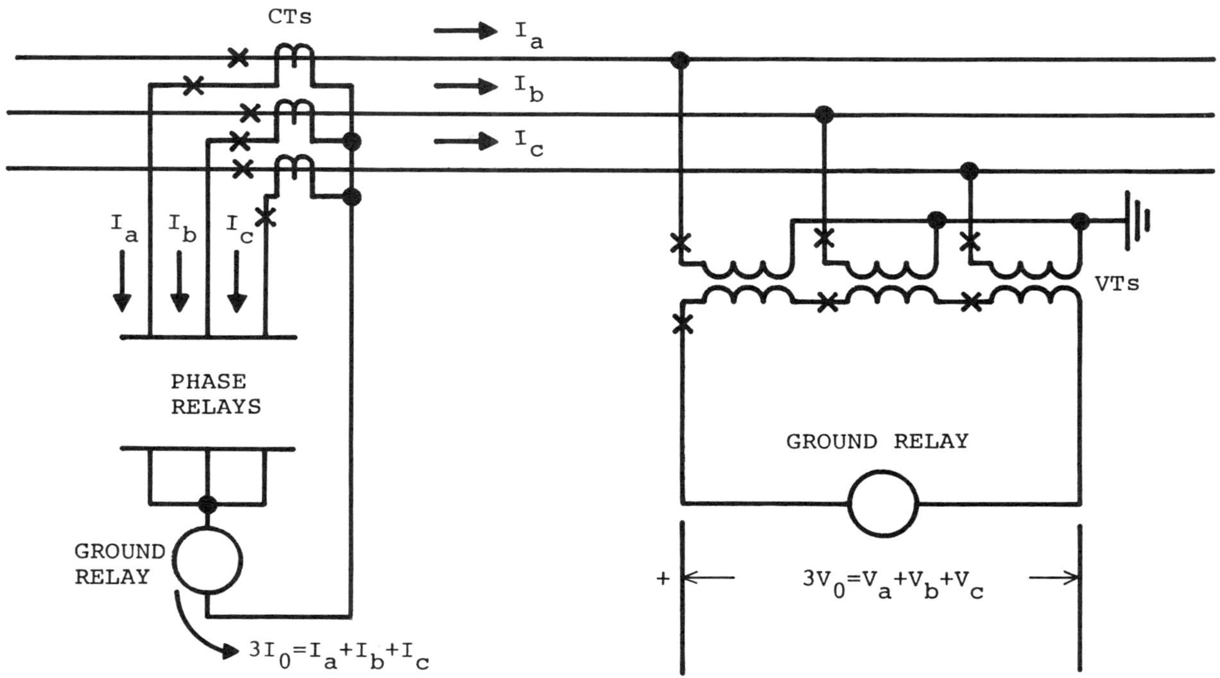

Fig. 4.4 Zero-sequence current and voltage networks used for ground-fault protection. See Figs. 3.9 and 3.10 for typical fault operation.

Before continuing with the sequence networks, a review of the sources of fault current is useful.

4.8 POSITIVE-SEQUENCE SOURCES

A single-line diagram of the power system or area under study is the starting point for setting up the sequence networks. A typical diagram for a section of a power system is shown in Fig. 4.5. In these diagrams, circles are used to designate the positive-sequence sources, which are the rotating machines in the system; generators, synchronous motors, synchronous condensers, and possibly induction motors. The symmetrical current supplied by these to power system faults decreases exponentially with time from a relative high initial value to a low steady-state value. During this transient period three reactance values are possible for use in the positive-sequence network and for the calculation of fault currents. These are the direct-axis subtransient reactance (X''_d), the direct-axis transient reactance (X'_d), and the unsaturated direct-axis synchronous reactance (X_d).

The values of these reactances vary with the designs of the machines, and specific values are supplied by the manufacturer. In their absence, typical values may be used and are available from many references (see, e.g., the books by Calabrese and Anderson in the Bibliography, and Westinghouse Electric Corp., *Electrical Transmission and Distribution Reference Book*). Very generally, X''_d in per unit of the machine kVA rating ranges from 0.7 to 3.0 pu with time constants of 0.035 to 0.05 s. Similarly, X'_d is about 1.5 to 2.5 X''_d with time constants on the order of 0.6 to 1.5 s. X_d is an unsaturated steady-state value for sustained faults about 4 to 15 times X''_d.

For system protection fault studies the almost universal practice is to use the subtransient (X''_d) for the rotating machines in the positive-sequence networks. This provides a maximum value of fault current useful for high-speed relaying. Although slower-speed protection may operate after the subtransient reactance has decayed into the transient reactance period, the general practice is to use X''_d except possibly for special cases where X'_d would be used. There are special programs to account for the decremental decay in fault current with time in setting the slower-speed protective relays, but these tend to be difficult and tedious, and may not provide significant advantages. A guide to aid in the understanding of the need for special considerations is outlined in Fig. 4.6. The criteria are very general and approximate.

Cases A and B are the most common situations, so that the use of X''_d has a negligible effect on the protection. Here the higher system Z_S tends to negate the source decrement effects.

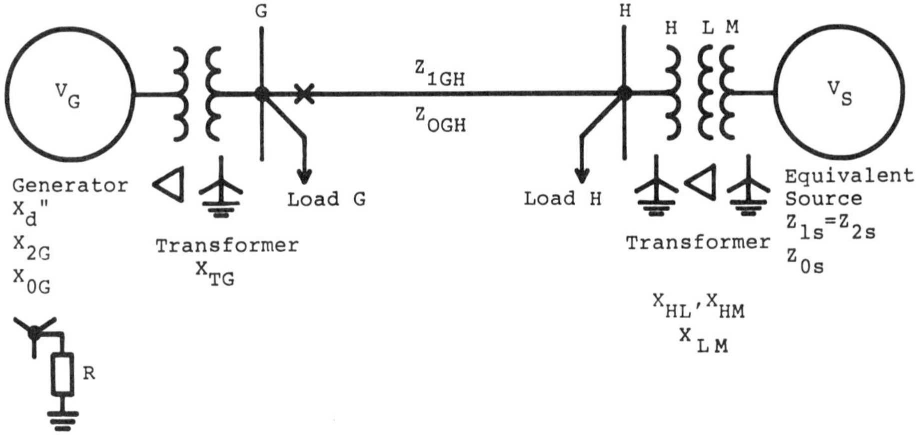

Fig. 4.5 Single-line diagram of a section of a power system.

Case C can affect the overall time operation of a slower-speed protection, but generally the decrease in fault current level with time will not cause coordination problems unless the time-current characteristics of various devices used are significantly different. Where Z_M predominates, the fault levels tend to be high and well above the maximum load current. The practice of setting the protection as sensitive as possible but not operating on maximum load (phase devices) should provide good protection sensitivity in the transient reactance period. If protection operating times are very long, so that the current decays into the synchronous reactance period, special phase relays are required, as discussed in Chapter 8.

In most cases induction motors are not considered sources of fault current for protection purposes (Case D). However, it must be emphasized that these motors must be considered in circuit breakers applications under the ANSI/IEEE standards. Without a field source, the voltage developed by induction motors decay rapidly within a few cycles, so generally have a negligible effect on the protection.

The dc offset that can result from sudden changes in current in ac networks is neglected in symmetrical components. It is an important consideration in all protection.

An equivalent source such as that shown in Fig. 4.5 represents the equivalent of all of the system not shown up to the point of connection to the part of the system under study. This includes one or many rotating machines together with any network of transformers, lines, and so on, that may be interconnected. In general, a network system can be reduced to two equivalent sources at each end of an area to be studied with an equivalent interconnecting tie between

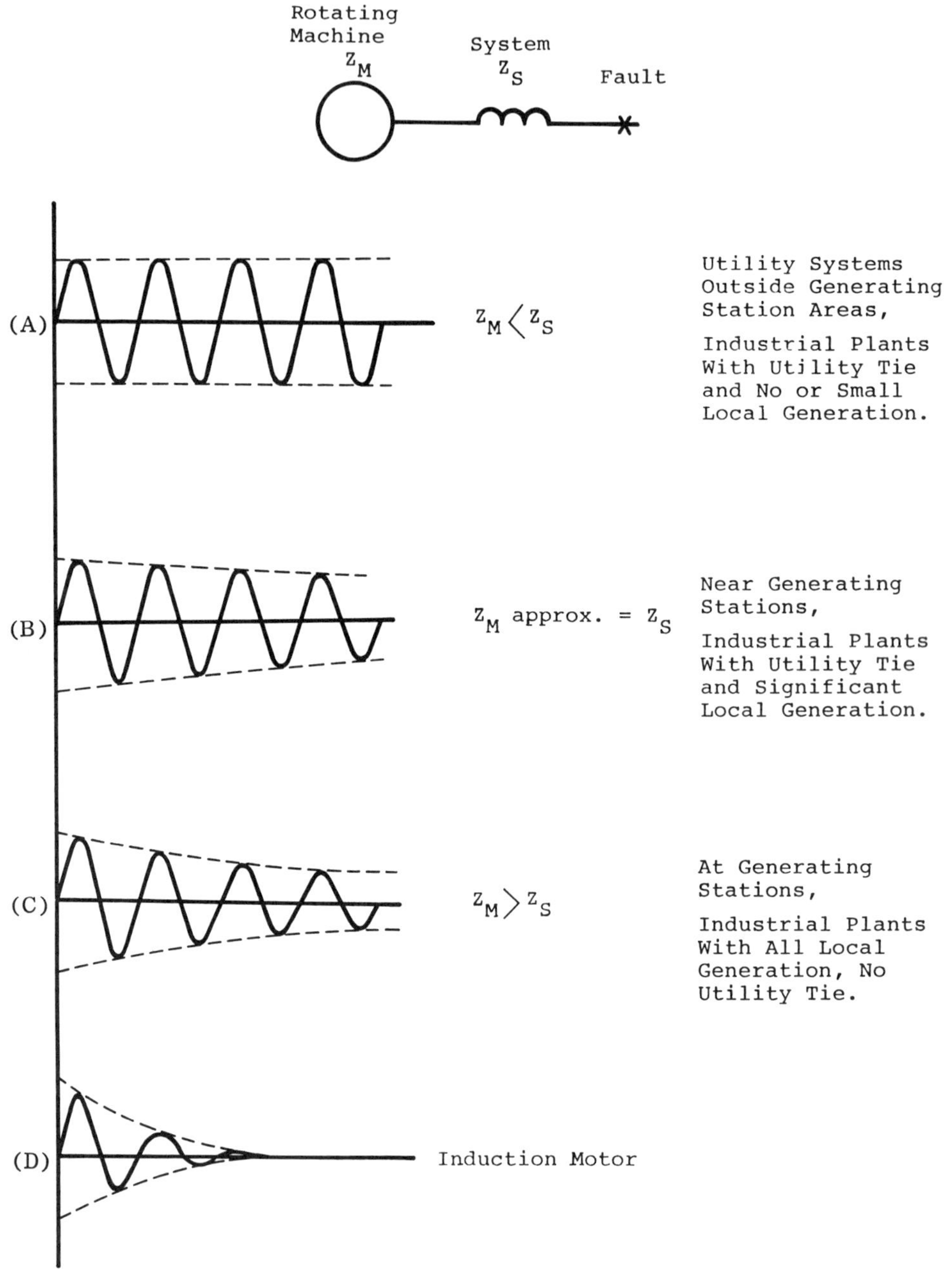

Fig. 4.6 Guide illustrating the effects of rotating machine decrements on the symmetrical fault current.

these two equivalent sources. When this equivalent tie is very large or infinite, indicating that little or no power is exchanged between the two source systems, it is convenient to express the equivalent source system up to a specified bus or point in short-circuit MVA (or kVA). Appendix 1 outlines this and the conversion to impedance or reactance values. In Fig. 4.5 the network to the right has reduced to a single equivalent impedance to represent it up to the M terminal of the three-winding transformer bank.

4.9 SEQUENCE NETWORKS

These represent one of the three-phase-to-neutral or ground circuits of the balanced three-phase power system, and documents how that sequence currents will flow if they can exist. These networks are best explained by an example, so consider the section of a power system of Fig. 4.5.

Reactance values only have been shown for the generator and the transformers. Theoretically, impedance values should be used, but the resistances of these units are very small and negligible for fault studies.

It is very important that all values be specified with a base [voltage if ohms are used, or MVA (kVA) and kV if per unit or percent impedances are used]. Before applying these to the sequence networks, all values must be changed to one common base. In most cases per unit (percent) values are used and a common base in practice is 100 MVA at the particular system kV.

4.9.1 Positive-Sequence Network

This is the usual line-to-neutral system diagram for one of the three symmetrical phases modified for fault conditions. The positive-sequence network for the system of Fig. 4.5 is shown in Fig. 4.7. V_G and V_S are the system line-to-neutral voltages. V_G is the voltage behind the generator subtransient direct-axis reactance X''_d, and V_S is the voltage behind the system equivalent impedance Z_{1S}.

X_{TG} is the transformer leakage impedance for the bank bus G, and X_{HM} is the leakage impedance for the bank at H between the H and M windings. More detail on these are given in Appendix 2. The delta winding L of this three-winding bank is not involved in the positive-sequence network unless a generator or synchronous motor is connected to this delta or unless a fault is to be considered in the L delta system. The connection would be as in Fig. A4.2-3.

For the line between buses G and H, Z_{1GH} is the line-to-neutral impedance of this three-phase circuit. For open-wire transmission lines an approximate estimating value is 0.8 Ω/mi for single conductor and 0.6 Ω/mi for bundled conductors. Typical values for shunt

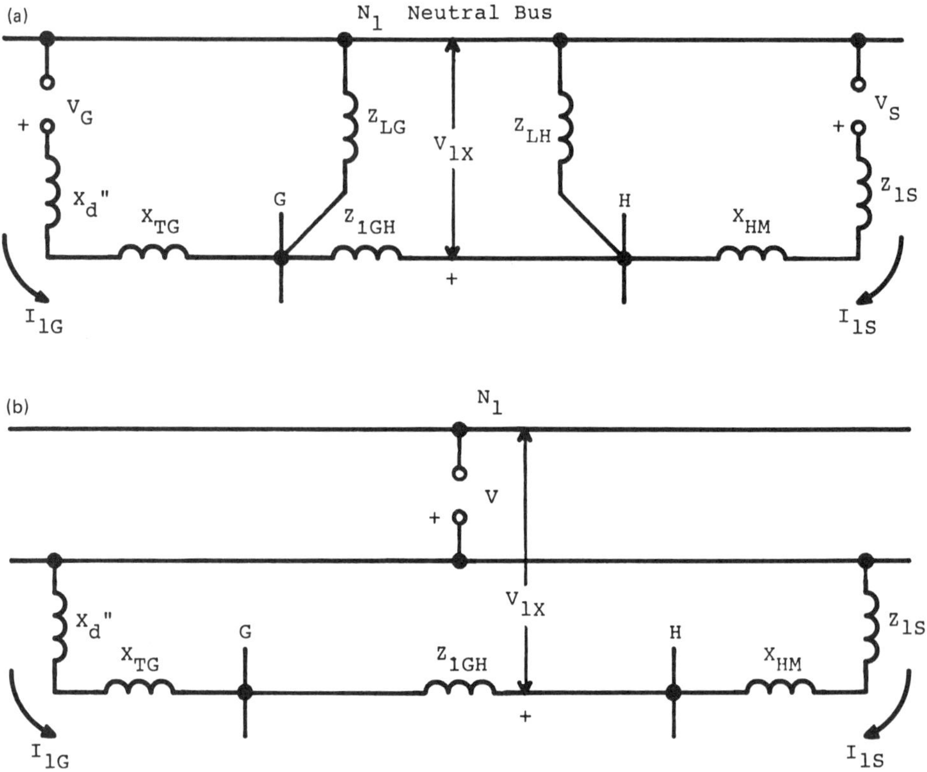

Fig. 4.7 Positive sequence networks for the system in Fig. 4.5: (a) network including loads; (b) simplified network with no load — all system voltages equal and in phase.

capacitance of these lines is 0.2 MΩ/mi for single conductor and 0.14 MΩ/mi for bundled conductors. Normally, this capacitance is neglected, as it is very high in relation to all other impedances involved in fault calculations. These values should be used for estimating or in the absence of specific line constants. The impedances of cables vary considerably, so specific data are necessary for these.

The impedance angle of lines can vary quite widely depending on the voltage and whether cable or open wire is used. In computer fault programs the angles are considered and included, but for hand calculation it is practical in most cases to simplify calculations by assuming that all the equipment involved in the fault calculation are at 90° or use reactance values only. Sometimes it may be preferred to use the line impedance values and treat them as reactances. Unless the network consists of a large proportion of low-angle circuits, the error of using all values as 90° will not be too significant.

Load is shown connected at buses G and H. Normally, this would be specified as kVA or MVA and can be converted into impedance:

$$I_{load} = \frac{1000\ MVA_{load}}{\sqrt{3}\ kV} \quad \text{and} \quad V_{LN} = \frac{1000\ kV}{\sqrt{3}}$$

$$Z_{load} = \frac{V_{LN}}{I_{load}} = \frac{kV^2}{MVA_{load}} = \Omega \text{ at kV} \tag{4.11}$$

Equation (4.11) is a line-to-neutral value and would be used for Z_{LG} and Z_{LH} representing the loads at G and H in Fig. 4.7a. If load is represented, the voltages V_G and V_S will be different in magnitude and angle, varying as the system load varies.

The value of load impedance ususally is quite large compared to the system impedances, so that load has a negligible effect on the faulted phase current. Thus it becomes practical and simplifies calculations to neglect load for shunt faults. With no load, Z_{LG} and Z_{LH} are infinite, V_G and V_S are equal and in phase and so are replaced by a common voltage V as in Fig. 4.7b. Normally, V is considered as 1 pu, the system-rated line-to-neutral voltages.

Conventional current flow is assumed to be from the neutral bus N_1 to the area or point of unbalance. With this the voltage drop V_{1x} at any point in the network is always

$$V_{1x} = V - \sum I_1 Z_1 \tag{4.12}$$

where V is the source voltage (V_G or V_S in Fig. 4.7a) and $\sum I_1 Z_1$ is the sum of the drops along any path from the N_1 neutral bus to the point of measurement.

4.9.2 Negative-Sequence Network

This network defines the flow of negative-sequence currents when they exist. The system generators do not generate negative sequence, but negative-sequence current can flow through their windings. Thus these generators and sources are represented by an impedance without voltage, as shown in Fig. 4.8. In the transformers, lines, and so on, the phase sequence of the current does not change the impedance encountered, so the same values are used as in the positive-sequence network.

A rotating machine can be visualized as a transformer with one stationary and one rotating winding. Thus dc in the field produces

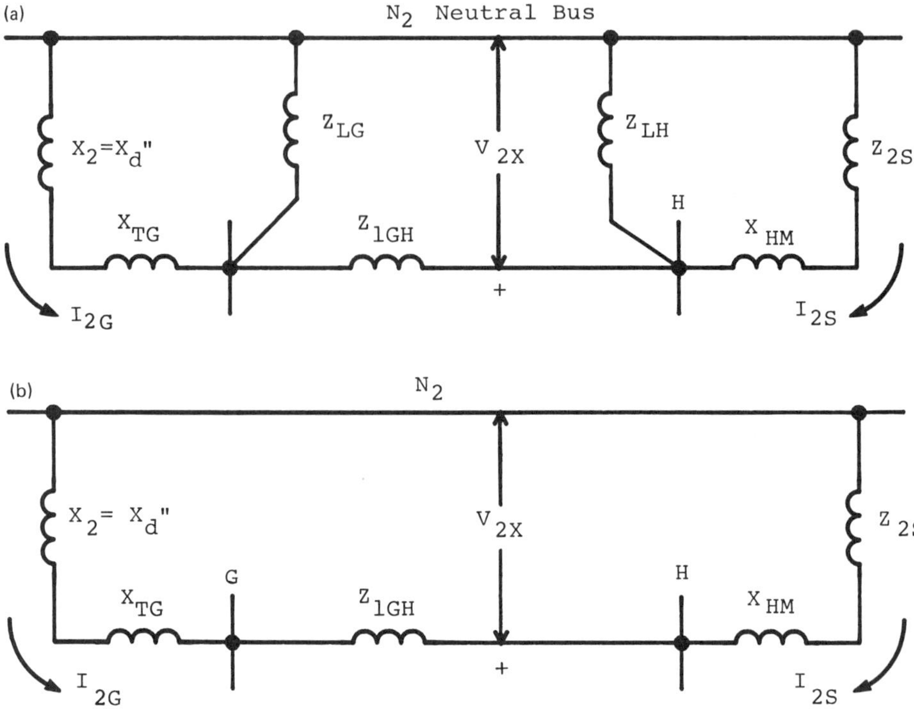

Fig. 4.8 Negative sequence networks for the system in Fig. 4.5: (a) network including loads; (b) network neglecting loads.

positive sequence in the stator. Similarly, the dc offset in the stator ac current produces an ac component in the field. In this relative motion model with the one winding rotating at synchronous speed, negative sequence in the stator results in a double-frequency component in the field. Thus the negative-sequence flux component in the air gap is alternately between and under the poles at this double frequency. One common expression for the negative-sequence impedance of a synchronous machine is

$$X_2 = \frac{1}{2}(X_d'' + X_q'') \tag{4.13}$$

or the average of the direct and quadrature axes subtransient reactances. For a round-rotor machine, $X_d'' = X_q''$, so that $X_2 = X_d''$. For salient-pole machines, X2 will be different, but this is frequently neglected unless calculating a fault very near the machine terminals. Where normally $X_2 = X_d''$, the negative-sequence network is equivalent to the positive-sequence network except for the omission of voltages.

Loads can be shown as in Fig. 4.8 and will be the same impedance as for positive sequence if they are static loads. Rotating loads such as those of induction motors have quite different positive- and negative-sequence impedances when running. This is discussed further in Chapter 11.

Again with load normally neglected, the network is as shown in Fig. 4.8 and is the same as the positive-sequence network (Fig. 4.7b), except that there is no voltage.

Conventional current flow is assumed to be from the neutral bus N_2 to the area or point of unbalance. With this the voltage drop V_{2x} at any point in the network is always

$$V_{2x} = 0 - \sum I_2 Z_2 \tag{4.14}$$

where $\sum I_2 Z_2$ is the sum of the drops along any path from the N_2 neutral bus to the point of measurement.

4.9.3 Zero-Sequence Network

The zero-sequence network is always different. It must satisfy the flow of equal and in-phase currents in the three phases. If the connections for this network are not apparent, or there are questions or doubts, these can be resolved by drawing the three-phase system to see how the equal in-phase zero-sequence currents can flow. For the example of Fig. 4.5, a three-phase diagram is shown in Fig. 4.9. The convention is that current always flows to the unbalance, so assuming an unbalance between buses G and H, the top left diagram shows I_{0G} flowing from the transformer at bus G. Zero sequence can flow in the grounded wye and to the fault since there is a path for it to flow in the delta. Thus X_{TG} is connected between the zero potential bus and bus G as shown in Fig. 4-10. This connection for the grounded-wye-delta transformer bank is also shown in Fig. A4.2-1.

Zero sequence impedance for transformer banks is equal to the positive and negative sequence and is the transformer leakage impedance. The exception to this is for three-phase core-type transformers, where the construction does not provide an iron flux path for zero sequence. For these the zero-sequence flux must pass from the core to the tank and return. Hence for these types X_0 usually is 0.85 to $0.9X_1$, and when known the specific value should be used.

The lower right-hand diagram of Fig. 4.9 is for the system connected to bus H (Fig. 4.5). Currents out of the three-winding transformer will flow as shown in the L and M windings. The three currents can flow in the M-grounded wye since the equivalent source is

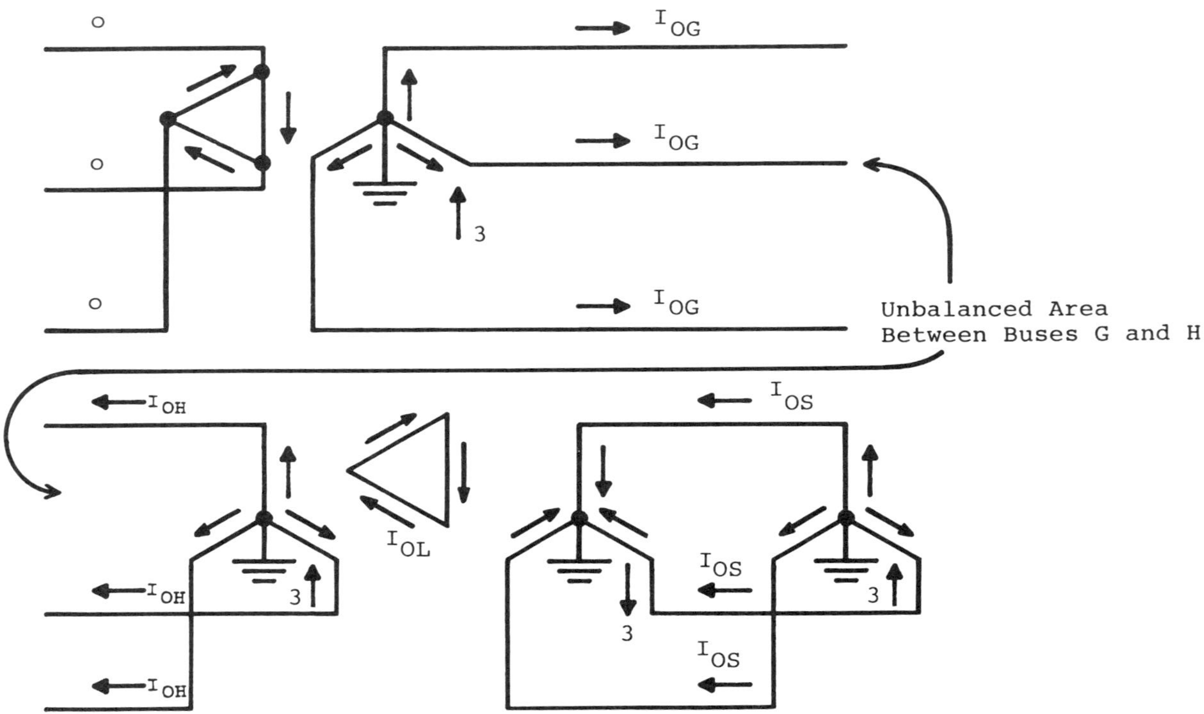

Fig. 4.9 Diagrams illustrating the flow of zero-sequence current as an aid in drawing the zero-sequence network. Arrows indicate current directions only, not relative magnitudes.

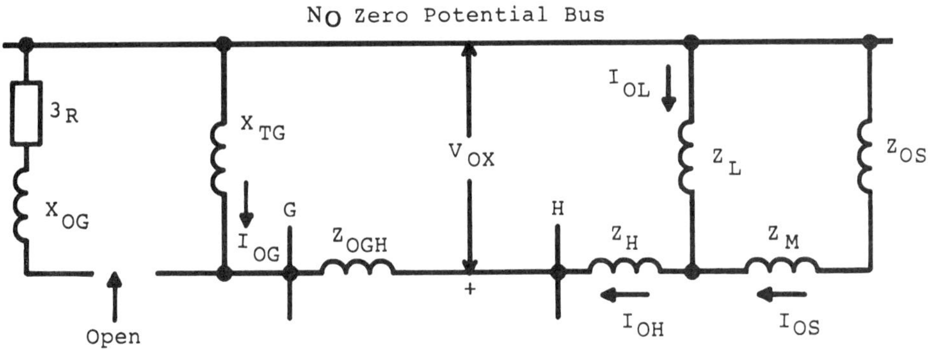

Fig. 4.10 Zero-sequence network for the system of Fig. 4.5.

shown grounded with Z_{0S} given. Thus the three-winding equivalent circuit is connected in the zero-sequence network (Fig. 4.10) as shown, which follows the connections documented in Fig. A4.2-3b.

Note that in the right-hand part of Fig. 4.9, if any of the wye connections were not grounded, the connections would be different. If the equivalent system, or the M winding were ungrounded, the network would be open between Z_M and Z_{0S}, as zero-sequence currents could not flow as shown. Loads, if desired, would be shown in the zero-sequence network only if they were wye grounded. Delta loads would not pass zero sequence.

Zero-sequence line impedance is always different, as it is a loop impedance; the impedance of the line plus a return path either in the earth, or in a parallel combination of the earth and ground wire, cable sheath, and so on. The positive-sequence impedance is a one-way impedance: from one end to the other end. As a result, zero sequence varies from 2 to 6 times X_1 for lines. For estimating open wire lines, a value of $X_0 = 3$ or $3.5X_1$ is commonly used.

The zero-sequence impedance of generators is low and variable, depending on the winding design. Except for very low voltage units, generators are never solidly grounded. This is discussed in Chapter 7. In Fig. 4.5, the generator G is shown grounded through a resistor R. Faults on bus G and in the system to the right do not involve the generator as far as zero sequence since the transformer delta blocks the flow of zero-sequence current as shown.

Conventional current flow is assumed to be from the zero-potential bus N_0 to the area or point of unbalance. Thus the voltage drop V_{0X} at any point in the network is always

$$V_{0x} = 0 - \sum I_0 Z_0 \tag{4.15}$$

where $\sum I_0 Z_0$ is the sum of the drops along any path from the N_0 bus to the point of measurement.

4.9.4 Sequence Network Reduction

For shunt fault calculations, the sequence networks can be reduced to a single equivalent impedance commonly designated as Z_1 or X_1, Z_2 or X_2, and Z_0 or X_0 from the neutral or zero-potential bus to the fault location. This is the Thevenin theorem equivalent impedance, and in the positive-sequence network, the Thevenin voltage. These values are different for each fault location. Short-circuit studies with computers use various techniques to reduce complex power systems and to determine fault currents and voltages.

For the positive-sequence network of Fig. 4.7b consider faults at bus H. Then by paralleling the impedances on either side, Z_1 becomes,

$$Z_1 = \frac{(X_d'' + X_{TG} + Z_{1GH})(Z_{1S} + X_{HM})}{X_d'' + X_{TG} + Z_{1GH} + Z_{1S} + X_{HM}}$$

Each term in parentheses in the numerator, divided by the denominator, provides a per unit value to define the portion of current flowing in the two parts of the network. These are known as distribution factors and are necessary to determine the fault currents in various parts of the system. Thus the per unit current through bus G is

$$I_{1G} = \frac{Z_{1S} + X_{HM}}{X_d'' + X_{TG} + Z_{1GH} + Z_{1S} + X_{HM}} \text{ pu} \qquad (4.16)$$

and the current through bus H is

$$I_{1S} = \frac{X_d'' + X_{TG} + Z_{1GH}}{X_d'' + X_{TG} + Z_{1GH} + Z_{1S} + X_{HM}} \text{ pu} \qquad (4.17)$$

The reduction of the positive-sequence network with load (Fig. 4.7a) requires determining the load current flow throughout the network before a fault, determining the open-circuit voltage (Thevenin voltage) at the fault point, and then the equivalent impedance looking into the network from the fault point with all voltages zero (Thevenin's impedance). After the fault currents are calculated, the total currents in the network are the sum of the prefault load and the fault currents.

The negative- and zero-sequence networks can be reduced in a similar manner to a single impedance to a fault point and with appropriate distribution factors. These three independent equivalent networks are shown in Fig. 4.11 with I_1, I_2, and I_0 representing the respective sequence currents in the fault, and V_1 V_2, and V_0 representing the respective sequence voltages at the fault.

Another approach is to generalize any power system and its sequence networks by the use of a box. These boxes represent the balanced or symmetrical part of the power system and its corresponding sequence networks up to a point or area of unbalance. This symbology is illustrated in Fig. 4.12.

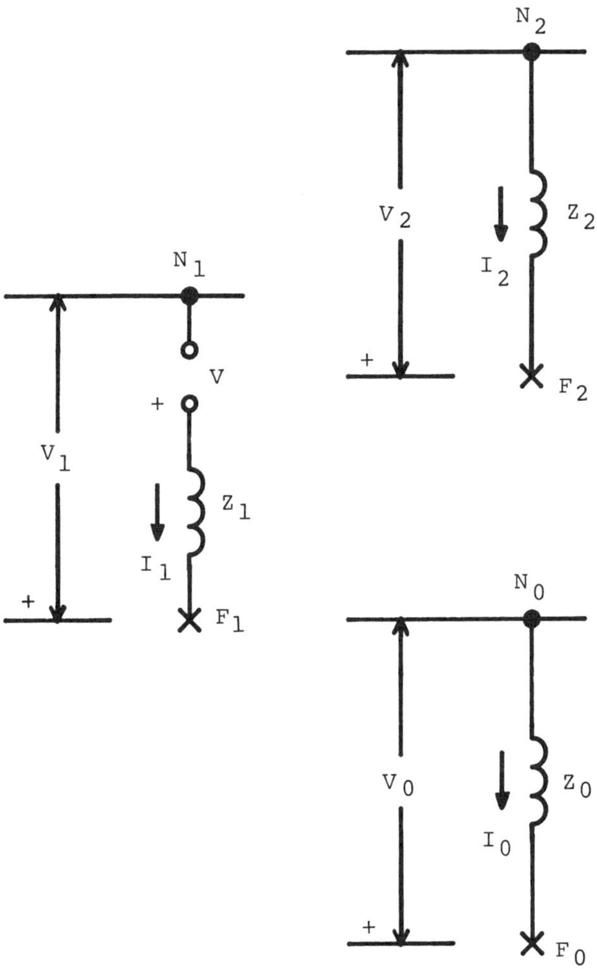

Fig. 4.11 Reduced sequence networks where Z_1, Z_2, and Z_0 are the equivalent impedances of the network to the fault point.

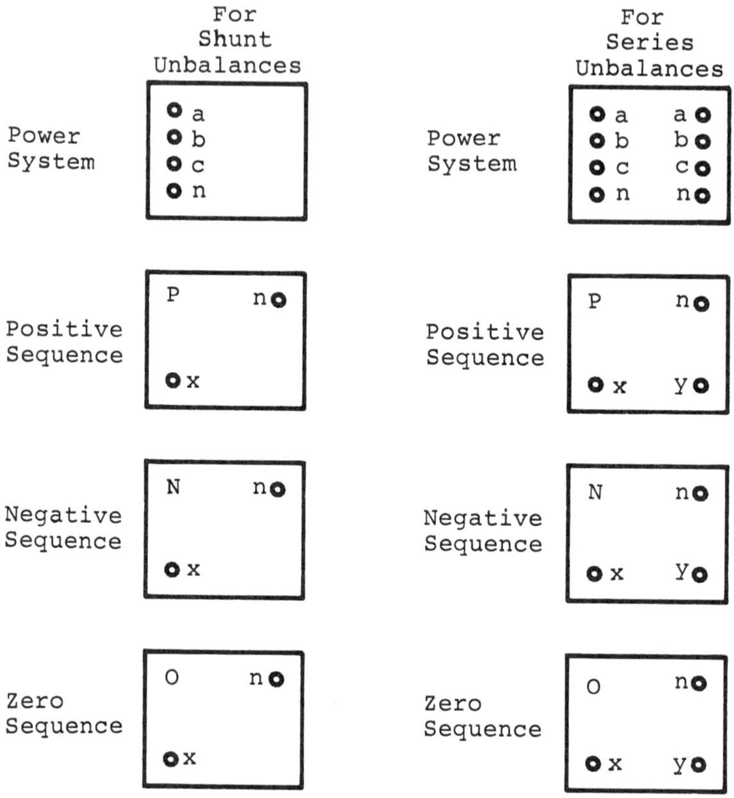

Fig. 4.12 Box symbols used to represent the symmetrical or balanced power system and their sequence networks up to a point or area of unbalance.

As indicated earlier, the sequence networks, such as shown in Fig. 4.11 and 4.12, are completely independent of each other. Next we discuss interconnections to represent faults and unbalances.

4.10 SHUNT UNBALANCE SEQUENCE NETWORK INTERCONNECTIONS

The principal shunt unbalances are faults: three-phase, phase to phase, two-phase to ground, and one-phase to ground.

4.10.1 Fault Impedance

Faults are seldom solid but involve varying amount of resistance. However, it is generally assumed in protective relaying and most fault studies that the connection and/or contact with the ground involves very low and generally negligible impedance. For the higher voltages of transmission and subtransmission this is essentially true. In distribution systems (34.5 kV and below) very large to basically infinite impedance can exist. This is true particularly at the lower voltages. Many faults are tree contacts, which can be high impedance and intermittent and variable. Conductors lying on the ground may or may not result in significant fault current and again can be highly variable. Many tests have been conducted over the years on wet soil, dry soil, rocks, asphalt, concrete, and so on, with quite variable and sometimes unpredictable results. Thus in most fault studies, the practice is to assume zero ground mat and fault impedance for maximum fault values. Protective relays are set as sensitively as possible, yet respond properly to these maximum values.

Thus while arcs are quite variable, a commonly accepted value for currents between 70 and 20,000 A has been an arc drop of 440 V per phase, essentially independent of current magnitude. Therefore,

$$Z_{arc} = \frac{440\,l}{I}\ \Omega \qquad (4.18)$$

where l is the arc length in feet and I the current in amperes. l/kV at 34.5 kV and higher is approximately 0.1 to 0.05. The arc essentially is resistance but can appear to protective relays as an impedance, with a significant reactive component resulting from out-of-phase contributions from remote sources. This is discussed in more detail in Chapter 12. In low-voltage (480-V) switchboard-type enclosures, typical arc voltages of around 150 V can be experienced. This is relatively independent of current magnitude.

It appears that since arcs are variable, their resistances tend to start at a low value and continue at this value for an appreciable time, then build up exponentially. Upon reaching a high value, an arc breaks over to shorten its path and resistance.

4.10.2 Substation and Tower Footing Impedance

Another highly variable factor, which is difficult both to calculate and measure, is the resistance between a station ground mat or line pole or tower and ground. In recent years a number of technical papers have been written and computer programs developed in this area, but there are still many variables and assumptions. All this is beyond the scope of this book. As indicated above, the general practice is to neglect these in most fault studies and relay applications and settings.

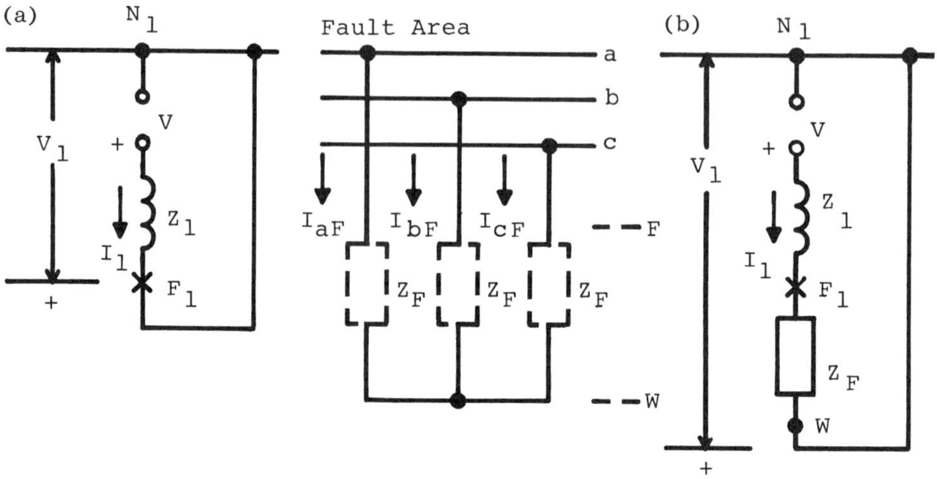

Fig. 4.13 System fault area (center and sequence connections for three-phase faults: solid fault (a); with fault impedance (b).

4.10.3 Sequence Interconnections for Three-Phase Faults

Three-phase faults are assumed to be symmetrical, so symmetrical components analysis is not necessary for their calculation. The positive-sequence network, which is the normal balanced diagram for a symmetrical system, can be used, and the connection is shown in Fig. 4.13. For a solid fault the fault point F_1 is connected back to the neutral bus (Fig. 4.13a); with fault impedance the connection includes this impedance, as shown in Fig. 4.13b. From these,

$$I_1 = I_{aF} = \frac{V}{Z_1} \text{ or } I_1 = I_{aF} = \frac{V}{Z_1 + Z_F} \tag{4.19}$$

and $I_{bF} = a^2 I_1$, $I_{cF} = aI_1$, according to Eq. (4.2).

Using the box symbolism of Fig. 4.12, balanced three-phase fault sequence connections are shown in Fig. 4.17b and c for solid faults, and in Fig. 4.17a for faults with impedance. These are the same as Fig. 4.13. Fig. 4.17a, b, and c show the negative- and zero-sequence networks and connections, but these are academic, as they have no driving voltage and are not interconnected with the positive-sequence network. Also, there is no difference between a solid three-phase fault (Fig. 4.17c) and a solid three-phase-to-ground fault (Fig. 4-17b).

4.10.4 Sequence Interconnections for Single-Phase-to-Ground Faults

A phase-a-to-ground fault is represented by connecting the three sequence networks together as shown in Fig. 4.14, with diagram a for solid faults, and b for faults with impedance. The same connections are shown in Fig. 4.17f and e, respectively. From these,

$$I_1 = I_2 = I_o = \frac{V}{Z_1 + Z_2 + Z_0} \quad \text{or}$$

$$I_1 = I_2 = I_0 = \frac{V}{Z_1 + Z_2 + Z_0 + 3Z_F} \tag{4.20}$$

$$I_{aF} = I_1 + I_2 + I_0 = 3I_1 = 3I_2 = 3I_0 \tag{4.21}$$

From Eq. (4.6) and (4.7) it can be seen that $I_{bF} = I_{cF} = 0$, which is correct in the fault. Also, $V_{aF} = 0$, which is supported by the sequence connections since $V_1 + V_2 + V_0 = 0$.

4.10.5 Sequence Interconnections for Phase-to-Phase Faults

For this type it is convenient to show the fault between phases b and c. Then the sequence connections are as shown in Fig. 4.15, with (a) for a solid fault and (b) for a fault with impedance. These are equivalent to the connections of Fig. 4.17h and g, respectively. From these,

$$I_1 = -I_2 = \frac{V}{Z_1 + Z_2} \quad \text{or} \quad I_1 = -I_2 = \frac{V}{Z_1 + Z_2 + Z_F} \tag{4.22}$$

From the fundamental equations (4.5) through (4.7),

$$I_{aF} = I_1 - I_2 = 0 \quad \text{as it should be in the fault}$$

$$I_{bF} = a^2 I_1 + aI_2 = (a^2 - a)I_1 = -j\sqrt{3}\, I_1 \tag{4.23}$$

$$I_{cF} = aI_1 + a^2 I_2 = (a - a^2)I_1 = +j\sqrt{3}\, I_1 \tag{4.24}$$

As is common, $Z_1 = Z_2$; then $I_1 = V/2Z_1$. disregarding $\pm j$ and considering only magnitude yields

$$I_{\phi\phi} = \frac{\sqrt{3}\,V}{2Z_1} = 0.866\,\frac{V}{Z_1} = 0.866\, I_{3\phi} \tag{4.25}$$

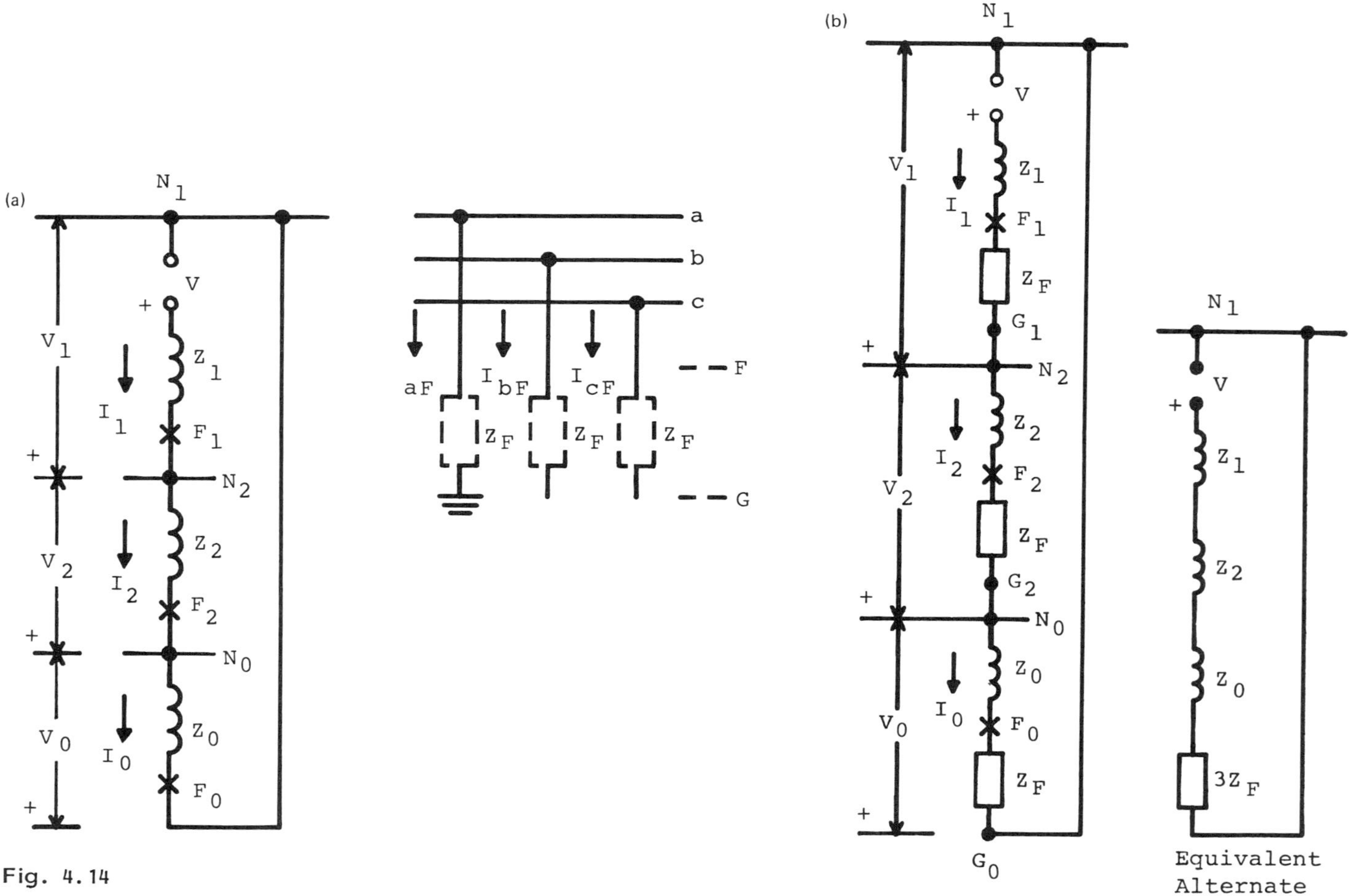

(a)
N 1
V
+
V 1
I 1
Z 1
F 1
N 2
V 2
I 2
Z 2
F 2
N 0
V 0
I 0
Z 0
F 0
a
b
c
aF
I bF
I cF
Z F
Z F
Z F
F
G
(b)
N 1
V
+
V 1
I 1
Z 1
F 1
Z F
G 1
N 2
V 2
I 2
Z 2
F 2
Z F
G 2
N 0
V 0
I 0
Z 0
F 0
Z F
G 0
N 1
V
+
Z 1
Z 2
Z 0
3Z F
Equivalent
Alternate

Fig. 4.14

Thus the solid phase-to-phase fault is 86.6% of the solid three-phase fault when $Z_1 = Z_2$.

4.10.6 Sequence Interconnections for Double-Phase-to-Ground Faults

The connections for this type are similar to those for the phase-to-phase fault, but with the addition of the zero-sequence network connected in parallel as shown in Fig. 4.16. Again a is for solid faults, b for faults with impedance. The corresponding connections in Fig. 4.17 are j and i, respectively. From these,

$$I_1 = \frac{V}{Z_1 + \dfrac{Z_2 Z_0}{(Z_2 + Z_0)}}$$

or

$$I_1 = \frac{V}{Z_1 + \dfrac{Z_F}{2} + \dfrac{\left(Z_2 + \dfrac{Z_F}{2}\right)\left(Z_0 + \dfrac{Z_F}{2} + 3Z_{FG}\right)}{Z_2 + Z_0 + Z_F + 3Z_{FG}}} \tag{4.26}$$

$$I_2 = -I_1 \frac{Z_0}{Z_2 + Z_0} \quad \text{and} \quad I_0 = -I_1 \frac{Z_2}{Z_2 + Z_0}$$

or

$$I_2 = -I_1 \frac{Z_0 + \dfrac{Z_F}{2} + 3Z_{FG}}{Z_2 + Z_0 + Z_F + 3Z_{FG}}$$

and

$$I_0 = -I_1 \frac{Z_2 + \dfrac{ZF}{2}}{Z_2 + Z_0 + Z_F + 3Z_{FG}} \tag{4.27}$$

Fig. 4.14 System fault area (center) and sequence connections for single-phase-to-ground faults: (a) solid fault; (b) with fault impedance.

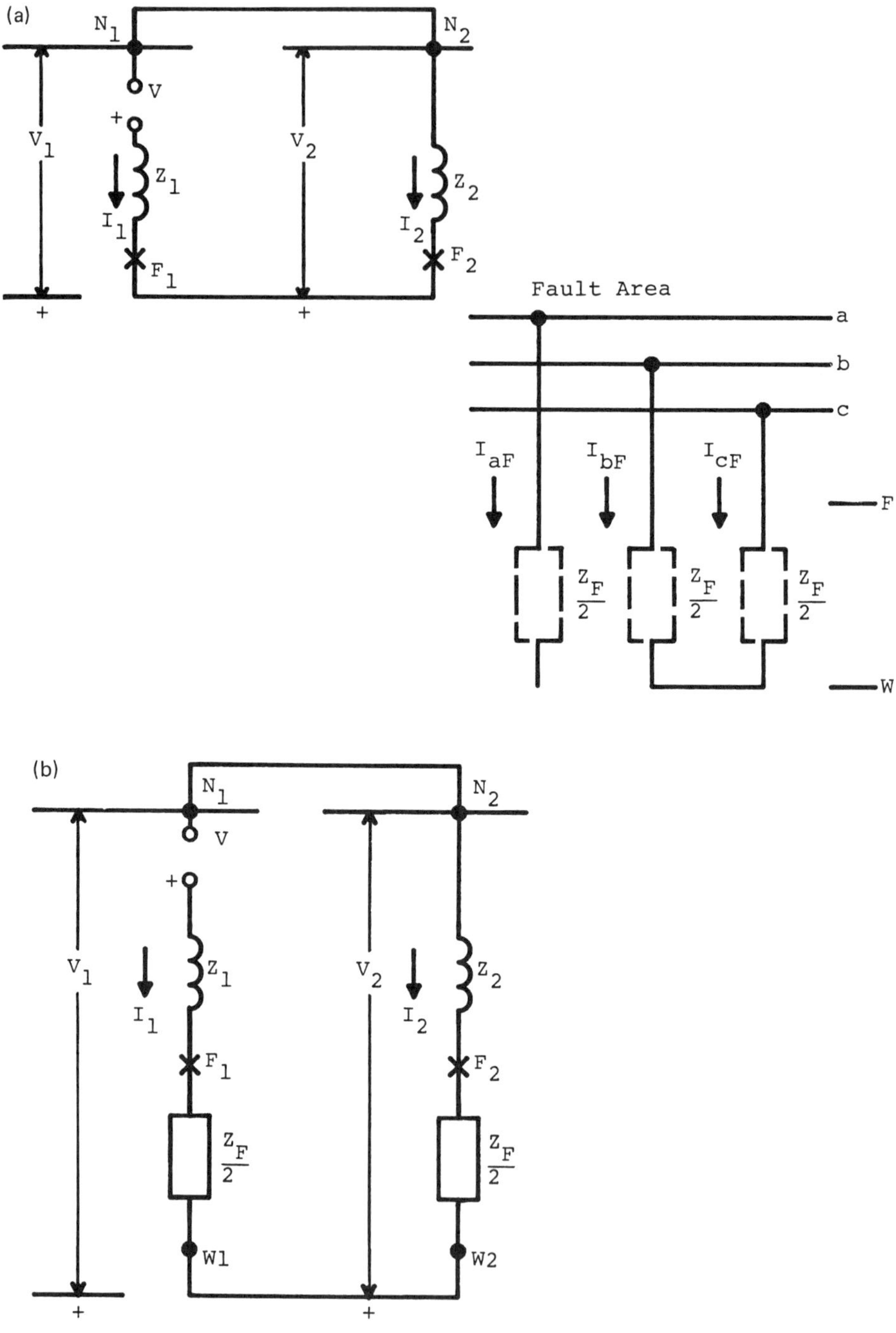

Fig. 4.15 System fault (right center) and sequence connections for phase-to-phase faults: (a) solid fault; (b) with fault impedance.

Equations (4.5) through (4.7) provide $I_{aF} = 0$, and fault magnitudes for I_{bF} and I_{cF}.

4.10.7 Other Sequence Interconnections for Shunt System Conditions

The impedances at the fault point in Fig. 4.13 through 4.16 were considered as resulting from the fault arc. However, they can also be considered as a shunt load, shunt reactor, shunt capacitor, and so on, connected at a given point to the system. Various types and their sequence interconnections are shown in Fig. 4.17. As an example, Fig. 4.17 can be used to determine the currents and voltages resulting from a shorted turn in a shunt reactor and the loss of a capacitor in a bank. Z_a was made less than the normal Z_b to model condition.

4.11 SERIES UNBALANCE SEQUENCE NETWORK INTERCONNECTIONS

A variety of possible series unbalances that might occur in the otherwise balanced system are documented in Fig. 4.18. A common one is a broken conductor or blown fuse. This is represented by Fig. 4.18j. this shows that the sequence networks are to be connected together at the x and y points on either side of the open area. This is applied to a typical system in Fig. 4.19. Load or the differences in the source voltages, such as in Fig. 4.5, must be included for these series unbalances. If Z_L, the load, is omitted in Fig. 4.19, as is common for shunt fault calculations, no current can flow. This makes sense, as opening an unloaded circuit does not change the current. Such opening will cause a transient arc as the system capacitance is interrupted.

The system of Fig. 4.19 is shown grounded on both sides of the open. If one side is ungrounded, it can be seen that the zero-sequence network is open, so no zero sequence current can flow in these cases. This is typical of a motor. This analysis is continued in Chapter 11.

4.12 SIMULTANEOUS UNBALANCES

A broken conductor falling to ground, or faults and/or opens in different parts of the system, can be represented by connecting the sequence networks as illustrated in Fig. 4.20. These require two different sequence network interconnections, so to provide isolation, ideal or perfect transformers can be used. These have no impedance and no losses. If the faults are related to the same phase, the ratios of the three ideal transformers are 1:1; if different phases are involved, the ratios include the phasors a and a^2 as shown.

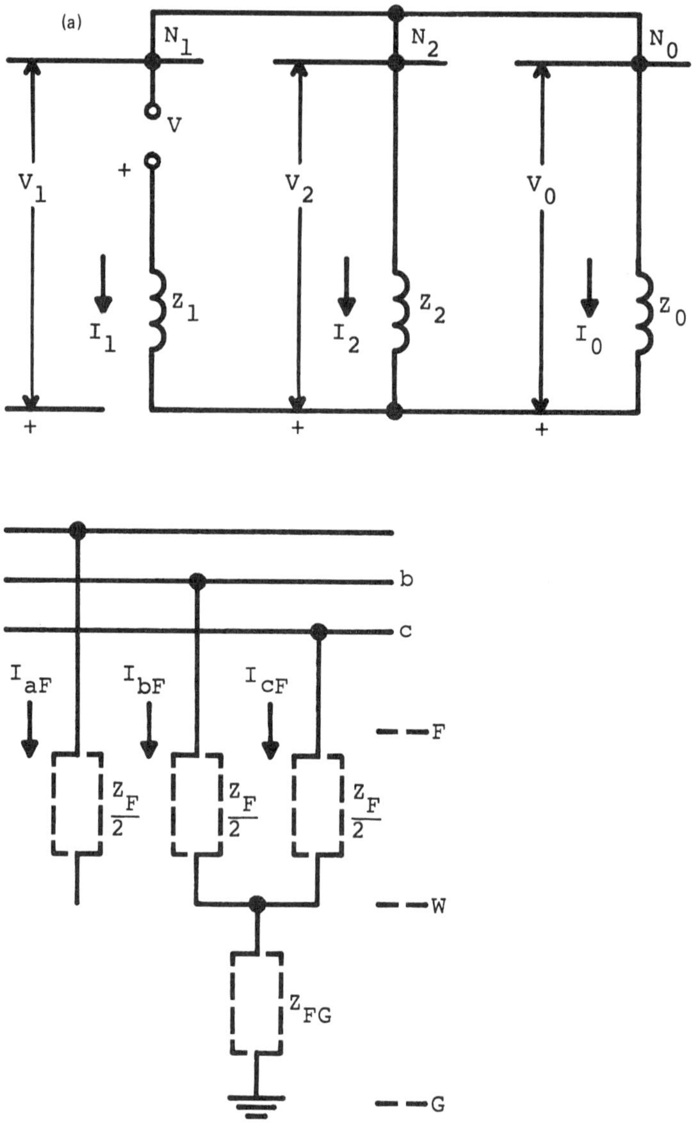

Fig. 4.16 System fault area (left bottom) and sequence connections for double-phase to ground faults: (a) solid fault; (b) with fault impedance.

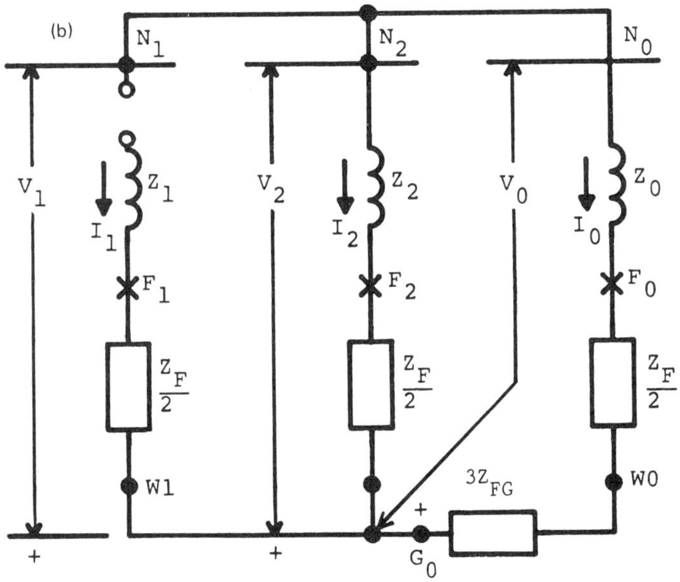

Fig. 4.16 (Continued)

An example of the connections for an open phase falling to ground on the load side is shown in Fig. 4.21. In this case, if Z_L the load is quite large compared to X_{TH}, the Z_L circuit can be omitted to simplify the calculations with negligible error. If the conductor falls to ground on the line side, the three ideal transformers would be connected to the y terminals rather than x. The load for this case cannot be neglected.

A set of simultaneous equations can be written and solved for these network interconnections. For computer studies, algorithms can be written to provide the equivalent.

4.13 EXAMPLE: FAULT CALCULATIONS ON A TYPICAL SYSTEM SHOWN IN FIG. 4.22

The system of Fig. 4.22 is similar to that shown in Fig. 4.5 but with typical constants for the various parts. These are on the bases indicated, so the first step is to transfer them to a common base, as discussed in Chapter 2. The positive- and negative-sequence network (negative the same as positive except for the omission of the voltage) is shown in Fig. 4.23. The conversion to a common base of 100 MVA is shown as necessary.

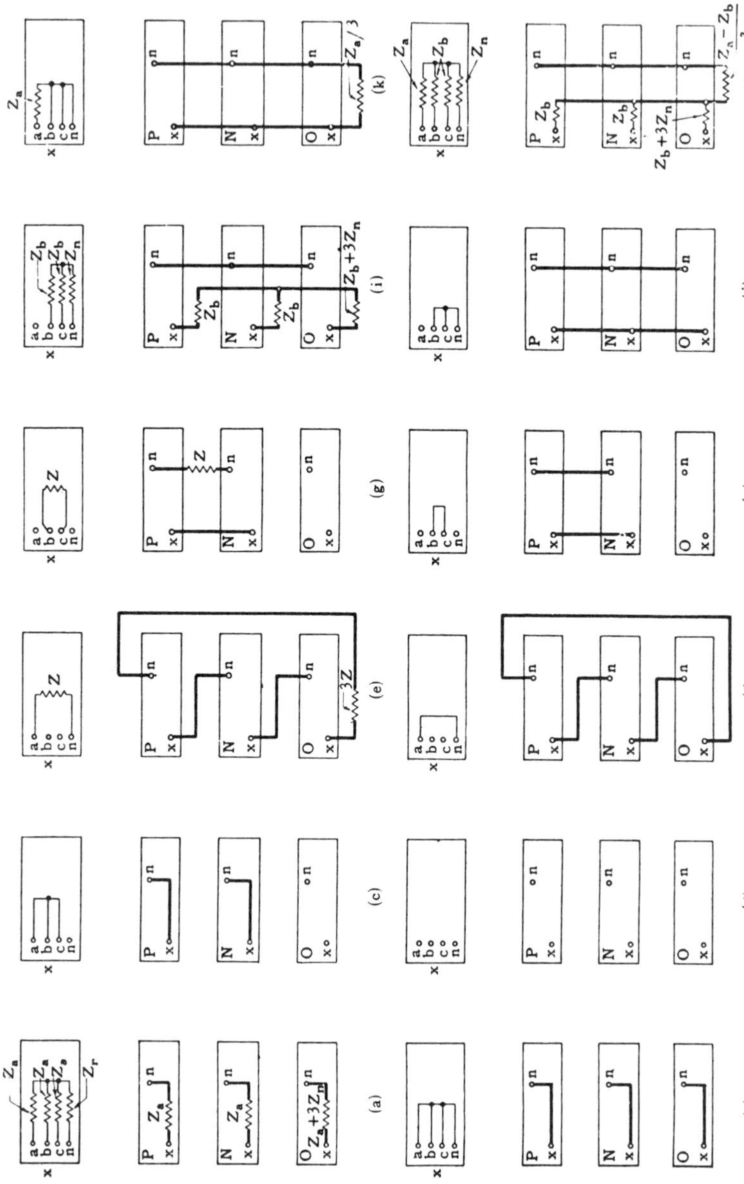

Fig. 4.17 Box sequence connections for shunt balanced and unbalanced conditions: (a) balanced load or three-phase-to-ground fault with impedances; (b) three-phase-to-ground fault; (c) three-phase fault; (d) shunt circuit open; (e) phase-to-ground fault through an impedance; (f) phase-to-ground fault; (g) phase-to-phase fault through impedance; (h) phase-to-phase fault; (i) two-phase-to-ground fault through impedance; (j) two-phase-to-ground fault; (k) three-phase-to-ground fault with impedance in phase a; (l) unbalanced load or three-phase-to-ground fault with impedance. (From E. L. Harder, Sequence Network Connections for Unbalanced Load and Fault Conditions, *The Electrical Journal*, December 1937.)

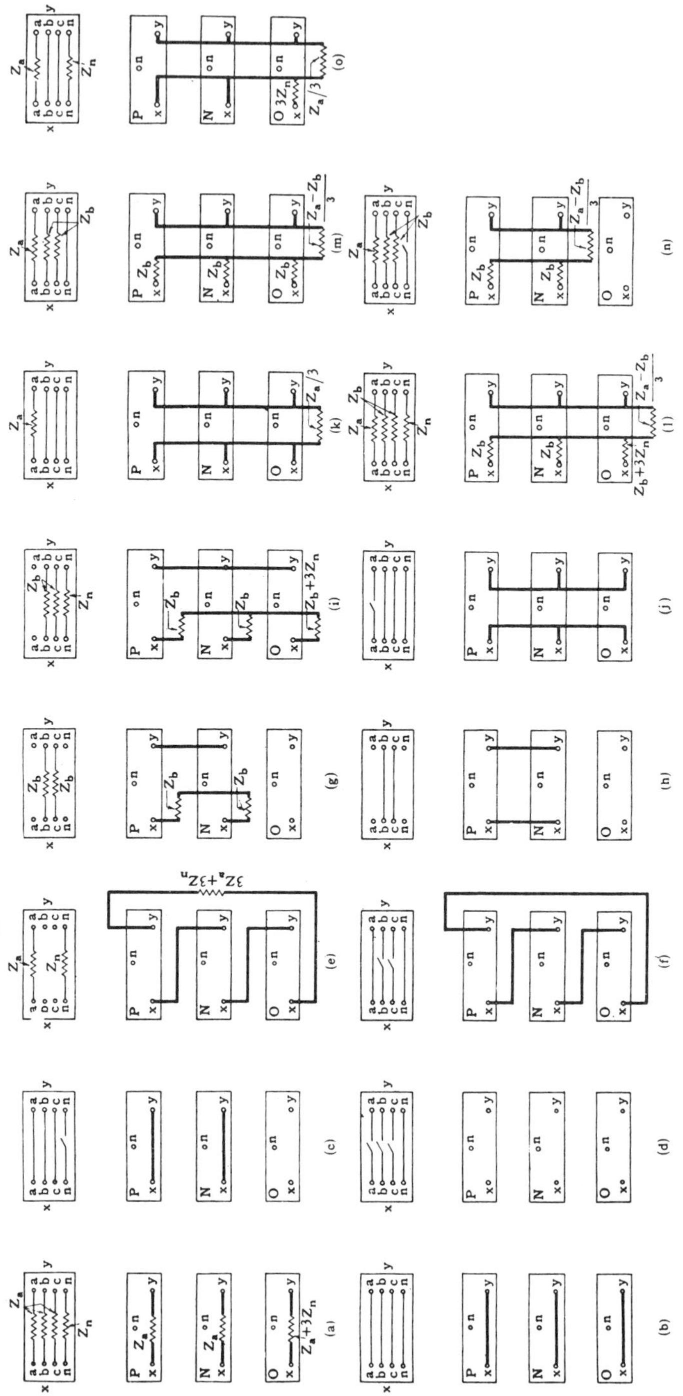
(a)
(b)
(c)
(d)
(e)
(f)
(g)
(h)
(i)
(j)
(k)
(l)
(m)
(n)
(o)

Fig. 4.18 Box sequence connections for series unbalanced conditions: (a) equal impedances in three phases; (b) normal balanced conditions; (c) neutral circuit open; (d) any three or four phases open; (e) phases b and c open and impedances in phase a and neutral; (f) phases b and c open; (g) phases a and neutral open and impedances in phases b and c; (h) phase a and neutral open; (i) phase a open and impedances in phases b, c, and neutral; (j) phase a open; (k) impedance in phase a; (l) equal impedances in phases b and c, impedance in neutral; (m) equal impedances in phases b and c; (n) equal impedances in phases b and c, neutral open; (o) impedances in phase a and neutral. (From E. L. Harder, Sequence Network Connections for Unbalanced Load and Fault Conditions, *The Electrical Journal*, December 1937.)

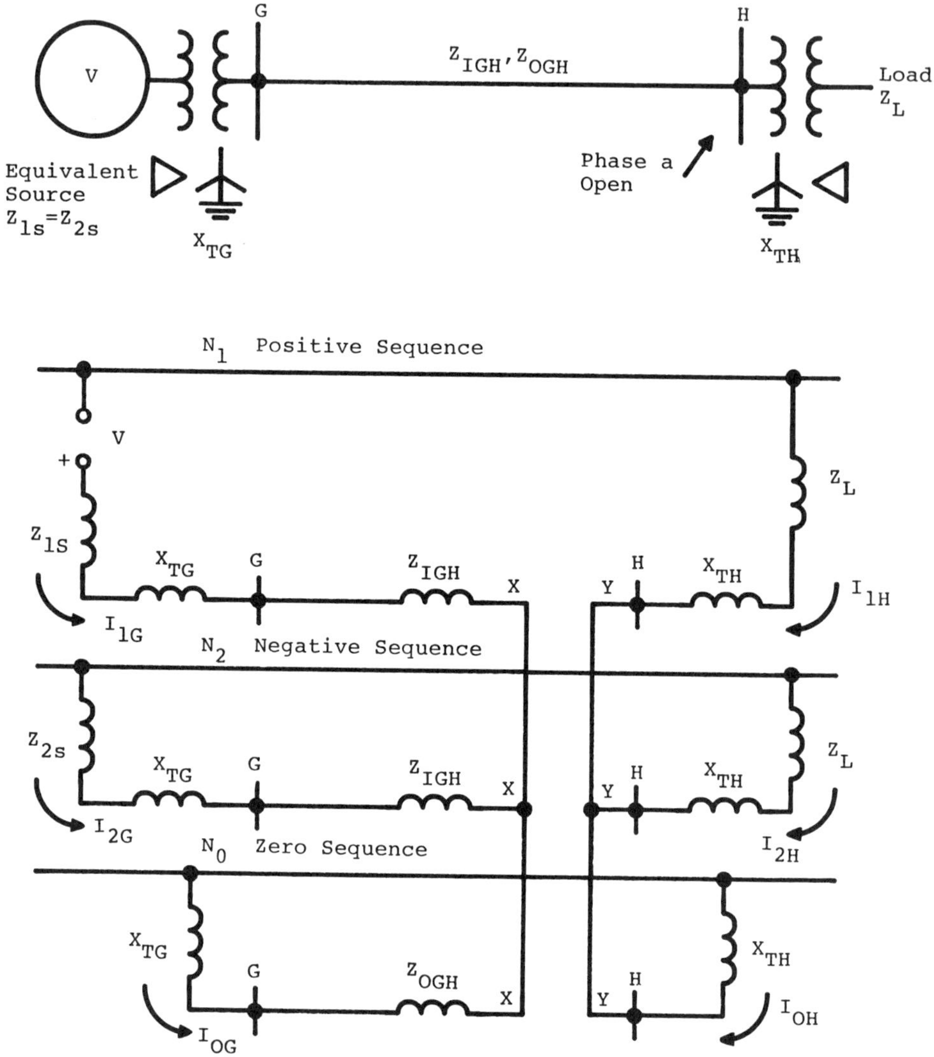

Fig. 4.19 Example of the sequence interconnections for phase a open at bus H for a typical power system.

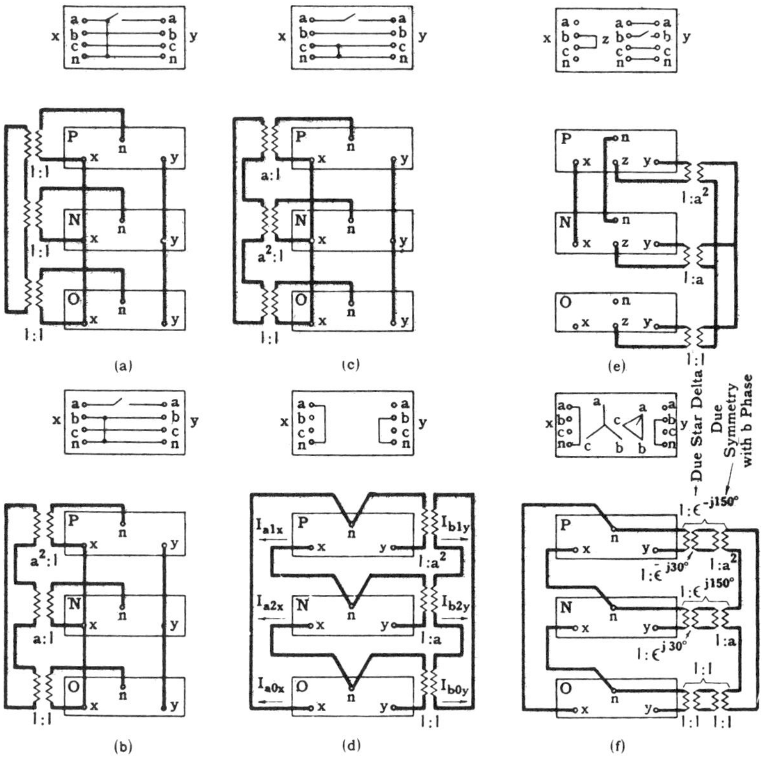

Fig. 4.20 Box sequence connections for simultaneous unbalanced conditions: (a) phase a open and phase-a-to-ground fault on the x side; (b) phase a open and a phase-b-to-ground fault; (c) phase a open and a phase-c-to-ground fault; (d) phase-a-to-ground fault at x and a phase-b-to-ground fault at y; (e) phase-b-to-c fault at x and phase b open z to y; (f) phase-a-to-neutral fault at x, phase-b-to-neutral fault on the other side of a wye-delta transformer bank at y, with x taken as the reference point. (From E. H. Harder, Sequence Network Connections for Unbalanced Load and Fault Conditions, *The Electrical Journal*, December 1937.)

For a fault at bus G, the right-hand impedances (j0.18147 + j0.03667 + j0.03 = j0.2481) are paralleled with the left-hand impedances (j0.20 + j0.1375 = j0.3375). Reactance values rather than impedance values are used, as is typical where the resistance is quite small relatively.

$$X_1 = X_2 = \frac{\overset{(0.5763)}{0.3375} \times \overset{(0.4237)}{0.2481}}{0.5856} = j0.1430 \text{ pu} \qquad (4.28)$$

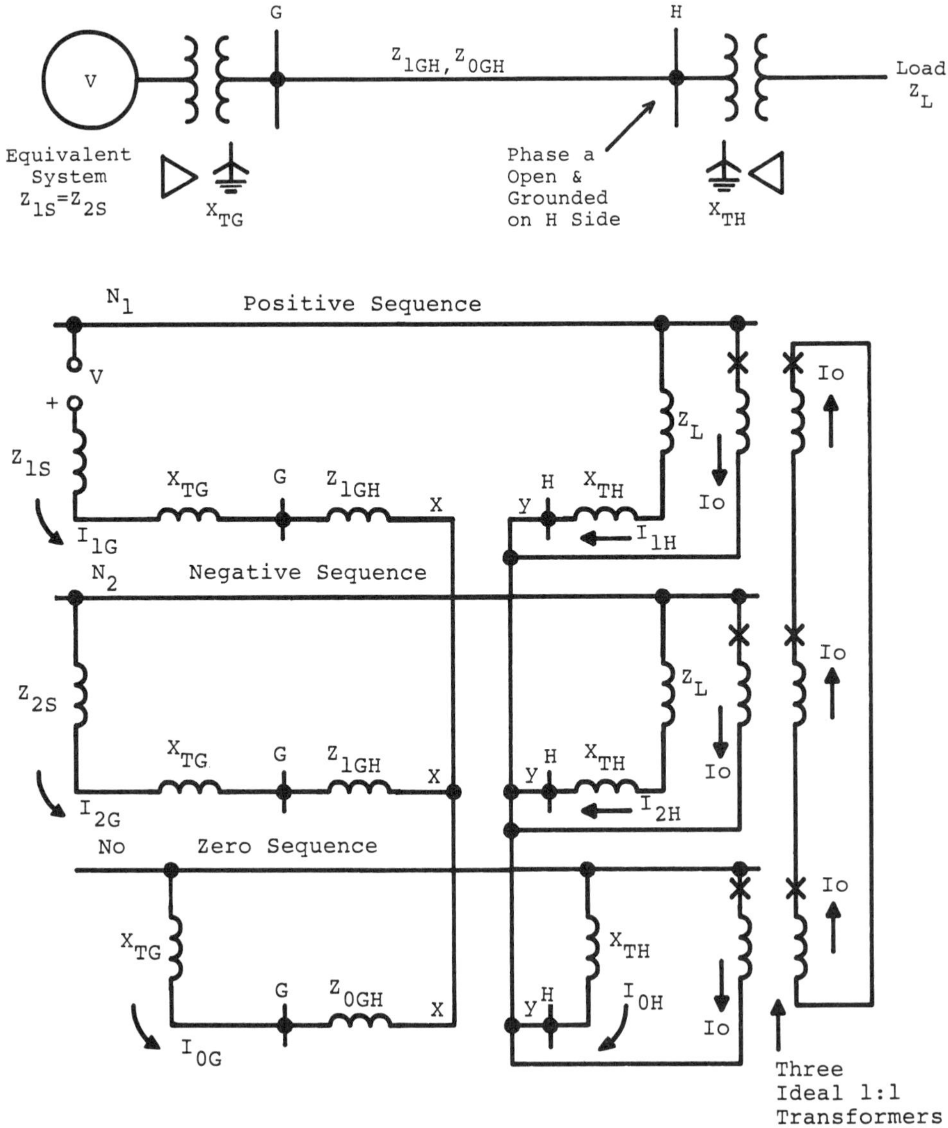

Fig. 4.21 Example of the sequence network interconnections for phase a open and grounded at bus H (broken conductor fallen to ground).

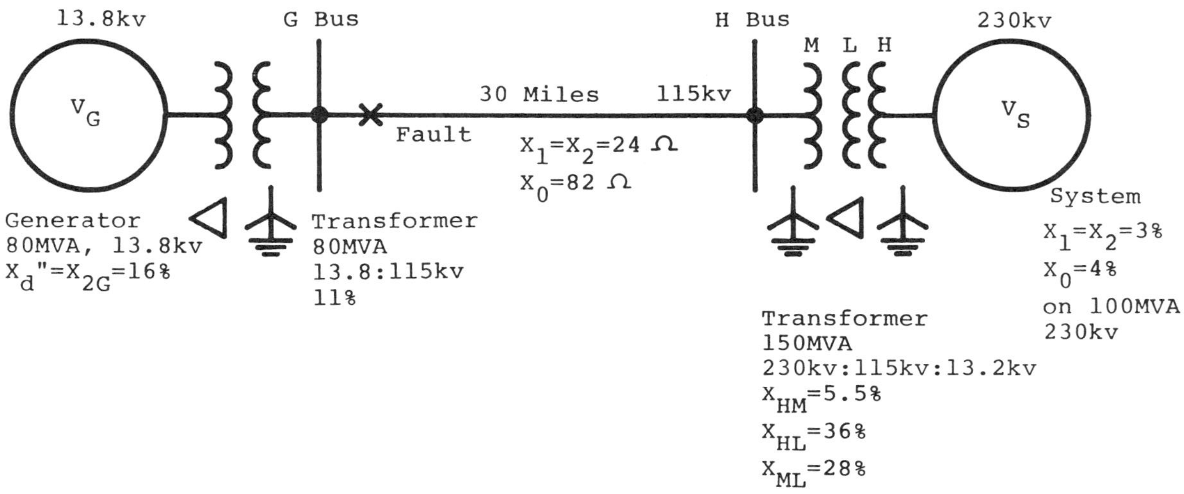

Fig. 4.22 Power system example for fault calculations.

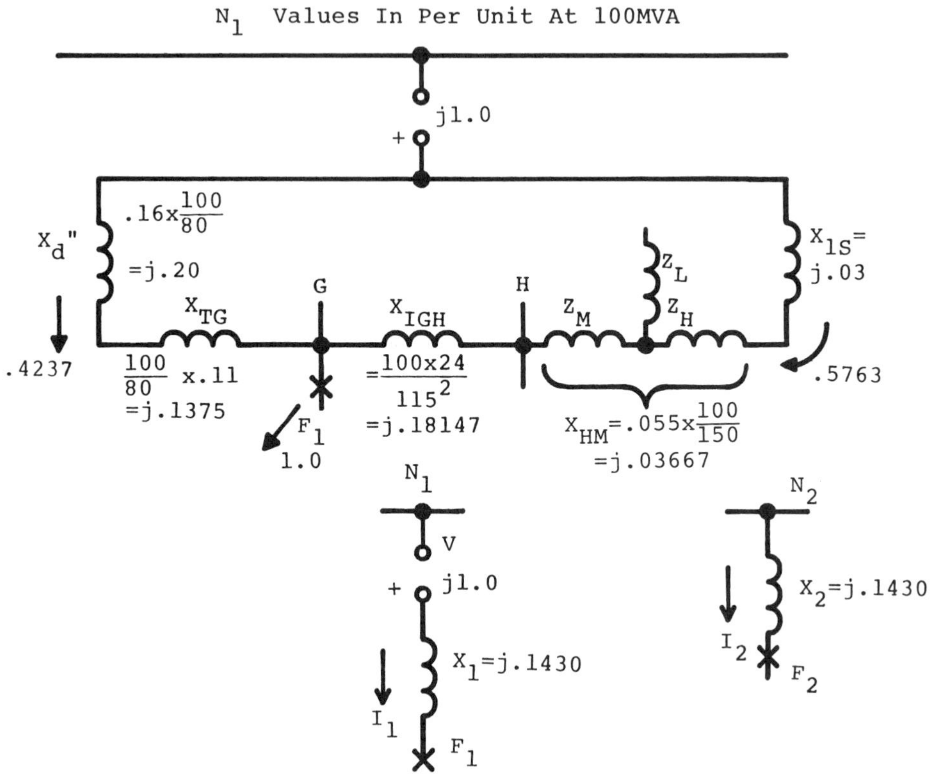

Fig. 4.23 Positive- and negative-sequence networks and their reduction to a single impedance for a fault at bus G in the power system of Fig. 4.22.

The division of 0.3375/0.5856 = 0.5763 and 0.2481/0.5856 = 0.4237 as shown provides a partial check, as 0.5763 + 0.4237 must equal 1.0 and are the distribution factors indicating the per unit current flow on either side of the fault. These values are added to the network diagram. Thus for faults at bus G, $X_1 = X_2 = j0.1430$ pu on 100-MVA base.

The zero-sequence network for Fig. 4.22 is shown in Fig. 4.24. Again the reactance values are converted to a common 100-MVA base. The three-winding bank connections are as indicated in Fig. A4.2-3b with $Z_{NH} = Z_{NM} = 0$ since the neutrals are shown solidly grounded.

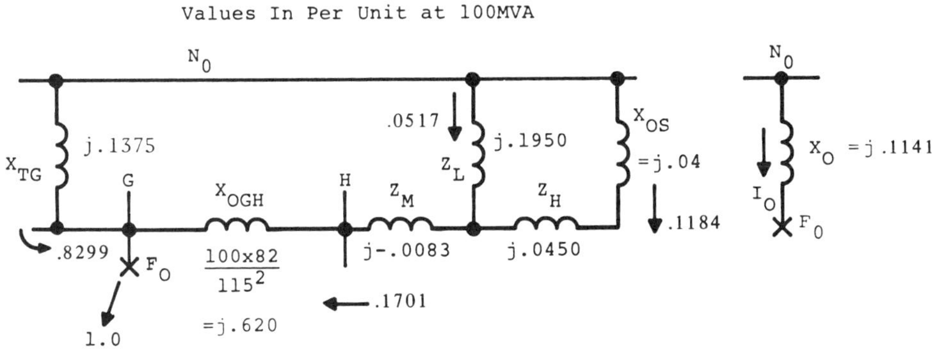

Fig. 4.24 Zero-sequence network and its reduction to a single impedance for a fault at bus G in the power system of Fig. 4.22.

The conversions to a common 100-MVA base are shown except for the three-winding transformer. For this bank,

$$X_{HM} = 0.055 \times \frac{100}{150} = 0.03667 \text{ pu}$$

$$X_{HL} = 0.360 \times \frac{100}{150} = 0.2400 \text{ pu}$$

$$X_{ML} = 0.280 \times \frac{100}{150} = 0.18667 \text{ pu}$$

and from Eq. (A4.2-13) through (A4.2-15),

$$X_H = \frac{1}{2}(0.03667 + 0.2400 - 0.18667) = 0.0450 \text{ pu}$$

$$X_M = \frac{1}{2}(0.03667 + 0.18667 - 0.240) = -0.00833 \text{ pu}$$

$$X_L = \frac{1}{2}(0.2400 + 0.18667 - 0.03667) = 0.1950 \text{ pu} \quad (4.29)$$

These are shown in Fig. 4.24.

This network is reduced for a fault at bus G by first paralleling $X_{OS} + Z_H$ with Z_L and then adding Z_M and X_{OGH};

$$\frac{\overset{(0.6964)}{0.1950} \times \overset{(0.3036)}{0.0850}}{0.280} = \begin{array}{rl} j0.0592 & \\ -j0.0083 & (Z_M) \\ \underline{j0.620} & (X_{OGH}) \\ j0.6709 & \end{array}$$

This is the right-hand branch. Paralleling with the left-hand branch,

$$X_0 = \frac{\overset{(0.8299)}{0.6709} \times \overset{(0.1701)}{0.1375}}{0.8084} = j0.1141 \text{ pu at 100 MVA} \tag{4.30}$$

The values (0.8299) and (0.1701) shown add to 1.0 as a check and provide the current distribution on either side of the bus G fault, as shown on the zero-sequence network. The distribution factor 0.1701 for the right side is further divided up by 0.6964 × 0.1701 = 0.1184 pu in the 230 kV system neutral, and 0.3036 × 0.1701 = 0.0517 pu in the three-winding transformer H neutral winding. These are shown on the zero-sequence network.

4.13.1 Three-Phase Fault at Bus G

For this fault,

$$I_1 = I_{aF} = \frac{j1.0}{j0.143} = 6.993 \text{ pu}$$

$$= 6.993 \frac{100{,}000}{\sqrt{3} \times 115} = 3510.8 \text{ A at 115 kV} \tag{4.31}$$

The divisions of current from the left (I_{aG}) and right (I_{aH}) are:

$$I_{aG} = 0.4237 \times 6.993 = 2.963 \text{ pu} \tag{4.32}$$

$$I_{aH} = 0.5763 \times 6.993 = 4.030 \text{ pu} \tag{4.33}$$

4.13.2 Single-Phase-to-Ground Fault at Bus G

For this fault,

$$I_1 = I_2 = I_0 = \frac{j1.0}{j(0.143 + 0.143 + 0.1141)} = 2.50 \text{ pu} \tag{4.34}$$

$$I_{aF} = 3 \times 2.5 = 7.5 \text{ pu at } 100 \text{ MVA}$$

$$= 7.5 \times \frac{100{,}000}{\sqrt{3} \times 115} = 3764.4 \text{ A at } 115 \text{ kV} \qquad (4.35)$$

Normally, the $3I_0$ currents are documented in the system, as these are used to operate the ground relays. As an aid to understanding, these are illustrated in Fig. 4.25 with the phase currents. Equations (4.5) through (4.7) provide the three phase currents. Since $X_1 = X_2$, so that $I_1 = I_2$, these reduce to, $I_b = I_c = -I_1 + I_0$ for the phase b and c currents, since $a + a^2 = -1$. The currents shown are determined by adding $I_1 + I_2 + I_0$ for I_a, $-I_1 + I_0$ for I_b and I_c, and $3I_0$ for the neutral currents.

In the 115-kV system the sum of the two neutral currents is equal and opposite to the current in the fault. In the 230-kV system the current up the neutral equals the current down the other neutral.

The calculations assumed no load, so prefault, all currents in the system were zero. With the fault involving phase a only, it will be observed that current flows in the b and c phases. This is because the distribution factors in the zero-sequence network are different from the positive- and negative-sequence distribution factors. On a radial system where positive-, negative-, and zero-sequence currents flow only from one source and in the same direction, the distribution factors in all three networks will be 1.0, although the zero-sequence impedances are different from the positive-sequence impedances. Then $I_b = I_c = -I_1 + I_0$ above becomes zero, and fault current only flows in the faulted phase. In this type $I_a = 3I_0$ throughout the system for a single-phase-toground fault.

4.14 EXAMPLE: FAULT CALCULATION FOR AUTOTRANSFORMERS

Autotransformers have become quite common in recent years. They provide some different and interesting problems. Consider a typical autotransformer in a system as shown in Fig. 4.26 and assume that a single-phase-to-ground fault occurs at the H or 345-kV terminal. The first step is to set up the sequence networks with all reactances on a common base, which will be 100 MVA. The 161- and 345-kV system values are given on this base. For the autotransformer, the equivalent

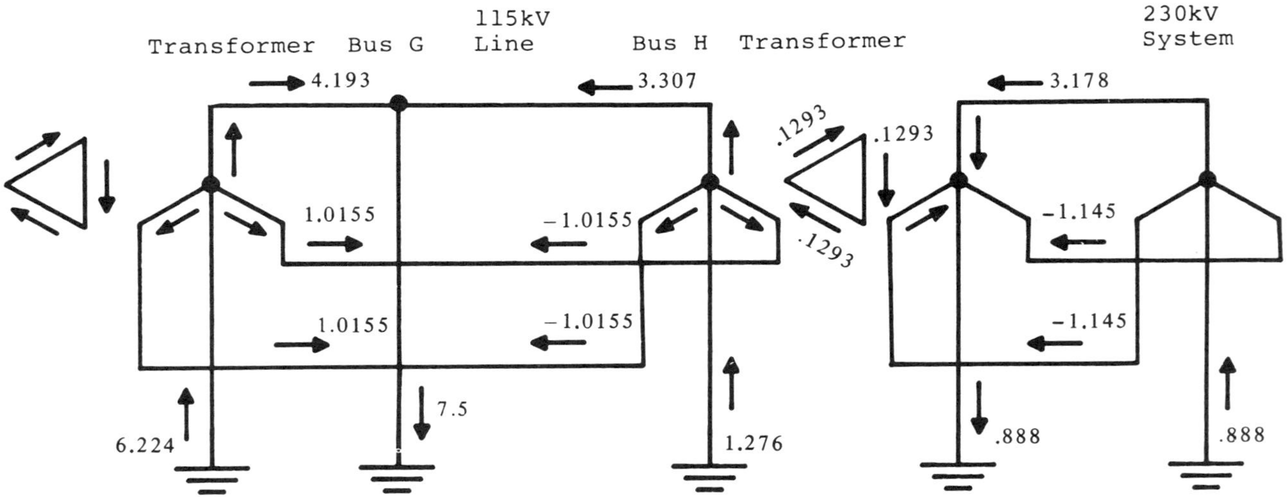

Per unit currents at 100 MVA.

Fig. 4.25 Phase and $3I_0$ current distribution for a single-phase-to-ground fault at bus G in Fig. 4.22.

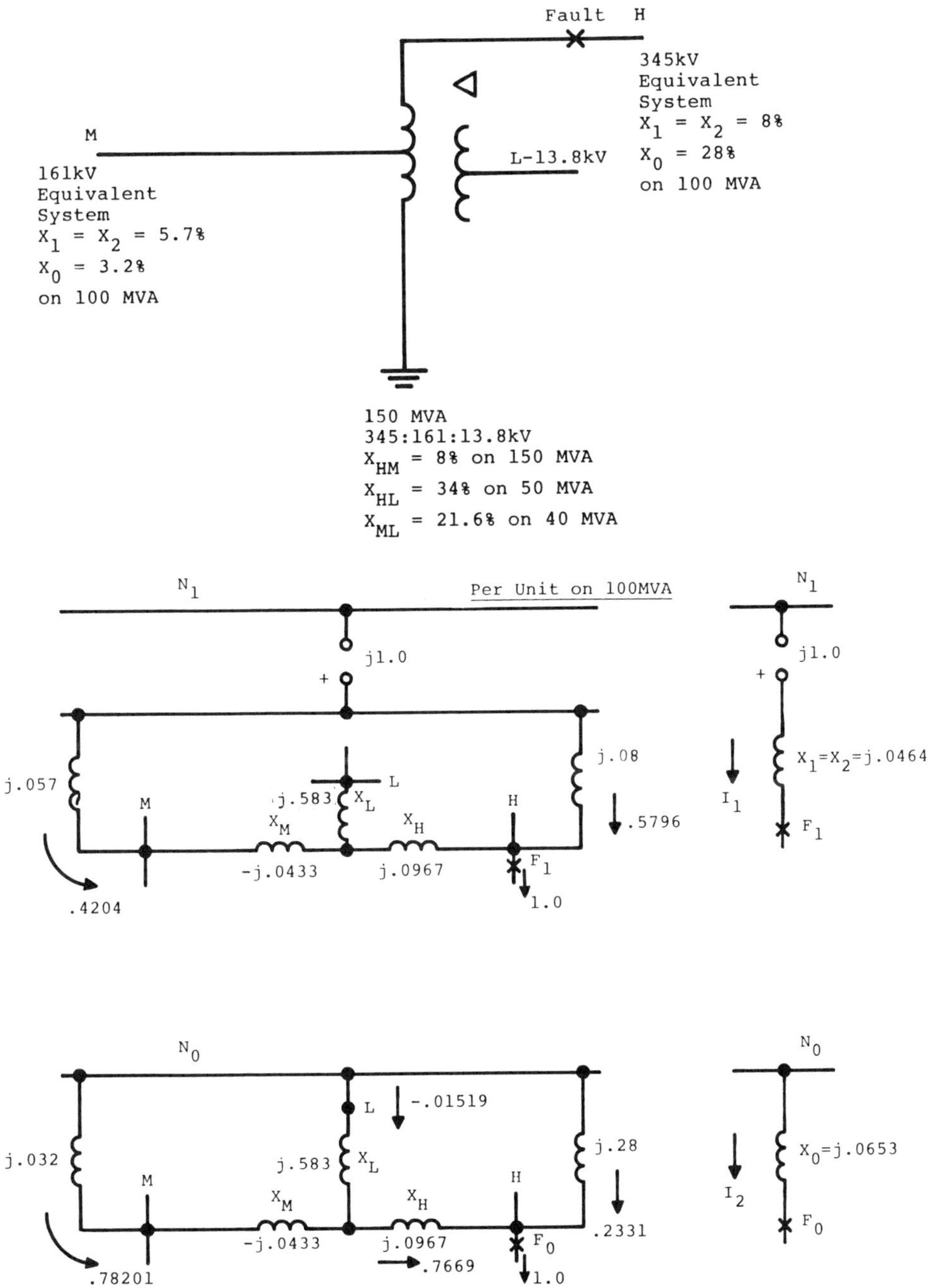

Fig. 4.26 Example of fault calculation for an autotransformer.

network is as in Fig. A4.2-3d, with values as follows: On a 100-MVA base,

$$X_{HM} \quad = \; = 8 \times \frac{100}{150} = 5.333\% = 0.05333 \text{ pu}$$

$$X_{HL} \quad = 34 \times \frac{100}{50} = 68\% = 0.68 \text{ pu} \tag{4.36}$$

$$X_{ML} \quad = 21.6 \times \frac{100}{40} = 54\% = 0.54 \text{ pu}$$

and from Eq. (A4.2-13) through Eq. (A4.2-15), page 146,

$$X_H = \frac{1}{2}(0.0533 + 0.68 - 0.54) = 0.09667 \text{ pu}$$

$$X_M = \frac{1}{2}(0.0533 + 0.54 - 0.68) = -0.04334 \text{ pu} \tag{4.37}$$

$$X_L = \frac{1}{2}(0.68 + 0.54 - 0.0533) = 0.58334 \text{ pu}$$

These values are shown on the sequence diagrams of Fig. 4.26. In the positive (and negative)-sequence networks for a fault at H,

$$X_1 = X_2 = \frac{\overset{(0.5796)}{(0.0533 + 0.057)}\overset{(0.4204)}{(0.08)}}{0.1903} = j0.04637 \text{ pu} \tag{4.38}$$

The zero-sequence network reduces as follows: first paralleling the left side,

$$= \frac{(0.032 - 0.0433)(0.5833)}{0.032 - 0.0433 + 0.5833} = \frac{\overset{(-0.0198)}{(-0.0113)}\overset{(1.0198)}{(0.5833)}}{0.5720}$$

$$= -0.01156 \text{ pu}$$

$$X_0 = \frac{(-0.01156 + 0.09667)(0.28)}{0.08511 + 0.28} = \frac{\overset{(0.2331)}{(0.08511)}\overset{(0.7669)}{(0.28)}}{0.36511} \tag{4.39}$$

$$= j0.06527 \text{ pu}$$

The current distribution factor through X_M path is $0.7669 \times 1.0198 = 0.78207$, and through X_L path is $0.7669 \times -0.0198 = -0.01519$. These current distributions are shown on the network.

4.14.1 Single-Phase-to-Ground Fault at H Calculation

$$I_1 = I_2 = I_0 = \frac{j1.0}{j(0.0464 + 0.0464 + 0.0653)} = \frac{1.0}{0.1580}$$

$$= 6.3287 \text{ pu}$$

$$= 6.3287 \times \frac{100{,}000}{\sqrt{3} \times 345} = 1059.1 \text{ A at 345 kV} \qquad (4.40)$$

$$I_{aF} = 3I_0 = 3 \times 6.3287 = 18.986 \text{ pu}$$

$$= 3 \times 1059.1 = 3177.29 \text{ A at 345 kV} \qquad (4.41)$$

It is recommended that amperes rather than per unit be used for fault current distribution, particularly in the neutral and common windings. The autotransformer is unique in that it is both a transformer and a direct electrical connection. Thus amperes at the medium-voltage base I_M are combined directly with amperes at the high-voltage base I_H for the common winding current I, or for the high-side fault,

$$I = I_H \text{ in amperes at } kV_H - I_M \text{ in amperes at } kV_M \qquad (4.42)$$

For the current in the grounded neutral,

$$3I_0 = 3I_{0H} \text{ in amperes at } kV_H - 3I_{0M} \text{ in amperes at } kV_M \qquad (4.43)$$

Both of the currents above are assumed to flow up the neutral and to the M junction point.

Correspondingly, for a fault on the M or medium-voltage system, the current flowing up the grounded neutral is,

$$3I_0 = 3I_{0M} \text{ in amperes at } kV_M - 3I_{0H} \text{ in amperes at } kV_H \qquad (4.44)$$

Thus these currents in the common winding and neutral are a mixture of high- and medium-voltage currents, so there is no base to which they can be referred. This makes per unit difficult, as it must have a base. When or if per unit must be used, a fictional base can be devised based on the transformer parts ratios. This is quite complex. Being fundamental, amperes are easy to handle and will be used in the following.

The sequence, phase a, and neutral currents are documented in Fig. 4.27 for the example of Fig. 4.26. There will be a current flowing

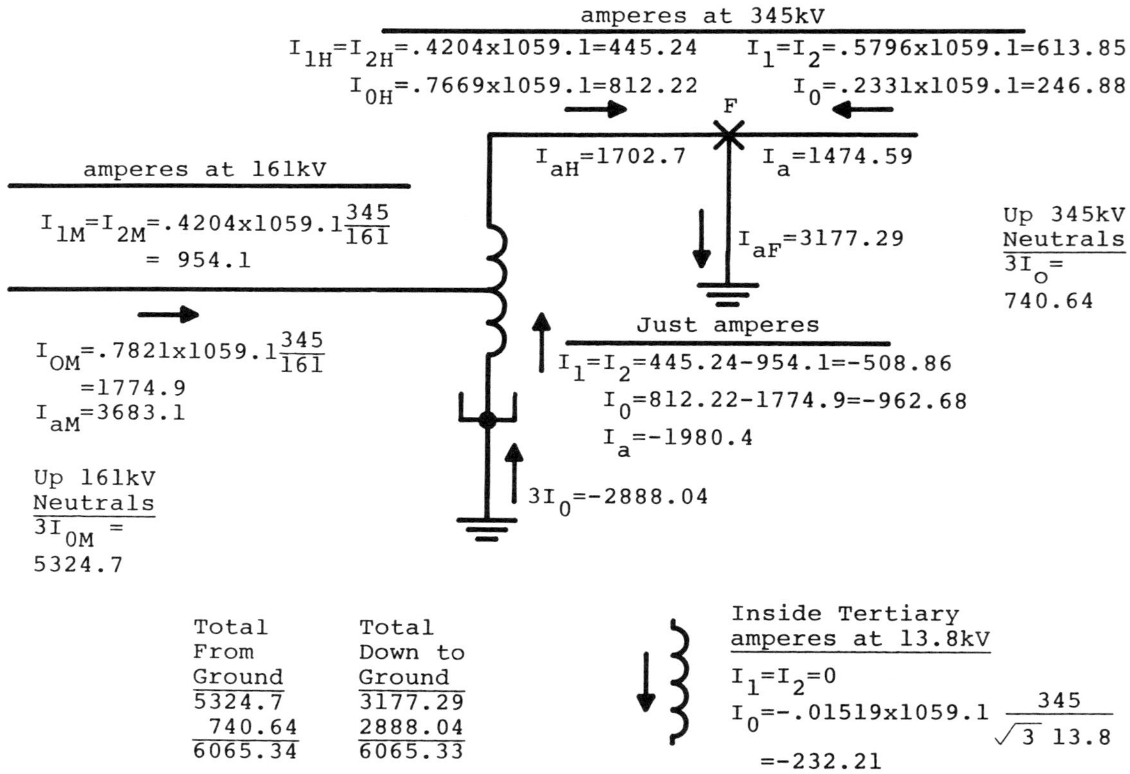

Fig. 4.27 Fault current distribution for the autotransformer of Fig. 4.26.

in phases b and c since the current distribution factors are different in the positive- and zero-sequence networks. These are not shown, as they are of little importance in protection.

The example indicates that current flows down the autotransformer neutral instead of up, as might be expected. Also in this case, the current in the delta has reversed because the negative branch of the transformer equivalent circuit is larger than the very solidly grounded 161-kV connected system. Both of these effects affect the protection, and this is discussed in Chapter 12.

There can be a question as to the direction of current in the tertiary. This can be checked by ampere-turns, as shown in Fig. 4.28. Arbitrarily, one per unit turn was assumed for the 161 kV winding, and the others were derived. Any winding or group could be used for the base, as convenient.

4.15 EXAMPLE: SERIES UNBALANCE CALCULATION

A typical system with load is shown in Fig. 4.29 with the three sequence networks. All constants on these networks are in per unit on

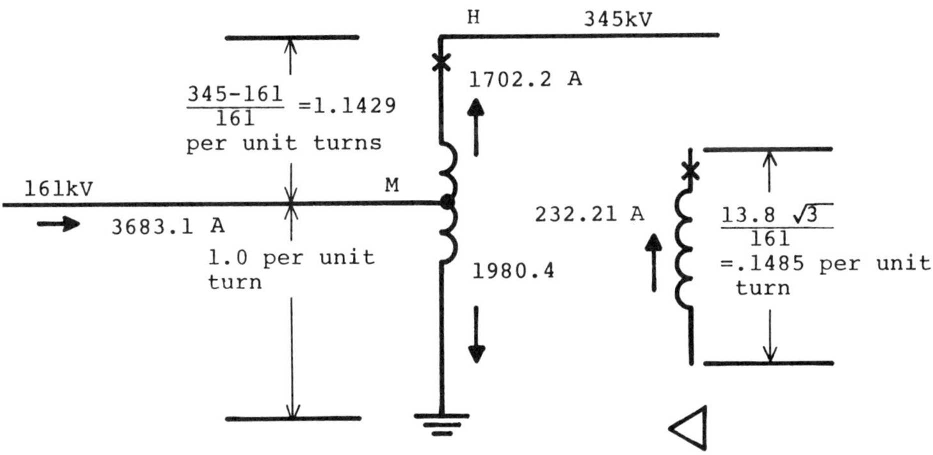

Ampere-turns Up = Ampere-turns Down

1702.7x1.1429 + 232.21x .1485 = 1980.4x1.0

1946.02 + 34.48 = 1980.4

1980.5 = 1980.4 Check.

Fig. 4.28 Ampere-turn check to confirm or establish the direction of current flow in the tertiary.

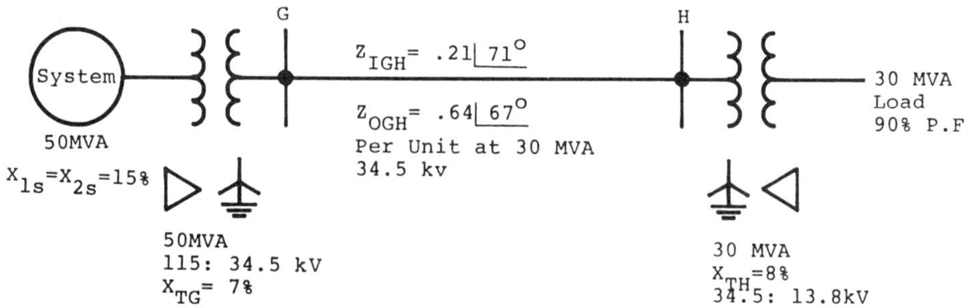

Fig. 4.29a Example for series unbalance calculation.

a 30-MVA base. With induction motors, the negative-sequence load impedance is less than the positive-sequence impedance. This is covered in Chapter 11. A series unbalance is assumed on the 34.5 kV line at bus H.

Assume that phase a is open at that point. The networks are interconnected as shown in Fig. 4.18j or as in Fig. 4.30. Z_2, which equals $Z_{2G} + Z_{2H}$, is in parallel with Z_0, which is $Z_{0G} + X_{TH}$. From the data,

$$Z_2 = 0.5932 + j0.7015 = 0.9187 \angle 49.78°$$

$$Z_0 = 0.250 + j0.711 = 0.7537 \angle 70.63°$$

$$\frac{Z_2 Z_0}{Z_2 + Z_0} = \frac{0.9187 \times 0.7537 \angle 49.78° + 70.63°}{0.8432 + j1.4125}$$

$$= 0.4209 \angle 61.245° = 0.2025 + j0.3690$$

The total impedance is,

$$Z_{1G} + \frac{Z_2 Z_0}{Z_2 + Z_0} + Z_{1H} = 1.171 + j1.216 = 1.688 \angle 46.08° \tag{4.45}$$

The equivalent system voltage to provide a voltage at the load of $1.0\angle 0°$ with I in the load of $1.0\angle -25.84°$ would be

$$-V + I(Z_{1G} + X_{TH}) + 1.0 = 0$$

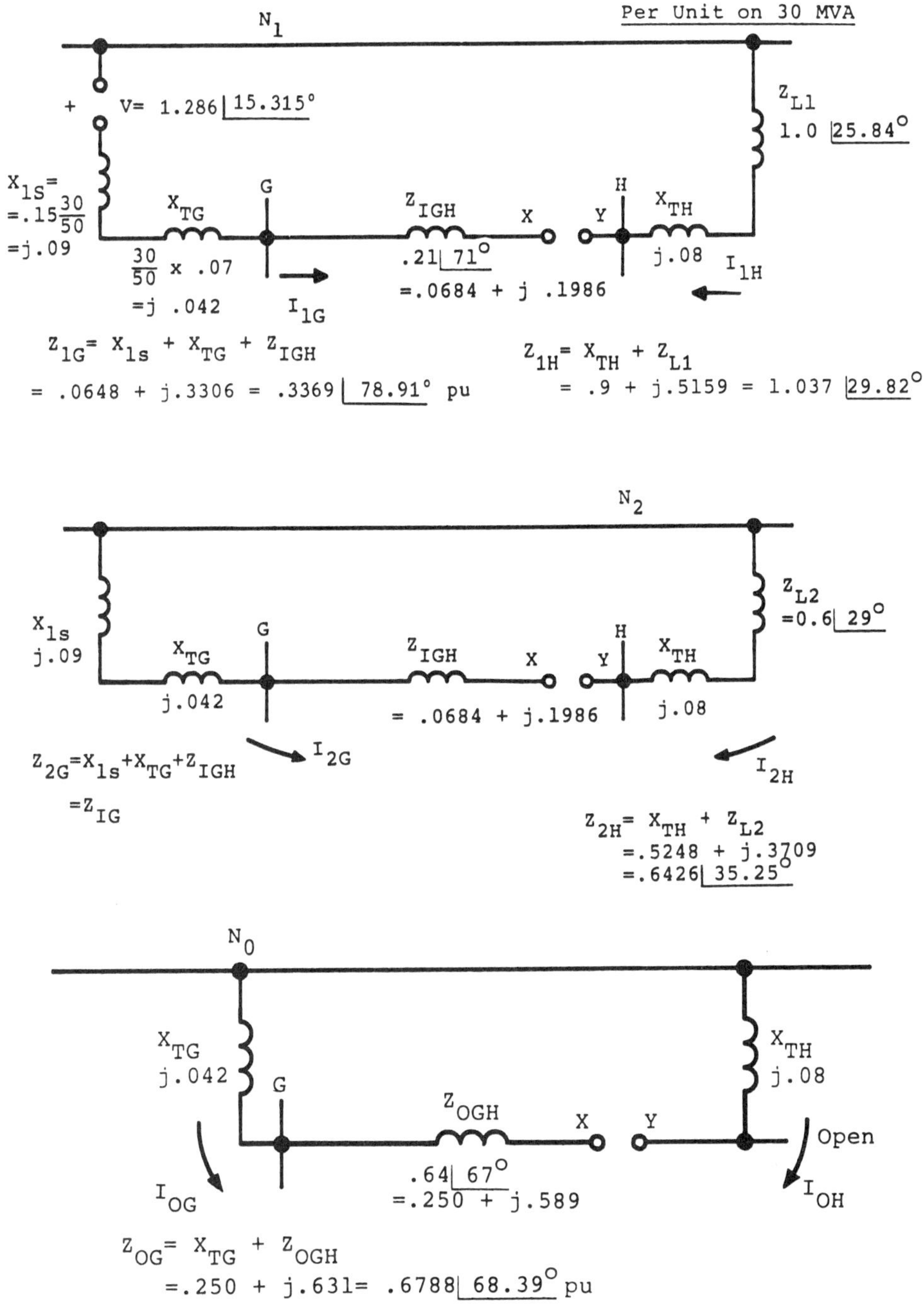

Fig. 4.29b The three sequence networks for the system of Fig. 4.29.

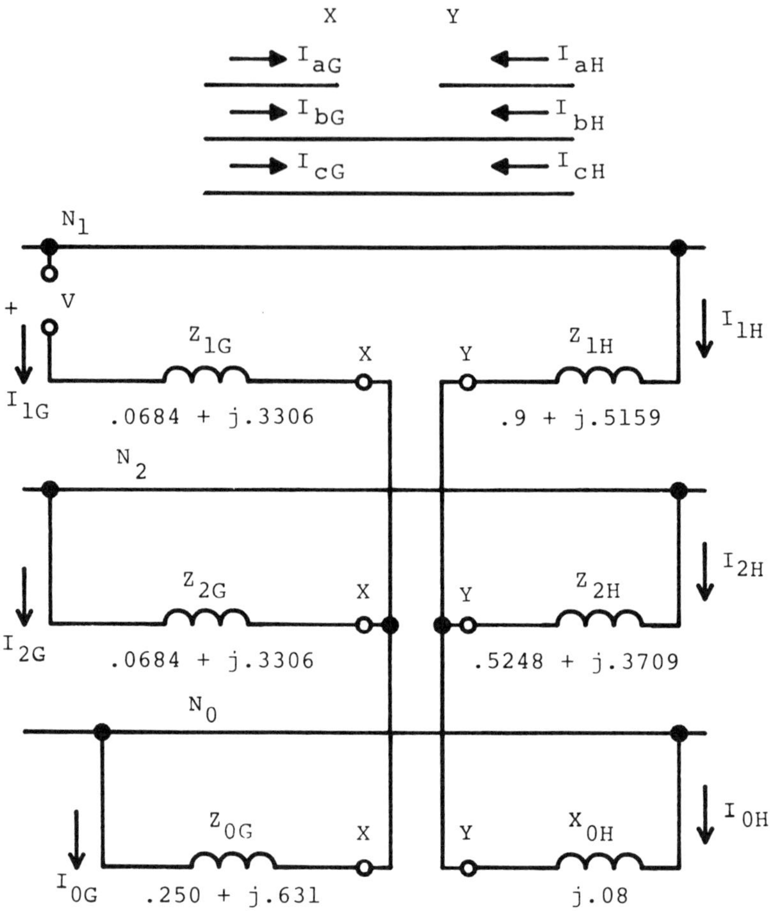

Fig. 4.30 Sequence interconnections for phase a open on the line at bus H of Fig. 4.29.

$$
\begin{aligned}
V &= 1.0 + 1.0\underline{/-25.84^\circ}\ (0.0684 + j0.4106) \\
&= 1.0 + 1 \times 0.4163\ \underline{/-25.84^\circ + 80.54^\circ} = 1 + .4163\ \underline{/54.7^\circ} \\
&= 1.286\ \underline{/15.315^\circ}\ \text{pu} \qquad (4.46)
\end{aligned}
$$

From Eq. (4.45) and (4.46),

$$
I_{1G} = \frac{1.286\ \underline{/15.315^\circ}}{1.688\ \underline{/46.08^\circ}} = 0.762\ \underline{/-30.765^\circ}\ \text{pu} = 0.655 - j0.390\ \text{pu} \qquad (4.47)
$$

$$I_{2G} = -I_{1G}\frac{Z_0}{Z_2 + Z_0} = -0.762\angle{-30.76°}\,\frac{0.7537\angle{70.63°}}{1.645\angle{59.165°}}$$

$$= 0.349\angle{160.70°} = -0.329 + j0.115 \text{ pu} \qquad (4.48)$$

$$I_{0G} = -I_{1G}\frac{Z_2}{Z_2 + Z_0} = -0.762\angle{-30.76°}\,\frac{0.9187\angle{49.78°}}{1.645\angle{59.165°}}$$

$$= 0.426\angle{139.85°} = -0.325 + j.274 \text{ pu} \qquad (4.49)$$

From these,

$$I_{aG} = I_{1G} + I_{2G} + I_{0G} = 0 \qquad (4.50)$$

$$I_{bG} = a^2 I_{1G} + aI_{2G} + I_{0G}$$

$$= 0.762\angle{209.24°} + 0.349\angle{280.7°} + 0.426\angle{139.85°}$$

$$= -0.925 - j0.441 = 1.025\angle{-154.53°} \text{ pu} \qquad (4.51)$$

$$I_{cG} = aI_{1G} + a^2 I_{2G} + I_{0G}$$

$$= 0.762\angle{89.24°} + 0.349\angle{40.7°} + 0.426\angle{139.85°}$$

$$= -0.050 + j1.264 = 1.265\angle{92.285°} \text{ pu} \qquad (4.52)$$

4.16 EXAMPLE: SIMULTANEOUS UNBALANCE CALCULATION

Consider an open phase with the conductor falling to ground on the load side in the system of Fig. 4.29. The sequence interconnections are illustrated in Fig. 4.31 per Fig. 4.20a. The load theoretically should be considered, but this requires more complex simultaneous equations. Since the load impedance is large compared to the transformer impedance at H (X_{TH}), neglecting the load simplifies the calculations with very small error. From the connections in Fig. 4.31, the following can be derived:

$$V_1 + V_2 + V_0 = 0 \qquad (4.53)$$

From the zero-sequence network,

$$3I_{1G}X_{TH} + V_0 = 0 \qquad (4.54)$$

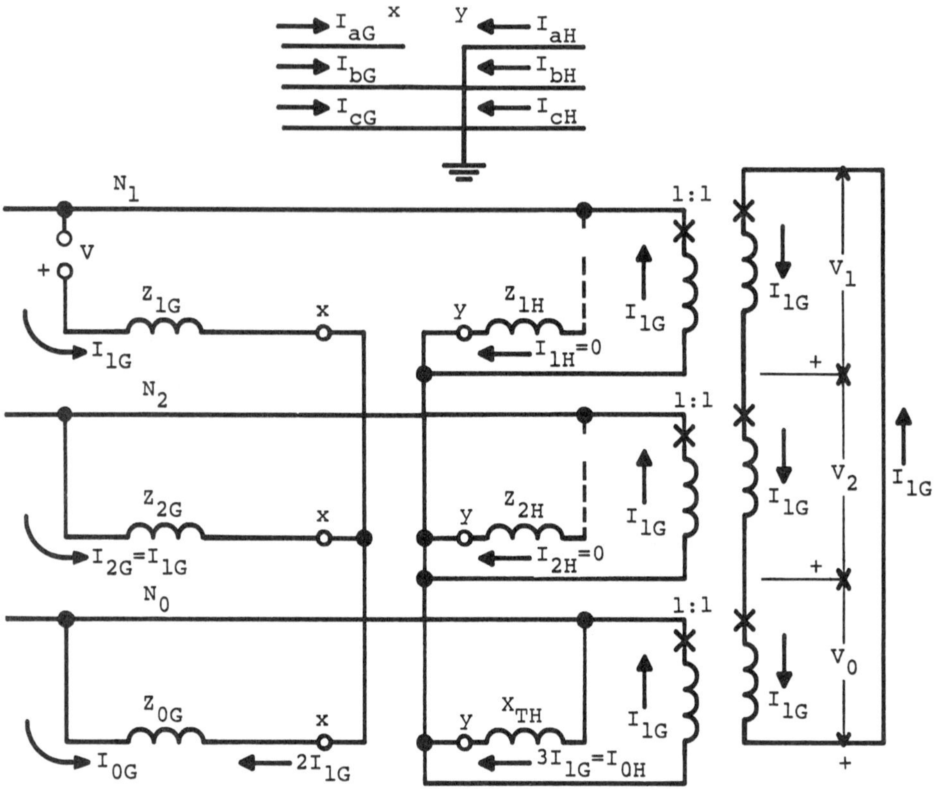

Fig. 4.31 Sequence interconnections for phase a open and grounded on the line at bus H of Fig. 4.29. For simplicity the current flow shown neglects the load.

Since the perfect or ideal transformers have no impedance, the voltage drop across them cannot be expressed by the current in their windings. Eq. (4.54) provides a value for V_0 in terms of the system. Subtracting Eq. (4.53) from Eq. (4.54) yields

$$3I_{1G}X_{TH} - V_1 - V_2 = 0 \tag{4.55}$$

Now the drops around the positive- and negative-sequence networks:

$$-V + I_{1G}Z_{1G} - I_{2G}Z_{2G} - V_2 + V_1 = 0$$

Since $I_{1G} = I_{2G}$ and $Z_{1G} = Z_{2G}$, the above becomes

$$-V - V_2 + V_1 = 0 \tag{4.56}$$

Subtracting (4.55) from (4.56) give us

$$-3I_{1G}X_{TH} - V + 2V_1 = 0 \tag{4.57}$$

Around the positive- and zero-sequence network loops,

$$-V + I_{1G}Z_{1G} + 2I_{1G}Z_{0G} + 3I_{1G}X_{TH} + V_1 = 0$$

Multiply by 2:

$$-2V + I_{1G}(2Z_{1G} + 4Z_{0G} + 6X_{TH}) + 2V_1 = 0 \tag{4.58}$$

Subtracting (4.57) from (4.58):

$$-V + I_{1G}(2Z_{1G} + 4Z_{0G} + 9X_{TH}) = 0$$

$$I_{1G} = I_{2G} = \frac{V}{2Z_{1G} + 4Z_{0G} + 9X_{TH}} \tag{4.59}$$

$$I_{0G} = -2I_{1G} \tag{4.60}$$

$$I_{1H} = I_{2H} = 0 \text{ and } I_{0H} = 3I_{1G} \tag{4.61}$$

with pu values from Fig. 4.29 of

$V = 1.0 \angle 0°$ for the no load assumption.

$Z_{1G} = 0.0684 + j0.3306$

$Z_{0G} = 0.250 + j0.631$

$X_{TH} = j0.08$

$$I_{1G} = I_{2G} = \frac{1.0 \angle 0°}{1.1368 + j3.905} = 0.2459 \angle -73.77° \tag{4.62}$$

$$I_{0G} = 0.4918 \angle 106.23^\circ \text{ pu} \tag{4.63}$$

$$I_{0H} = 0.7376 \angle -73.72^\circ \text{ pu} \tag{4.64}$$

The current flow is shown in Fig. 4.32 for this open phase and ground. It will be noted that the current in the transformer bank at G is reversed, flowing down instead of up the neutral. Also, the fault current is low relative to the bank neutral currents. If the conductor on the right or line side had fallen to ground, quite different results may be expected.

4.17 OVERVIEW

Faults and the sequence quantities can be visualized and perhaps better understood by an overall view in contrast to the specific representations and calculations. Accordingly, several overviews are presented next.

4.17.1 Voltage and Current Phasors for Shunt Faults

The first overview is a review of shunt faults, which are the common types experienced on a power system. These are illustrated in Fig. 4.33, with (a) showing the normal balanced voltage and load current phasors. Load is slightly lagging, normally from unit power factor to around 30° lag. With capacitors at light load, the currents may slightly lead.

When faults occur the internal voltage of the generators does not change, that is, unless the fault is left on too long and the voltage regulators attempt to increase the fault-reduced machine terminal voltage.

A three-phase fault (b) reduces all three voltages and causes a large increase and usually highly lagging fault current symmetrically in all three phases. The angle of lag is determined by the system as discussed above and will usually vary from about 30 to 45° lag up to nearly 90° as shown.

The single-phase-to-ground fault (c) is the most common. The faulted phase voltage collapses and its current increases as shown. Load current is neglected, as it usually is quite small relatively, and $I_b = I_c = 0$. As has been indicated above, fault current will flow in the unfaulted phases on loop systems where the distribution factors for the three sequence networks are different. Again the fault current will lag normally. It is shown at 90° in Fig. 4.33.

The phase-to-phase fault is (d). Neglecting load, for a b-to-c phase fault, V_a is normal, $I_a = 0$. V_b and V_c collapse from their

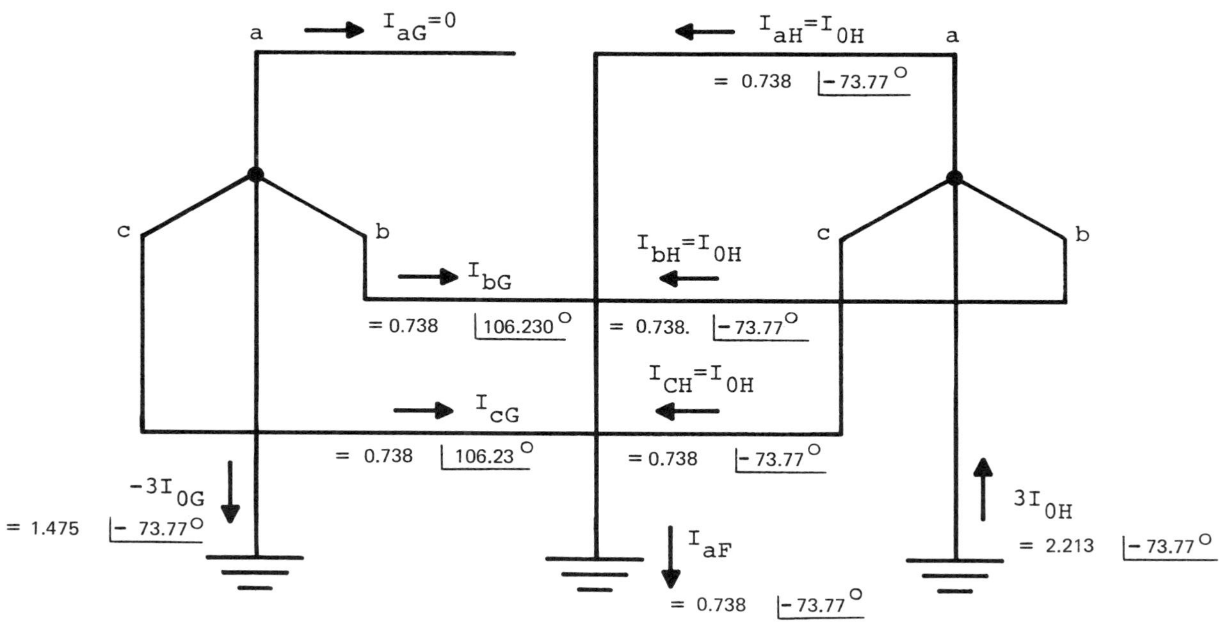

Fig. 4.32 Current flow in per unit at 30-MVA base for the open phase and ground fault for the power system of Fig. 4.29.

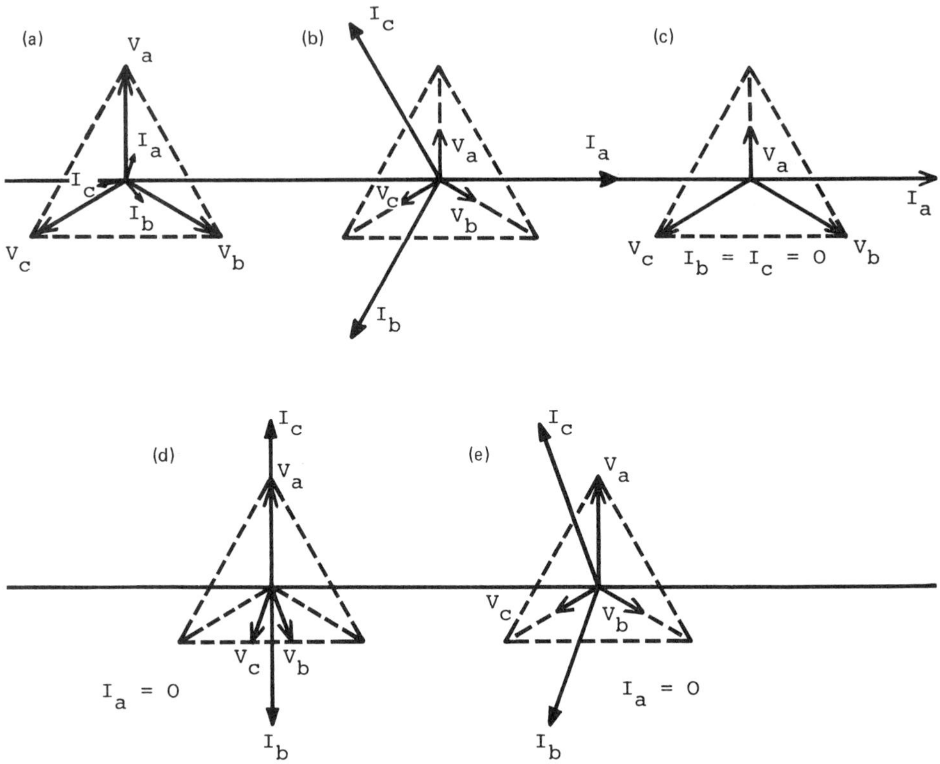

Fig. 4.33 Typical current and voltage phasors for common shunt faults. Currents shown for 90° lagging or for a power system with Z = X, neglecting resistance. Load current is assumed negligible. (a) Normal phasors for a balanced system; (b) three-phase fault; (c) single-phase-to-ground fault (a-gnd.); (d) phase-to-phase fault (bc fault); (e) two-phase-to-ground fault (bc-gnd.).

normal position to vertical phasors at a solid fault point where V_{bc} = 0. I_b and I_c are equal and opposite and lagging normally, 90° in Fig. 4.33.

The two-phase-to-ground fault (e) results in the faulted phase voltages collapsing along their normal position until for a solid fault they are zero. Thus at the fault, $V_b = V_c = 0$, which is not true for the phase-to-phase fault (d). I_b and I_c will be in the general area as shown. An increasing amount of zero-sequence current will cause I_b and I_c to swing closer to each other; contrarily, a very low zero-sequence current component will result in the phasors, to approach the phase-to-phase fault of (d). This can be seen from the sequence

sequence network connections of Fig. 4.16. If Z_0 becomes infinite (essentially ungrounded system), the interconnection becomes as in Fig. 4.15 for a phase-to-phase fault. On the other hand, for a very solidly grounded system where Z_0 approaches zero relatively, the negative network becomes shorted, and this fault becomes like the three-phase fault of Fig. 4.13.

In some parts of a loop network it is possible for the zero-sequence current to flow opposite to the positive- and negative-sequence currents. In this area I_c may lag the V_a phasor rather than lead it as shown, with I_b correspondingly leading the position shown.

4.17.2 System Voltage Profiles During Faults

The trends of the sequence voltages for the various faults of Fig. 4.33 are illustrated in Fig. 4.34. Only the phase a sequence voltages are shown for an ideal case where $Z_1 = Z_2 = Z_0$ — an impossibility, but it makes the presentation less complex and does not affect the trends shown.

With the common assumption of no load, the system voltage is equal throughout the system, as indicated by the dashed lines. When a solid three-phase fault occurs, the voltage at the fault point becomes zero, but as indicated above, does not change in the source until the regulators act to change the generator fields. By this time the fault should have been cleared by protective relays. Thus the voltage profile is shown in (a).

For phase-to-phase faults (b), the positive-sequence voltage drops to half value ($Z_1 = Z_2$). This unbalance fault is the source of negative sequence and the V_2 drops are as shown, being zero in the generators.

For two-phase-to-ground faults (c) with $Z_1 = Z_2 = Z_0$, the positive-sequence voltage at the fault drops to one-third of V_1. The fault now generates both negative and zero sequence, which flows through the system, producing voltage drops as shown. V_2 becomes zero in the generators, while V_0 is zero at the grounded transformer neutral point.

The fault voltage for a phase-a-to-ground solid fault is zero and as phase a is documented in Fig. 4.34d, the sum of the positive, negative, and zero voltage components at the fault add to zero. Thus the positive-sequence voltage drops to $2/3V_1$ when $Z_1 = Z_2 = Z_0$ at the fault point, where $-1/3V_2$ and $-1/3V_0$ are generated. Again they drop to zero in the generator or source for the negative sequence and to zero at the grounded transformer bank neutral.

The fundamental illustrated in Fig. 4.34 is that *positive-sequence voltage is always maximum at the generators and minimum at the fault. The negative- and zero-sequence voltages are always maximum at the fault and minimum at the generator or grounded neutral.*

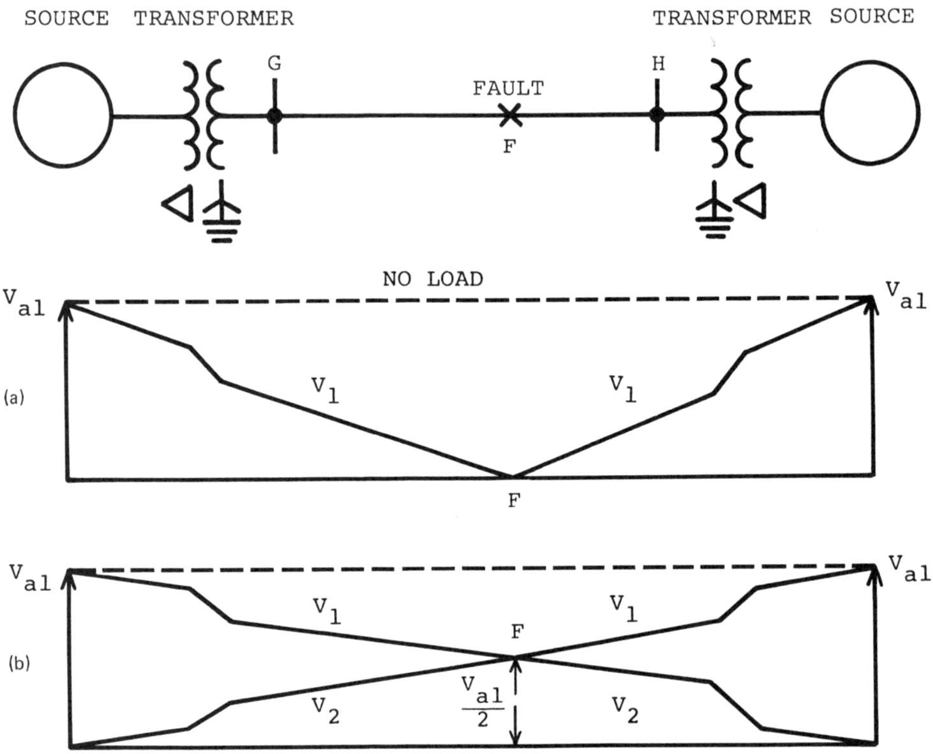

Fig. 4.34 System sequence voltage profiles during shunt faults: (a) three-phase fault; (b) phase-to-phase (bc) fault; (c) two-phase-to-ground (bc-G) fault; (d) single-phase-to-ground (a-G) fault.

It is common to refer to the grounded-wye-delta or similar banks as "ground sources." This is really a misnomer, as the source of zero sequence is the unbalance, the ground fault. However, so designating these transformers as ground sources is practical, as by convention ground ($3I_0$) current flows up the grounded neutral, through the system, and down the fault into ground.

4.17.3 Voltage and Current Fault Phasors for All Combinations of the Different Faults

Another look at the sequence phasors is presented in Figs. 4.35 and 4.36. The voltages and currents generated by the sources can be only positive sequence by design, nothing else. Yet the unbalanced faults require unbalanced quantities. How can this difference be resolved to satisfy both requirements: balanced quantities by the generators and

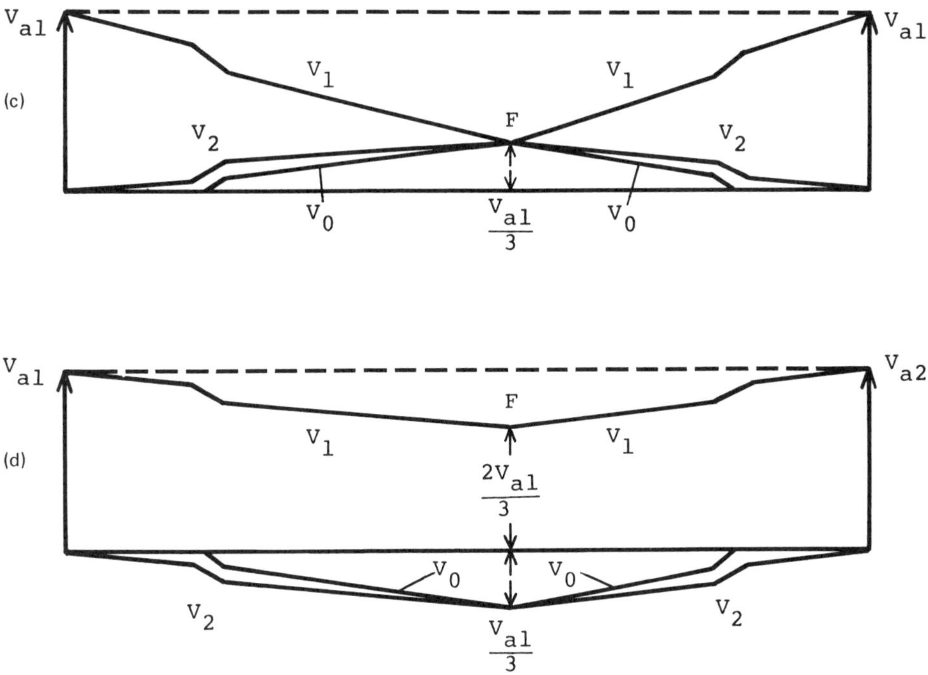

Fig. 4.34 (Continued)

unbalanced quantities at the faults? The resolution can be considered as the function of the negative-sequence quantities and for ground faults the zero-sequence quantities. This can be seen from Figs. 4.35 and 4.36. Considering the voltages of Fig. 4.35, positive-sequence voltage developed by the source or generator is the same for all faults. For three-phase faults no transition help is required, as these faults are symmetrical, hence no negative or zero sequences. For the phase-to-phase faults, negative sequence appears to provide the transition. Note that for the several combinations, ab, bc, and ca phases, the negative sequence is in different positions to provide the transition. The key is that for, say, an ab fault, phase c will be essentially normal, so V_{c1} and V_{c2} are basically in phase to provide this normal voltage. Correspondingly, for a bc fault, V_{a1} and V_{a2} are essentially in phase, and so on.

The two-phase-to-ground faults are similar; for ab-G faults, the uninvolved phase c quantities V_{c1}, V_{c2}, V_{c0} combine to provide the uncollapsed phase c voltage. In the figure these are shown in phase and half magnitude. In actual cases there will be slight variations, as the sequence impedances do not have the same magnitude or phase angle.

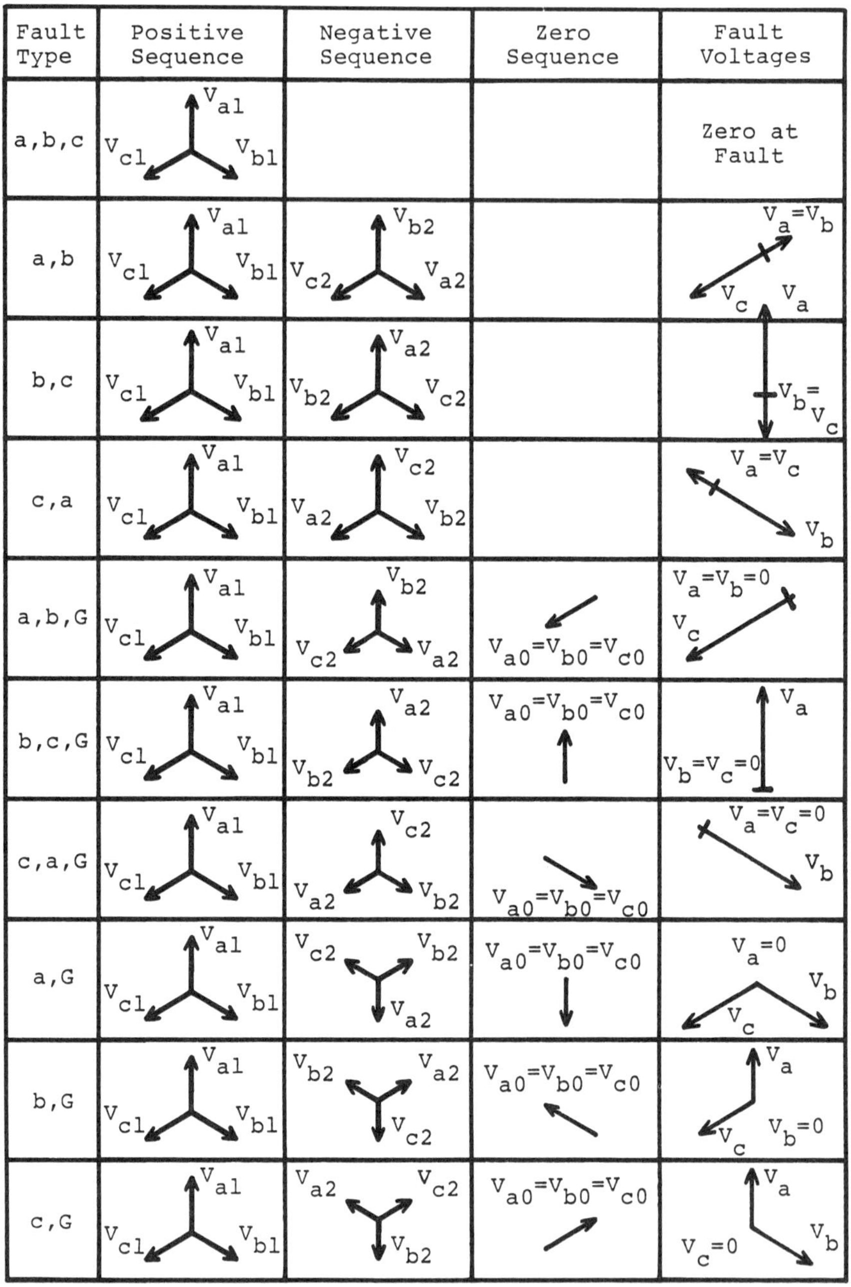

Fig. 4.35 Sequence voltages and the voltage at the fault point for the various fault types. Solid faults with $Z_1 = Z_2 = Z_0$ for simplicity. Magnitudes not to scale.

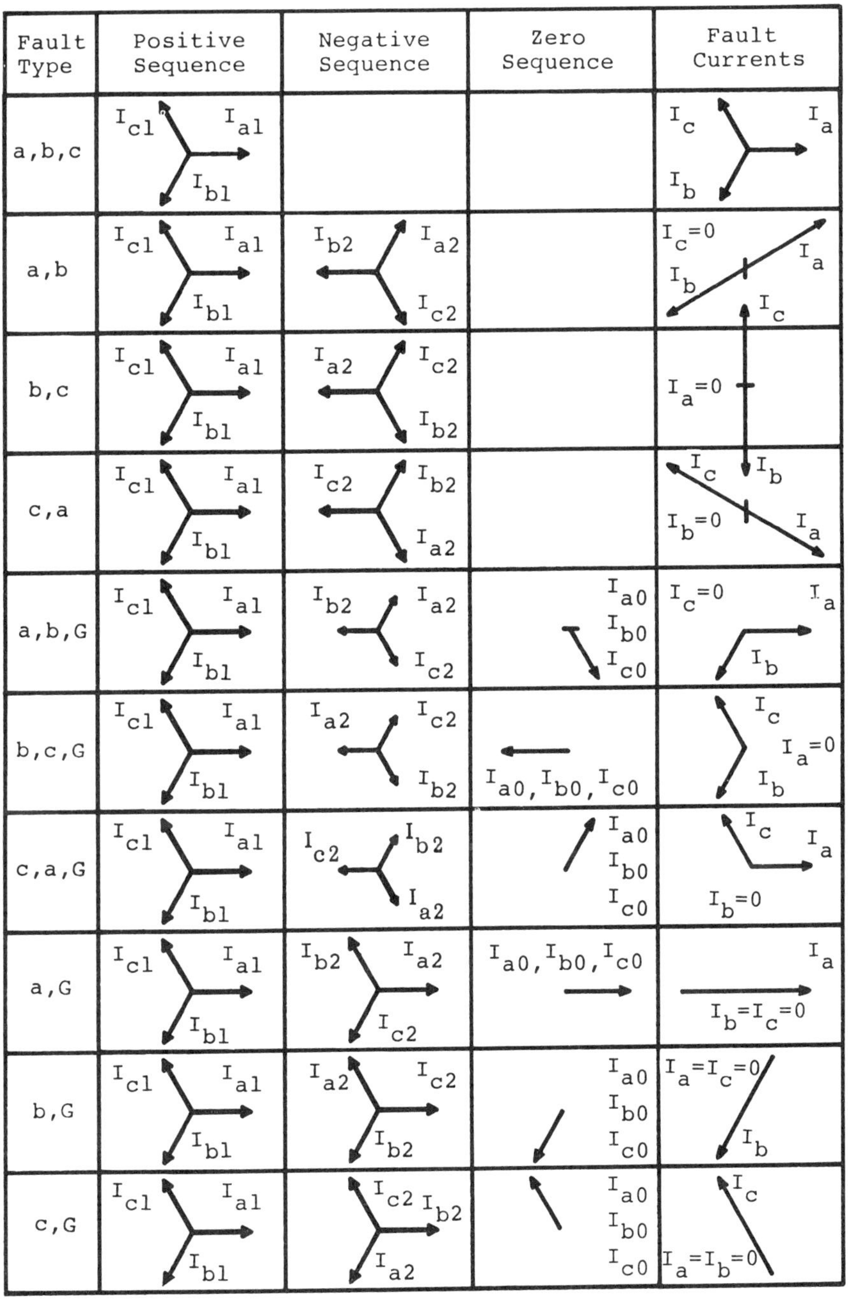

Fig. 4.36 Sequence currents and the fault current for the various fault types. Solid faults with $Z_1 = Z_2 = Z_0$ for simplicity. Magnitudes not to scale.

For single-phase-a-to-ground faults, the negative (V_{a2})- and zero (V_{a0})- sequence voltages add to cancel the positive- sequence V_{a1}, which will be zero at a solid fault. Correspondingly, for a phase b fault, V_{b2} and V_{b0} oppose V_{b1}, and similarly for the phase c fault.

The same concept applies to the sequence currents, as shown in Fig. 4.36. The positive-sequence currents are shown the same for all faults and for 90° lag (X-only system) with respect to the voltages of Fig. 4.35. These will vary depending on the system constants, but the concepts illustrated are valid. Again for three-phase faults, no transition help is required; hence there is no negative- or zero-sequence involvement.

For phase-to-phase faults negative sequence provides the necessary transition, with the unfaulted phase sequence currents in opposition to provide zero or low current. Thus for the ab fault, I_{c1} and I_{c2} are in opposition.

Similarly for two-phase-to-ground faults; for an ab-G fault, I_{c2} and I_{c0} tend to cancel I_{c1}, and so on. For single-phase-to-ground faults the faulted phase components tend to add to provide a large fault current, since $I_{a1} + I_{a2} + I_{a0} = I_a$.

4.18 SUMMARY

A question often asked is: Are the sequence quantities real or only useful mathematical concepts? This has been debated for years, and in a sense they are both. Yes, they are real; positive sequence certainly since it is generated, sold, and consumed; zero sequence because it flows in the neutral and ground and deltas; and negative sequence because it can cause serious damage to rotating machines. Negative sequence, for example, cannot be measured directly by an ammeter or voltmeter. Networks are available and commonly used in protection to measure V_2 and I_2, but these are designed to solve the basic equations for those quantities.

In either case, symmetrical components analysis is an extremely valuable and powerful tool. Protection engineers automatically tend to think in its terms when evaluating and solving unbalanced situations in the power system.

It is important to remember always that any sequence quantity cannot exist in only one phase; this is a three-phase concept. If any sequence is in one phase, it *must* be in all three phases according to the fundamental definitions of Eq. (4.2) through (4.4).

BIBLIOGRAPHY

See the Bibliography at the end of Chapter 1 for additional information.

Anderson, P. M., *Analysis of Faulted Power Systems*, Iowa State University Press, Ames, IA, 1973.

Calabrese, G. O., *Symmetrical Components Applied to Electric Power Networks*, The Ronald Press Company, New York, 1959.

Clarke, Edith, *Circuit Analysis of A-C Power Systems*, Vols. I and II, General Electric Co., Schenectady, NY, 1943, 1950.

Guide for Calculation of Fault Currents for Application of AC High Voltage Circuit Breakers Rated on A Total Current Basis, ANSI/IEEE Standard C37.5-1979, IEEE Service Center.

Harder, E. L., Sequence Network Connections for Unbalanced Load and Fault Conditions, *The Electric Journal*, December 1937, pp. 481–488.

IEEE Recommended Practice for Electric Power Distribution for Industrial Plants, IEEE Red Book, Standard 141-1976, IEEE Service Center.

Wagner, C. F., and R. D. Evans, *Symmetrical Components*, MGraw-Hill Book Company, 1933 (Available in reprint from R. E. Krieger Publishing Co., Inc., Melbourne, FL, 1982).

APPENDIX 4.1: SHORT-CIRCUIT MVA AND EQUIVALENT IMPEDANCE

Quite often short-circuit MVA data are supplied for three-phase and single-phase-to-ground faults at various buses or interconnection points in a power system. The derivation of this and conversion into system impedances is as follows:

Three-Phase Faults

$$MVA_{SC} = 3\phi \text{ fault-short-circuit MVA} = \frac{\sqrt{3}\ I_{3\phi}\ kV}{1000} \qquad \text{(A4.1-1)}$$

where $I_{3\phi}$ is the total three-phase fault current in amperes and kV is the system line-to-line voltage in kilovolts. From this,

$$I_{3\phi} = \frac{1000\ MVA_{SC}}{\sqrt{3}\ kV} \qquad (A4.1\text{-}2)$$

$$Z_{\Omega} = \frac{V_{LN}}{I_{3\phi}} = \frac{1000\ kV}{\sqrt{3}\ I_{3\phi}} = \frac{kV^2}{MVA_{SC}} \qquad (A4.1\text{-}3)$$

Substituting Eq. (2.15), which is

$$Z_{pu} = \frac{MVA_{base}\ Z_{\Omega}}{kV^2} \qquad (2.15)$$

the positive-sequence impedance to the fault location is

$$Z_1 = \frac{MVA_{Base}}{MVA_{SC}}\ pu \qquad (A4.1\text{-}4)$$

$Z_1 = Z_2$ for all practical cases. Z_1 can be assumed to be X_1 unless X/R data are provided to determine an angle.

Single-Phase-to-Ground Faults

$$MVA_{\phi GSC} = \phi G \text{ fault short-circuit MVA} = \frac{\sqrt{3}\ I_{\phi G}\ kV}{1000} \qquad (A4.1\text{-}5)$$

where $I_{\phi G}$ is the total single-line-to-ground fault current in amperes and kV is the system line-to-line voltage in kilovolts.

$$I_{\phi G} = \frac{1000\ MVA_{\phi GSC}}{\sqrt{3}\ kV} \qquad (A4.1\text{-}6)$$

However,

$$I_{\phi G} = I_1 + I_2 + I_0 = \frac{3V_{LN}}{Z_1 + Z_2 + Z_0} = \frac{3V_{LN}}{Z_g} \qquad (A4.1\text{-}7)$$

where $Z_g = Z_1 + Z_2 + Z_0$. From Eqs. (A4.1-3) and (A4.1-7),

$$Z_g = \frac{3\ kV^2}{MVA_{\phi GSC}} \text{ in } \Omega \qquad (A4.1\text{-}8)$$

$$Z_g = \frac{3\ MVA_{Base}}{MVA_{\phi GSC}} \text{ pu} \qquad (A4.1\text{-}9)$$

Then $Z_0 = Z_g - Z_1 - Z_2$, or in most practical cases, $X_0 = X_g - X_1 - X_2$, since the resistance is usually very small in relation to the reactance.

Example

A short-circuit study indicates that at bus X in the 69-kV system,

$$MVA_{SC} = 594 \text{ MVA}$$

$$MVA_{\phi GSC} = 631 \text{ MVA}$$

on a 100-MVA base.

Thus the total reactance to the fault is

$$X_1 = X_2 = \frac{100}{594} = 0.1684 \text{ pu}$$

$$X_g = \frac{300}{631} = 0.4754 \text{ pu}$$

$X_0 = 0.4754 - 0.1684 - 0.1684 = 0.1386$ pu, all values on a 100-MVA 69-kV base.

APPENDIX 4.2 IMPEDANCE AND SEQUENCE CONNECTIONS FOR TRANSFORMER BANKS

Two Winding Transformer Banks

Typical banks are shown in Fig. A4.2-1. H is the high-voltage winding and L the low-voltage winding. These designations can be interchanged as required. Z_T is the transformer leakage impedance between the two windings. It is normally designated in per unit or percent by the manufacturer and stamped on the transformer nameplate. Unless otherwise specified, this value is on the self-cooled kVA or MVA rating at the rated voltages.

It can be measured by shorting one winding and applying voltage to the other winding. This voltage should not cause the transformer to saturate. From Figure A4.2-2,

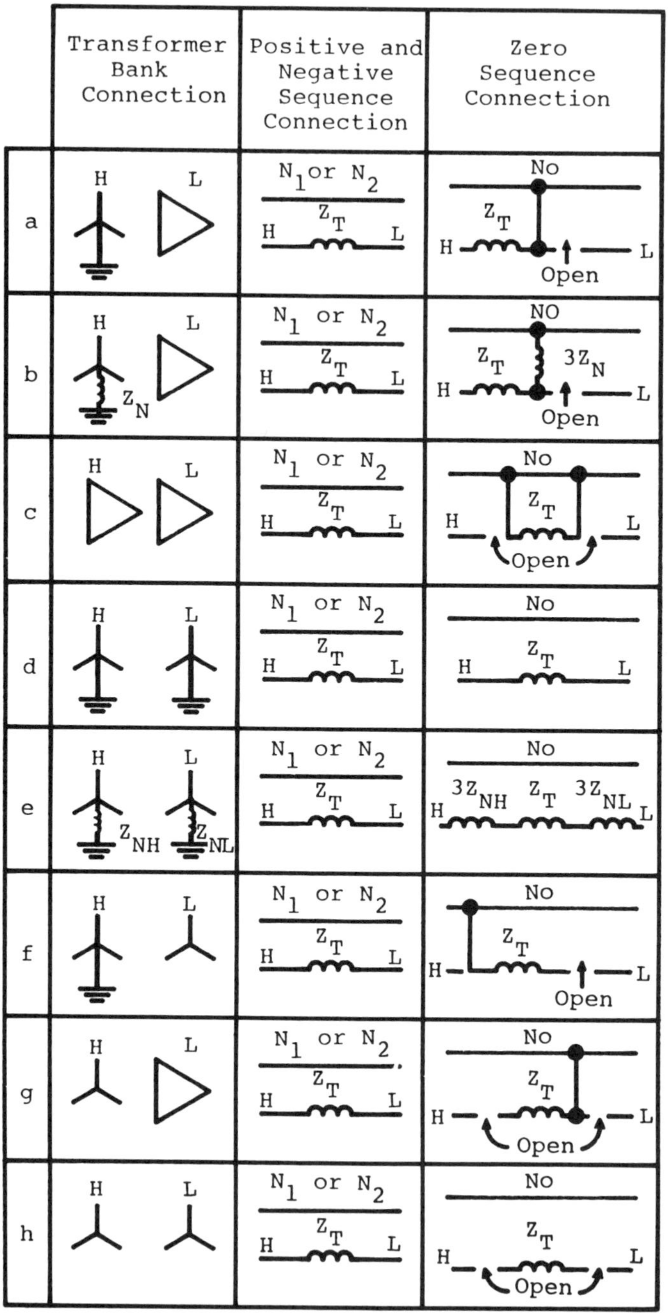

Fig. A4.2-1 Sequence connections for typical two-winding transformer banks.

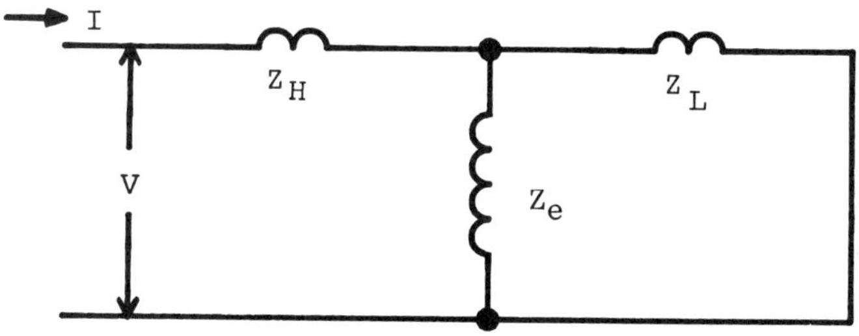

Fig. A4.2-2 Simplified equivalent diagram for a transformer. Z_H and Z_L are the components of the transformer leakage impedance and Z_e is the exciting impedance. All values in per unit or primary H side ohms.

$$Z_T = \frac{V}{I} = Z_H + \frac{Z_L Z_e}{Z_L + Z_e} \qquad \text{(A4-2-1)}$$

Since Z_e unsaturated is very large compared to Z_L, the term $Z_L Z_e / Z_L + Z_e$ approaches and is approximately equal to Z_L, so that for practical purposes

$$Z_T = \frac{V}{I} = Z_H + Z_L \qquad \text{(A4-2-2)}$$

Z_T is measured in practice by circulating rated current (I_R), through one winding with the other shorted and measuring the voltage (V_W) required to circulate this rated current. Then,

$$Z_T = \frac{V_W}{I_R} \text{ ohms.} \qquad \text{(A4-2-3)}$$

This test can be done for either winding as convenient. On the measured side, the base impedance will be

$$Z_B = \frac{V_R}{I_R} \text{ ohms.} \qquad \text{(A4-2-4)}$$

where V_R and I_R are the rated voltage and current respectively.

Then the per unit impedance from equation 2-1 is;

$$Z_T = \frac{Z_T \text{ in ohms}}{Z_B \text{ in ohms}} = \frac{V_W I_R}{I_R V_R} = \frac{V_W}{V_R} \text{ per unit} \qquad \text{(A4-2-5)}$$

For three-phase type transformer units, the nameplate should specify this Z_T, usually in percent, on the three-phase kVA (MVA) rating, and the kV line-to-line voltages. Where several kVA (MVA) ratings are specified, the normal rating without fans, pumps, etc. (lowest) rating should be used as one of the impedance bases.

For individual single-phase transformers, the transformer impedance normally is specified on a single-phase kVA (MVA) and the rated winding voltages (kV) of the transformer. When three such units are used in three-phase systems, then the three-phase kVA (MVA) and line-to-line voltages (kV) bases are required, as outlined in Chapter 2.

Thus, when three individual single-phase transformers are connected in the power system, the individual nameplate percent or per unit impedance will be the Z_T leakage impedance, but on the three-phase kVA (MVA), base and the system line-to-line kV.

Example: Impedances of Single-Phase Transformers in Three-Phase Power Systems

Consider single-phase transformers, each with a nameplate rating of 20 MVA, 66.5 kV:13.8 kV, X = 10%. Considering the individual transformer alone, its leakage reactance is

$$X_T = 0.10 \text{ p.u. on 20 MVA, 66.5 kV}$$

$$\text{or } X_T = 0.10 \text{ p.u. on 20 MVA, 13.8 kV.} \qquad \text{(A4-2-6)}$$

Converting these to actual ohms using equation A4-2-5); $V_{WH} = 0.10V_{RH} = 0.10 \times 66{,}500 = 6650$ volts on the high side, where $I_{RH} = 20{,}000/66.5 = 300.75$ amperes primary.

Then from equation A4-2-3,

$$X_{TH} = \frac{6650}{300.75} = 22.11 \text{ ohms primary} \qquad \text{(A4-2-7)}$$

or on the secondary side, $V_{WL} = 0.10 \times 13{,}800 = 1380$ volts, and $I_{RL} = 20{,}000/13.8 = 1449.28$ amperes secondary.

$$X_{TL} = \frac{1380}{1449.28} = 0.952 \text{ ohm secondary} \qquad \text{(A4-2-8)}$$

Check:

$$\left(\frac{66.5}{13.8}\right)^2 \times 0.952 = 22.11 \text{ ohms primary.}$$

Now consider two possible applications of three of these individual transformers to a power system. These are intended to demonstrate the fundamentals and do not consider if the transformer windings are compatable or suitable for the system voltages shown.

Case 1

Connect the high voltage windings in wye to a 115 kV system and the low voltage windings in delta to a 13.8 kV system. As indicated previously, the leakage impedance of this transformer bank for this application is

$$X_T = 0.10 \text{ p.u. on } 60 \text{ MVA, } 115 \text{ kV}$$

$$X_T = 0.10 \text{ p.u. on } 60 \text{ MVA, } 13.8 \text{ kV} \qquad \text{(A4-2-9)}$$

Now to check this. From equation 2.17,

$$X_{TH} = \frac{115^2 \times 0.10}{60} = 22.11 \text{ ohms primary}$$

$$X_{TL} = \frac{13.8^2 \times 0.10}{60} = 0.317 \text{ ohm secondary} \qquad \text{(A4-2-10)}$$

It will be noted that the individual transformer reactance per equation A4.2-8 is 0.952 ohm, but this is the reactance across the 13.8 kV because of the delta connection. The equivalent wye impedance can be determined by the product of the two-delta branch on either side of the desired wye branch divided by the sum of the three-delta branches. Thus the wye equivalent is

$$\frac{(.952)\ .952}{(3)\ .952} = \frac{.952}{3} = 0.317 \text{ ohms as above.}$$

Check

$$\left(\frac{115}{13.8}\right)^2 \times 0.317 = 22.1 \text{ ohms primary}$$

Case 2

Connect the high voltage windings in delta to a 66.5 kV system, and the low voltage side in wye to a 24 kV system. Now the transformer bank impedance for this system application is

$$X_T = 0.10 \text{ p.u. on } 60 \text{ MVA, } 66.5 \text{ kV}$$

$$\text{or } X_T = 0.10 \text{ p.u. on } 60 \text{ MVA, } 24 \text{ kV} \qquad \text{(A4-2-11)}$$

Now to check this by converting to ohms, using equation 2.17

$$X_{TH} = \frac{66.5^2 \times 0.10}{60} = 7.37 \text{ ohms primary} \qquad \text{(A4-2-12)}$$

Now this primary winding of 22.11 ohms (Eq. A4.2-7) is connected across the 66.5 kV system because of the delta. Accordingly the equivalent wye reactance is $22.11 \times 22.11/3 \times 22.11 = 7.37$ ohms line to neutral on the primary side.

On the secondary side, $X_{TL} = 24^2 \times 0.10/60 = 0.952$ ohm secondary as Eq. A4.2-8.

Check

$$\left(\frac{66.5}{24}\right)^2 \times 0.952 = 7.37 \text{ ohms primary}$$

The connections of two-winding transformers in the sequence networks are documented in Figure A4.2-1. It will be noted that the connections for positive and negative sequences are all the same and independent of the bank connections. This is not true for the zero sequence with different connections for each type of bank.

Neutral impedance is shown for several connections. If the banks are solid grounded, the neutral impedance is zero and the values shown are shorted out in the system and zero sequence diagrams.

Three-Winding and Auto Transformer Banks

Typical banks are shown in Figure A4.2-3. H, M, and L are the high, medium, and low-voltage windings. These designations can be interchanged as required. Normally, the manufacturer provides the leakage impedances between the windings as Z_{HM}, Z_{HL}, and Z_{ML}, usually on different kVA or MVA ratings and at the rated winding voltages.

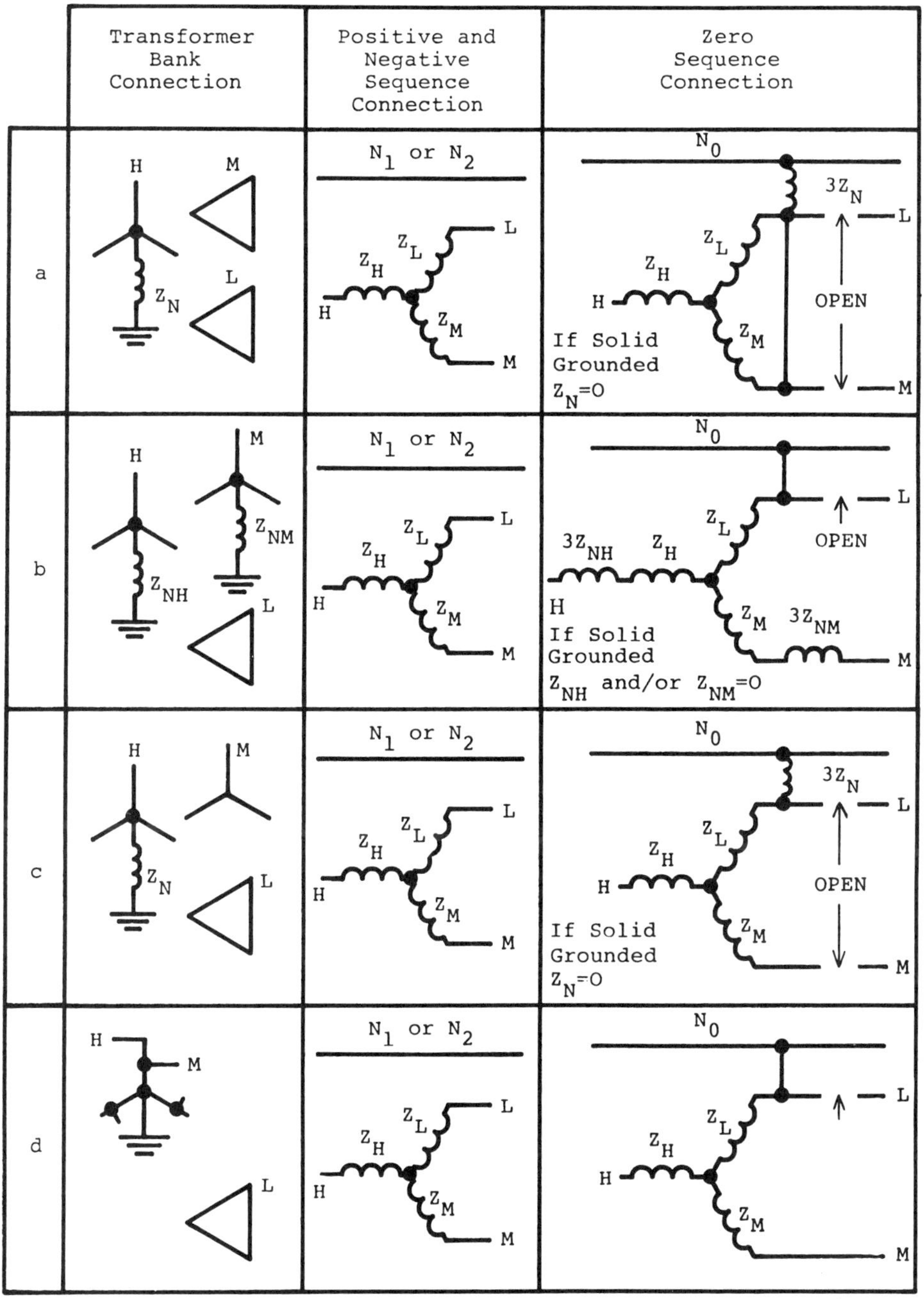

Fig. A4.2-3 Sequence connections for typical three-winding and autotransformer banks.

In order to use these impedances in the sequence networks, they must be converted to an equivalent star type network as shown. This conversion is

$$Z_H = \frac{1}{2}(Z_{HM} + Z_{HL} - Z_{ML}) \qquad \text{(A4-2-13)}$$

$$Z_M = \frac{1}{2}(Z_{HM} + Z_{ML} - Z_{HL}) \qquad \text{(A4-2-14)}$$

$$Z_L = \frac{1}{2}(Z_{HL} + Z_{ML} - Z_{HM}) \qquad \text{(A4-2-15)}$$

It is easy to remember this conversion as the equivalent star value is always half the sum of the leakage impedances involved less the one not involved. For example, Z_H is half of Z_{HM} and Z_{HL}, both involving H, minus Z_{ML} that does not involve H.

After determining Z_H, Z_M, and Z_L, a good check is to see if they add up as $Z_H + Z_M = Z_{HM}$, etc.

If these values are not available, they can be measured as described above for a two-winding transformer. For the three-winding or auto transformers: Z_{HM} is the impedance looking into H winding with M shorted, L open; Z_{HL} is the impedance looking into H winding with L shorted, M open; Z_{ML} is the impedance looking into M winding with L shorted, H open.

This equivalent star or wye is a mathematical network representation valid for determining currents and voltages at the transformer terminals or in the associated system. The wye or star point has no physical meaning. Quite often, one of the values will be negative and should be used as such in the network. It does not represent a capacitor.

The positive and negative sequence connections are all the same and independent of the actual bank connections. However, the connections for the zero sequence network are all different and depend on the transformer bank connections. If the neutrals are solidly grounded, then the Z_N and $3Z_N$ components shown are shorted out in the system and sequence circuits.

APPENDIX 4.3: SEQUENCE PHASE SHIFTS THROUGH WYE-DELTA TRANSFORMER BANKS

As has been indicated positive and negative sequences pass through the transformer bank, and in the sequence networks the impedance is the same independent of the bank connection. This is shown in Figs. A4.2-1 and A4.2-3. In these networks the phase shift is ignored, but if currents and voltages are transferred from one side of the transformer bank to the other, these phase shifts must be taken into account. This appendix will document these relations. For this the standard ANSI connections are shown in Fig. A4.3-1.

From Fig. A4.3-1a, all quantities are phase-to-neutral values, and in amperes or volts; for per unit, N = 1, n = $1/\sqrt{3}$.

$$I_A = n(I_a - I_c) \text{ and } V_a = n(V_A - V_B)$$

For positive sequence [Eq. (4.2)],

$$I_{A1} = n(I_{a1} - aI_{a1}) = n(1 - a)I_{a1}$$

$$= \sqrt{3}\, nI_{a1} \underline{/-30°} = NI_{a1} \underline{/-30°} \quad \text{(A4.3-1)}$$

$$V_{a1} = n(V_{A1} - a^2V_{A1}) = n(1 - a^2)V_{A1}$$

$$= \sqrt{3}\, nV_{A1} \underline{/+30°} = NV_{A1} \underline{/+30°} \quad \text{(A4.3-2)}$$

For negative sequence, [Eq. (4.3)],

$$I_{A2} = n(I_{a1} - a^2I_{a1}) = n(1 - a^2)I_{a2}$$

$$= \sqrt{3}\, nI_{a2} \underline{/+30°} = NI_{a2} \underline{/+30°} \quad \text{(A4.3-3)}$$

$$V_{a2} = n(V_{A2} - aV_{A2}) = n(1 - a)V_{A2} \quad \text{(A4.3-4)}$$

$$= \sqrt{3}\, nV_{A2} \underline{/-30°} = NV_{A2} \underline{/-30°} \quad \text{(A4.3-5)}$$

Now consider the connections in Fig. A4.3-1b. Again all values are in phase-to-neutral amperes or volts; for per unit, N = 1, n = $\sqrt{3}$.

$$I_a = \frac{1}{n}(I_A - I_B) \text{ and } V_A = \frac{1}{n}(V_a - V_c)$$

For positive sequence [Eqs. (4.2)],

$$I_{a1} = \frac{1}{n}(I_{A1} - a^2I_{A1}) = \frac{1}{n}(1 - a^2)I_{A1}$$

$$= \frac{\sqrt{3}}{n} I_{A1} \underline{/+30°} = \frac{1}{N}I_{A1} \underline{/+30°} \quad \text{(A4.3-6)}$$

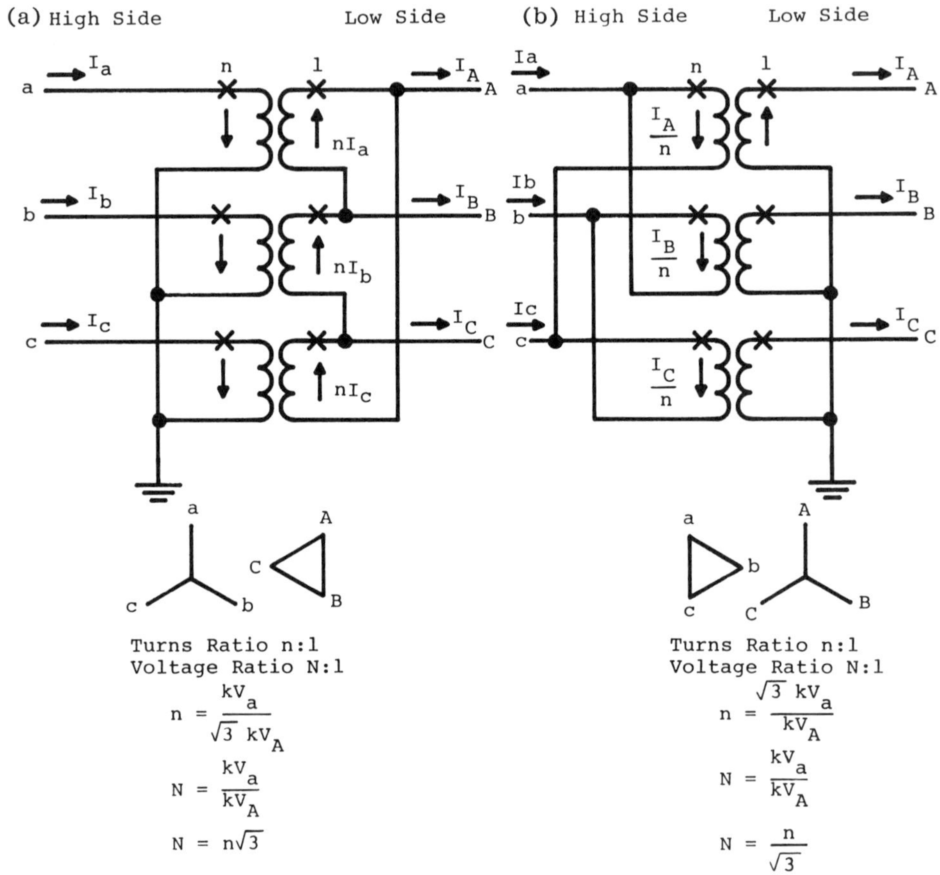

Fig. A4.3-1 ANSI connected wye-delta transformer banks. The high-voltage-side phase a leads the low-voltage-side phase a for both connections illustrated: (a) wye (star) on high side; (b) delta on high side.

$$V_{A1} = \frac{1}{n}(V_{a1} - aV_{a1}) = \frac{1}{n}(1 - a)V_{a1}$$

$$= \frac{\sqrt{3}}{n} V_{a1} \angle -30^\circ = \frac{1}{N} V_{a1} \angle -30^\circ \qquad \text{(A4.3-7)}$$

For negative sequence [Eqs. (4.3)],

$$I_{a2} = \frac{1}{n}(I_{A2} - aI_{A2}) = \frac{1}{n}(1 - a)I_{A2}$$

$$= \frac{\sqrt{3}}{n}I_{A2} \angle -30° = \frac{1}{N}I_{A2} \angle -30° \qquad \text{(A4.3-8)}$$

$$V_{A2} = \frac{1}{n}(V_{a2} - a^2V_{a2}) = \frac{1}{n}(1 - a^2)V_{a2}$$

$$= \frac{\sqrt{3}}{n}V_{a2} \angle +30° = \frac{1}{N}V_{a2} \angle +30° \qquad \text{(A4.4-9)}$$

Summary

An examination of the equations above shows that for ANSI standard connected wye-delta transformer banks: (a) if both the positive-sequence current and voltage on one side lead the positive-sequence current and voltage on the other side by 30°, the negative-sequence current and voltage correspondingly will both lag by 30°; and (2) similarly, if the positive-sequence quantities lag in passing through the bank, the negative-sequence quantities correspondingly will lead 30°. This fundamental is useful in transferring currents and voltages through these banks.

Zero sequence is not phase shifted *if* it can pass through and flow in the transformer bank. The zero-sequence circuits for various transformer banks are shown in Figs. A4.2-1 and A4.2-3.

5

Relay Input Sources

5.1 INTRODUCTION

Protective relays require reasonably accurate reproduction of the normal, tolerable, and intolerable conditions in the power system for correct sensing and operation. This information input from the power system is through current and voltage transformers in most cases. An exception is for temperature-type relays, which receive their information from thermocouples or temperature indicators.

These devices—current transformers (CTs), voltage transformers (VTs, formerly PTs), and coupling capacitor voltage transformers (CCVTs)—provide insulation from the higher system voltages and a reduction of the primary current and voltage quantities. Thus the transformer primary is connected into or to the power system with insulation compatible with that of the system. The secondary is standardized for the convenience of application and relay design. These relay input sources are an important member of the protection "team." Typical units are illustrated in Figs. 5.1 through 5.5 and in Fig. 1.8.

There are other means of providing primary information to the protective relays, by electronic, light pipe, and fiber optic devices, for example, but so far these are primarily research projects and none are known in practical operation or production. They do hold promise for the future.

The intolerable system problems may be transient or permanent, and involve a change in the primary current or voltage over a wide range. This is true particularly for the current, which can suddenly change from a few amperes to several hundred amperes secondary when a fault occurs. Voltages during faults can collapse from the rated value down to zero.

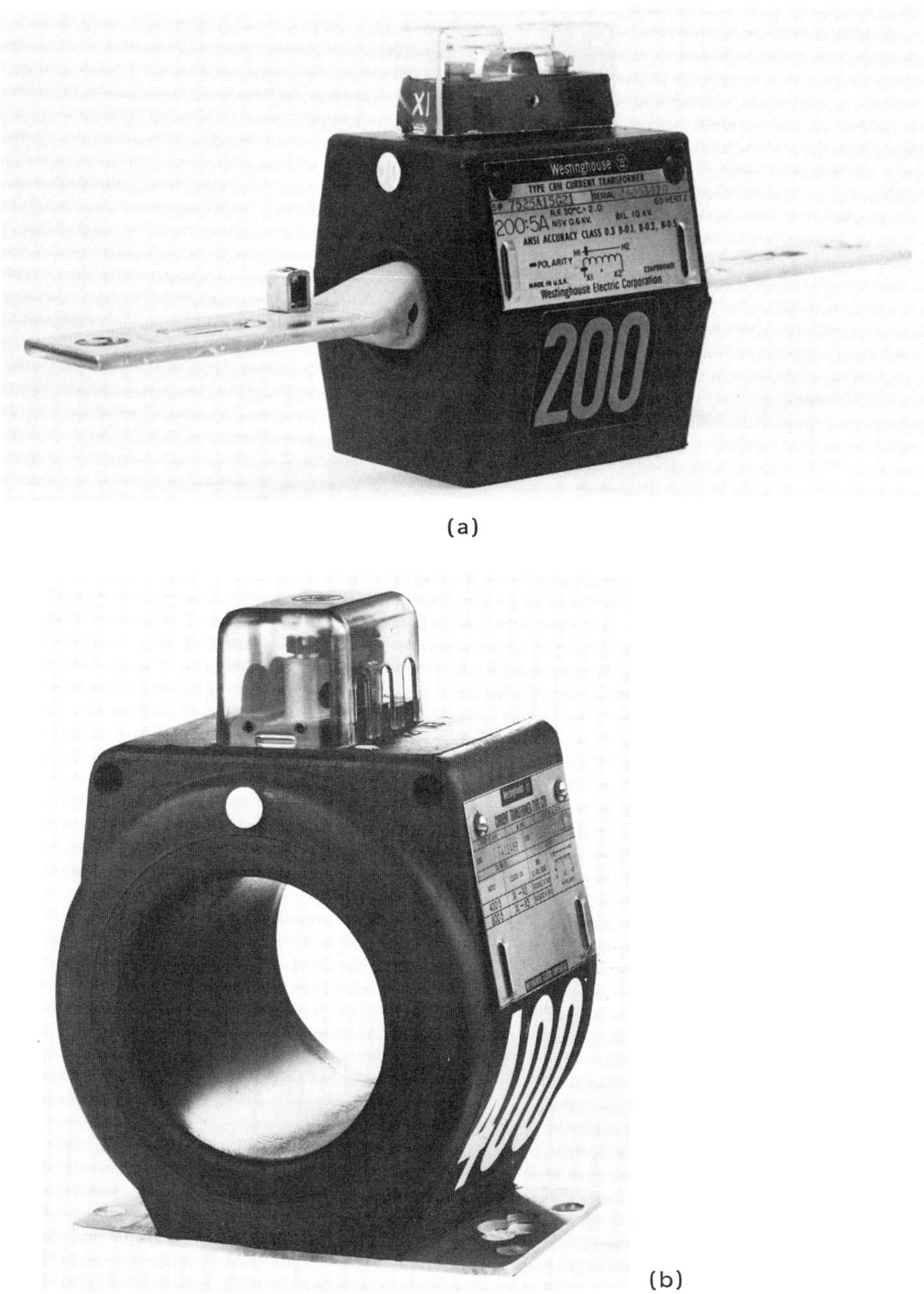

(a)

(b)

Fig. 5.1 Typical low-voltage current transformers: (a) bar type, with the bar as the primary; (b) through type, where the primary cable is passed through the CT opening. (Courtesy of Westinghouse Electric Corp.)

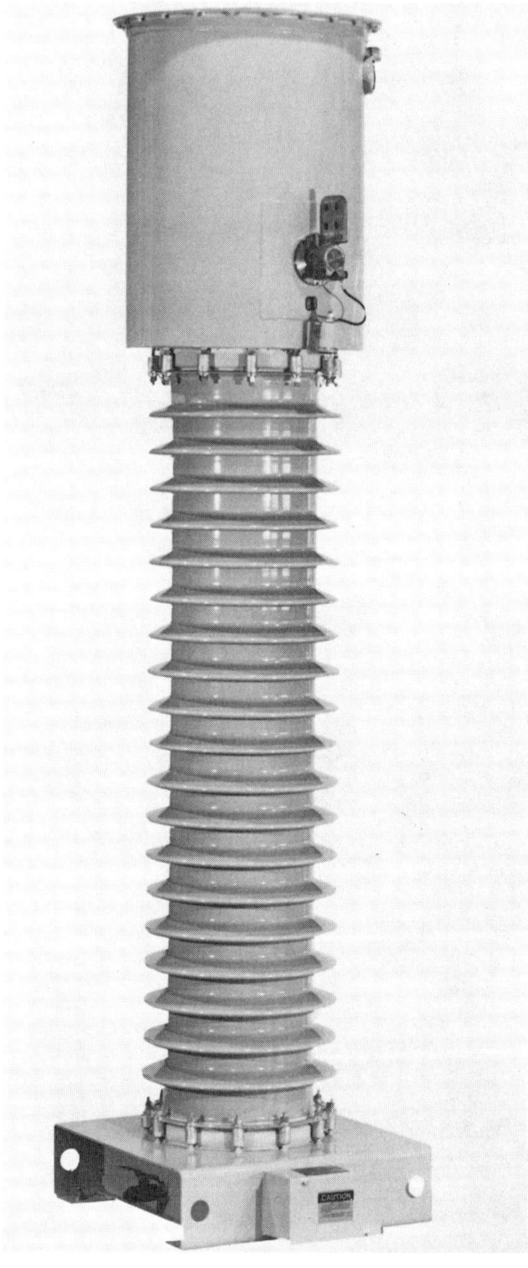

Fig. 5.2 A 230-kV oil-filled current transformer. The unit is approximately 10 ft high and 2.5 ft in diameter. (Courtesy of Westinghouse Electric Corp.)

Fig. 5.3 Voltage transformer for use in low-voltage power systems. (Courtesy of Westinghouse Electric Corp.)

Fig. 5.4 Typical 115-kV switchyard with three 115-kV voltage transformers in the foreground, mounted on top of a 10-ft pipe. Behind are oil circuit breakers and in the right rear, a 325-MVA 230:115-kV autotransformer bank. (Courtesy of Puget Sound Power & Light Co.)

5.2 EQUIVALENT CIRCUITS OF CURRENT AND VOLTAGE TRANSFORMERS

The equivalent diagrams for an instrument transformer are shown in Fig. 5.6. The exciting (magnetizing) impedance Z_e in Fig. 5.6(a) is shown in two parts. Z_e' is that associated with the leakage flux within the transformer core and its related leakage reactance X. Z_e'' is that associated with the flux that does not reach the core. X_p is the leakage reactance from this flux that does not cut the transformer core. R_p and R_s are the resistances of the primary and secondary windings, respectively.

For voltage transformers, the value of $(R_p + R_s) + j(X_p + X)$ is kept low to minimize the loss of voltage and shift of the phase angle from primary to secondary. Current transformers are of two types; those with significant leakage flux in the core (Fig. 5.6a), and those with negligible leakage flux in the core (Fig. 5.6b). In either type the

Fig. 5.5 A 500-kV coupling capacitor voltage transformer undergoing factory test and calibration. (Courtesy of Westinghouse Electric Corp.)

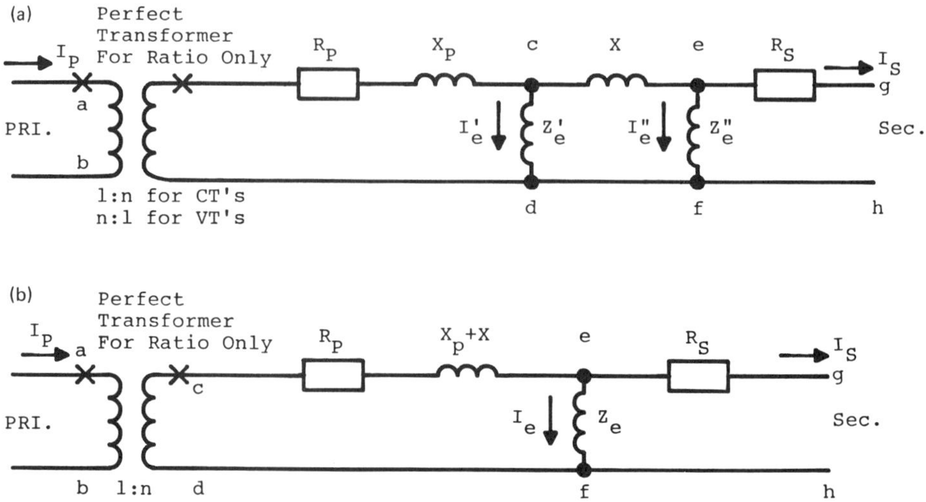

Fig. 5.6 Equivalent diagrams for instrument transformers. With perfect (ideal) transformers for ratio primary to secondary as shown, the values are in secondary ohms. n = turns. With per unit, omit the perfect transformers. (a) For significant leakage flux in transformer core – class T CTs and voltage transformers. (b) For negligible leakage flux in transformer core – class C CTs.

Z_e shunt impedance(s) are kept high to minimize current loss from the primary to the secondary.

The perfect or ideal transformers shown in the diagrams are to provide the necessary ratio change; they have no losses or impedance. Although shown in the primary, they may be connected in the secondary instead. As shown, all the impedances are on a secondary basis. With per unit nomenclature, the perfect transformers are omitted, as they are not required.

The primary quantities are reduced by the turns ratio n to provide a secondary current or voltage to energize protective relays and other equipment. The impedances of these loads are commonly called "burden." The term can refer to individual devices or to the total load connected, including the instrument transformer secondary impedance where that is significant. For the devices the burden is often expressed in volt-amperes at a specified current or voltage. Thus the burden impedance Z_B is

$$Z_B = \frac{VA}{I^2} \text{ ohms for CTs} \quad \text{or} \quad = \frac{V^2}{VA} \text{ ohms for VTs} \tag{5.1}$$

where VA is the volt-ampere burden and I or V the amperes or volts at which the burden was measured or specified.

5.3 CURRENT TRANSFORMERS FOR PROTECTION APPLICATIONS

In the United States, current transformers almost universally have 5-A secondary ratings. Other ratings, such as 1A, exist but are not common, although they are used in other countries. Advantages for this lower rating may exist where unusually long secondary leads are required between the CTs and the relays, such as in high-voltage (HV) installations. However, changing the current transformer rating does not necessarily reduce the energy required for relay operation. With a constant VA, lower current means higher voltage and more insulation between the primary and the secondary. For the most part the advantages appear to be offset by the disadvantages. Today and in the future, the trends to solid-state relays with very low burdens makes the need for another standard less useful.

The measure of a current transformer performance is its ability to reproduce accurately the primary current in secondary amperes both in wave shape and magnitude. There are two parts: (1) the performance on the symmetrical ac component, and (2) the performance on the offset dc component. Modern CTs do a remarkable job of reproducing wave shapes as long as they do not saturate.

5.4 CURRENT TRANSFORMER PERFORMANCE ON SYMMETRICAL AC COMPONENT

For the symmetrical component the performance is determined by the highest current that can be reproduced without saturation to cause large ratio errors. Phase-angle errors are not usually critical for relaying.

If the CT does not saturate, it is practicable to assume that I_e is negligible.

$$I_S = \frac{I_P}{R_c} \text{ amperes} \quad \text{or} \quad I_S = I_P \text{ per unit} \tag{5.2}$$

where R_c is the current transformer ratio and equivalent to n in Fig. 5.6.

For CTs connected in the phase leads of the power system with load passing through them, the ratio R_c is selected so that the maximum anticipated secondary current does not exceed 5 A secondary.

This comes from the long-standing practice of indicating instrument movements being standardized at 5 A independent of the primary ampere scale markings. By selecting a ratio to give a little less than 5 A at the maximum load, any instrument connected in the circuit would not go off scale. Instruments may or may not be connected in the relay circuits, but this practice continues, with the continuous ratings of the CTs and relays often based on 5 A.

However, I_e, the CT exciting current, is never zero if the CT is energized either by the primary or on the secondary. Thus it must be checked to assure that it is negligible. This can be done by one of three methods: (1) classical transformer formula, (2) CT performance curves, and (3) ANSI/IEEE accuracy classes for relaying.

5.4.1 Performance by Classical Analysis

The classical transformer formula is

$$V_{ef} = 4.44fNA\beta_{max} \times 10^{-8} \text{ V} \tag{5.3}$$

where f is the frequency in hertz, N the number of secondary turns, A the iron-core cross-sectional area in square inches, and β_{max} the iron-core flux density in lines per square inch. However, most of these quantities are not normally available, so this method is used primarily by CT designers. V_{ef} is the voltage the current transformer can develop to drive secondary current through the load.

This load on the CTs consists of its secondary resistance R_S, the impedance of the connecting leads Z_{ld}, and the equipment (relays, etc.) Z_r. The voltage required by the burden (load) is

$$V_{ef} = I_S(R_S + Z_{ld} + Z_r) \quad \text{V} \tag{5.4}$$

5.4.2 Performance by CT Characteristic Curves

The calculation of the performance with the equivalent circuit of Fig. 5.6a is difficult even where the value of X is known. The ANSI/IEEE standard (C57.13) for instrument transformers recognizes this and classifies CTs that have significant leakage flux within the transformer core as class T (class H before 1968). Wound CTs, those that have one or more primary winding turns mechanically encircling the core, are usually class T. Their performance is best determined by test.

CTs constructed to minimize the leakage flux in the core, such as the through, bar, and bushing types (Fig. 5.1), can be represented by the modified equivalent circuit of Fig. 5.6b. Effectively, the leakage reactance X is ahead of the exciting branches, and these branches can be paralleled to Z_e. With this the performance can be calculated. These are designated as class C (class L before 1968).

Evaluation by performance curves requires the availability of test curves such as shown in Fig. 5.7 for class T type, or excitation curves such as shown in Fig. 5.8 for the Class C type. The B-1 through B-6 indications on Fig. 5.7 represent standard burdens. The number indicates the impedance of the burden in ohms at 0.5 power factor. Thus B-1 is a burden of 1Ω, B-8 a burden of 8Ω, and so on.

The "knee" or effective point of saturation is defined by the ANSI/IEEE standard as the intersection of the curve with a 45° tangent line. However, the IEC (International Electrotechnical Commission) defines the knee as the intersection of straight lines extended from the nonsaturated and saturated parts of the exciting curve. The IEC knee is at a higher voltage than the ANSI knee.

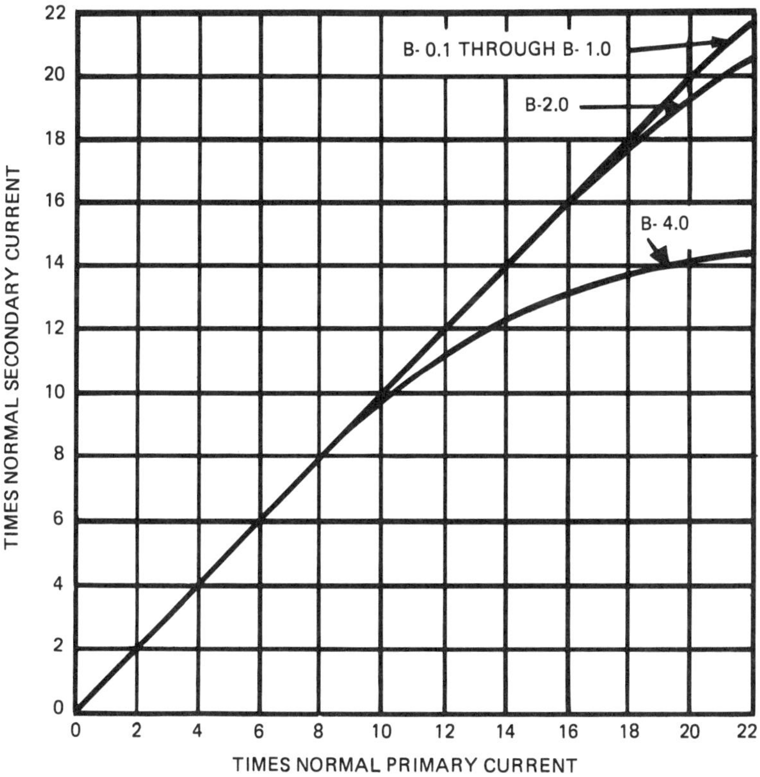

Fig. 5.7 Typical overcurrent ratio curves for class T current transformers. (Figure 5 of ANSI/IEEE Standard C37.13-1978.)

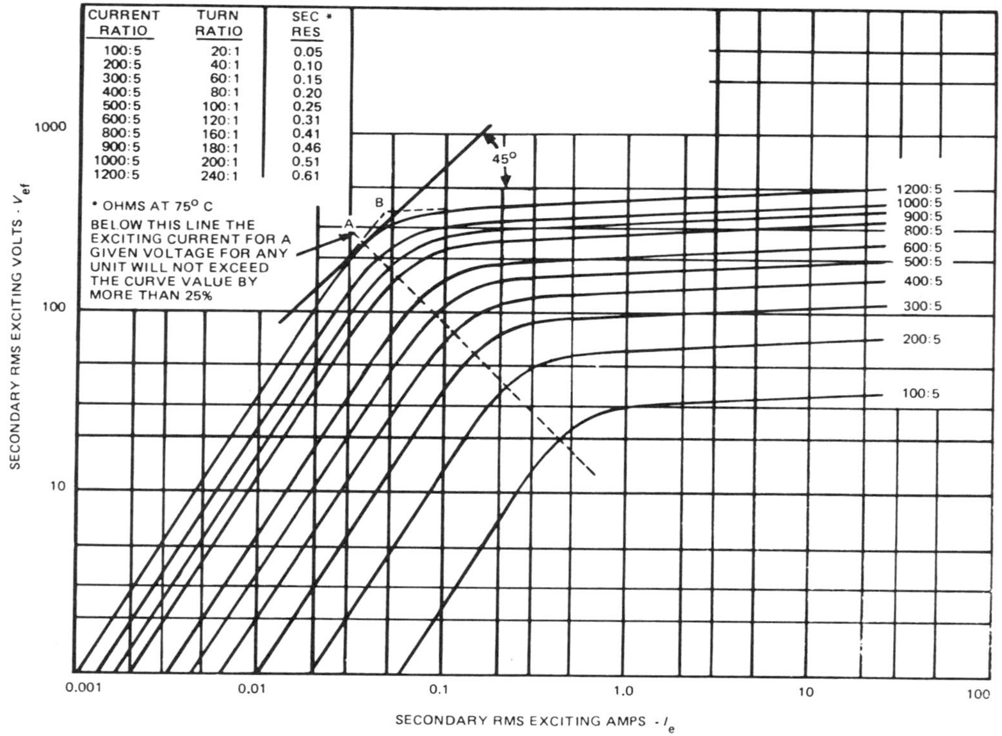

Fig. 5.8 Typical excitation curves for a multiratio class C current transformer. Point A is the ANSI knee. Above this line the voltage for a given exciting current for any unit will not be less than 95% of the curve value. Point B is the IEC knee. (Figure 4 of ANSI/IEEE Standard C57.13-1978.)

5.4.3 Performance by ANSI/IEEE Standard Accuracy Classes

In many applications the use of the ANSI/IEEE accuracy class designations is adequate to assure satisfactory relay performance. As indicated, there are two standard classes; class T, where performance is not easy to calculate, so manufacturer's test curves must be used, and class C, where the performance can be calculated. These designations are followed by a number indicating the secondary terminal voltage (V_{gh}) which the transformer can deliver to a standard burden at 20 times rated secondary current without exceeding 10% ratio correction. This 10% will not be exceeded at any current from 1 to 20 times rated secondary current at the standard burden or any lower standard burden. For relaying the voltage classes are 100,200, 400, and 800, corresponding to standard burdens of B-1, B-2, B-4, and B-8, respectively. These burdens are at 0.5 power factor. The burden numbers are ohms, obtained by dividing the voltage rating by 20 times rated secondary current. Thus with the 800 V rating and its associated B-8 burden: 8 Ω x 5 A x 20 = 800 V.

If the current is lower, the burden can be higher in proportion; however, this does not necessarily apply to higher currents since the internal impedance basically neglected in the standard can affect performance. Thus a 400-V rated CT will pass 100 A (20 times rated) with a burden of 4 Ω or less with not more than 10% ratio error. Correspondingly, it can pass 50 A with an 8-Ω burden with not more than the 10% error.

For these voltage ratings, terminal voltage is used rather than the excitation voltage of Eq. (5.4), or

$$V_{gh} = I_S(Z_{ld} + Z_r) \quad \text{V} \tag{5.5}$$

The lower voltage classes of 10, 20, and 50 with standard burdens of B-0.1, B-0.2, and B-0.5 at 0.9 power factor are primarily for metering service and should be used cautiously for protection.

Two CTs with the same ratio and their secondaries connected in series will increase the accuracy capability. Each CT thus develops one-half of the voltage required by the burden. Two CTs with primaries in series and secondaries in parallel can provide an overall lower ratio with higher-ratio individual CTs and their correspondingly higher accuracy rating.

It should be appreciated that the ANSI classifications merely indicate that the ratio correction or error will not exceed 10%; they do not provide information as to the actual value, which may be any value but not exceeding 10%. It is also very important to note that these accuracy class values apply only to the full winding, and are reduced proportionally where lower taps are available and used. Many type C transformers are bushing multiratio type with five secondary taps, providing ratios such as those shown on Fig. 5.8. Performance on these lower taps is significantly reduced and limited. As a general rule, the use of the lower ratio taps should be avoided, and when used, the performance should be checked.

According to the standards, the manufacturer is to supply application data for relaying service current transformers of (1) the accuracy class, (2) short-time mechanical and thermal ratings, (3) resistance(s) of the secondary winding, and (4) typical curves, such as Fig. 5.7 for class T and Fig. 5.8 for class C.

5.5 SECONDARY BURDENS DURING FAULTS

The burden imposed on CTs for various fault and connections is documented in Fig. 5.9. Z_B is the sum of the leads or connecting circuit between the CTs and the relays plus the relays and other equipment. It is assumed that the burden in each phase is the same, but this may not always be correct. Where the CTs are connected in delta (Fig. 5.9b), the burden on the phase A current transformer is $(I_A - I_B)Z_B - (I_C - I_A)Z_B$. This reduces to $[2I_A - (I_B + I_C)]Z_B$. For a three-phase fault, $I_A + I_B + I_C = 0$, so $(I_B + I_C) = -I_A$, and substituting this, the burden reduces to $3I_AZ_B$.

Since phase relays are set on three-phase fault currents, and ground relays set on single-phase-to-ground fault currents, the phase-to-phase fault diagrams are not generally used. They do show that the burden is the same as for three-phase faults, and with less phase-to-phase fault current, the three-phase fault cases are the maximum.

5.6 SUMMARY OF CT STEADY-STATE PERFORMANCE EVALUATION

First assume a negligible exciting current requirement and determine the maximum secondary current from Eq. (5.2) for the maximum fault for which the protection is expected to perform. Determine the total burden that is connected to the CTs.

If an estimation of performance is to be made using the ANSI standards, determine the maximum voltage from Eq. (5.5) that the CT must develop and compare with the specified CT rating. For more accurate determination with type T CTs, determine the performance from the maximum primary current and burden using curves such as Fig. 5.7 supplied with the CTs.

For the C types, calculate the voltage that must be developed at maximum fault using Eq. (5.4) and determine the magnitude of excitation curve supplied with the CT, such as Fig. 5.8. Comparison of I_e with the required secondary current indicates the performance.

If the voltage V_{ef} is less than the capability of the CT, then I_e will be small and have a minimum effect on I_S. On the other hand, if V_{ef} exceeds the CT capability, I_e will be significant, and I_S for the relays will be less than expected or required and the transformer performance impaired. The waveform distortion is high when CTs operate over the knee of the saturation curve. This may impair operation of some designs of relays.

In many cases it is practical to add the burden impedances and the currents algebraically, but theoretically they should be combined phasorally. If the performance is marginal, combine phasorally; otherwise, direct addition is satisfactory and simpler. The burdens are generally near unity power factor while I_e lags 90°, so combining I_e and I_S at right angle is a good approximation.

Some relays, such as those of the electromechanical time-overcurrent type, tend to saturate at the higher currents, so their internal impedance decreases and becomes more resistive. For these relays the manufacturer provides burden data at several levels of current. For example, one type of relay on the most sensitive 0.5-A tap has an impedance of 9.5 Ω at 0.5 A, 9.3 Ω at 1.5 A, 5.3 Ω at 5 A, and 3.5 Ω at 10 A (20 times tap). Such reduced burdens aid CT performance.

5.7 EXAMPLES OF STEADY-STATE CT PERFORMANCE

Consider C100, 600:5 multiratio CTs to be used with overcurrent relays for the protection of a feeder with a maximum three-phase fault current of 2500 A primary and a minimum fault current of 350 A primary. The maximum load on the feeder is 90 A.

With this load it would have been preferable to have selected lower-ratio CTs, but often the CTs are selected before adequate system data are known. Choosing a high-ratio multitap CT appears to cover a wide range of possible system operation. However, it can give protection problems.

A first choice would appear to select a 100:5 (20:1) ratio tap. This gives a maximum load current of 90/20 = 4.5 A secondary. For the phase-fault relays, select tap 5, which is above the maximum load as required to avoid operation on load. At this relay tap the burden is 2.64 VA. The lead resistance from the relays to the CTs is 0.4 Ω.

With the exciting current assumed negligible or zero, the relay would just operate (pick up) at 5 x 20 = 100 A primary on tap 5.

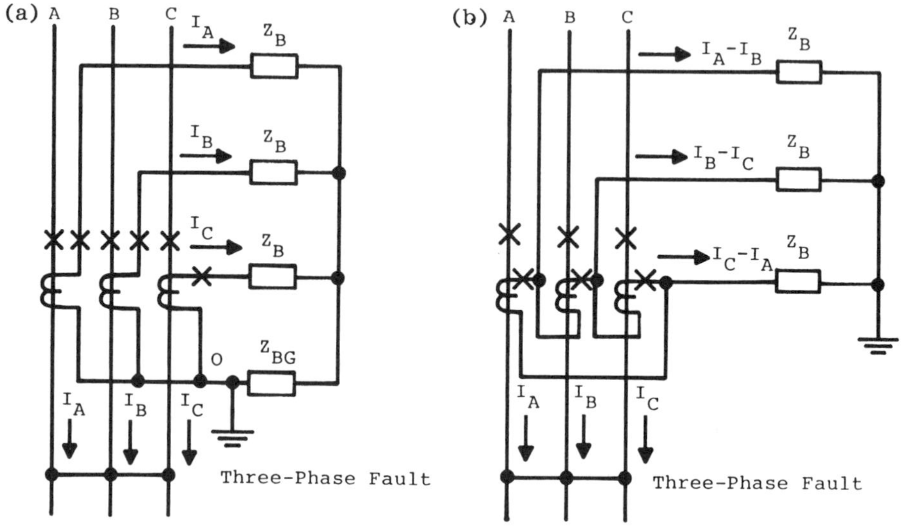

Fig. 5.9 Burdens on CTs for various types of CT connections and faults. The unexcited CT load is neglected.

$Z_B = Z_{leads} + Z_{phase\ relays}$ and $Z_{BG} = Z_{lead} + Z_{ground\ relay}$.

(a) Each CT has Z_B burden. (b) Each CT has $3Z_B$ burden. (c) B & C CT have Z_B burden each. (d) B & C have $3Z_B$ burden each. (e) A CT has $Z_B + Z_{BG}$ burden. (f) A CT has $2Z_B$ burden.

However, considering the exciting current, the primary current pickup will be higher. The total secondary impedance is, added directly:

Relay burden, $2.64/5^2$	0.106 Ω	
Lead resistance	0.40 Ω	
CT secondary resistance	0.082 Ω	(Fig. 5.10)
Total on CT secondary	0.588 Ω	

The voltage required to drive 5 A through the secondary is 5 × 0.588 = 2.94 V. From the CT characteristic of Fig. 5.10, I_e for 2.94 V on the 100/5 curve is about 0.22 A.

If I_S is directly added to I_e, the total is 5.22 and the primary current just to operate the relay would be 20 x 5.22 or 104.4 A primary.

If I_S and I_e are added at 90°, the total is

$$\sqrt{5^2 + 0.22^2}$$

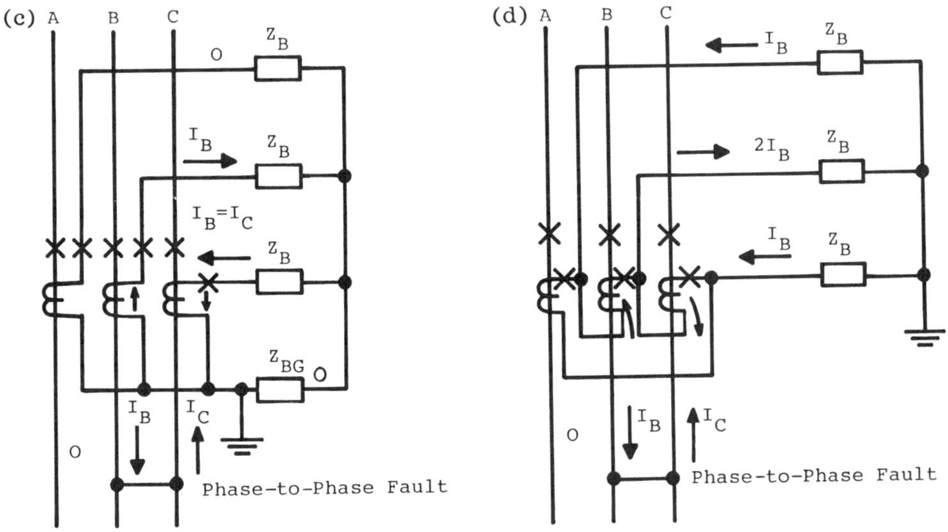

Fig. 5.9 (continued)

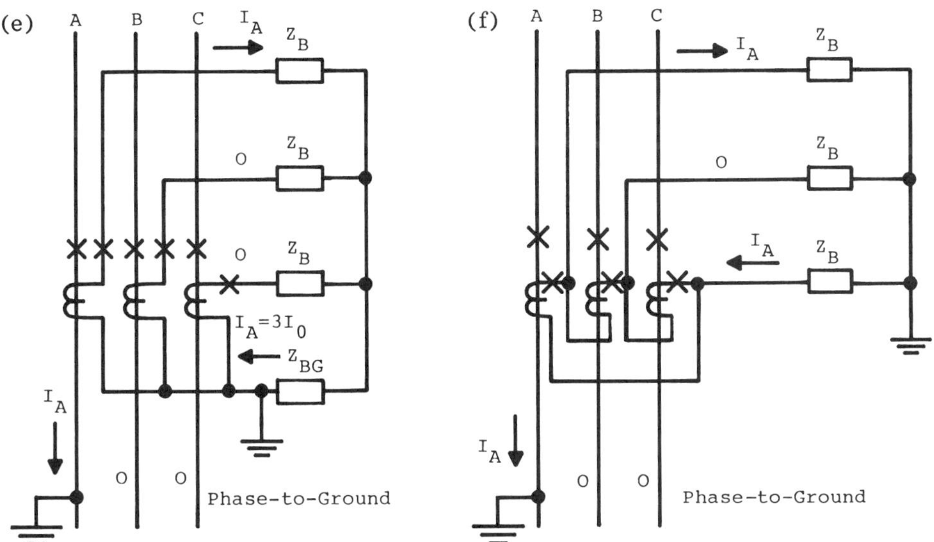

Fig. 5.9 (continued)

or 5.005, giving a primary pickup of 20 × 5.005 = 100.10 A. With a minimum fault of 350 A, this is 350/104.4 = 3.35 or 350/100.10 = 3.5 times the relay pickup, which is quite acceptable.

However, for maximum fault, the secondary current neglecting I_e should be 2500/20 = 125 A. At 20 times tap the relay burden was given as 580 VA, so for the higher current the total burden is:

Relay burden, $580/100^2$	0.058 Ω
Lead resistance	0.400 Ω
CT secondary resistance	0.082 Ω
Total on CT secondary	0.540 Ω

and the voltage required to pass 125 A through the relay would be 125 × 0.54 = 67.5 V. As can be seen from Fig. 5.10, this is not possible; the CTs on the 100:5 tap can produce only 12 to 15 V before saturation, and from the curve a maximum of around 20 to 25 V. Thus this combination cannot be used.

Consider now using the 400:5 CT tap. With this the load is 90/80 = 1.13A secondary. Choose the 1.5 A relay tap, where the burden is 3.5 VA. With this tap and assuming no exciting current, the primary pickup would be 1.5 × 80 = 120 A. The total burden is:

Relay burden, $3.5/1.5^2$	1.556 Ω
Lead resistance	0.40 Ω
CT secondary resistance	0.211 Ω (Fig. 5.10)
Total on CT secondary	2.167 Ω

The voltage required to develop 1.5 A in the relays is V_{ef} = 1.5 × 2.167 = 3.25 V, and from Fig. 5.10, I_e = 0.024 A. Directly adding 1.5 + .024 for the worst case, the primary pickup would be 80 × 1.524 = 121.92 A. This is well below the 350 A minimum fault current; this fault is around 2.9 times the relay pickup current.

For the maximum fault, the secondary current neglecting I_e would be 2500/80 = 31.25 A, and assuming that the relay burden at this level is not significantly different, the secondary voltage V_{ef} required is 31.25 × 2.167 = 67.72 V. While this is around the knee of the saturation curve, I_e for 68 V is about 0.16 A, so does not significantly decrease the fault current to the relay, which is around 20 times the relay pickup.

If a 1.5-A tap were not available, the 2.0 tap could be used with the 400:5 CT tap. With this and neglecting I_e, the primary pickup would be 80 × 2 = 160 A. Assuming the same relay burden as that of the 1.5 tap, the total burden is:

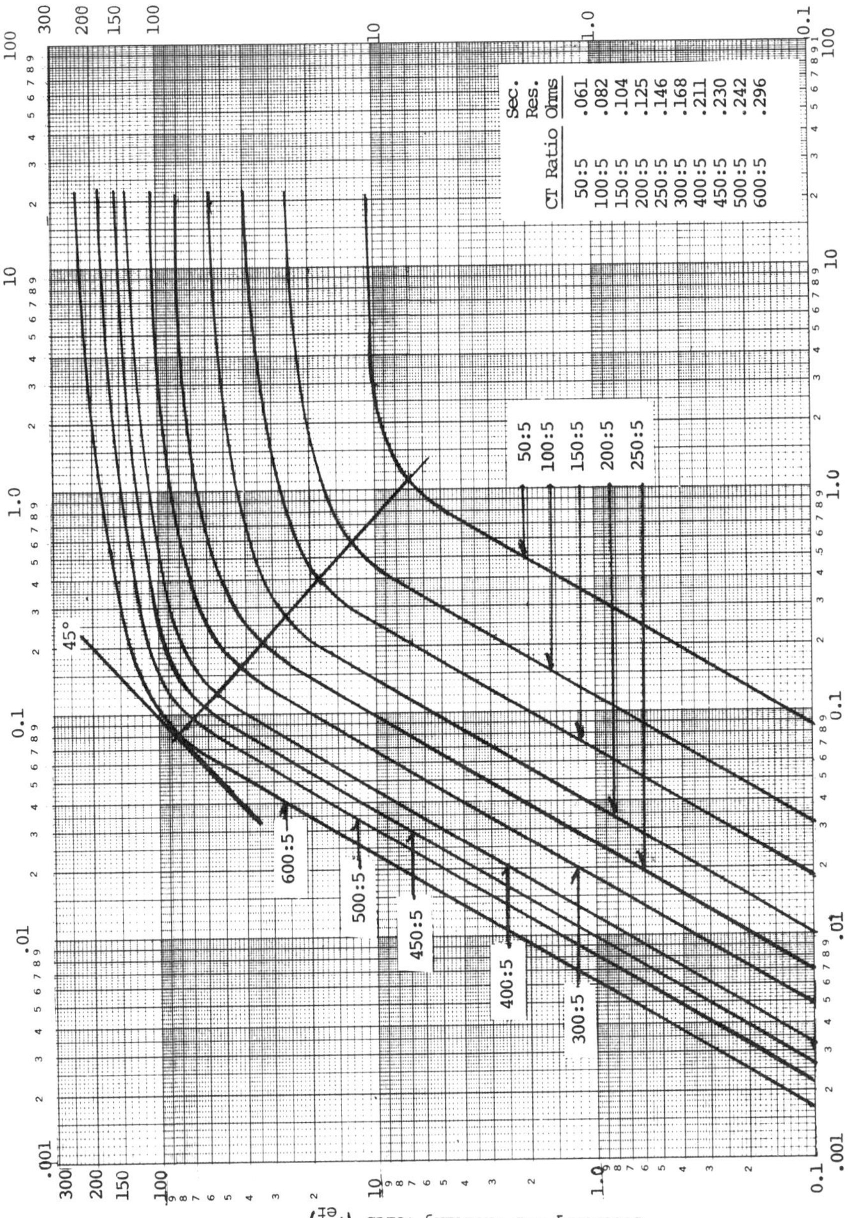

Fig. 5.10 Typical excitation curves for a 600:5 multiratio class C100 current transformer.

Relay burden, $3.5/2^2$	0.875 Ω
Lead resistance	0.400 Ω
CT secondary resistance	0.211 Ω
Total on CT secondary	1.486 Ω

$V_{ef} = 2 \times 1.486 = 2.972$ V, where $I_e = 0.022$ A, so with direct addition, the primary pickup would be 80 x 2.022 = 161.76. This is satisfactory since the minimum fault of 350 A is 2.16 times the pickup.

For the maximum fault of 2500/80 = 31.25 secondary amperes, $V_{ef} = 31.25 \times 1.486 = 46.438$ V, where $I_e = 0.11$ A, and this does not decrease the fault current to the relay significantly.

Evaluation by the ANSI standard provides equivalent results. With the 100:5 tap and relay tap 5, the terminal voltage from the previous analysis would be, from Eq. (5.5) for the maximum fault,

$$V_{gh} = \frac{2500}{20}(0.058 + 0.4) = 57.25 \text{ V}$$

but the C100 CT on the 100:5 tap can only supply 100 (100/600) = 16.7 V, and since it should supply 57.25 V, the CT will be severely saturated, as shown previously.

Using the 400:5 tap, the C100 CT can supply only 100 (400/600) = 66.67 V. However, the voltage required is:

$$1.5 \text{ relay tap: } V_{gh} = \frac{2500}{80}(1.56 + 0.40) = 61.25 \text{ V}$$

$$2.0 \text{ relay tap: } V_{gh} = \frac{2500}{80}(0.875 + 0.4) = 39.84 \text{ V}$$

Both of these are within the 66.67-V rating, so the ratio error will be not more than 10%.

5.8 EFFECT OF UNENERGIZED CTS ON PERFORMANCE

Current transformers with their secondaries interconnected but with the primaries passing zero or negligible current are excited from the secondary and thus will require I_e current. This can occur when paralleling CTs in a differential circuit, or during single-phase-to-ground faults. An example of the later case is illustrated in Fig. 5.11. Fault I_A flows in that phase CT, but the currents in the unfaulted phases B and C are zero. To emphasize the effect, assume that the 100:5 tap of a C100 600:5 multiratio CT is used. The secondary resistance of the CT, leads, and the phase relay is given at 0.63 Ω. The ground relay has 16 Ω on its 0.5-A tap at 68° lag. In order to pass pickup current through the ground relay, 0.5 × 16 = 8 V is required. This voltage,

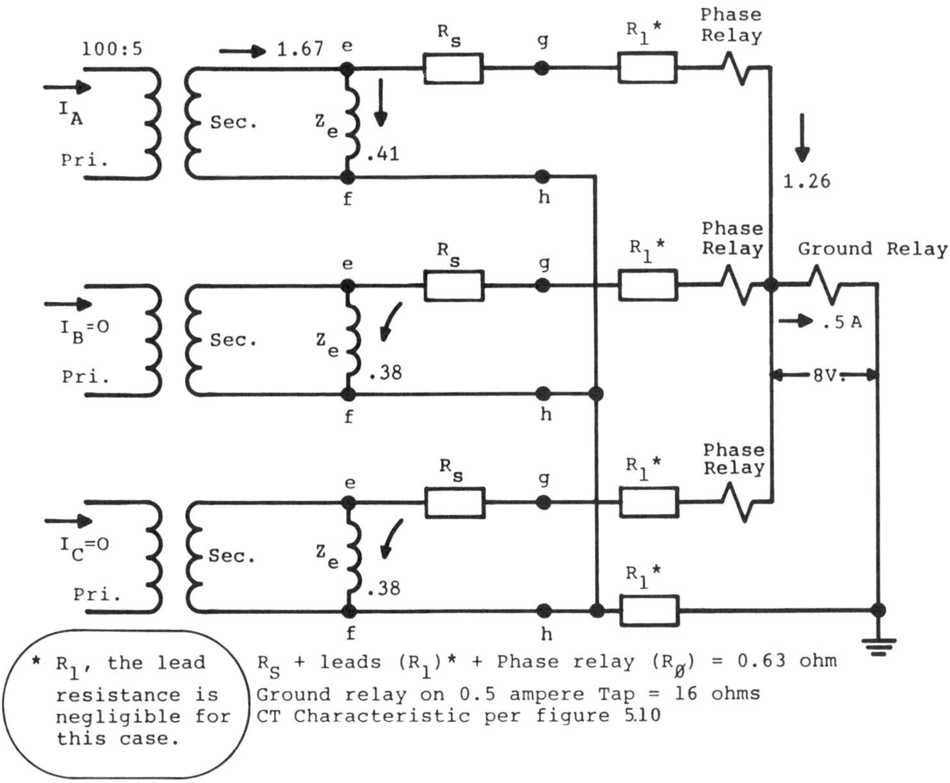

Fig. 5.11 Equivalent diagram for a three-phase and ground relay connection. Current shown for a phase-A-to-ground fault.

less the small drop through the phase relay circuit, will appear across phase B and C current transformer secondaries to excite them. The voltage V_{ef} depends on the current, which in turn depends on the voltage, so the exact determination is a "cut and try" process. As the first try, assume that V_{ef} = 8 V. From the CT characteristic, (Fig. 5.10) I_e for 8 V = 0.39 A. This current through phase circuit impedance results in a drop to make V_{ef} = 8 - 0.39 x 0.63 = 7.75 V, where I_e = 0.37. Another iteration provides the I_e = 0.38 A needed to excite both the phase B and C current transformers. Any primary current would help offset this.

Thus the current in the phase A circuit is the sum of the ground relay pickup and the phase B and C exciting currents. By direct addition this would be 0.50 + 0.38 + 0.38 = 1.26 A. By phasor addition it would be essentially $0.5\angle 68^\circ + j0.38 + j0.38 = 1.24\angle 81.30^\circ$, so the difference is not significant. The exciting voltage for the phase A current transformer V_{ef} = 8.0 + 1.26 × 0.63 = 8.79 V, where from Fig. 5.10, I_e = 0.41 A. Directly adding, the total is 1.46 + 0.41 = 1.67 secondary amperes or 20 × 1.67 = 33.4 primary amperes just to pick up the ground relay. This is in contrast to the 20 × 0.5 = 10 primary amperes required just to pick up the ground relay if the exciting currents for the three CTs were neglected.

It should be recognized that these CTs are not adequate for fault protection on the CT and ground relay taps used. As explained in Section 5.4.3, higher taps can improve performance and would decrease the shunting effect described above. This effect should be considered, especially where a number of CTs are paralleled with only one or two carrying current, as in a differential scheme, as described later.

5.9 TOROIDAL FLUX SUMMATION CURRENT TRANSFORMER

This type, also known as the "ring type," consists of a magnetic toroid core around which a distributed secondary winding is wound. The power conductors are passed through the center opening. It is very useful in protection at the lower voltages.

When the three-phase conductors are passed through the opening, the secondary measures $I_a + I_b + I_c = 3I_0$, the ground current. When the same phase conductor on the two ends of a device are passed through the opening, the net current for a load or fault current passing through the device is zero. For an internal fault with one or both supplying current of different magnitude or phase angle, the net or sum equals the fault current.

This is a flux summation rather than individual summing of separate transformer secondary currents. The advantages are that the CT ratio is independent of the load current or kVA of the circuit, and that it avoids the possible difficulties of unequal individual CT saturation or performance with paralleled CTs. The disadvantage is the limitation of the size of conductors that can be passed through the opening. A typical ratio for this CT is 50:5, and the maximum opening is around 8 in. in diameter.

The CT commonly is used with an 0.25-A instantaneous overcurrent unit. The combination provides a primary pickup of 5 A rather than 2.5 A if the exciting current were negligible. Specific applications are discussed in later chapters.

Metallic sheath or shielded cables passed through the toroidal CT can result in cancellation of the fault current. This is illustrated in Fig. 5.12. This applies either to three-phase cable as shown, or to single-phase cables. The cancellation may be partial or complete, depending on the sheath grounding. This sheath component of fault current can be removed from passing through the CT by connecting a conductor as shown.

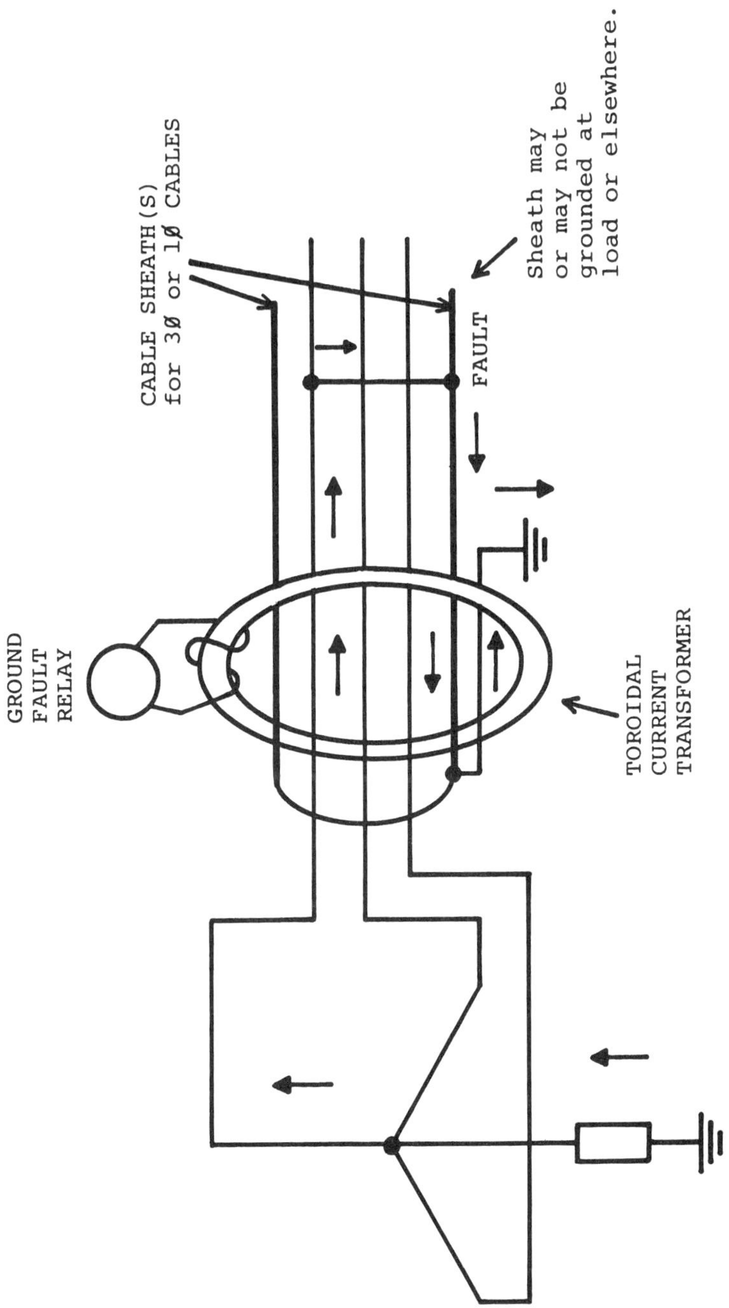

Fig. 5.12 Typical application of the toroidal current transformer for ground-fault protection with metallic sheath conductors.

5.10 CURRENT TRANSFORMER PERFORMANCE ON THE DC COMPONENT

Since transformers are paralyzed by direct current, CT performance is affected significantly by the dc component of the ac current. When a current change occurs in the primary ac system, one or more of the three-phase currents will have some dc offset, although none may be maximum and one could not have any offset. This dc results from the necessity to satisfy two conflicting requirements that may occur: (1) in a highly inductive network of power systems, the current wave must be near maximum when the voltage wave is at or near zero; and (2) the actual current at the time of change is that determined by the prior network conditions. For example, energizing a circuit with current being zero prior to closing the circuit at the instant when the voltage wave is zero presents a problem. By requirement (1) the current should be at or near maximum at this time. Thus a countercurrent is produced to provide the zero required by (2). This is the dc component equal and opposite to the required ac current by (1), with the two adding to zero at the instant of closing the circuit.

Having provided this function, the dc is no longer required, but it can disappear only by decaying according to the L/R time constant of the power system. This decaying dc acts somewhat like a low-frequency ac current in passing through the current transformer. It can saturate the iron so that the secondary reproduction of the primary current can be severely limited and distorted. This is illustrated in Fig. 5.13 for a 20-times-rated fully offset current with resistive burden. This type of burden causes a sharp drop-off of the secondary current during each cycle.

After saturation occurs, the decay of the dc component results in the CT recovering so that during each subsequent cycle the secondary current more nearly approaches the primary. As the dc disappears, the secondary is again a reproduction of the primary. This assumes no ac saturation. It is possible but rarely occurs that the secondary current may be practically zero for a few cycles in very severe cases.

Inductance in the burden results in a more gradual drop-off, while a lower burden reduces the distortion. These several effects are shown in Fig. 5.13. As shown, this saturation does not occur instantly, so initially the secondary current follows the primary current, before it is reduced and distorted by saturation.

The time to saturate and the calculation of the secondary current are complex and depend on many factors: the nature of the fault current, the constants and design of the current transformer, and the burden connected. A simplified method for estimating the performance of ring-core CTs is available from the IEEE report "Transient Response of Current Transformers." From a practical standpoint and as a general rule of thumb, the CTs used for relaying can be expected to

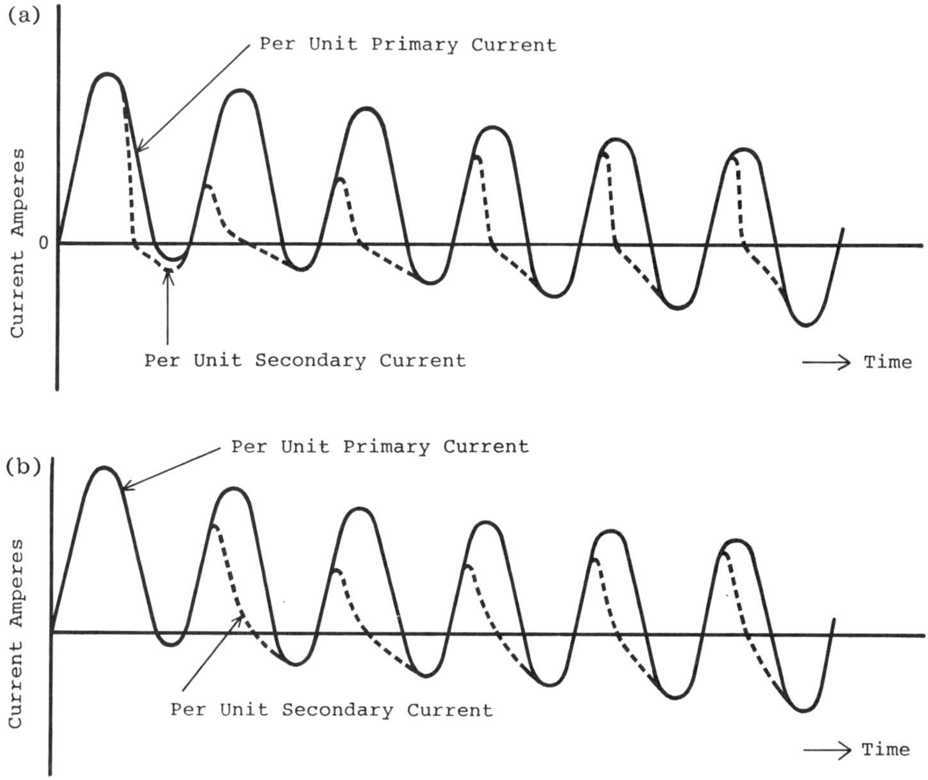

Fig. 5.13 Typical possible distortion in CT secondary current resulting from dc saturation: (a) large resistive burden; (b) smaller resistive burden. (Figure 3 of IEEE 76-CH1130-4, PWR Transient Response of Current Transformers.)

reproduce with reasonable accuracy the primary current for about one-half cycle or more before significant dc saturation.

5.11 CURRENT TRANSFORMER PERFORMANCE EVALUATION

As a generality, CT performance is most critical at the relay decision point. In current differential protection this decision point is at the CTs nearest an external fault, and frequently this is a very high value fault, the maximum fault. The nearest CT must reproduce the high current for comparison with the secondary currents of other CTs, which may carry only a part of the total current. Thus these schemes require close attention to CT performance.

Since dc saturation is likely to occur for those heavy external faults for which the relays are not to operate, and it is not practical to avoid it by CT design, various differential relay designs use several techniques to prevent misoperation.

In most other applications it is not likely to be too severe and/or to significantly inhibit the protection from a practical standpoint. However, it should always be considered and checked.

Most faults tend to occur near maximum voltage where the prefault current is low in the inductive power system. This minimizes the dc offset, so it is seldom at its maximum possible value. In many parts of the power system the time constant will be short, so that when the dc offset occurs, it rapidly decays. Also, faults other than differential protection are not at the maximum value at the critical decision point of the protection.

For example, in line protection, the relay decision point is remote from the CTs, so that the fault current is often lower, and line resistance is available to help moderate the effect. Also, the decision point may not be too critical since time is often involved to clear remote faults. Generally, the close-in high-current faults will be many times greater than the relay pickup current and with high-speed relays operation may take place before dc current transformer saturation occurs. Should saturation occur before the line protection relays can operate, generally a delay in operation occurs until the CTs recover sufficiently to permit operation. Thus the tendency for this type of protection usually is to "under reach" momentarily rather than "over reach."

Saturation on ac should always be avoided, especially at the relay decision point. Again, the effect of a reduced secondary current is to reduce the reach of the relay. This may or may not be critical in some applications. For the heavy close-in faults, some saturation may not be a serious problem as long as it is not severe enough to prevent or significantly delay relay operation. Waveform distortion generally has not been too critical for electromechanical relay designs. This may or may not be true for solid-state designs.

5.12 CURRENT TRANSFORMER RESIDUAL FLUX AND SUBSIDENCE TRANSIENTS

When a current transformer is energized by a load for the first time, the excursion on the hysteresis loop is symmetrical, with a flux variation such as rr' in Fig. 5.14. Faults with higher current produce increased flux and a wider excursion. When the fault is cleared and the primary current becomes zero, a unidirectional transient current can flow in the secondary. This is the trapped exciting current, which is

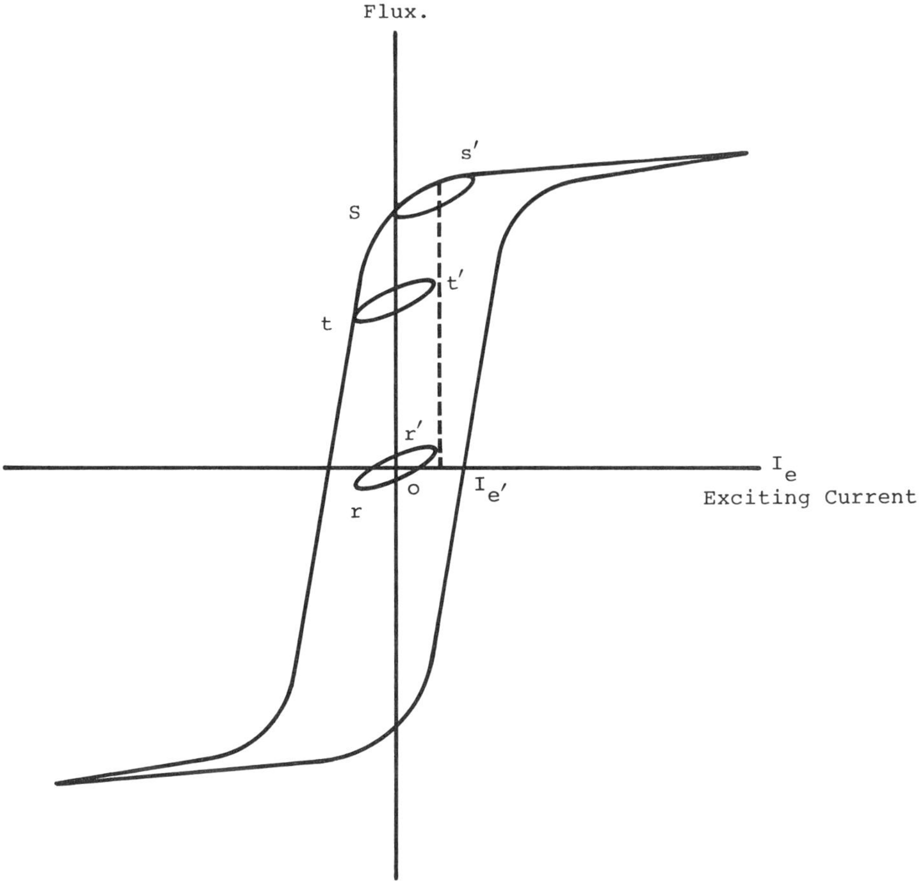

Fig. 5.14 Typical hysteresis loops and residual flux in current transformers.

out of phase with the fault primary-secondary current through the resistive-type burden before interruption. The time constant for this is usually short with resistive burdens unless air gaps are utilized in the CT core. This transient current may delay the dropout of very sensitive high-speed overcurrent relays utilized in breaker failure protective schemes, and may cause misoperation for the short times programmed in these schemes, particularly for HV system protection.

Interruption of the fault and the decay of I_e to zero still leaves flux in the CT. This is called residual flux, such as at point s in Fig. 5.14. Now if the CT is reenergized with the original load current flowing, the flux excursion would resume, but from the residual flux level s with a loop as ss', where the flux variation of ss' = rr'. However, it cannot continue in loop ss' because this would require direct current to maintain it in this offset position. So it shifts down to a symmetrical position tt', where the variation tt' = ss' = rr'. During this shift a small dc current flows in the secondary circuit according to the burden secondary time constant. Until the load changes or another fault occurs, the flux will vary in this tt' loop indefinitely. If the circuit is opened to deenergize the primary current to zero, the residual flux would be the value existing at the moment of interruption and somewhere within loop tt'. Thus current transformers, once energized, will have residual flux anywhere between zero and the saturated value in either the positive or negative loop.

For a later fault the residual can either add or reduce the CT capabilities. In the example, residual flux points s or t are much closer to the saturation level for flux excursions to the right but are far away from saturation for excursions to the left. Thus the performance depends on in which half cycle the next fault occurs. Since this is not predictable in theory, the residual flux can cause saturation and problems in protection. However, the general experience in the United States has not indicated this to be a serious problem. Only a very few cases have been cited where this might have caused problems.

Air-gap CTs have been used to minimize this residual; however, their performance for general application is not nearly as good. While they do not saturate as rapidly as the nongap designs, their exciting current is much higher; there is a loss of steady-state accuracy and of transformation of the dc transient. The unidirectional CT current after a fault clears decays out much more slowly with air gaps and so is more likely to cause problems in breaker failure relaying, as outlined above. At present, air-gap CTs have very minimum use in the United States.

5.13 AUXILIARY CURRENT TRANSFORMERS IN CT SECONDARY CIRCUITS

Auxiliary current transformers are sometimes required to provide (1) different ratios than would otherwise be available, (2) phase shift in the current and/or (3) circuit isolation. They should be used to step down the current to the burden wherever possible, to minimize the impedance loading on the main current transformer. This is illustrated in Fig. 5.15. Z'_B is the impedance at the secondary of the main CT, which is the reflected impedance Z_B or load connected to the auxiliary CT secondary. Neglecting the loss of the auxiliary CT yields

$$Z'_B = \frac{Z_B}{N^2} \tag{5.6}$$

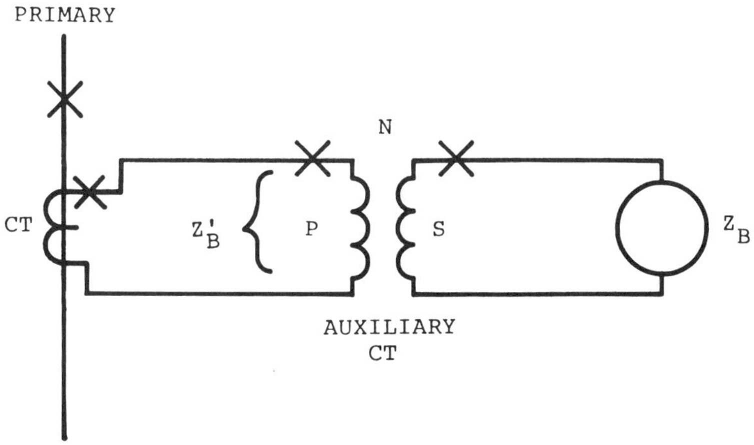

Fig. 5.15 Auxiliary current transformers applied to change overall current transformer ratio.

where N is the ratio of the auxiliary CT. Thus for a step-down connection with P:S of 10:5, N = 2 and $Z'_B = 0.25Z_B$. However, with a step-up connection, where P:S is 5:10, $Z'_B = 4.0Z_B$. With high lead resistance an auxiliary step-down CT near the main CT can be used to reduce this lead resistance load. The auxiliary CT does add its losses to the total load on the main CT.

5.14 VOLTAGE TRANSFORMERS FOR PROTECTIVE APPLICATIONS

Voltage transformers have wound primaries that are either connected directly to the power system (VTs) or across a section of a capacitor string connected between phase and ground (CCVTs). Typical units are illustrated in Fig. 5.3 through 5.5, with connection schematics shown in Fig. 5.16.

Protective relays utilizing voltage are usually connected phase to phase, so the transformers are normally rated 120 V line to line. Taps may be provided to obtain either 69.3 V or 120 V line to neutral. When available, double secondaries provide the means of obtaining zero-sequence voltage for ground relays. This is shown in Fig. 5.16a. If only a single transformer secondary winding is available, an auxiliary wye grounded-broken delta auxiliary voltage transformer can be connected to the secondary a, b, c bus of Fig. 5.16a for $3V_0$, similar to the connections shown. A typical example is shown in Fig. 1.10. CCVTs commonly have double secondaries for both phase and $3V_0$ voltages.

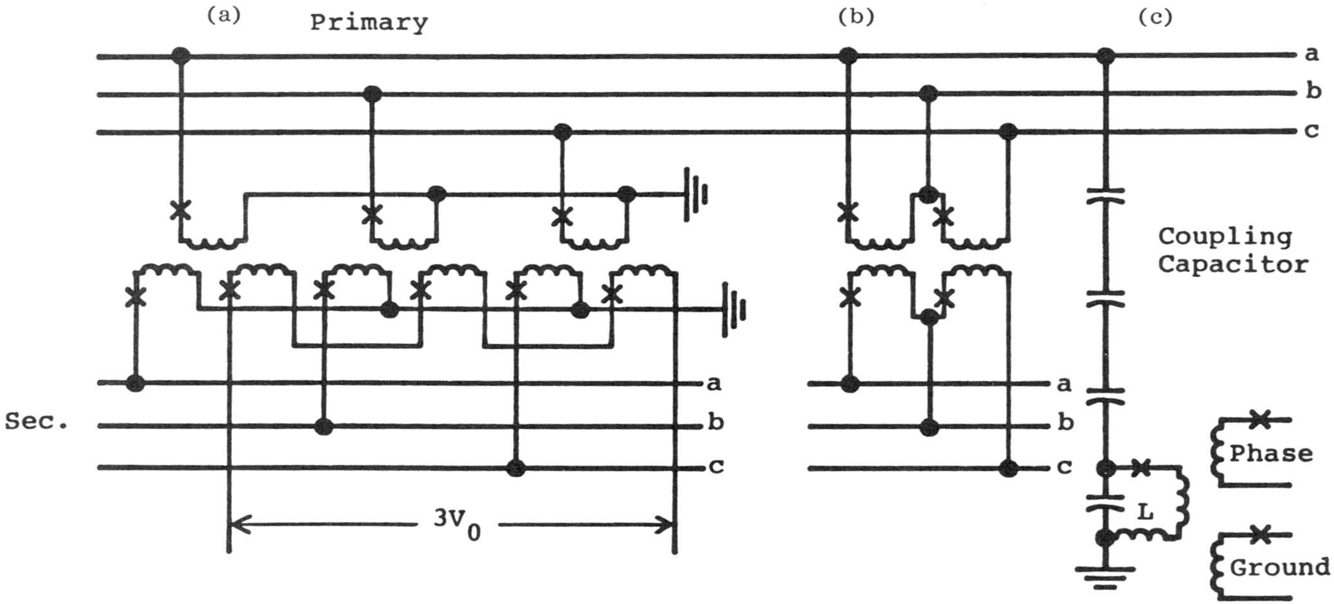

Fig. 5.16 Typical voltage sources for relays. The secondary circuits for the coupling capacitor voltage transformer (CCVT) device are simplified schematics, for concept only. (a) Secondary phase and ground voltage with three double secondary VTs connected phase-to-ground. (b) Secondary phase voltage with two single secondary VTs connected open delta. (c) Secondary phase and ground voltage with three CCVTs connected phase-to-ground. (Only one phase shown, b and c phases duplicate with secondaries connected as in a).

Three VTs or three CCVTs, such as shown in Fig. 5.16a and c, pass positive-, negative-, and zero-sequence voltage. The open-delta connection of Fig. 5.16b will pass both positive- and negative-sequence voltage, but not zero-sequence voltage.

VTs are used at all power system voltages and are usually connected to the bus. At around 115 kV, the CCVT type become applicable and generally more economical than VTs at the higher voltages. Usually, the CCVTs are connected to the line rather than to the bus, since the coupling capacitor device may also be used as a means of coupling radio frequencies to the line for use in pilot relaying. This is discussed in Chapter 13.

Either type of transformer provides excellent reproduction of primary voltage, both transient and steady state, for protection functions. Saturation is not a problem since power systems should not be operated above normal voltage, and faults result is a collapse or reduction in voltage. Both have ample capacity and are highly reliable devices. VTs are normally installed with primary fuses, which are not necessary with CCVTs. Fuses are also used in the secondary. A common practice is to use separate secondary fusing for voltage supply to different groups of relays used in the total protection. Fuses are a hazard. A loss of potential via a fuse may result in unwanted, incorrect relay operations. In some cases overcurrent fault detectors are used to minimize this possibility.

Some CCVTs may exhibit a subsidence transient when the system voltage is suddenly reduced so that the secondary voltage momentarily is not a replica of the primary. This is caused by the trapped energy ringing in the secondary compensating or tuning reactor (L) and the associated circuit. This transient can be at a different frequency from that of the system frequency, or unidirectional. This has not been a problem for electromechanical relays but may cause problems for solid-state types. Modern-design CCVTs are available to eliminate this problem.

BIBLIOGRAPHY

See the Bibliography at the end of Chapter 1 for additional information.

Connor, E. E., E. C. Wentz, and D. W. Allen, Methods for Estimating Transient Performance of Practical Current Transformers for Relaying, *IEEE Trans. Power Appar. Syst.*, PAS 94, January 1975, pp. 116–122.

Iwanusiw, O. W., Remenant Flux in Current Transformers, *Ontario Hydro Res. Q.*, 1970, pp. 18–21.

Requirements for Instrument Transformers, ANSI/IEEE Standard C57.13-1978, IEEE Service Center.

Transient Response of Coupling Capacitor voltage Transformers, IEEE Power System Relaying Committee Report, *IEEE Trans. Power Appar. Syst.*, PAS 100, December 1981, pp. 4811–4814.

Transient Response of Current Transformers, Power System Relaying Committee Report 76-CH1130-4 PWR, IEEE Special Publication. A summary report and discussion, *IEEE Trans. Power Appar. Syst.*, PAS 96, November-December 1977, pp. 1809–1814.

Wentz, E. C., and D. W. Allen, Help for the Relay Engineer in Dealing with Transient Currents, *IEEE Trans. Power Appar. Syst.*, PAS 101, March 1982, pp. 517–525.

6

Protection Fundamentals

6.1 INTRODUCTION

The best protection technique now and for more than 50 years is that known as differential protection. Here the electrical quantities entering and leaving the protected zone or area are compared via current transformers. If the net between all the various circuits is essentially zero, no fault or intolerable problem is assumed to exist. However, if the net is not zero, an internal problem exists and the difference current can operate the associated relays. In general, internal faults provide significant operating current, even for fairly light faults.

Differential protection is applicable universally to all parts of the power system: generators, motors, buses, transformers, lines, capacitors, reactors, and sometimes combinations of these. As the protection of each part of the power system is discussed, invariably, differential protection is the first consideration, and often is the choice for the primary protection.

6.2 THE DIFFERENTIAL PRINCIPLE

This fundamental technique is illustrated in Fig. 6.1 and for simplicity only two circuits to the protection zone are shown. Multiple circuits may exist but the principle is the same. The sum of the currents flowing essentially equals the sum of the currents flowing out during normal operation. The voltage differential system is similar and is discussed in Chapter 10.

For normal operation and all external faults (the through condition), the secondary currents in Fig. 6.1a in the protective relay is the difference in the exciting currents of the differentially connected current transformers. Per unit current distribution is shown. For example, I_p is the primary current in the lines entering or leaving the protected area. $I_p - I_e$ is the secondary ampere current and is equal

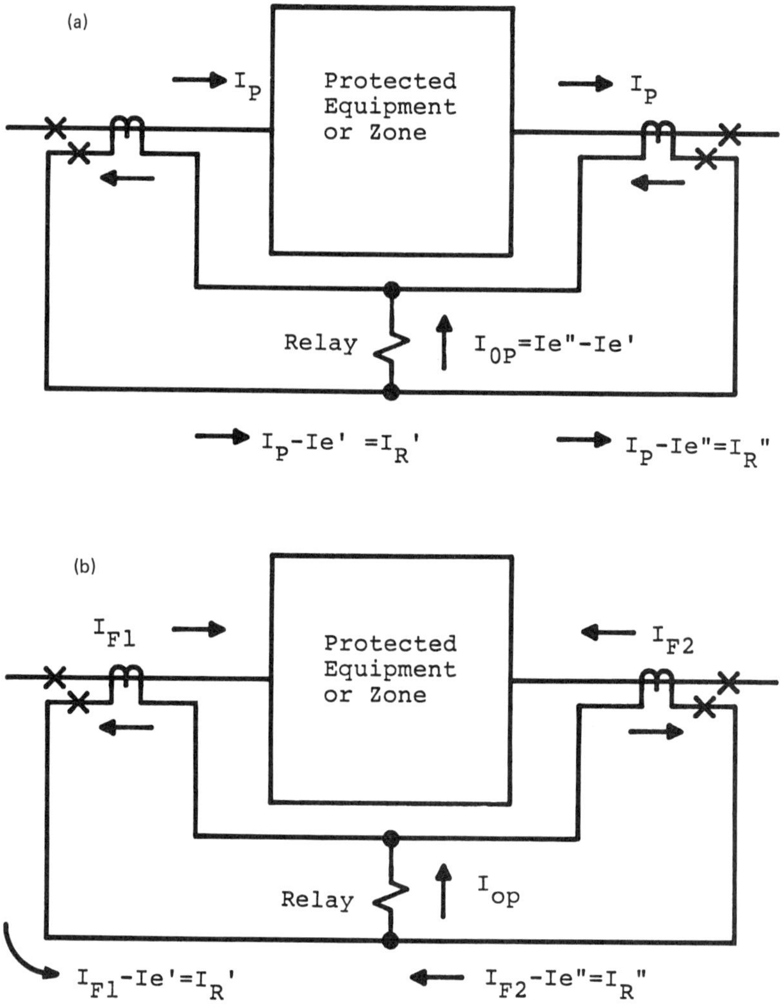

Fig. 6.1 Basic current differential scheme illustrated for the protection of a zone with two circuits: (a) normal conditions, $I_{op} = I_e'' - I_e'$; (b) internal fault, $I_{op} = I_{F1} + I_{F2} - (I_e' + I_e'')$.

to primary current divided by the current transformer ratio minus the secondary exciting current. Even with exactly the same ratio and type of current transformer, the relay current I_{op} will be small but never zero. This is because of the losses within the protected area and small differences between the same CTs. This assumes that no current transformer significantly saturates for the maximum symmetrical ac through currents. With different CTs and ratios, larger differences will exist which must be minimized and/or the pickup of the relay set not to operate on any through condition.

During external faults the transient performance of the several CTs resulting from the sudden increase in current and the associated offset (dc component) can produce rather large transient operating currents. Thus it is difficult and impractical to apply an instantaneous relay. Time-delay relays can be used with care.

For internals faults, Fig. 6.1b shows that the differential relay operating current essentially is the sum of the input currents feeding the fault. This is the total fault current on a secondary ampere basis. Except for very light internal faults, good discrimination is available to detect problems (faults) within the differential zone. For the differential relay to operate, it is not necessary for all the circuits to supply fault current if the circuits are not supplying current to the fault.

To provide high sensitivity to light internal faults with high security (high restraint) for external faults, most differential relays are of the percentage differential type. Fig. 6.2 is a simplified schematic of this type of relay for two circuits, as shown in Fig. 6.1. The secondaries of the CTs are connected to restraint windings (R). Currents in these inhibit operation. Associated with these restraint windings is the operating winding (OP). Current in this winding tends to operate the relay. Differential relays may be of either fixed or variable percentage, and typical characteristics are illustrated in Fig. 6.3. The abscissa is the restraint current. This can be either the smaller current (I''_R) or the larger current (I'_R), depending on the design. The ordinate is the current (I_{OP}) required to operate the relay. Fixed percentage relays exist between 10 and 50% and may or may not have taps to change the percentage.

Thus with a 50% characteristic, an external or through current of 10 A would require a difference or operating current of 5 A or more for the relay to operate. With a 10% type, and 10 A through current, 1 A or more difference current would produce relay operation.

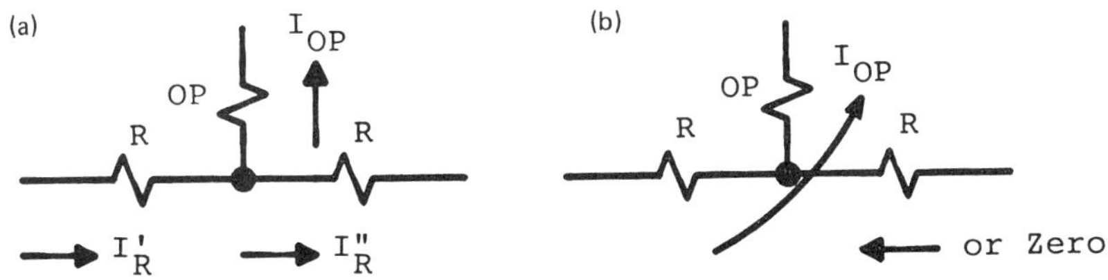

Fig. 6.2 Percentage differential relay: (a) external faults; (b) internal faults.

Fig. 6.3 Typical through current characteristics of various differential type relays.

The variable percentage types do not have percentage taps. At low through currents the percentage is low since at these levels the current transformer performance is usually quite good. At high through-fault currents, where the CT performance may not be as good, a high percentage characteristic is provided. This gives increased sensitivity with higher security.

It is important to recognize that characteristics such as those shown in Fig. 6.3 apply only for external faults or through current flow. Differential relays are quite sensitive to internal faults where the currents in the restraint windings are in opposite directions or one restraint current is zero, as in Fig. 6.2. These relays are calibrated with current through one restraint and the operating windings with no current through the other restraint(s). Typical pickup currents for differential relays are on the order of 0.14 to 3.0 A, depending on the type, tap, and application.

As has been seen, the differential principle compares the outputs of the current transformers in all the circuits into and out of the protected area or zone. For equipment such as generators, buses, transformers, motors, and so on, the CTs usually are all in the same general area, so that it is not too difficult to interconnect their secondaries with the relays. For lines where the terminals and CTs are separated by considerable distances, it is not possible practically to use differential relays as described above. Still the differential principle provides the best protection and is still widely used. This is true particularly at the higher voltages. A communication channel such as a pilot wire, power line carrier (radio frequency), audio tones over wire, or microwave is used for information comparison between the various terminals. These systems are described later.

6.3 THE BASIC PROTECTION PROBLEM

Where the differential principle is not applicable or not practical, the protection problem is as illustrated in Fig. 6.4. The relays at station G for the protection of the line GH should operate fast for all faults in the area between the two line terminals. This is the primary protection zone for the relays at G and the similar relays at H. Fault F_1 is in this primary zone, but faults F and F_2 are external and should be cleared by other protection. However, for the relays at G, the currents are the same since the distances between these three faults is very small and negligible. Thus practically, $I_F = I_{F1} = I_{F2}$. Therefore, the relays at G cannot determine by current (or voltage) magnitude if the remote fault is at F1, where desirably, they should operate fast, or at F or F2, where they should delay. Hence the problem is how to distinguish the internal fault F1 from the external faults F and F2. There are two possible solutions: (1) time or (2) communication.

6.3.1 The Time Solution

The time solution delays the operation of the relays at G for the

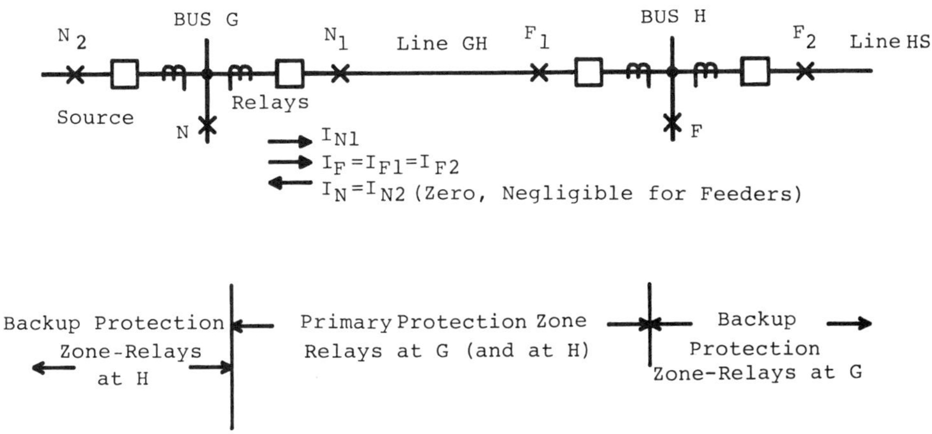

Fig. 6.4 Protection problem for protective relays at station G for line GH.

faults near or at bus H. This delay is to permit the primary relays for the bus H and for the line(s) to the right of H to clear faults such as F and F2. Unfortunately, this means that the internal faults on line GH near bus H, such as F1, will be delayed.

Setting the relays, either phase or ground, for this time solution is called coordination or selectivity. Basically, this technique attempts to set the primary relays to provide fast operation for close-in faults (N_1), yet to delay, for coordination with the relays at H, for faults at or near bus H. This is accomplished with inverse time-overcurrent relays, where the operating time increases as the current magnitude decreases, or with instantaneous relays and constant (fixed) time delays. Coordination is discussed in more depth later, particularly when we discuss line protection in Chapter 12.

6.3.2 The Communication Solution

This second solution is a type of differential protection, as outlined above. The relays at H for the protection of line GH indicate by the direction of the power flow or the relative phase angle information whether the fault is internal (F1) or external (F and F2). This information is communicated via a channel to the relays at G. Similarly, the relays at G provide information which is communicated to H. If the fault is in the primary zone (faults N1 to F1), both G and H operate together at high speed. For the external faults (on the bus at G or H, to the left of G, and to the right of H), the relays at both G and H do not operate. This is an elementary introduction to pilot relaying, which is expanded in later chapters.

6.4 TIME-OVERCURRENT RELAYS

This type of relay, one of the first protective relays, developed some 60 to 70 years ago, is still widely used in many applications throughout the power system. Originally, it was basically a watthour meter with contacts and restricted disk travel. Today, as it has been for many years, the design is entirely different, except that it uses the fundamental induction disk principle. One typical type is illustrated in Fig. 6.5. Alternating current or voltage applied to the main coil produces magnetic flux, most of which passes through the air gap and disk to the magnetic keeper. This returns through the disk to the two side legs of the electromagnet. The shorted turns of the lag coil on one side leg cause a time and phase shift in the flux through that side of the disk to produce rotation of the disk. This rotation is damped by a permanent magnet. The spiral spring provides reset of the contacts after the operating quantity is removed or reduced below pickup value. Contacts attached to the disk shaft can normally be open or closed in

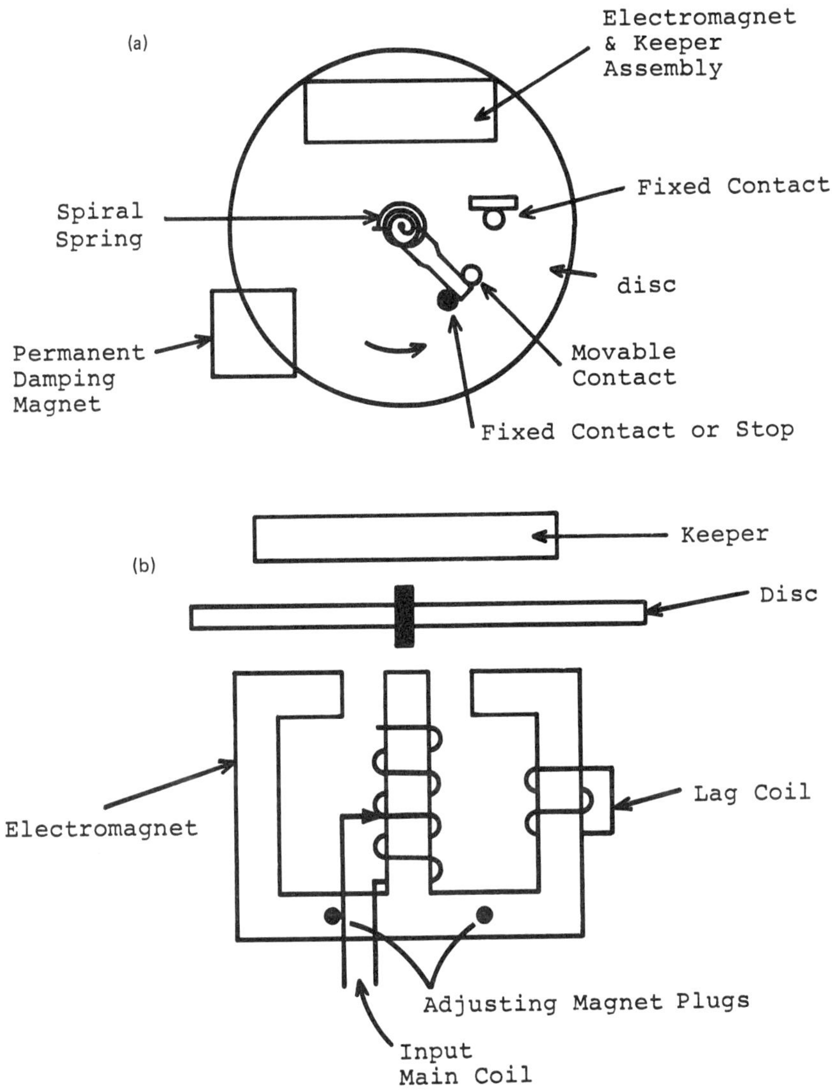

Fig. 6.5 Typical induction disk inverse-time-overcurrent or voltage relay: (a) top view; (b) side view.

the deenergized state. This combination produces fast operation at high current and slow operation at light current, hence an inverse time characteristic. Over the years a number of shapes of time curves have evolved, which are shown in Fig. 6.6 for general comparison.

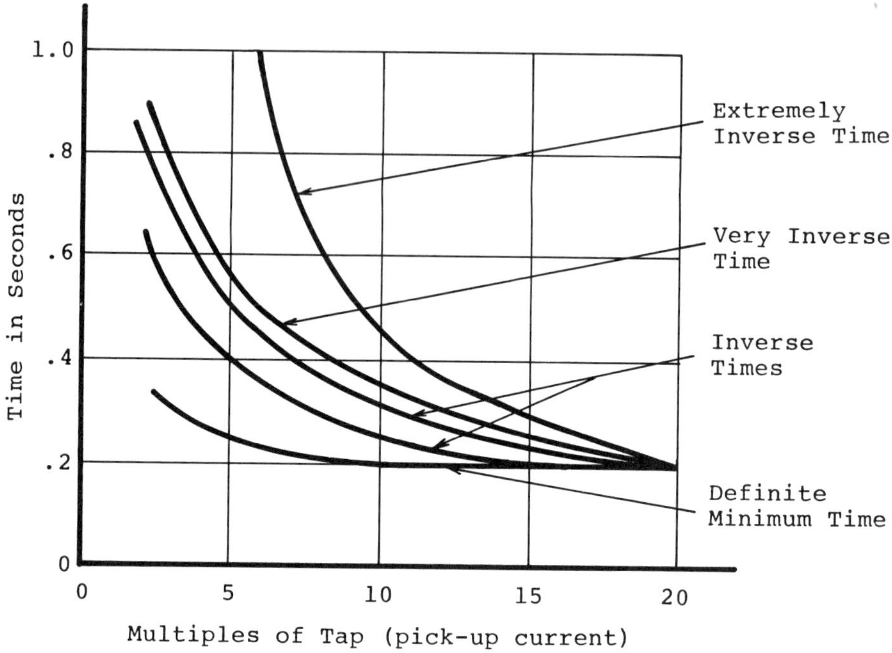

Fig. 6.6 Typical inverse-time-overcurrent relay characteristics. For general comparison the curves are fixed a 0.2 s at 20 times minimum pickup current.

Solid-state versions of these relays are available which essentially duplicate these curves and general characteristics, with lower burdens, wider application ranges, and possibly adjustable time characteristics. Although they will find increasing applications, the electromechanical type described will continue in service for many years.

All relays have a number of taps, each of which represents the minimum current (or voltage) at which the unit will start to operate. This is the minimum pickup value. Thus a current relay set on tap 2 will begin to operate at 2.0 A ± the manufacturer's tolerances. At this current the time will be very long and difficult to check unless the current is maintained at an extremely accurate value. Any small or transient deviation at this level will result in a significant time change. Hence manufacturers generally do not show their time curves below 1.5 to 2 times minimum pickup. Practically, this is not a usable part of the curve for protection.

The abscissa of the characteristic curves is shown in multiples of tap or pickup current. This is for the convenience of providing one scale for all taps. For example, with tap 5, a multiple of 5 on the curve represents 25 A, with tap 2, the 5 multiple represents 10 A, and so on.

In addition to the taps, the spacing for the contact travel is adjustable and marked by a scale known originally as a "time lever" and currently as a "time dial." This provides different operating times at the same operating current level, a family of curves not illustrated in Fig. 6.6. Thus for each relay type, typical time-current characteristic curves are available, usually with curves from 1/2 through 11 time dial settings. These are available on semilog or on log-log fuse-type coordinate paper.

These relays have overtravel and reset times, which can be important in some applications. Overtravel in the electromechanical units is the travel of the contacts after the current drops below the pickup value. Typical values are on the order of 0.03 to 0.06 s generally negligible in most applications. Reset time can be important in coordination with fast reclosing or with fast repetitive faults. It is a function of the time dial setting and the design. Data on reset are available from the manufacturers. The values are generally negligible for solid-state types. Fast reset may or may not be advantageous when coordinating with fuses and relays that do not have fast-reset characteristics.

In a protective relay, the opening of the circuit breaker generally reduces the current through the relay to zero. This may not be true when a relay is used as a fault detector and set to operate for faults well beyond the primary protection zone. Most induction disk time overcurrent relays will not start to reset until the current drops below about 60% of the pickup current.

The relays described above are nondirectional; that is, they can operate independent of the direction of the current. Where this is not desired, a separate directional element such as that discussed in Chapter 3 is used. The induction disk unit provides rotational torque by lag loops on the electromagnet to provide a flux shift. "Torque control" varieties have this lag circuit available for external use, or the directional unit contacts or output are internally connected in this circuit. As long as the lag coil or its equivalent is open, no amount of current will cause operation. When the directional unit closes the circuit to the lag coil or equivalent, operation is as in the characteristic curves. This is directional torque control. The induction disk unit and its solid-state equivalent are used for overcurrent, overvoltage, undervoltage, combined over and undervoltage, and power, as well as in other types of designs.

6.5 INSTANTANEOUS CURRENT-VOLTAGE RELAYS

These relays are used in many areas of protection as overcurrent or over- and/or undervoltage units to trip directly or as fault detectors

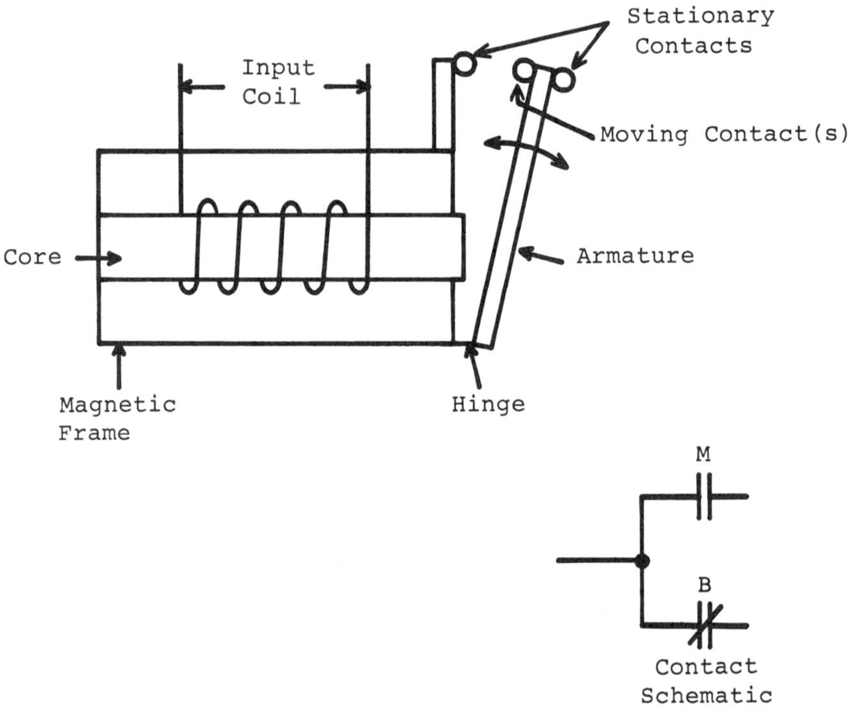

Fig. 6.7 Typical electromechanical clapper or telephone relay.

for security. Typical types in general use are the clapper or telephone relay (Fig. 6.7), solenoid or plunger relay (Fig. 6.8), and induction cup or induction cylinder relay (Fig. 6.9).

The term "telephone relay" dates from its very wide application originally in telephone exchange systems. Today this use is superseded by modern solid-state electronic switching. However, relays of this type are still used in many auxiliary applications, both ac and dc. It is a common output relay for many solid-state protective relays.

An example of the solenoid type as a dc type is its use as a seal-in contact switch (CS) of Fig. 1.9. With an ac coil and constructions, the solenoid unit serves as an instantaneous trip unit (IT and IIT).

The operation of the first two is basic; current or voltage applied to the coil produces flux, which attracts the armature or plunger. Contacts on the moving member are thus operated. Multiple contacts are possible, especially on the telephone types.

The ac types have taps or other means to change the pickup value. For trip service, dropout is seldom a problem, and many of these units do not drop out after closing their contacts upon operation

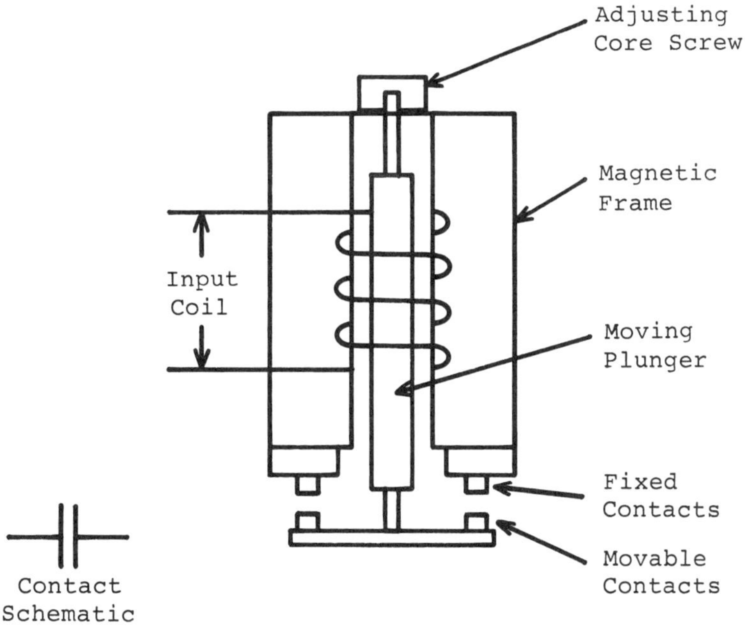

Fig. 6.8 Typical electromechanical plunger relay.

until the current (voltage) drops to the order of 60% of the pickup value. Where this is a problem, high-dropout models of the Fig. 6.8 type are available. These will reset at 90% of the pickup value or higher. In general, fault detectors should be of the high-dropout type.

The ac induction cup or induction cylinder unit of Fig. 6.9 basically is a "two-phase motor" with the two coils wound as shown on the four poles of the electromagnet. In the center is a magnetic core. Around or over this is the moving cup or cylinder, with the moving contacts and spring to provide reset. When the fluxes of coils 1 and 2 are in phase, no rotational torque exists. As an instantaneous overcurrent unit, a phase shift is designed in one coil circuit, so that operating torque is produced when the current is above the pickup value. The rotation is limited to a few millimeters, enough to close the contacts. Typical times of operation are in the order of 16 to 20 ms.

6.6 DIRECTIONAL-SENSING POWER RELAYS

The induction cup and induction cylinder units are used to indicate the direction of power flow and magnitude. This first application for

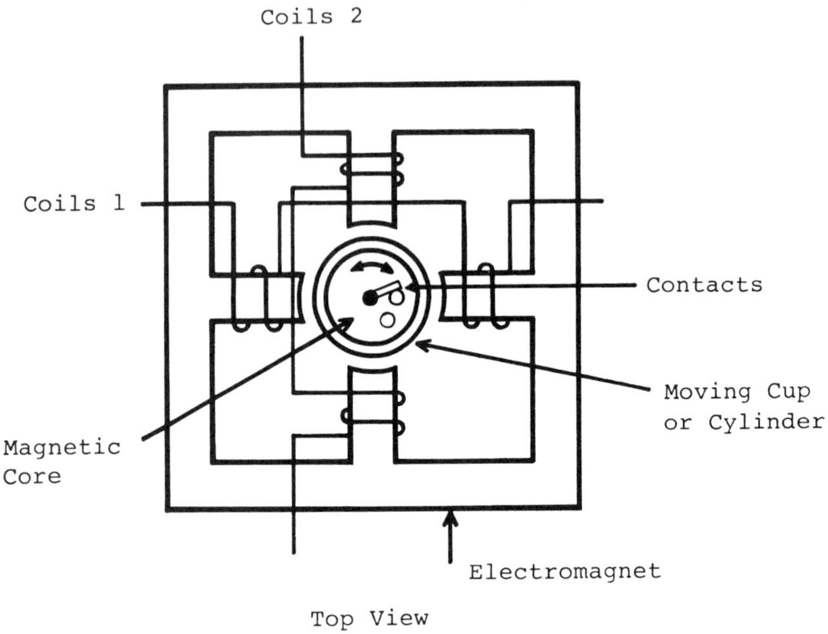

Fig. 6.9 Typical electromechanical induction cup or cylinder relay.

directional sensing was discussed in Chapter 3. Typical characteristics are shown in Fig. 3.7. The operating current is passed through one set of windings, and a reference voltage or current through the other. When the phase relations are as indicated, the unit operates. Since these units as directional units are very sensitive, they are used in almost all applications with fault-sensing units such as the time-overcurrent or instantaneous-overcurrent units above. Other types have tapped windings to operate when the power level exceeds a preset value.

6.7 DISTANCE RELAYS AND THE R-X DIAGRAM

Fundamentally, distance relays compare the power system voltage and current. They operate when the ratio is less than its preset value. For balanced conditions and for all phase faults, the ratio of the voltage to current applied to the relay is the impedance of the circuit since $V/I = Z$. Thus these relays are set as a function of the fixed impedance of the power system for the zone they are to protect. An early design (no longer manufactured) provides a good basic understanding of the principle and appreciation of common terms currently

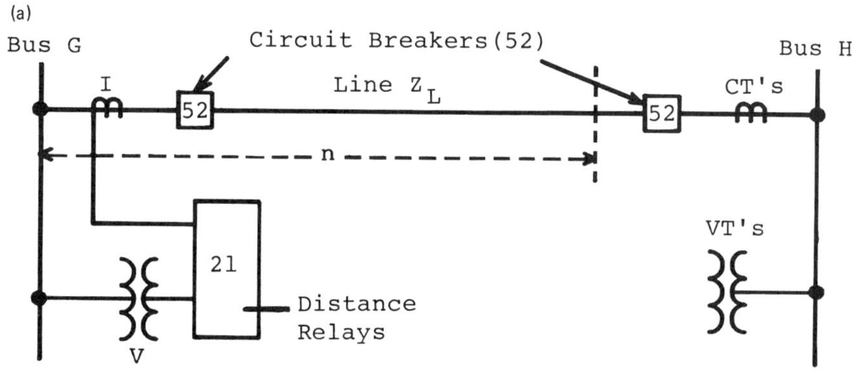

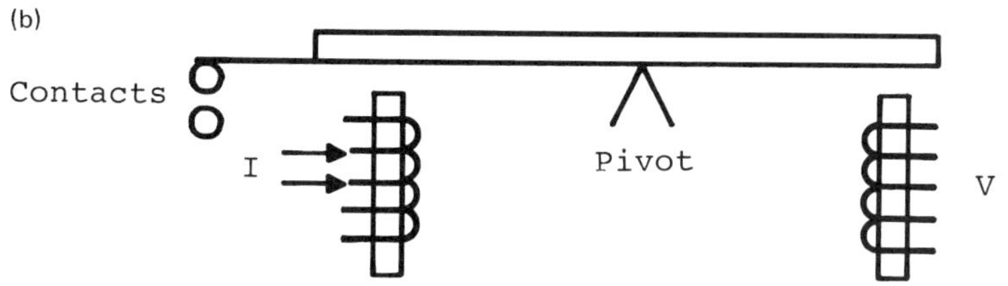

Fig. 6.10 Distance relay operating principles explained fundamentally by the balanced-beam impedance unit: (a) distance relay application to line GH; (b) simplified explanation diagram with a beam unit.

used. This early type is illustrated in Fig. 6.10. A balanced beam has a voltage-energized electromagnet to restrain its movement, and a current-operated electromagnet to close its contacts. By design and setting the voltage restraint force can be made to equal the current operating force for a solid zero-voltage three-phase fault at the set point shown as nZ_L. This threshold point is known as the "balance point," "operating threshold," or "decision point" of the unit. For a fault between the relay and point n, the current I will be larger and V will decrease or remain approximately the same relative to the values for the fault at n. Thus the increased current causes the beam to tip at the left end to close the contacts.

For an external fault to the right of point n, the current will be less than for the fault at n, and the voltage will be higher. Thus the torque or pull of the voltage coil is greater than that of the current coil for restraint or no operation.

With the solid three-phase fault at the balance point n, the voltage at n will be zero. Then the voltage at the relay location will be the drop along the circuit, or InZ_L. Dividing this voltage by the current, the unit responds to impedance:

$$Z_R = \frac{V}{I} = \frac{InZ_L}{I} = nZ_L \tag{6.1}$$

Thus the setting and operation are a function of the impedance from the relay voltage measurement point to the balance or set point. By design and connections, this reach is the same for all phase faults: three-phase, phase to phase, and two-phase to ground. It is fixed for a given setting and it is independent over very wide range of fault currents, and so provides an instantaneous fixed reach protective relay.

Distance relays as currently available, both electromechanical and solid-state types, operate by comparison of the power system voltages and currents but with various techniques, depending upon the particular manufacturer. Discussion of these are beyond the scope of this book. Distance relays for single-phase-to-ground fault protection are available but are more complex. Their applications are covered later.

The characteristics of distance relays are shown most conveniently on an impedance R-X diagram, where the resistance R is the abscissa and the reactance X is the ordinate. Typical characteristics on these axes are shown in Fig. 6.11. For any given discussion the origin is the relay location, with the operating area generally in the first quadrant. Whenever the ratio of the system voltage and current falls within the circle shown or in the crosshatched area, the unit operates.

The elementary type discussed in Fig. 6.10 provided an impedance characteristic such as that shown in Fig. 6.11a. This obsolete design was independent of the phase relation of the voltage and current, so operated in all four quadrants. Thus a separate directional-sensing unit was necessary to prevent operation for faults in the system to the left of bus G (Fig. 6.10).

All of the characteristics shown are in practical application with the "mho" type of Fig. 6.11b as the most universal and common. The application for the primary protection of line GH of Fig. 6.10a would require two units at station G and two at station H. Typical settings are shown in Fig. 6.12 for the relays at G looking toward bus H. The zone 1 unit operates instantaneously and n is less than 1, typically 0.9. The zone 2 unit is set with n greater than 1.0, so a time delay is required for coordination with the protection at station H and to the right.

Many phase distance relays are of the single-phase type, requiring three separate units to cover the various faults. In these the characteristics, such as the mho circle of Fig. 6.11b, are valid for any

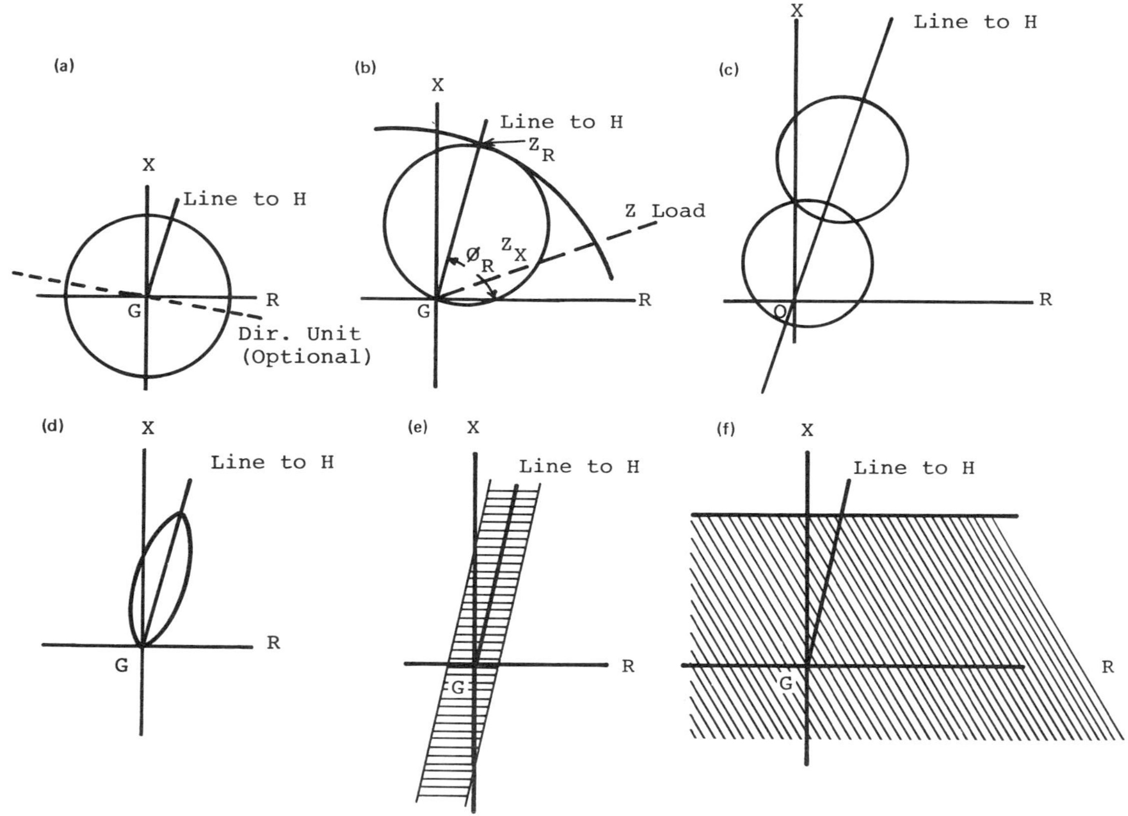

Fig. 6.11 Distance relay operating principles explained fundamentally by the balanced-beam impedance unit: (a) impedance; (b) MHO; (c) offset MHO's; (d) lens; (e) double blinders; (f) reactance.

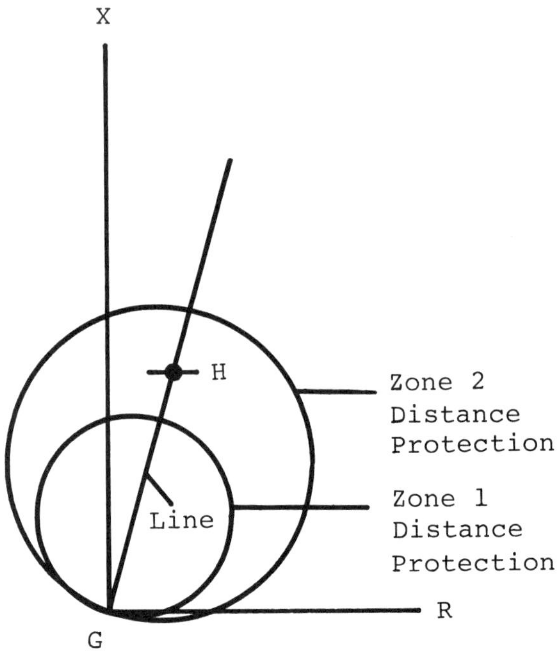

Fig. 6.12 Distance mho protective relays applied at station G for the protection of line GH of Fig. 6.10.

unit only for faults of specific phases; that is, one unit operates as shown for phase ab and ab-ground faults, one for bc and bc-ground faults, and the third for ca and ca-ground faults. All three operate for three-phase faults.

Another type is three-phase, so that one relay with two operating units operates for all combinations of phase faults independent of the particular faulted phases. One of the units (three phase) of this type has a mho circle characteristic as in Fig. 6.11b and operates for all three-phase and some phase-to-phase-to-ground faults that occur within the circle. The other unit (phase to phase) has the arc characteristic shown in Fig. 6.11b and will respond only to all phase-to-phase faults (ab, bc, ca) and corresponding, phase-to-phase-to-ground faults not covered by the three-phase unit. The operating area is only in the first quadrant for the faults within the sector between the line vector, the relay characteristic, and the R axis. Thus it is directional, as is the mho unit. This phase-to-phase unit will not respond to balanced currents so does not operate for three-phase faults, load, swings, and so on. It can be set independent of load and swings, while the mho and similar units must be set with regard to maximum load and swings.

If load were flowing from G to H (Fig. 6.10a), it would appear to the relay at G as

$$Z_{load} = \frac{V_{LN}}{I_{load}} = \frac{120}{\sqrt{3}\,5} = 13.86\ \Omega \text{ secondary} \tag{6.2}$$

based on 5 A load and 120 V line to line at the secondaries of the CTs and VTs, respectively. In most cases load into the line would be between 0° and about 30° lagging current. On an R-X diagram, such as Fig. 6.12, this load would be a phasor in the first quadrant, 0° to about 30° counterclockwise from the +R axis. For heavy load, large I, the magnitude of Z_{load} is small. Thus for heavy loads the impedance phasor moves toward the origin; for light loads it moves to the right or away from the origin.

Load flowing from H to G, or out of G, will appear as a phasor in the second quadrant, or to the left of G on the R-X diagram and above the -R axis for lagging load.

The operating points on the mho circle characteristic can be determined by

$$Z_X = Z_R \cos(\phi_R - \phi_X) \tag{6.3}$$

Where Z_X is the impedance from the origin to any point on the circle at angle ϕ_X, and Z_R is the relay reach at ϕ_R. As an example, what is the reach of a mho unit along a 75° angle line if the maximum load into the line is 5 A secondary at 30° lagging? From Eq. (6.2), the load impedance is 13.86 Ω secondary. This is Z_X in Eq. (6.3) with $\phi_X = 30°$. Thus

$$13.86 = Z_R \cos(75° - 30°), \text{ and solving yields}$$

$$Z_R = 19.60\ \Omega \text{ secondary} \tag{6.4}$$

This can be translated into primary line ohms by the basic formula

$$Z_{R(Sec)} = \frac{Z_{R(Pri)} R_c}{R_v} \tag{6.5}$$

where R_c and R_v are the CT and VT ratios (Chapter 5). This line reach on a 115-kV line with 600:5 CTs would be

$$Z_{R(Pri)} = \frac{Z_{R(Sec)} R_v}{R_c} = \frac{19.6\ (1000)}{120} = 163.3\ \Omega \text{ primary} \tag{6.6}$$

On the basis of the typical 0.8 Ω/mi, the 163.3-Ω reach is around 204 m, a very long line at this voltage. It is interesting to note that one type of distance relay has its maximum specified reach at 20 secondary ohms, derived from Eq. (6.4).

The MVA represented by the 5-A load is

$$\text{MVA} = \frac{\sqrt{3}\,\text{kV} \times I}{1000} = \frac{\sqrt{3}\,(115)\,(5)\,(120)}{1000} = 119.5\ \text{MVA} \tag{6.7}$$

When distance relays are set to overreach several lines or are applied to long lines, usually EHV lines, it is quite possible for maximum load impedance to fall within the operating circle. In these applications various relay characteristics, as shown in Fig. 6.11c—e, are used to restrict relay operation in the load area, but provide long line reaches.

The distance relay characteristic of Fig. 6.11f is a reactance type. These require additional units, to avoid the reactance unit operating on loads and for faults behind the relay. The double blinder characteristic of Fig. 6.11e is composed of two reactance characteristic with phase shifts, as shown. The left unit operates over a wide area to the right, while the right unit operates over a wide area to the left. Relay operation occurs only when both units operate, and operation is in the area between them (shown crosshatched). Specific applications of distance relays are discussed in various chapters.

6.8 BACKUP PROTECTION: REMOTE VERSUS LOCAL

The importance of backup and redundancy in system protection has been indicated in Chapter 1 and is discussed again in Chapters 8 through 13. Backup is defined as "protection that operates independently of specified components in the primary protective system." It may duplicate the primary protection or may be intended to operate only if the primary protection fails or is temporarily out of service (IEEE 100-1984). The various types can be illustrated with reference to the protection at station G for line GH of Fig. 6.4. Relays applied at G as shown are principally to provide primary protection for line GH. For faults on the line, generally more than one of the several primary relays may operate. This provides primary backup through redundancy. For very important circuits or equipment, especially at HV and EHV levels, completely separate protection, operating from different CTs (and sometimes different VTs, separate dc supplies, and operating different trip circuits on the breakers, is used commonly. One protective system is designated as the primary one, and the other as secondary — somewhat of a misnomer since usually they operate together at high speed.

In the simple two-line system of Fig. 6.4, it was seen that the relays at G must be set to operate for the external faults F, F2, and others out on line HS in order to provide protection for fault F1. Thus relays G provide primary protection for line GH, and backup protection for bus H and line HS. This is remote backup. If the F, F2, and so on, faults are not cleared by their primary relays and associated circuit breaker(s), relays at G should operate and remove the G source from the fault. Similarly, all other sources supplying current to the uncleared fault should be cleared by the backup operation of the relays at their remote terminals.

In recent years it has been desirable to have the backup protection at the local station and to open all the breakers around the bus rather than at the remote terminals. This is local backup, also associated with breaker failure. In this type of application, the breaker at H on line GH rather than the breaker at G would be relayed to clear faults F2, and so on, should the primary relays and/or the H breaker on line HS fail to operate.

For local backup there should be a separate, independent set of relays, as always exists for remote backup. This is available with the independent primary and secondary relay systems indicated above applied principally at high voltages. This independence may not exist in lower-voltage protection systems.

6.9 SUMMARY

This chapter has presented the fundamentals of system protection and very briefly outlined various basic designs in wide use in these systems throughout the United States. The aim is to provide a background for the later chapters on the protection aspects of the various power system components.

BIBLIOGRAPHY

See the Bibliography at the end of Chapter 1 for additional information.

Lewis, W. A, and L. S. Tippett, Fundamental Basis for Distance Relaying on a Three Phase System, *AIEE Trans.* 66, 1947, pp 694–708. The original was presented but not published at an AIEE meeting in 1932.

Sonnemann, W. K., A Study of Directional Elements for Phase Relays, *AIEE Trans.*, 69, P. II, 1950, pp. 1438–1451.

Van, C. Warrington, A.C., Application of the OHM and MHO Principle to Protective Relaying, *AIEE Trans.*, 65, 1946, pp. 378–386, 490.

7

System Grounding Principles

7.1 INTRODUCTION

Power system grounding is very important, particularly since the large majority of faults involve ground. Thus it has a significant effect on the protection of all the components of the power system. The principal purposes of grounding are to minimize potential transient overvoltages, to comply with local, state, and national codes for personnel safety requirements, and to assist in the rapid detection and isolation of the trouble or fault areas.

A basic review of system grounding is in order together with its fundamental technology and a general evaluation of the methods. There are four types: (1) ungrounded, (2) high impedance, (3) low impedance, and (4) effective or solid grounding. Each has its application in practice, together with advantages and disadvantages. The recommendations are based on general practices plus some personal preferences. It should be recognized that there are many factors in each specific system or application which can well justify variations or a different approach. As relaying is highly influenced by personality, as indicated earlier, so to a degree is system grounding.

7.2 UNGROUNDED SYSTEMS

These are power systems with no intentional applied grounding. However, they are grounded by the natural capacitance of the system to ground. Thus the fault current level is very low, so that equipment damage is minimum and it is not necessarily essential that the faulted area be isolated rapidly. This is an advantage, so it is sometimes used in industrial plant systems where a high continuity of service is important to minimize costly production process interruptions. However, ungrounded systems are subject to high, destructive transient

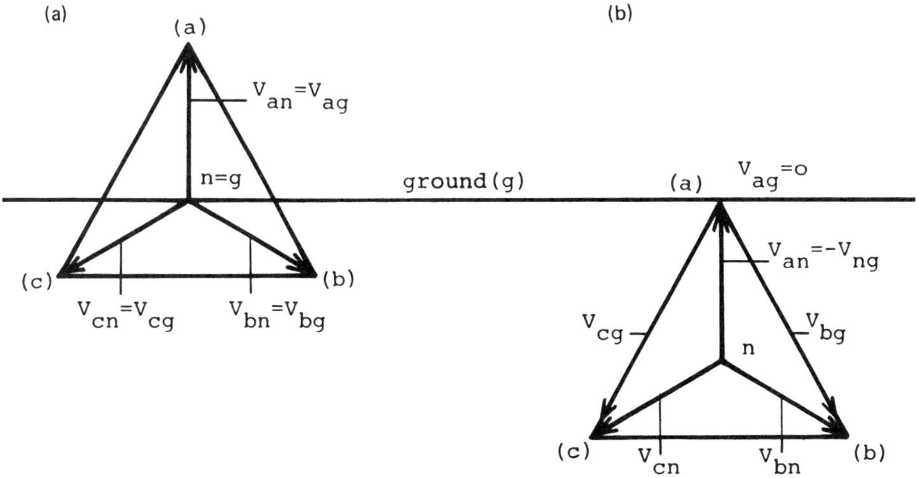

Fig. 7.1 Voltage shift for a phase-a-to-ground fault on an ungrounded system: (a) normal balanced system; (b) phase a solidly grounded.

overvoltages and so are always potential hazards to equipment and personnel. Thus as a generality they are not recommended, although, as indicated, are used.

Phase-to-ground faults on an ungrounded system essentially shift the normal balanced voltage triangle as shown in Fig. 7.1. The small currents flowing through the series phase impedances will cause a very slight distortion of the voltage triangle, but practically, it is as shown in Fig. 7.1b. A typical circuit is illustrated in Fig. 7.2 showing the current flow, and the sequence networks are shown in Fig. 7.3. The distributed capacitive reactance values X_{1C}, X_{2C}, and X_{0C} are very large, while the series reactance (or impedance) values X_{1S}, X_T, X_{1L}, X_{0L}, and so on, are relatively very small. Thus practically, X_{1C} is shorted out by X_{1S} and X_T in the positive-sequence network, and similarly for the negative-sequence network. Since these series impedances are very low, X_1 and X_2 approach zero relative to the large value of X_{0C}. Therefore,

$$I_1 = I_2 = I_0 = \frac{V_S}{X_{0C}} \tag{7.1}$$

and

$$I_a = 3I_0 = \frac{3V_S}{X_{0C}} \tag{7.2}$$

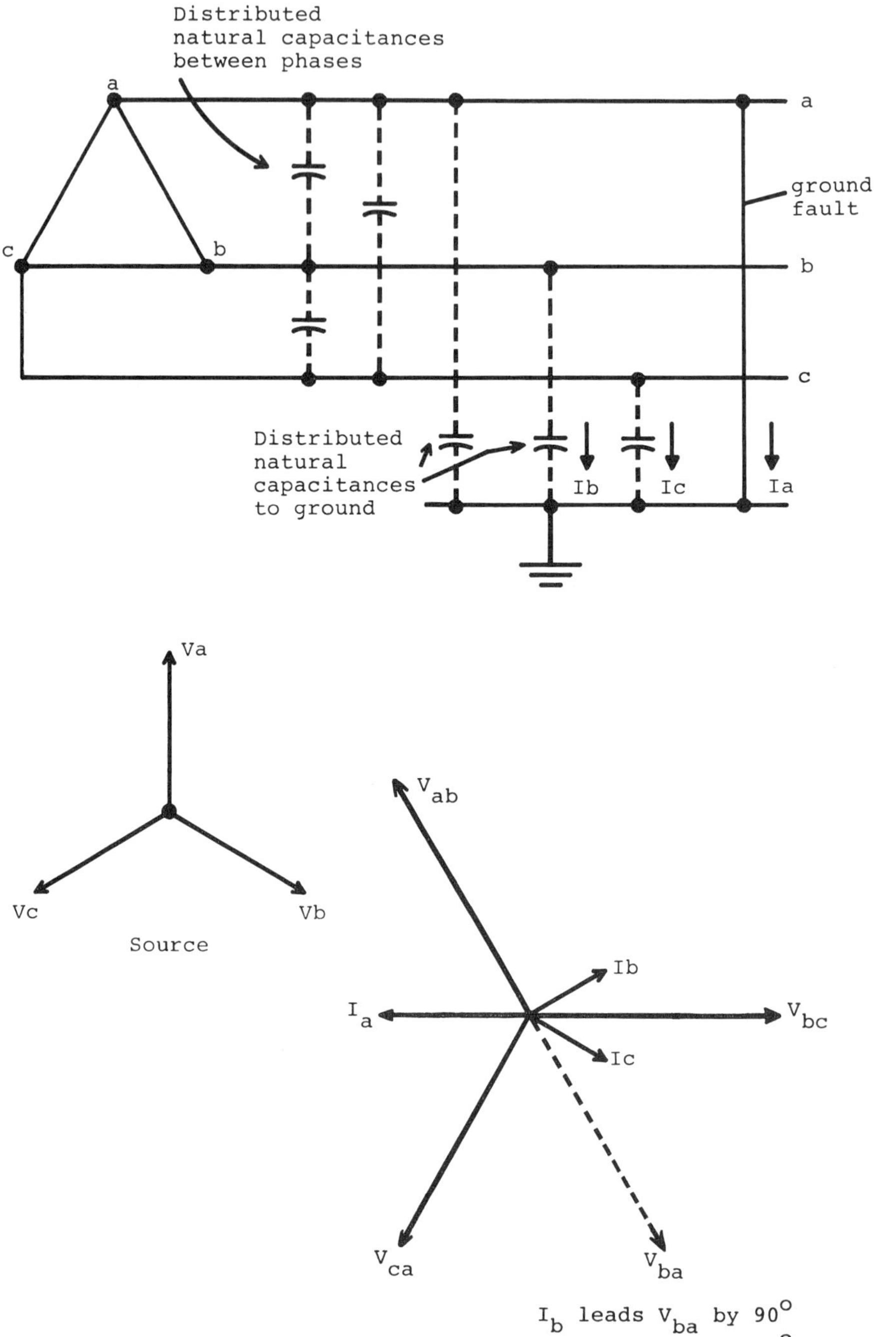

I_b leads V_{ba} by 90°

I_c leads V_{ca} by 90°

$I_a = -I_b - I_c = 3I_0$

Fig. 7.2 Phase-to-ground fault on an ungrounded system.

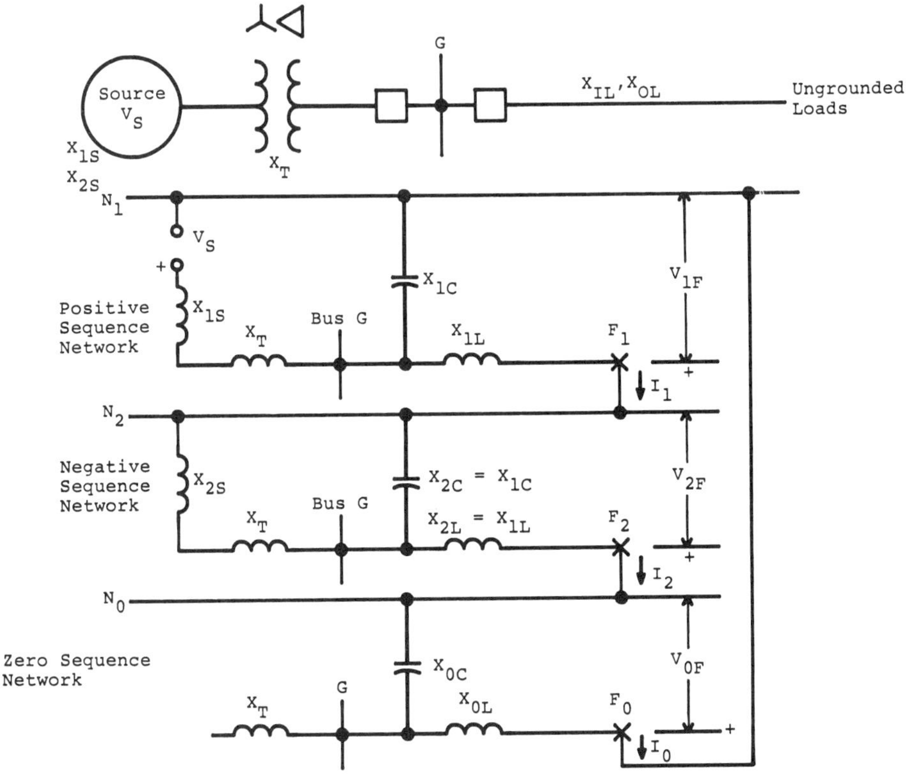

X_{1C}, X_{2C}, X_{0C} are lumped equivalents of the distributed capacitance between phases to network and to ground.

Fig. 7.3 Sequence networks and interconnections for a phase-a-to-ground fault on an ungrounded system.

This calculation can be made in per unit or amperes, remembering that V_S and all the reactances (impedances) are line-to-neutral quantities.

The unfaulted phase b and c currents will be zero when determined from the sequence currents of Eq. (7.1). This is correct for the fault itself. However, throughout the system the distributed capacitance X_{1C} and X_{2C} is actually paralleled with the series reactances X_{1S}, X_T, and so on, so that in the system I_1 and I_2 are not quite equal to I_0. Thus I_b and I_c exist and are small but necessary as the return paths for I_a fault current. This is shown in Fig. 7.2. If $I_a = -1$ pu, then $I_b = 0.577 \angle +30°$ and $I_c = 0.577 \angle -30°$ pu.

In industrial applications where ungrounded systems might be used, the X_{0C} is equal practically to $X_{1C} = X_{2C}$ and is equivalent to the charging capacitance of the transformers, cables, motors, surge-supression capacitors, local generators, and so on, in the ungrounded circuit area. Various reference sources provide tables and curves for typical charging capitances per phase of the power system components. In an existing system the total capacitance can be determined by dividing the measured phase charging current into the line-to-neutral voltage.

Note that as faults occur in different parts of the ungrounded system, X_{0C} does not change significantly. Since the series impedances are quite small in comparison, the fault currents are the same practically and independent of the fault location. This makes impractical selective location of faults on these systems by the protective relaying.

When a phase-to-ground fault does occur, the unfaulted phase-to-ground voltages are increased essentially by $\sqrt{3}$, as shown in Fig. 7.1b. Thus these systems require line-to-line voltage insulation.

In the normal balanced system (Fig. 7.1a), $V_{an} = V_{ag}$, $V_{bn} = V_{bg}$, and $V_{cn} = V_{cg}$. When a ground fault occurs, the phase-to-neutral voltages and the phase-to-ground voltages are quite different. The neutral "n" or "N" is defined as "the point that has the same potential as the point of junction of a group (three for three-phase systems) of equal nonreactive resistances if connected at their free ends to the appropriate main terminals (phases of the power system)" (IEEE 100-1984). This is the "n" shown in Fig. 7.1b.

From this figure, the voltage drop around the right-hand triangle is

$$V_{bg} - V_{bn} - V_{ng} = 0 \tag{7.3}$$

and around the left triangle,

$$V_{cg} - V_{cn} - V_{ng} = 0 \tag{7.4}$$

Also,

$$V_{ng} + V_{an} = 0 \tag{7.5}$$

From the basic equations,

$$V_{ag} + V_{bg} + V_{cg} = 3V_o \tag{7.6}$$

$$V_{an} + V_{bn} + V_{cn} = 0 \tag{7.7}$$

Subtracting Eq. (7.7) from (7.6), substituting Eqs. (7.3) through (7.5), and with $V_{ag} = 0$:

$$V_{ag} - V_{an} + V_{bg} - V_{bn} + V_{cg} - V_{cn} = 3V_o$$

$$V_{ng} + V_{ng} + V_{ng} = 3V_o$$

$$V_{ng} = V_o \qquad (7.8)$$

Thus the neutral shift is the zero-sequence voltage. In the balanced system of Fig. 7.1a, n = g and V_o is zero and there is no neutral shift.

7.3 TRANSIENT OVERVOLTAGES

Restricting arcs after current interruption in the breaker or in the fault can result in large destructive overvoltages in ungrounded systems. This phenomenon is illustrated in Fig. 7.4. In the capacitive system the current leads the voltage by nearly 90°. When the current is interrupted or the arc extinguished at or near its zero value, the voltage will be at or near its maximum value. With the open this voltage remains on the capacitor to decay at the time constant of the capacitive system. In the source system it continues as shown for V_S. Thus in a half cycle, the voltage across the open is almost twice the normal peak value. If a restrike occurs (switch closed in Fig. 7.4), the essentially +1 pu voltage of the capactive system will go to the system voltage of −1 pu, but because of the system inductance and inertia it will overshoot to a maximum possibility of −3 pu. If the arc goes out again near current zero (switch open) but restrikes (switch closed) again, the system voltage will try to go to +1 pu but again will overshoot, this time to a potential maximum of +5 pu. This could continue to −7 pu, but by this time the system insulation no doubt would break down, causing a major fault. Thus ungrounded systems should be used with caution, and applied at the lower voltages (below 13.8 kV), where the system insulation levels are higher.

If used, prompt attention is important to locate and correct the ground fault. As the fault current is very low, it is easy to ignore the fault and continue operation. However, with the fault the other phases are operating at essentially 1.73 times normal line-to-ground voltage. If an insulation deterioration caused the first ground fault, the higher voltages might accelerate breakdown of the unfaulted phases, to result in a double line-to-ground or three-phase fault. Then high fault currents would result, requiring fast shutdown and instant loss of production.

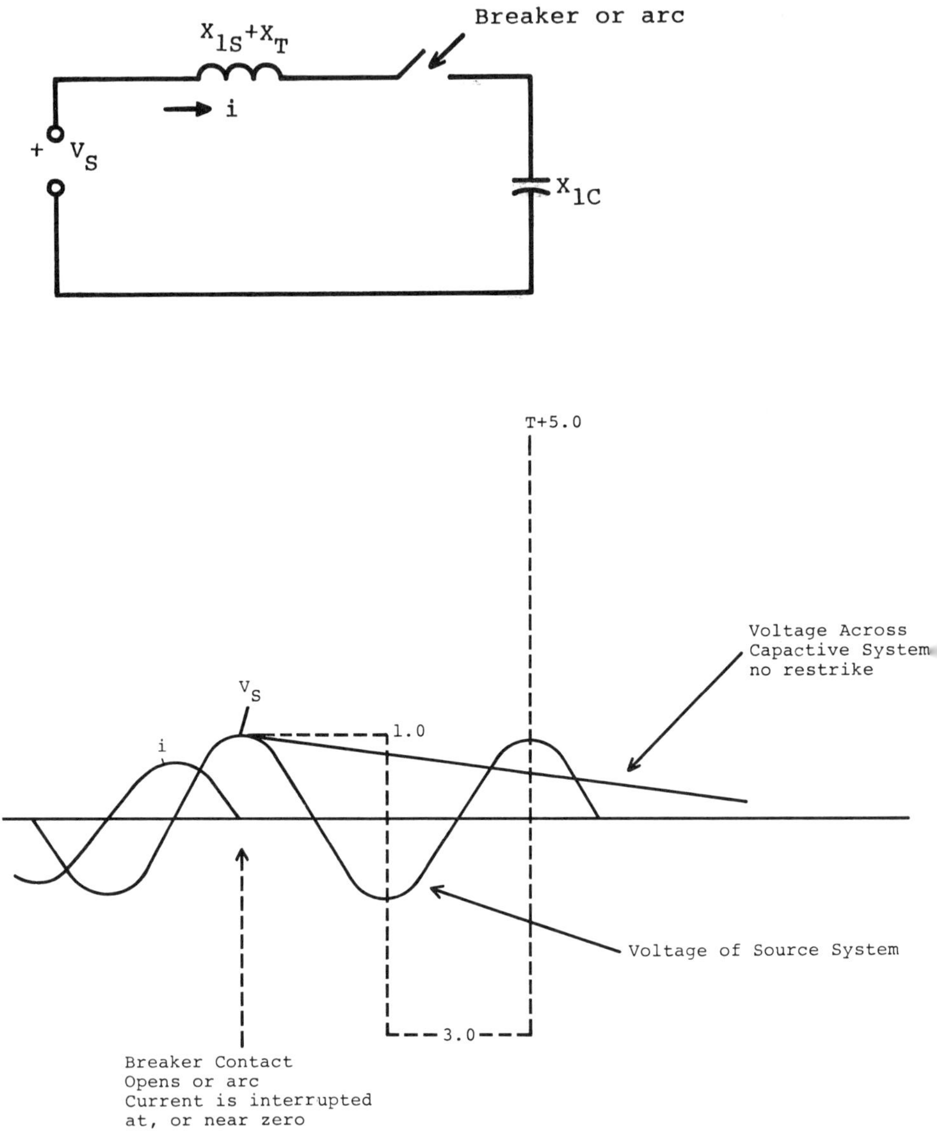

Fig. 7.4 Transient overvoltage on an ungrounded system.

In actual practice there are no totally ungrounded systems. As soon as a fault detector is applied using one or three voltage transformers, the system is grounded through the high impedance of these devices. The resistance of the relays and associated ballast resistors helps limit the transient overvoltages, so that very few cases of overvoltage actually exist.

7.4 GROUND-DETECTION METHODS FOR UNGROUNDED SYSTEMS

Voltage provides the best indication of a ground fault since the current is very low and basically does not change with the fault location. The two methods used are shown in Fig. 7.5. These indicate that a ground fault exists but not where it is in the primary system. Wye grounded-broken delta voltage transformer connections are preferred (Fig. 7.5a). Ballast resistors are used to reduce the shift of the neutral from either unbalanced excitation paths of the voltage transformers or from ferroresonance between the inductive reactance of the voltage transformers and relays and the capactive system. The single voltage transformer of Fig. 7.5b is especially subject to possible ferroresonance without adequate resistance in the secondary. Without this resistance,

$$V_{bg} = \frac{\sqrt{3}\, V_{LL}}{3 - \dfrac{X_C}{X_e}} \tag{7.9}$$

If the distributed system capacitance X_C divided by the transformer exciting reactance X_e equals 3, then theoretically, V_{bg} is infinite. Saturation of the voltage transformer would prevent this, but it is quite possible for the voltage triangle abc to have its ground point well outside this triangle. This is called "neutral inversion," where substained phase-to-ground voltages of almost four times normal phase-to-neutral voltages have been experienced. Also, the interaction of the variable transformer exciting reactance with system capacitance can produce ferroresonance with very high and distorted waveforms. This application of the single VT is not recommended.

The voltage for the relay in Fig. 7.5a, from Fig. 7.1b, is

$$V_{pq} = 3V_o = V_{ag} + V_{bg} + V_{cg}$$

$$= (\sqrt{3}\, V_{LN} \cos 30°) \times 2 = 3V_{LN} \tag{7.10}$$

Thus the voltage available to the relay for a phase-to-ground fault on the ungrounded system is three times the line-to-neutral normal voltage. Usually, the VT ratio of primary V_{LN}: 69.3 V is used so that the

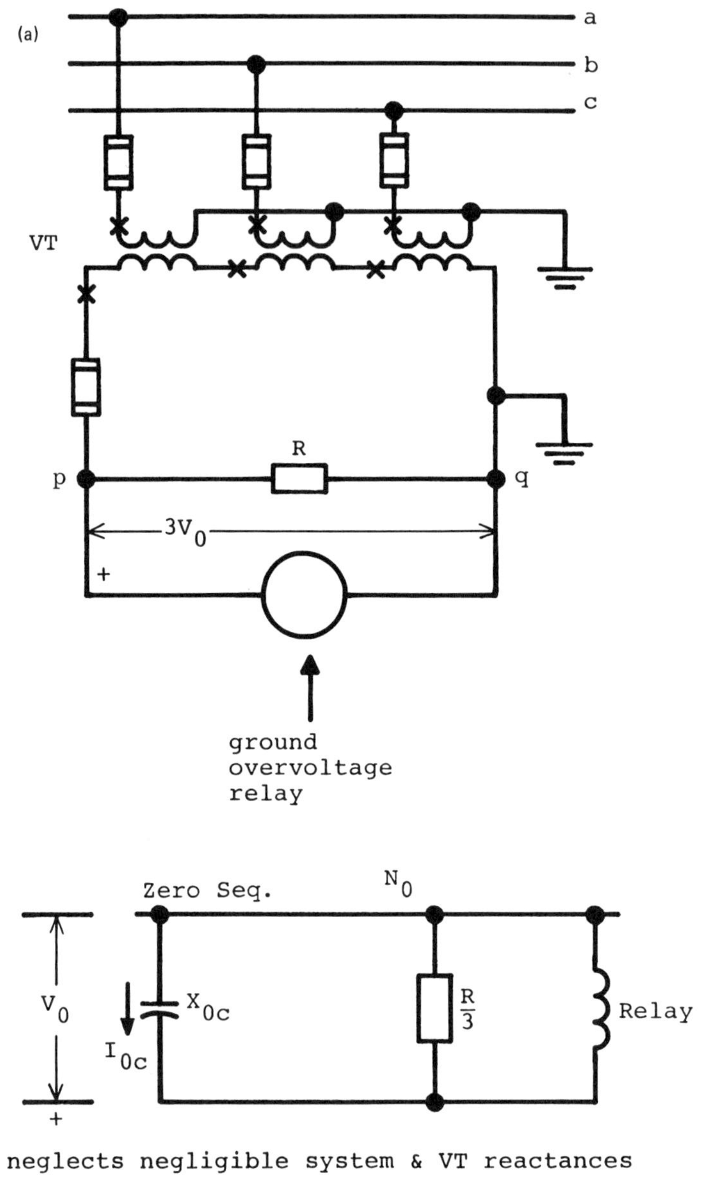

Fig. 7.5 Voltage ground-fault detection methods: (a) three voltage transformers; (b) single voltage transformers.

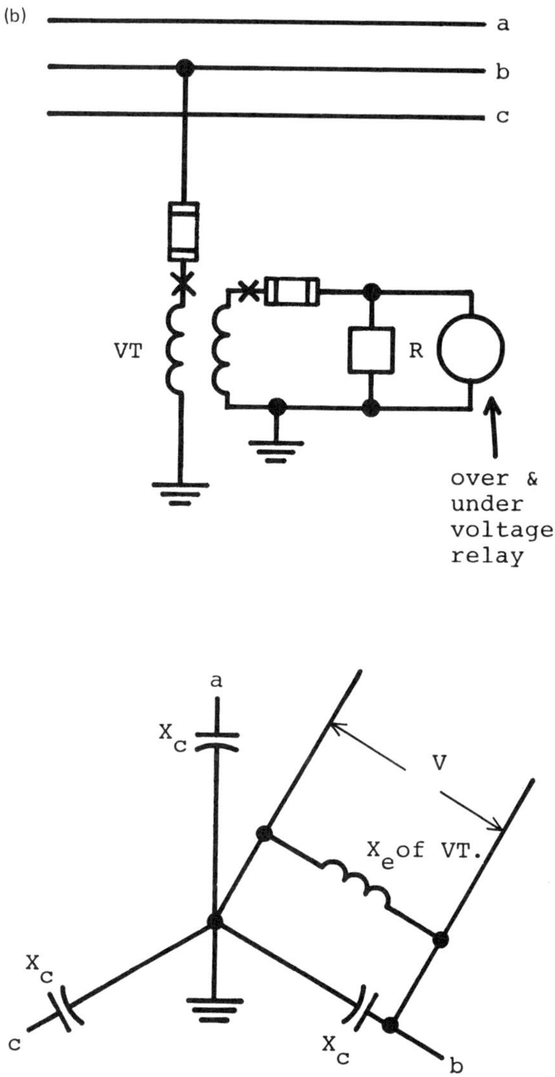

Fig. 7.5 (continued)

maximum solid ground relay voltage would be 3 × 69.3 = 208 V. Since the relay will be used to alarm, its continuous voltage rating should be greater than or equal to this 208-V value. Otherwise, an auxiliary step-down voltage transformer must be used.

Fig. 7.5a is simplified. Usually, the voltage transformer will be wye grounded-wye grounded and an auxiliary wye grounded-broken delta transformer used. In some cases the main voltage transformer will have a double secondary, one of which can be connected broken delta. Lamps can be connected across each broken-delta secondary winding to provide visual indications. Typical resistance values across the secondary derived from experience are shown in Table 7.1.

The ground detection scheme of Fig. 7.5b must be used with caution to avoid "neutral" inversion and ferroresonance, as outlined above. The voltage relay is set to have its contacts held open for the normal line-to-ground secondary voltage. When a ground fault occurs on phase b, the voltage collapses and the voltage relay resets to close the undervoltage contacts. If a phase a or c ground fault occurs, the relay voltage increases by about 1.73 to cause the relay to operate on overvoltage. Either under- or overvoltage operation usually operates an alarm to alert the operators of a ground fault so that they can arrange an orderly or convenient shutdown.

7.5 HIGH-IMPEDANCE GROUNDING SYSTEMS

There are two types of high-impedance grounding systems: resonant grounding and high-resistance grounding. The first has very limited use in the United States, although it is used elsewhere. The major American use is for generator grounding in the New England area. High-resistance grounding is widely used for generators and in industrial plants. These applications are reviewed later.

Table 7.1

Nominal system voltage (kv)	VT ratio	Resistor R	
		Ohms	Watts at 208 V
2.4	2400:120	250	175
4.16	4200:120	125	350
7.2	7200:120	85	510
13.8	14,400:120	85	510

7.5.1 Resonant Grounding

These are also known as ground-fault neutralizer or "Petersen coil" systems. The total system capacitance to ground is canceled by an equal inductance connected in the neutral (Fig. 7.6). If the neutral reactor is tuned exactly to the total system capacitance, the current in the fault is zero. Actually, in practice, taps on the reactor permit close tuning, so that the fault current is very small and the fault arc will not maintain itself. The circuit is a parallel resonance circuit, and the very low fault currents cause minimum fault damage. As the fault arc is extinguished, the reactor capacitance continues to produce essentially a voltage equal to the system line-to-neutral voltage, so that the voltage across the arc is small and the potential for restrike is negligible.

When this system is used on distribution systems, it becomes difficult always to provide a good match with system changes and switching. Further, these systems are seldom transposed and may have many single-phase taps, all of which result in small zero-sequence voltages generated by the normal load currents. These voltages act in a series resonance circuit with the grounding reactor and shunt capacitance and may produce a high voltage across the reactor and capacitance. The system must have full line-to-line insulation. Experience has shown that a larger number of line-to-line faults may occur, and there can be a higher incidence of simultaneous faults. When used, a very sensitive overcurrent relay provides an alarm and after some 10 to 20 s, if the fault still exists, the reactor is shorted out. This provides a high ground-fault current to operate other relays to isolate the fault. As indicated, it is not used in the United States, although there may be one or two older applications not yet converted to other grounding methods.

The application of this technique to grounding unit generators is more favorable because of the short distances involved and the fixed value of system capacitance involved. With the very low fault current, it is promoted as permitting the continued operation of the generator with a ground fault until an orderly convenient shutdown can be arranged. After a number of years of experience on a number of unit generators, several faults external to the generators have been detected successfully, but fortunately, no internal fault experience within the generators has occurred.

7.5.2 High-Resistance Grounding

In this system the power system is grounded through a resistor, and the accepted practice is to use a value of resistance equal or slightly less than the total system capacitance to ground. This provides a low fault current to minimize damage, yet limits the potential transient overvoltages to less than 2.5 times the normal crest value to ground.

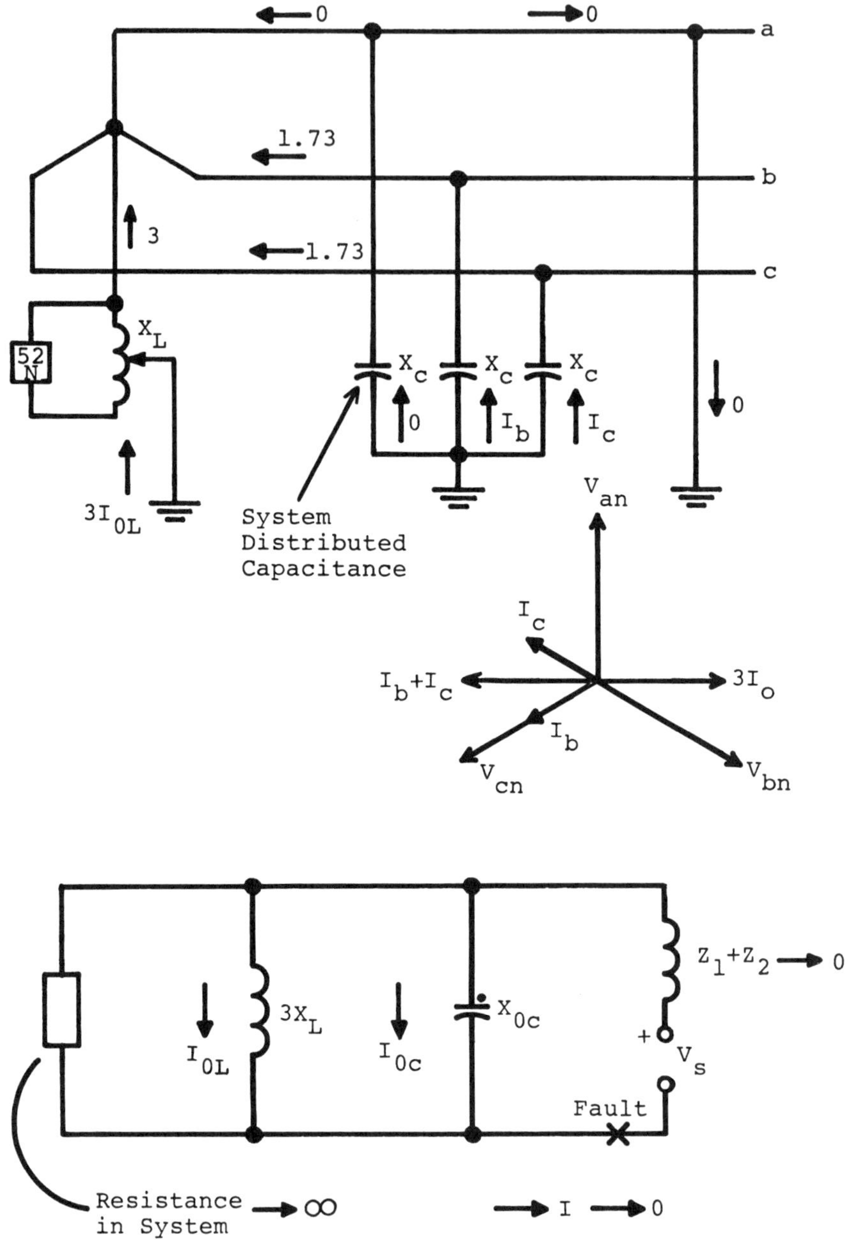

Fig. 7.6 Resonant grounding.

The fault current range normally encountered with this method is between 1 and 25 A primary, usually between 1 and 10 A.

The grounding resistor may be connected in the neutral of a generator or power transformer (Fig. 7.7), or across the broken delta of line-to-ground-connected distribution transformers (Fig. 7.8). With the resistor in the neutral as in Fig. 7.7, a solid ground fault can produce a maximum V_0 equivalent to the phase-to-neutral voltage as illustrated in Fig. 7.1. Thus a line-to-neutral-rated distribution transformer is normally used, although line-to-line ratings have also been used. For the grounding system of Fig. 7.8, a solid ground fault can raise the voltage on two of the distribution transformers to line-to-line equivalent, as shown in Fig. 7.1. Thus line-to-line ratings are suggested for this application, especially if the protection system is used for alarm rather than direct trip.

The neutral connection (Fig. 7.7) is used for unit generator applications and in industrial systems with a single power transformer supply. For multiple generators connected to a common bus or for systems with several power sources, the grounding system of Fig. 7.8 may be preferable. These two methods are best documented by typical examples.

7.5.3 Example: Typical High-Resistance Neutral Grounding

This type of grounding is applied to a 160-MVA 18-kV unit generator, as shown in Fig. 7.9. The area of ground protection is the generator to the low-voltage winding of the power transformer and to the high-voltage winding of the unit auxiliary transformer. In this area the following capacitances to ground (microfarads per phase) that must be considered:

Generator windings	0.24
Generator surge capacitor	0.25
Generator to transformer leads	0.004
Power transformer low-voltage winding	0.03
Station service transformer high-voltage winding	0.004
Voltage transformer windings	0.0005
Total capacitance to ground	0.5285

$$X_C = -j\frac{10^6}{2\pi fC} = -j\frac{10^6}{2(3.1416)(60)(0.5285)} = 5019.08\ \Omega/\text{phase} \tag{7.11}$$

This capacitive reactance, in per unit on a 100-MVA 18-kV base, is from

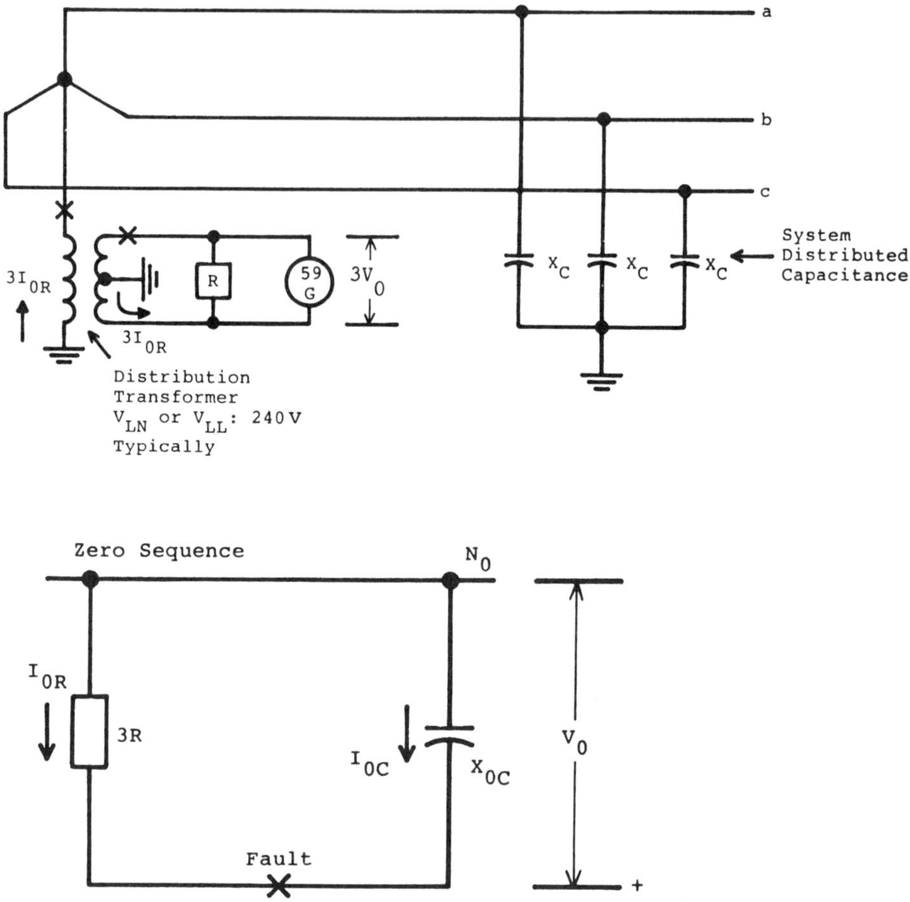

Fig. 7.7 High-resistance grounding with resistor in the neutral.

Eq. (2.15),

$$\frac{100(5019)}{18^2} = 1549.1 \text{ pu} \tag{7.12}$$

or on the generator MVA base, $160(5019)/18^2 = 2478.56$ pu. Selecting the grounding resistor to be equal to the capactive reactance and using the convenient 100-MVA base, 3R in the zero-sequence network would be 1549.1 pu. For a solid fault in this area,

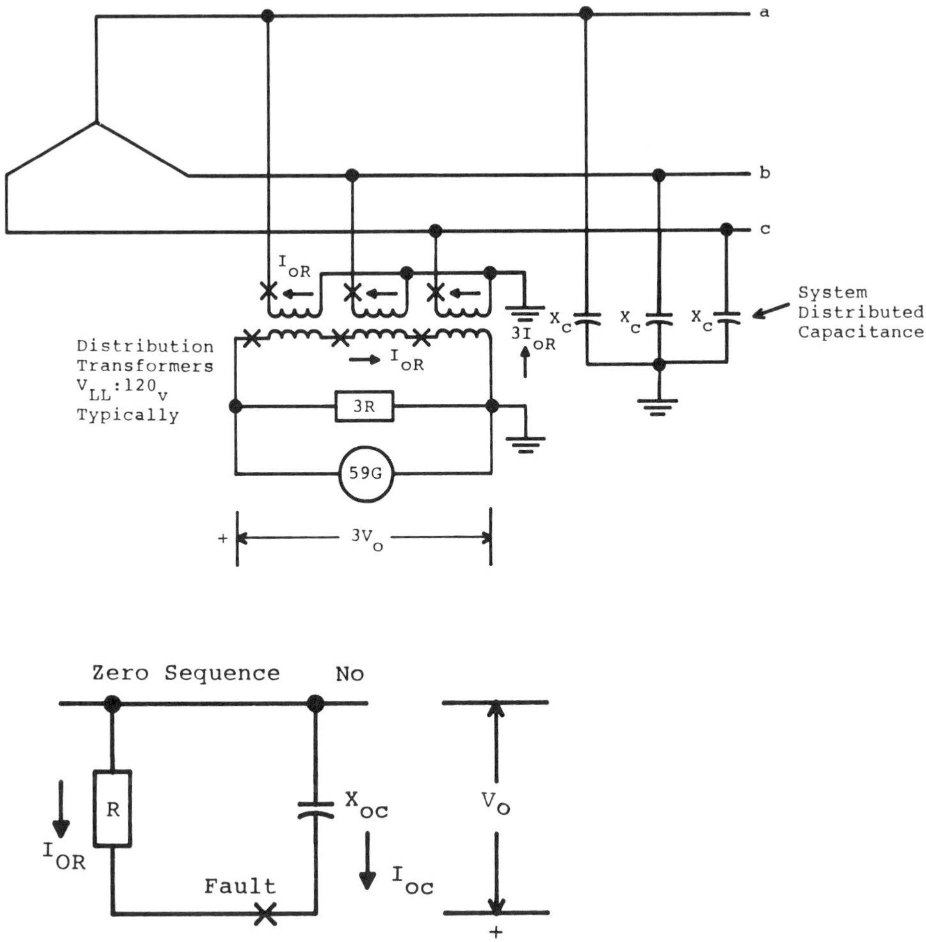

Fig. 7.8 High-resistance grounding with resistor across distribution transformer secondaries.

$$Z_0 = \frac{1549.1\ (1549.1\ \angle -90^\circ)}{1549.1 - j1549.1} = 1095.38\ \angle -45^\circ \text{ pu}$$

In contrast, the positive- and negative-sequence reactance for this system is j0.066 pu and so quite negligible. From Eq. (7.1) and (7.2),

$$I_1 = I_2 = I_0 = \frac{1.0}{1095.38\ \angle -45^\circ} = 0.00091\ \angle 45^\circ \text{ pu} \qquad (7.13)$$

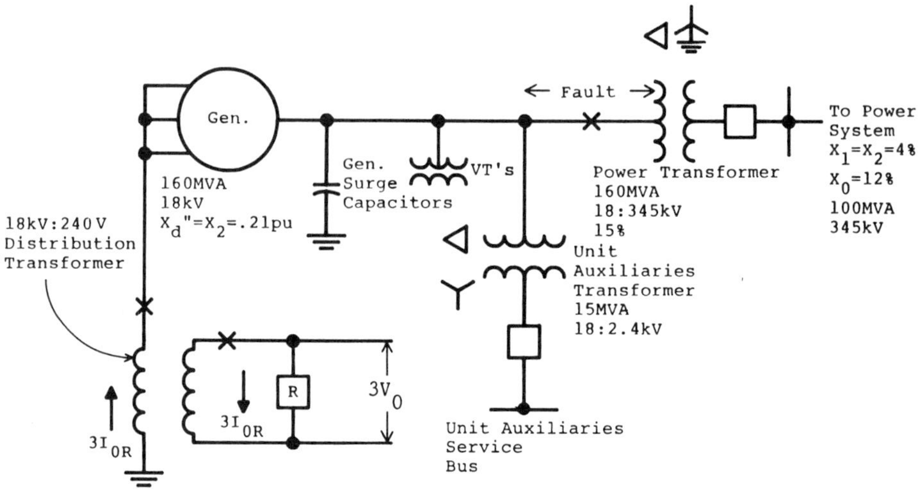

(a)

Fig. 7.9 Typical example of high-resistance grounding with neutral resistor: (a) unit generator system; (b) sequence networks, values in per unit at 100 MVA, 18 kV; (c) fault current distribution.

$$1.0 \text{ pu } I = \frac{100{,}000}{\sqrt{3} \times 18} = 3207.5 \text{ A at 18 kV} \tag{7.14}$$

so the fault currents are

$$I_1 = I_2 = I_0 = 0.00091\,(3207.5) = 2.92 \text{ A at 18 kV} \tag{7.15}$$

$$I_a = 3I_0 = 3(2.92) = 8.76 \text{ A at 18 kV} \tag{7.16}$$

The distribution of these fault currents is shown in Fig. 7.9c. The resistor selected with its primary resistance 3R equal to X_C provides a value of 5019.08/3 = 1673.03 Ω at 18 kV. The actual resistor value connected to the secondary of the distribution transformer will be

$$R = 1673.03\left(\frac{240}{18{,}000}\right)^2 = 0.2974\ \Omega \tag{7.17}$$

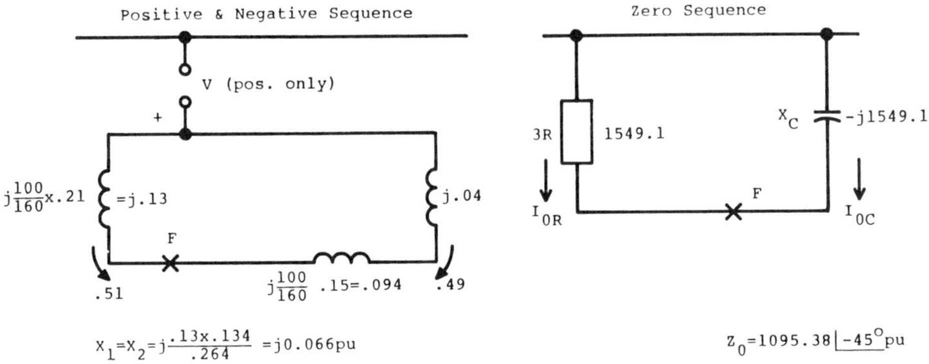

(b)

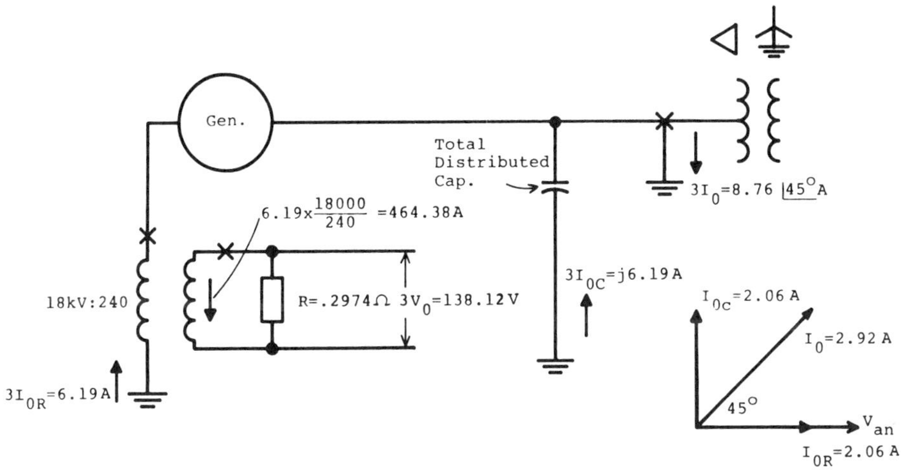

(c)

Fig. 7.9 (continued)

With a secondary current of 6.19(18,000/240) = 464.38 A in the distribution transformer secondary, the V_0 available for a primary line-to-ground fault will be

$$3V_0 = (464.38)(0.2974) = 138.12\ \text{V} \tag{7.18}$$

The watts in the resistor during the fault is

$$\frac{(464.38)^2(0.2974)}{1000} = 64.14 \text{ kW} \tag{7.19}$$

Similarly, the distribution transformer kVA is

$$6.19\left(\frac{18}{\sqrt{3}}\right) = 64.33 \text{ kVA}$$

equal within decimal-point accuracy. When this grounding is used for generator units, tripping the unit is recommended, so these ratings can be short-time rather than continuous ratings.

The normal charging current for this system would be

$$I_C = \frac{18{,}000}{\sqrt{3} \times 5019} = 2.07 \text{ A/phase at 18 kV} \tag{7.20}$$

The use of a distribution transformer and a secondary resistor rather than a resistor directly connected in the neutral is an economic consideration. With high-resistance grounding it is generally less expensive to use the resistor in the secondary, as shown.

The flow of current through the system for a ground fault is sometimes hard to visualize from the zero-sequence quantities, such as shown in Fig. 7.9. Although positive- and negative-sequence impedances are quite negligible in high-resistance grounded system, the three sequence currents are equal at the fault [Eq. (7.13)] and flow through the system. Since there is a positive-sequence source at either end of Fig. 7.9, the positive- and negative-sequence currents divide as shown by the 0.51 and 0.49 distribution factors in the positive- and negative-sequence networks. The approximate currents flowing through this system are documented in Fig. 7.10. They are approximate since the capacitance normally distributed is shown lumped. Before the fault a charging current of 2.07 A [Eq. (7.20)] flows symmetrically in the three phases. Since this is of the same order of magnitude as the fault currents, Thevenin's theorem and superposition must be used to determine the currents flowing during the ground fault. Thus, in phase a from the generator to the lumped capacitance, I_{a1} is the sum of the prefault charging current plus the fault component, or $2.07\,\underline{/90^\circ} + 0.51 \times 2.92\,\underline{/45^\circ}$ which is $I_{a1} = 3.29\,\underline{/71.4^\circ}$. Similarly, $I_{b1} = 2.07\,\underline{/-30^\circ} + 0.51 \times 2.92\,\underline{/-75^\circ} = 3.29\,\underline{/-48.7^\circ}$ and $I_{c1} = 2.07\,\underline{/210^\circ} + 0.51 \times 2.92\,\underline{/165^\circ} = 3.29\,\underline{/191.41^\circ}$. The negative- and zero-sequence components are as normally determined by the fault. In the lumped shunt capacitance, the charging current of 2.07 A cancels the zero-sequence phase a component of 2.07 A to give zero current, since this branch is essentially shorted

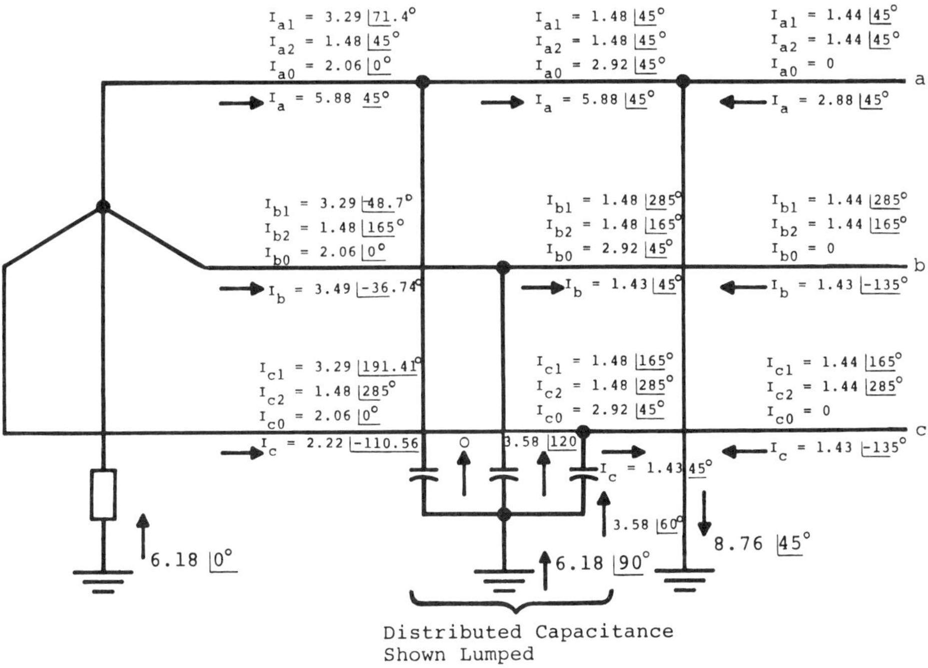

Fig. 7.10 Three-phase and sequence current distribution for the system of Fig. 7.9 during a solid phase-a-to-ground fault.

out by the solid phase-a-to-ground fault. In the unfaulted phases the charging currents add to the zero- sequence component, providing currents as shown in Fig. 7.10. Fig. 7.10 is similar and consistent with Fig. 7.2. The source voltage for Fig. 7.2 is 1 $\angle 90°$, while for Fig. 7.10 it is 1 $\angle 0°$.

Again this assumes that none of the distributed capacitance is in the area to the right of the fault location. As has been indicated, the total fault value would not change for different fault locations; similarly, the distribution will not change essentially, as no series impedance is considered between the generator and the power transformer.

7.5.4 Example: Typical High-Resistance Grounding with Three Distribution Transformers

An industrial plant 13.8-kV system is illustrated in Fig. 7.11. The main source is the utility, but the plant has a small local generator. Either the supply transformer or the generator could be grounded with

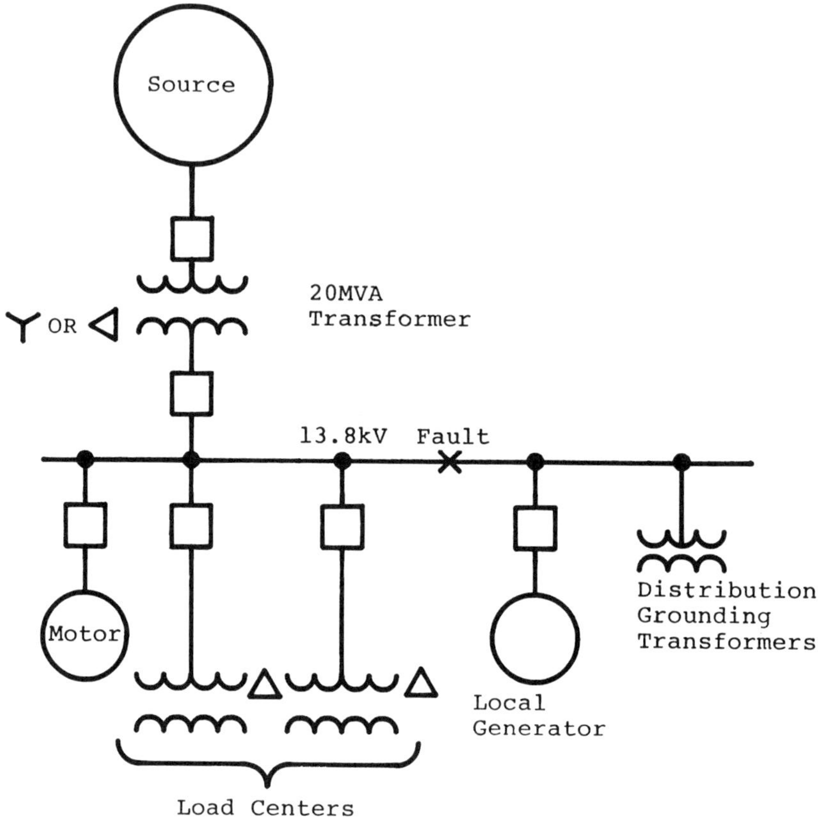

(a)

Fig. 7.11 Typical example of high-resistance grounding with three distribution transformers: (a) industrial system; (b) zero sequence network, values in per unit on 20 MVA, 13.8 kV; (c) ampere currents flowing for the ground fault.

a resistor in the neutral (if the supply transformer is wye secondary), but it is possible that either the local generator or the utility supply may be out of service. Thus this system is to be grounded using the method shown in Fig. 7.8.

From estimating data or specific test, we have the following capacitances to ground (microfarads per phase):

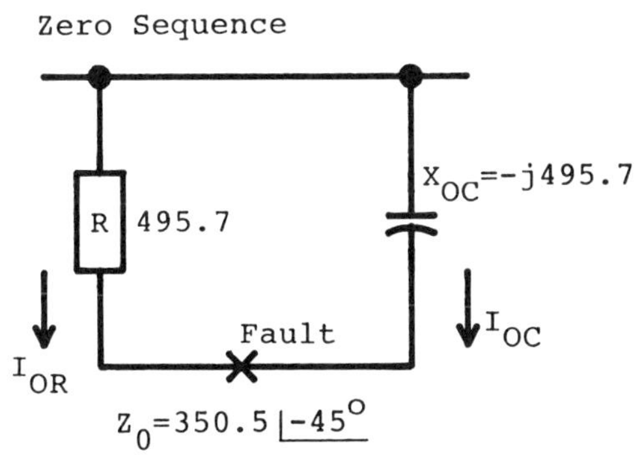

(b)

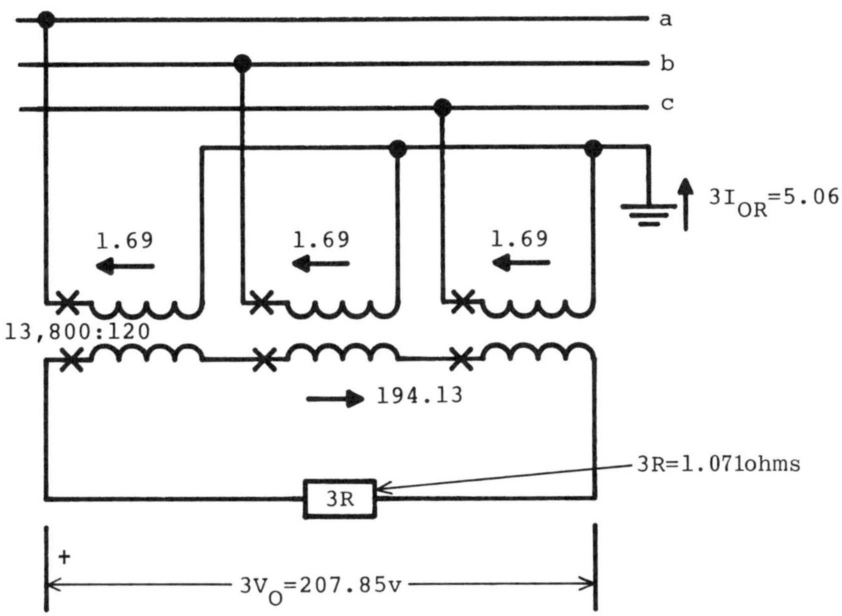

(c)

Fig. 7.11 (continued)

Source transformer	0.004
Local generator	0.11
Motor	0.06
Power center transformers	0.008
Total connecting cables	0.13
Surge capacitor	0.25
Total capacitance to ground	0.562

$$X_C = -j\frac{10^6}{2\pi fC} = -j\frac{10^6}{2(3.1416)(60)(0.0562)} = 4719.9\ \Omega/\text{phase} \quad (7.21)$$

Thus the charging current of this 13.8 kV system is

$$I_C = \frac{13{,}800}{\sqrt{3} \times 4719.9} = 1.69\ \text{A/phase at } 13.8\ \text{kV} \quad (7.22)$$

The total capacitance in per unit on a 20 MVA 13.8 kV base is from Eq. (2.15)

$$X_C = \frac{20\ (4719.9)}{13.8^2} = 495.68\ \text{pu} \quad 7.23)$$

For high-resistance grounding per Fig. 7.8, $R = X_{0C}$, so R in the zero-sequence network is 495.68 pu. For the system,

$$Z_0 = \frac{(495.7)(-j495.7)}{495.7 - j495.7} = 350.5\ \angle{-45^\circ}\ \text{pu} \quad (7.24)$$

For a line-to-ground fault the positive- and negative-sequence values of the system are very small and can be ignored. Thus for a line-to-ground fault on this 13.8 kV system,

$$I_1 = I_2 = I_0 = \frac{1.0}{350.5\ \angle{-45^\circ}} = 0.00285\ \angle{45^\circ}\ \text{pu} \quad (7.25)$$

The base per unit current is

$$I_{base} = \frac{20{,}000}{\sqrt{3} \times 13.8} = 836.74\ \text{A at } 13.8\ \text{kV} \quad (7.26)$$

Thus

$$I_1 = I_2 = I_0 = 0.00285(836.74) = 2.39 \text{ A at } 13.8 \text{ kV} \tag{7.27}$$

$$I_a = 3I_0 = 0.00856 \text{ pu} = 7.16 \text{ A at } 13.8 \text{ kV} \tag{7.28}$$

$$I_{0R} = 0.00285 \cos 45° = 0.00202 \text{ pu} = 1.69 \text{ A at } 13.8 \text{ kV} \tag{7.29}$$

The three distribution transformers have the ratio 13.8 kV:120 = 115. Thus the secondary current for the ground fault is

$$I_{0R(\text{Sec})} = 1.69(115) = 194.13 \text{ A} \tag{7.30}$$

The resistor was sized at 495.68 pu or 4719.9 Ω at 13.8 kV. Reflected to the secondary, the resistor value becomes

$$3R = 3 \times \frac{4719.9}{115^2} = 1.071 \ \Omega \text{ secondary} \tag{7.31}$$

This is the resistance value for installation. Alternatively, this value can be calculated directly from the system values:

$$3R = \frac{(\sqrt{3}\, V_{\text{sec}})^2}{2\pi f C (V_{\text{pri LL}})^2} = \frac{(\sqrt{3} \times 120)^2}{377\ (0.562)(13.8)^2}$$

$$= 1.071 \ \Omega \text{ secondary} \tag{7.32}$$

$$3V_0 = 194.13(1.071) = 207.85 \text{ V secondary} \tag{7.33}$$

The continuous resistor and transformer ratings are:

$$\text{Resistor:} \quad I^2(3R) = \frac{(194.13)^2(1.071)}{1000} = 40.36 \text{ kW} \tag{7.34}$$

$$\text{Transformer:} \quad VI = \frac{1.69(13{,}800)}{1000} = 23.3 \text{ kVA} \tag{7.35}$$

The line-to-line voltage was used as during a fault; this voltage appears essentially across the primary winding. If relays are used to trip, short-time ratings may be used for the resistor and transformers.

7.6 SYSTEM GROUNDING FOR MINE OR OTHER HAZARDOUS TYPE APPLICATIONS

A medium high resistance grounding system was developed originally for underground coal mining systems and is being used more and more for all mining and hazardous type applications. In view of the great hazardous conditions often encountered in mining these systems emphasize personnel safety and limit the fault current to 25 to 50 primary amperes. A typical system is shown in Fig. 7.12. All portable equipment in the mine is fed by a separate feeder that is resistance grounded separately at a location at least 50 ft from the supply substation. Many applications limit the ground current to 25 primary amperes. This separate ground is carried as a fourth wire with its impedance not to exceed 4 Ω. A pilot conductor is also carried in the supply cable with a monitoring system to assure continuity of this ground. For a ground fault to frame at the portable equipment 25 A flowing over the 4-Ω safety ground wire would produce a maximum of 100 V across the operator. Very fast sensitive relay is used to detect a ground fault and trip off instantly this feeder without regard to fault location. With high hazard potentials, safety concerns override selectivity and continuity of service. All essential loads, such as lighting, fans, and vital support services, are fed from normal feeders from the substation.

7.7 LOW-IMPEDANCE GROUNDING

This type of grounding limits line-to-ground fault currents to approximately 50 to 600 A primary. It is used to limit the fault current, yet permit selective protective relaying by magnitude differences in fault current by the power system impedances. There are also cost advantages since line-to-neutral equipment insulation can be utilized, as the unfaulted phase voltages are not increased significantly by the ground faults.

Most typically, this type of grounding is accomplished by a reactor or resistor in the system neutral (Fig. 7.13). In a distribution station it would be in the neutral of the delta-wye supply transformer. Several generator units that are connected to a common bus may be grounded in this manner. The zero-sequence network for this is also shown in Fig. 7.13.

Where a delta-connected transformer exists or a system neutral is not available, this type of grounding can be accomplished either by a shunt connection of a wye-grounded-delta or by a zigzag transformer. The wye-grounded-delta transformer could be applied only for grounding purposes and not for transmitting power. The grounding would be as indicated above and shown in Fig. 7.13.

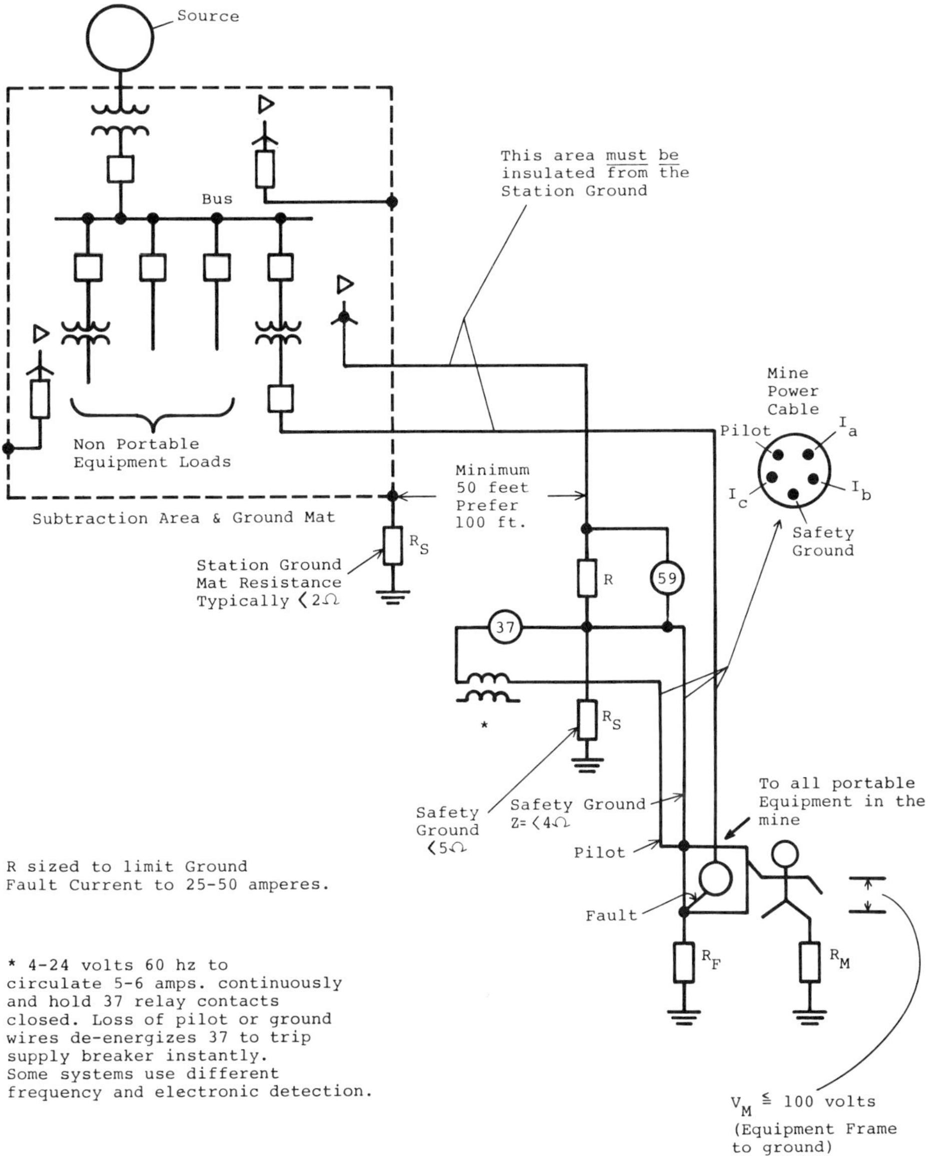

Fig. 7.12 Typical system grounding for mining or other hazardous applications.

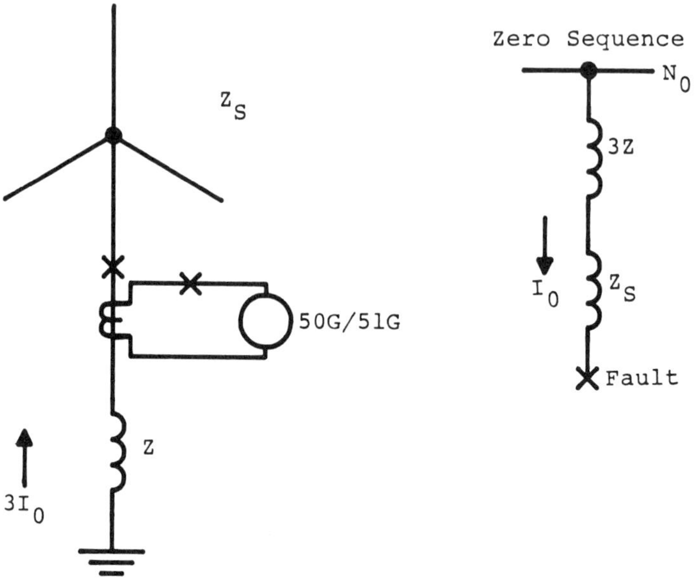

Fig. 7.13 Low-impedance grounding with impedance in the system neutral.

The zigzag transformer is illustrated in Fig. 7.14. Basically, this unit consists of three 1:1 ratio transformers interconnected to pass zero-sequence current only. With the transformer polarities shown and since $I_{ao} = I_{bo} = I_{co} = I_o$, zero-sequence current can flow, but positive- and negative-sequence currents cannot since $I_{a1} \neq I_{c1} \neq I_{b1}$ and $I_{a2} \neq I_{b2} \neq I_{c2}$. The impedance of the zero-sequence path is the leakage impedance of the transformer, X_T. With a line-to-neutral voltage of 1 pu, the voltage across each winding is 0.866 pu. The zigzag transformer grounding is reactance, as the transformer resistance is very low. If the zigzag X_T is too low for the desired fault limiting, a resistor (or reactor) can be added as shown.

7.7.1 Example: Typical Low-Resistance Neutral Reactor Grounding

The grounding reactor in the typical system of Fig. 7.15 is to be applied to limit the maximum line-to-ground fault current to 400 A primary. For convenience a 20-MVA base is used for the calculations. From Eq. (A4.1-4):

$$\text{Source:}\quad X_1 = X_2 = \frac{\text{MVA}_{\text{Base}}}{\text{MVA}_{\text{SC}}} = \frac{20}{3200} = j0.00625\ \text{pu} \qquad (7.36)$$

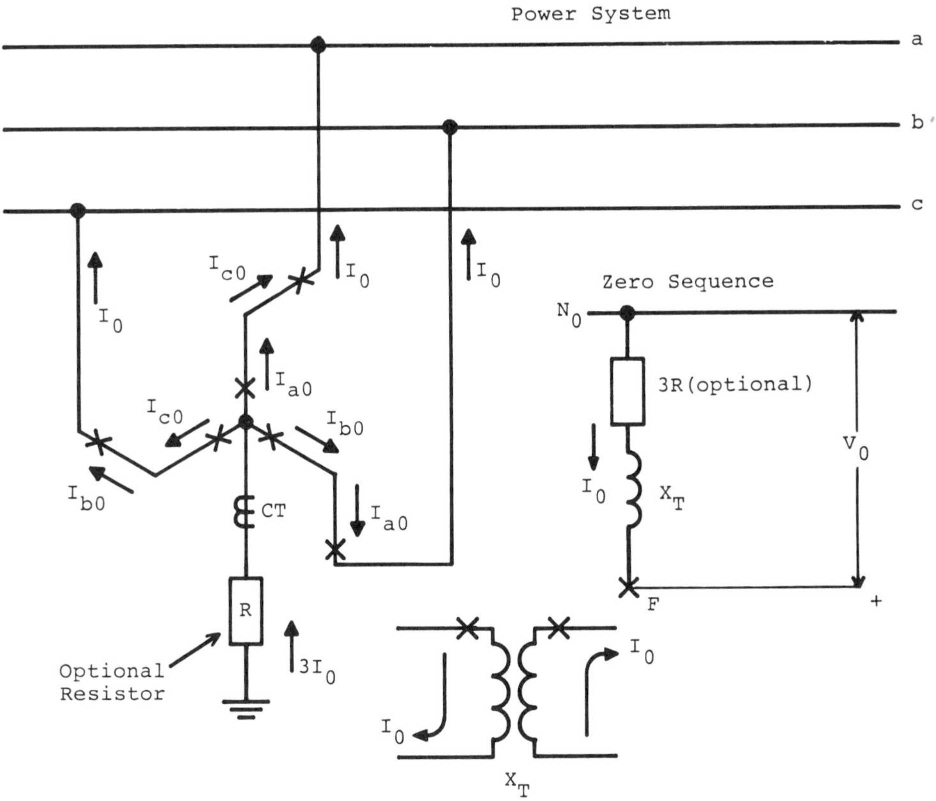

Fig. 7.14 Low-impedance system grounding with a zigzag grounding transformer.

$$\text{Transformer: } X_T = j0.052 \text{ pu} \tag{7.37}$$

$$\text{Total } X_1 = X_2 = j(0.0063 + 0.052) = j\,0.0583 \text{ pu} \tag{7.38}$$

$$\text{Total } X_0 = j(0.052 + 3X) \text{ pu} \tag{7.39}$$

For a 400-A primary fault:

$$I_1 = I_2 = I_0 = \frac{400}{3} = 133.33 \text{ A at } 13.8 \text{ kV} \tag{7.40}$$

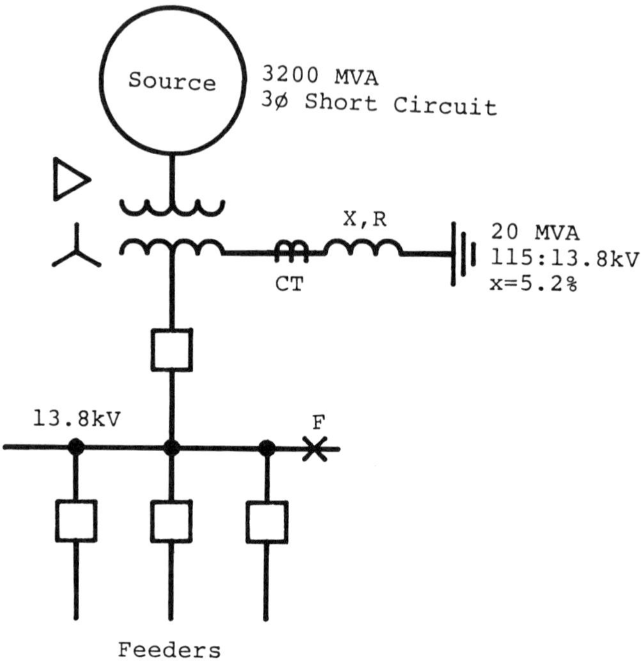

Fig. 7.15 Typical example of low-impedance grounding.

$$I_{Base} = \frac{20,000}{\sqrt{3} \times 13.8} = 836.74 \text{ A} \tag{7.41}$$

$$I_1 = I_2 = I_0 = \frac{133.33}{836.74} = 0.159 \text{ pu} \tag{7.42}$$

$$X_1 + X_2 + X_0 = j(0.1685 + 3X) \tag{7.43}$$

$$0.159 = \frac{j1.0}{j(0.1685 + 3X)} \tag{7.44}$$

$$X = 2.036 \text{ pu}$$

$$= \frac{13.8^2(2.036)}{20} = 19.38 \ \Omega \text{ at } 13.8 \text{ kV} \quad \text{(Eq. 2.17)} \tag{7.45}$$

7.7.2 Example: Typical Low-Resistance Neutral Resistance Grounding

The grounding resistor is to be applied in the typical system of Fig. 7.15 to limit the maximum line-to-ground fault current to 400 A primary. As for the previous example, $X_1 = X_2 = j0.0583$, but with a resistor

$$Z_0 = 3R + j0.052 \text{ pu}$$

Also as before, 400 A for a line-to-ground fault represents an $I_1 = I_2 = I_0$ of 0.159 pu from Eq. (7.42).
Thus

$$I_1 = I_2 = I_0 = 0.159 \angle ?° = \frac{j1.0}{3R + j0.1685} \tag{7.46}$$

This is best solved by the sum of the squares, so rearranging yields

$$3R + j0.1685 = \frac{j1.0}{0.159\angle ?°}$$

$$(3R)^2 + (0.1685)^2 = (6.29)^2$$

$$R = 2.09 \text{ pu} = \frac{13.8^2(2.09)}{20} = 19.91\ \Omega \text{ at } 13.8 \text{ kV} \tag{7.47}$$

Comparing the values of 19.38 [Eq. (7.45)] for a reactor, and 19.91 [Eq. (7.47)] for a resistor shows that in many cases with resistor grounding the angle can be ignored and the resistor added arithmetically rather than vectorally. This simplifies the calculations, generally with little error. Thus from a practical standpoint, where the resistor value is large compared to the system reactance or impedance, its value can be calculated directly as,

$$R = \frac{V_{LN}}{I} = \frac{13,800}{\sqrt{3} \times 400} = 19.92\ \Omega \tag{7.48}$$

7.8 SOLID (EFFECTIVE) GROUNDING

Effective grounding is defined by ANSI/IEEE standards (IEEE 100-1984) when the power system constants are

$$\frac{X_0}{X_1} \leqq 3.0 \quad \text{and} \quad \frac{R_0}{X_1} \leqq 1.0$$

where X_0 and R_0 are the zero-sequence reactance and resistance and X_1 the positive-sequence reactance of the power system. From a practical standpoint, this means that there can be no impedance between the system neutral and ground; hence these systems are solidly grounded. Usually, this is accomplished by connecting the neutral of the wye windings of the power transformer(s) to the station ground mat and ground. In the diagrams this would be shown as in Figs. 7.13 or 7.15, with X, R omitted. As a result the ground-fault currents can vary considerably, from very small currents to currents greater than the three-phase fault value. The magnitudes depend on the power system configuration and constants, location of the fault, and the fault resistance which may or may not be significant. Since the current level can vary with the fault location, it becomes easier to locate the fault and selectively isolate the trouble area by protective relays. The various techniques utilized are covered in the various chapters on equipment protection. The CTs shown in the grounded neutral in Figs. 7.13, 7.14 and 7.15 are used to operate time-overcurrent relays set sensitively and with time to coordinate with the various line, feeder, and so on, relays that they "overreach." Hence this serves as backup — "last resort" protection for ground faults around the area that are not properly cleared by their primary and associated backup protection.

7.8.1 Example: Solid Grounding

Assume that the transformer bank of the system of Fig. 7.15 is solidly grounded (X, R = 0). For a fault at F, $X_1 = X_2 = j0.0583$ pu [Eq. (7.38)], and $X_0 = j0.052$ pu, all on a 20 MVA 13.8 kV base. Thus $X_1 + X_2 + X_0 = j0.1685$ pu and

$$I_1 = I_2 = I_0 = \frac{j1.0}{j0.1685} = 5.934 \text{ pu}$$

$$= 4965.8 \text{ A at } 13.8 \text{ kV} \tag{7.49}$$

$$I_a = 3I_0 = 17.8 \text{ pu}$$

$$= 14{,}897.5 \text{ A fault current at } 13.8 \text{ kV} \tag{7.50}$$

This is over 37 times larger than the 400-A low-impedance grounded example.

For a three-phase fault at F:

$$I_1 = \frac{j1.0}{j.0583} = 17.17 \text{ pu}$$

$$= 14{,}364.6 \text{ A at } 13.8 \text{ kV}$$

Thus the ground-fault current is larger than the current for a three-phase fault. The difference in this example is small because the source is quite large compared to the supply transformer. If the source impedance were larger, the two fault currents would be lower, but the ground fault would be a larger percentage of the three-phase fault.

7.9 FERRORESONANCE IN THREE-PHASE POWER SYSTEMS

At present, ferroresonance appears to be occurring more frequently in power systems, especially in distribution systems. Thus a review is in order since it relates to power system grounding. This is a complex nonlinear phenomenon occasioned by the system capacitance resonating with the nonlinear magnetizing (exciting) reactance of connected transformer(s). It is characterized by sustained but variable overvoltages with very irregular wave shapes and appreciable harmonics. The magnitudes are sufficient to damage the equipment connected.

A detailed discussion of this phenomenon is beyond our scope, so the intent is to present a brief overview of typical possibilities that can lead to potential hazards. The possibility of ferroresonance with voltage transformers in ungrounded power systems has been outlined in Section 7.4.

In source-grounded distribution systems it is general practice to connect the load transformers with their primaries ungrounded. A typical system is shown in Fig. 7.16a. Often, the distribution circuit is overhead with fuse cutout or single-phase disconnect switches on the pole. From this point cable is run underground to a pad-mounted transformer near the utilization point. Several possiblities exist in these circuits for ferroresonance at light load or no load on the secondary.

In Fig. 7.16b, energizing phase a before phases b and c provides a path for current flow, as indicated by the arrows. The nonlinear exciting reactance of the transformers is in series with the system capacitance to ground. With the common delta primary, transformers ab and ac are energized at 0.577 rated voltage, so that exciting current or, if energized at or near the zero point on the voltage wave, magnetizing inrush currents flow. With residual flux in the transformer core, this

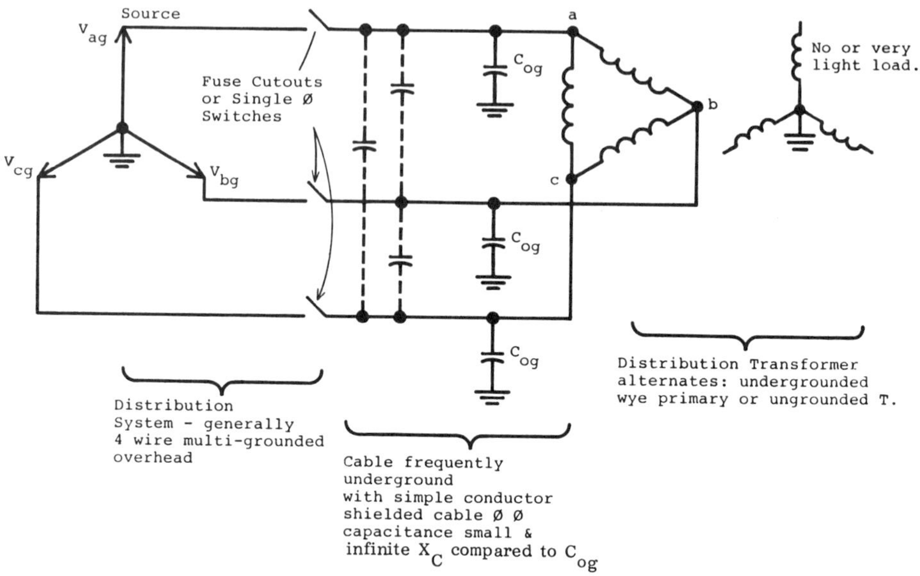

(a)

Fig. 7.16 Ferroresonance possibilities with ungrounded transformers connected to grounded power system: (a) typical distribution system; (b) ferroresonance current paths when one phase (phase a) is energized before the other phases or if two phases are opened by fuse, breaker, or switch operation; (c) ferroresonance current paths when one phase (phase a) opens by a blown fuse, broken conductor, defective breaker or switch, etc.

can result in a large current. At the end of the half wave the transformer core drops out of saturation but a trapped charge or voltage is left on the cable capacitance. During the next half cycle the polarity of the source voltage and trapped charge of the cable capacitance are the same and can force the core into saturation in the opposite direction.

As the core goes into and out of saturation in either a random or a periodic manner, high overvoltages appear between the phases and phase to ground. These can be in the order of 5 pu or higher. When a second phase is energized, the overvoltages can continue and may become larger. Energizing the third phase eliminates the single-phase condition and usually the ferroresonance.

Similarly, the capacitance-to-ground and exciting reactances have the possibility of being in resonance for a broken fuse, broken conductor, and so on, where one phase is open as illustrated in Fig. 7.16c.

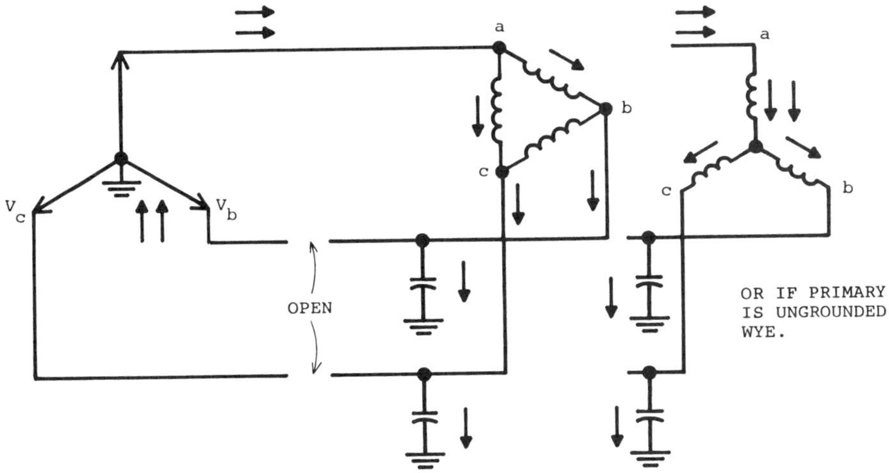

(b)

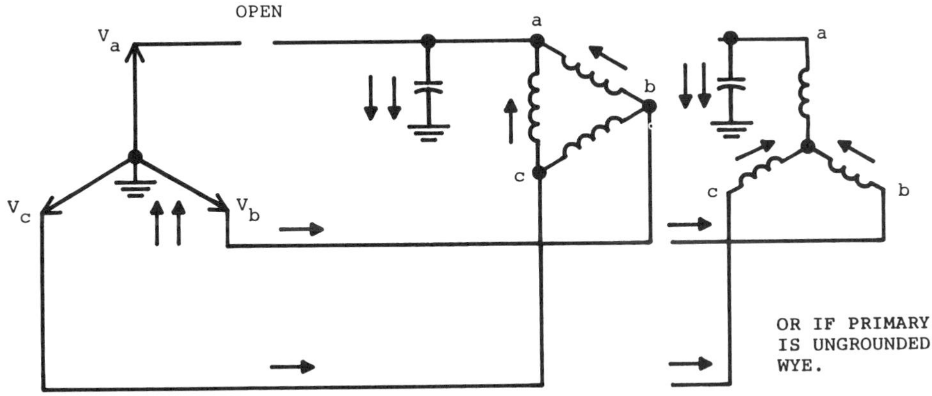

(c)

Fig. 7.16 (continued)

To limit the voltage on the open phase to around 1.25 pu or less,

$$\frac{X_C}{X_e} \geqq K \tag{7.51}$$

where X_C is the equivalent capacitive reactance per phase and X_e is the equivalent exciting reactance. K has been indicated to be 40, although some feel it should be in the range 5 to 10. In terms of the system, Eq. (7.51) can be expressed as

$$\frac{X_C \text{ kVA}}{10^5 \text{ kV}^2}(I_e) \geqq K \tag{7.52}$$

From this it can be observed that, in general, ferroresonance is more likely to occur

1. With smaller size transformers. Ferroresonance can occur in small transformers in 25-kV and 35-kV systems, resulting from the internal capacitance of the transformer and the transformer exciting reactance.
2. At higher voltages, around 15 kV and above.
3. With long cable runs or circuits with high capacitance.

For single conductor cables with shields, typical capacitance values are about 0.25 to 0.75 (average 0.5) μF/mi compared to open-wire line values of 0.01 μF/mi. Thus the capacitive reactance of underground cables is only around 2% of the capacitive reactance of an overhead circuit.

Whereas the probability of ferroresonance is low for systems operating at 15 kV or less, cases have been reported of ferroresonance in 15-kV systems with long cable runs. It is more probable in systems of 25 kV and 35 kV using cable. It is generally unlikely to occur if the primary windings are grounded-wye or grounded tee, especially if the transformers are three independent single-phase units or of three-phase triplex type, where there is no magnetic coupling or negligible coupling between the phase windings, and with single-conductor shielded cable, where the interphase capacitances are very large. However, ferroresonance can occur with grounded-wye transformers, as shown in Fig. 7.17, either through the interphase capacitance on a long line or three-conductor cable and/or through ungrounded shunt capacitor banks.

It can also occur when four- and five-legged core transformer banks (Fig. 7.18) are used. There is magnetic and capacitive coupling between the three phases. This magnetic coupling between the phases with the phase-to-ground capacitive reactance on the open phases provides a series-parallel circuit that can resonate in a nonlinear manner. The magnitude of the overvoltages reported for this type are less than 2.5 pu, in contrast with 5 pu and higher for the ungrounded transformers.

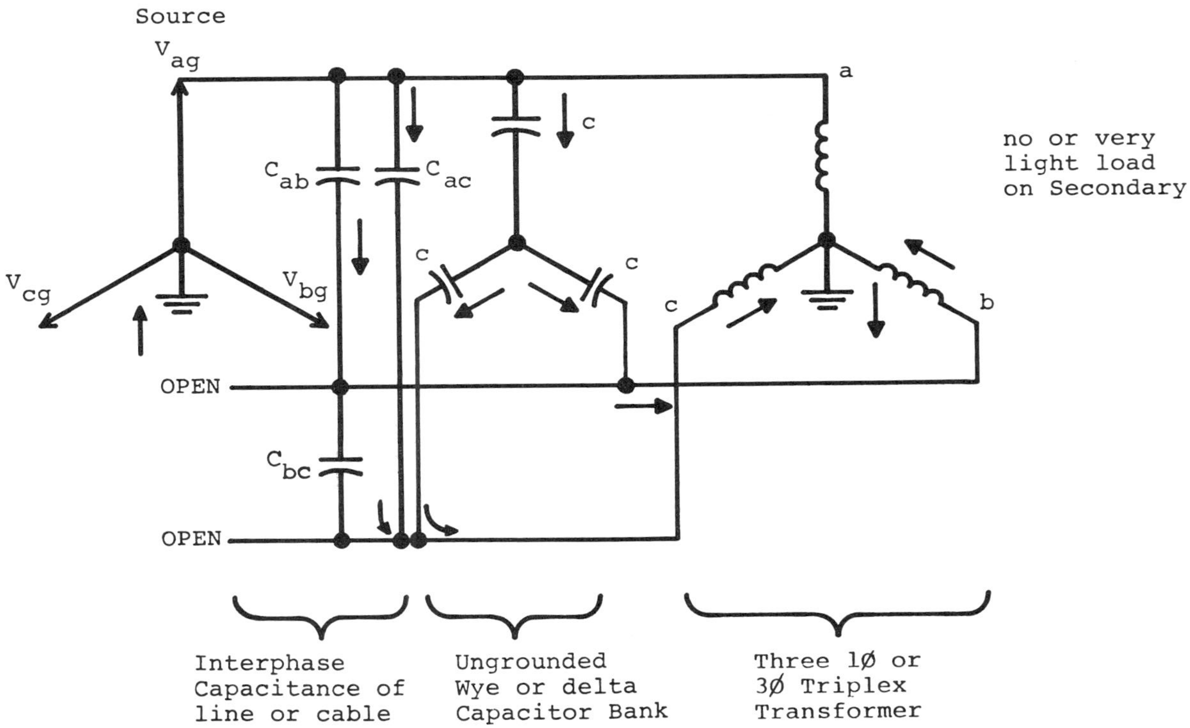

Fig. 7.17 Ferroresonance possibilities with grounded transformers resulting from interphase capacitance on long circuits and/or with ungrounded shunt capacitor banks.

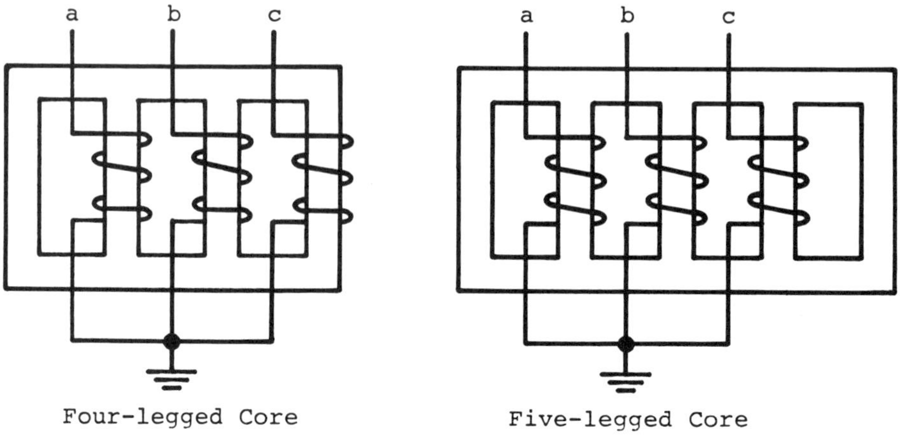

Fig. 7.18 Three-phase transformer banks with four- and five-legged cores provide magnetic and capacitive coupling paths between the three phases that can contribute to ferroresonance.

7.9.1 General Summary for Ferroresonance for Distribution Systems

This is far from an exact science, so the discussion above is very general. In practice, there are many circuits with so little capacitive reactance or such high capacitive reactances relative to the exciting reactances that ferroresonance and high voltages will not occur when one or two conductors are open. Also, resistive load on the secondary of the transformers, on the order of 5 to 15% of the transformer rating, should prevent ferroresonance.

Where ferroresonance may occur with ungrounded transformers that are supplied by cable, possible solutions are three-phase switching or switching closer or at the bank. Both of these can be expensive and/or impractical in many applications. Switching only with load can help, but again may not always be practical and does not prevent the blown-fuse or broken-conductor problems that could occur at light load. The application of transformers with grounded primaries and minimum interphase coupling generally should provide a solution.

7.9.2 Ferroresonance at High Voltages

Ferroresonance is not limited to distribution but can occur almost anywhere in the power system. As an example, sustained 60- and 20-Hz ferroresonance has been experienced in a parallel 13-km 500-kV line where one end of each line connects without a breaker to separate 750-MVA grounded wye-delta tertiary autotransformer banks. This randomly occurred in the line when it was deenergized by opening the

500-kV breaker at one end and the autotransformer secondary 230-kV breaker at the other end with the parallel line still energized. The ferroresonance circuit is the 500-kV line capacitive reactance to ground and the exciting reactance of the 500-kV autotransformer winding to ground. The mutual coupling between the two lines, which are on the same tower, provide the voltages for this phenomenon. The tertiary was essentially unloaded, so the solution to avoid future problems was to load down the tertiary. Calculations indicated that about 590 kW was supplied to the ferroresonant circuit from the energized line and that about 250 kW additional via the 28 kV tertiary was sufficient to damp the oscillations.

7.10 SAFETY GROUNDING

Power stations and substations for either utilities or industrial plants are built on ground mats. These are carefully designed to provide minimum voltage drops across the mat in all directions (step and touch potentials) and minimum impedance between the mat and true earth or remote ground (ground potential rise). The primary aim is to reduce and minimize electric shock potentials for personnel safety. These designs are a specialized field and beyond our scope here. Standard IEEE 80-1976 is the basic guide for this area.

All equipment frames within the ground-mat area in these stations must be solidly bonded to the mat. This includes all exposed metallic parts of relays, relay switchboards, fences, secondary wiring, and so on. Thus all secondary circuits from the CTs and VTs are grounded. There should be only *one* ground in the circuit, and the general practice favors grounding at the switchboard or relay house. Multiple grounds may short-out relay(s) and prevent proper clearing of a fault, and may cause secondary wiring damage. A ground in the yard and another in the switchhouse put the secondary wiring in parallel with the ground mat, so that part of the heavy fault current can flow in the secondary winding directly, to either damage or cause misoperation. Only one ground in the circuit is sufficient to minimize any electrostatic potential.

If any equipment cannot be properly grounded, it should be carefully isolated from all contact with personnel. Special care must be taken for equipment associated with both the station and remote ground (transfer potential). Communication channels can fall into this category and are discussed in greater detail in Chapter 13.

Electromagnetic induction should be minimized by design with the station. Two grounds are required to reduce this hazard; again, the technique is discussed in Chapter 13.

In areas involving electrical equipment where a ground mat is not possible or practical, safety must be carefully examined. The fundamentals of the problem are illustrated in Fig. 7.19. Although many

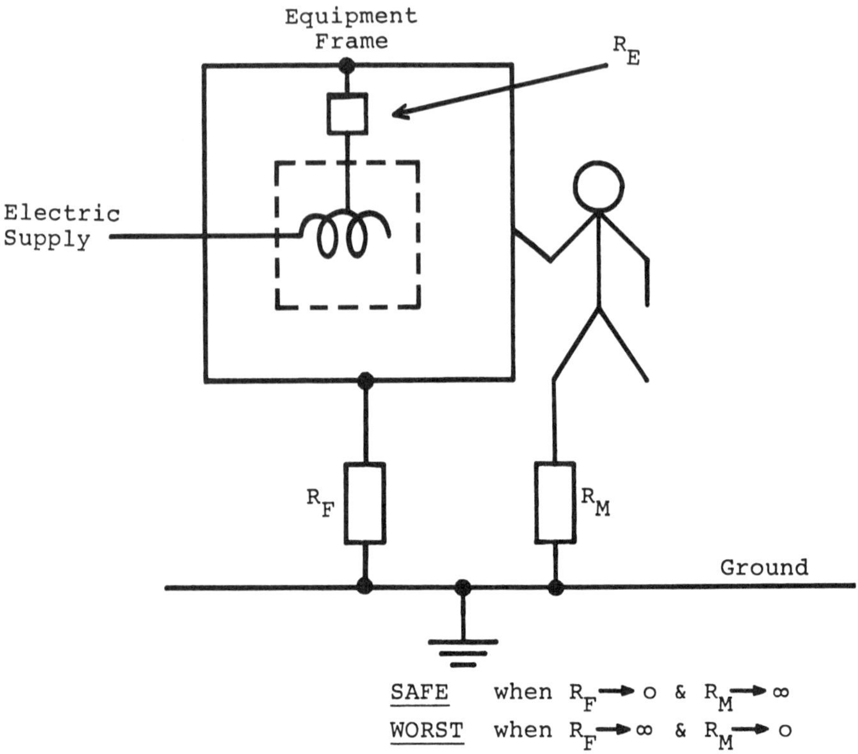

Fig. 7.19 Basic fundamentals of safety grounding.

diverse factors are involved, it appears that the average or reasonable resistance of a human being is from 1000 to 2000 Ω foot to foot and 500 to 1000 Ω foot to arm, although these limits are general. R_F can be made small by a low-impedance ground wire (fourth wire or third wire in a single phase) that is effectively and adequately connected to ground with minimum impedance.

R_M can be made large by high insulation to ground (nonwet) and by avoiding moist earth or possible contact with metallic surfaces, such as water pipes, that are connected to ground. Also sensitive, fast ground protection is helpful.

An application of this is in system grounding for mines (Section 7.6, Fig. 7.12). In the home, the ground wire system has been used, but the modern approach is to make R_E approach infinity through double insulation and other techniques, thus making ground faults more improbable.

Table 7.2 Types of System Grounding and General Recommendations

Types	Approximate fault current primary amperes	Industrial and utility station service systems	Utility Transmission subtransmission distribution	General comments
Ungrounded	Very low	Not generally recommended but used for high service continuity	Not recommended	1. Faults easy to detect but difficult to selectively locate 2. Fault current low, minimum damage 3. High potential of transient over-voltages 4. Ferroresonance and neutral VT inversion possible 5. Fault maintenance important
High resistance	1 to 10	Recommended for high service continuity	Not recommended	1. Same as above 2. Fault current low, low damage 3. Transient overvoltage limited to $2.5V_{LN}$
Low impedance	50 to 600	Recommended[a]	Not recommended	1. Easier to detect and locate faults selectively
Effective—solid	Low to very high	Recommended[a]	Recommended	1. Easy to detect and selectively locate faults.

[a]Up to 1000 V, solid grounding; 1000 V to 15 kV, low-impedance grounding; above 15 kV solid grounding.

7.11 GROUNDING SUMMARY AND RECOMMENDATIONS

A summary of the various grounding methods with suggested recommendations is given in Table 7.2. One column is for industrial and station service electrical systems for large generators in utility power systems. The other column is for all other parts of utility systems. As indicated, there are always many factors that influence the grounding choice, so any recommendations can be changed with very valid logic. Thus the recommendations in Table 7.2 are a guide. Although they reflect the personal opinions of the author, they also generally reflect actual practice in the United States.

For utility systems, only solid grounding is recommended. The logic here is that many distribution circuits extend through public areas and can be fairly long, with taps, branches, and so on. Thus while the line-to-ground fault at the substation bus will be greater than the three-phase fault, this reverses quickly, as the high zero-sequence impedance soon becomes predominate. These lower fault currents for faults out on the feeders, along with the probably of high earth contact resistance at the lower voltages, often makes ground-fault detection very difficult if not impossible. These currents are quite hazardous because of the involvement and/or presence of the public. Any fault limiting by neutral impedance tends to increase significantly the difficulty of detecting light faults out on the feeders.

For very short line distribution circuits, fault limiting may be desirable and is generally recommended for industrial systems from 1000 V to 15 kV. Here the distances between supply and utilization are usually quite short and are over or in an area not accessible to the public. Since most faults involve ground, it is desirable to limit their current magnitude and possible damage, which can be high because of the short distances of the utilization equipment from the supply source.

Solid grounding is used in most applications at utilization voltages below 1000 V. In many cases this is required by local, state, or national codes. Above 15 kV, the general practice is solid grounding.

BIBLIOGRAPHY

See the Bibliography at the end of Chapter 1 for additional information.

Cottrell, R. G. and W. D. Niebahn, An Analysis of Potential Ferroresonance at 138 kV Distribution Substations, McGraw-Edison's The Line 80/2 pp. 9-12, 17-18.

Forster, R. S. and L. R. Beard, Solving the Ferroresonance Problem, Transmission and Distribution, Dec. 1984. pp. 44, 46, 48.

Gleason, L. L., Neutral Inversion of a Single Potential Transformer Connected to Ground on an Isolated Delta System, *AIEE Trans.*, 70, P. 1, 1951, pp. 103–111.

Guide for Safety in AC Substation Grounding, IEEE Standard 80-1976, IEEE Service Center.

Hopkinson, R. H., Ferroresonant Overvoltage Control Based on TNA Tests on Three-Phase Delta-Wye Transformer Banks, IEEE Transactions, Vol. PAS-86, No. 10, Oct. 1967, pp. 1258-1265.

Hopkinson, R. H., Ferroresonant Overvoltage Control Based on TNA Tests on Three-Phase Wye-Delta Transformer Banks, IEEE Transactions, Vol. PAS-87, No. 2, Feb. 1965, pp. 352-361.

Locke, P., Check Your Ferroresonance Concepts at 34 kV Transmission and Distribution, April 1978, pp. 32–34, 39.

Price, E. D., Voltage Transformer Ferroresonance in Transmission Substations, Texas A&M University 30th Annual Protection Relay Engineers Conference, April 25-27, 1977.

Recommended Practice for Grounding of Industrial and Commercial Power Systems, IEEE Green Book, Standard 142-1982, IEEE Service Center.

Rudenberg, Reinhold, *Transient Performance of Electric Power Systems*. The MIT Press, Cambridge. MA, 1969.

Smith, D. R., S. R. Swanson and J. D. Bout, Overvoltages with Remotely-Switched Cable Fed Grounded Wye-Wye Transformer. IEEE Transaction Vol. PAS-75, Sept.-Oct. 1975, pp. 1843–1853.

8

Generator Protection

8.1 INTRODUCTION AND POTENTIAL PROBLEMS

Generators are subject to a number of possible hazards for which protection should be considered and given careful attention. These problems can be categorized as (1) internal faults within the protection zone, and (2) abnormal operating and/or abnormal system conditions. The protection discussed is primarily for the electrical power generator, which is essentially independent of the prime mover. Thus whether the prime mover is hydro, coal, oil, gas, or nuclear, the protection of the generator itself is basically the same.

Generators vary widely in size and are located in stations or plants generally near or convenient to the prime-mover supply, and/or major utilization areas as possible. Typical hydro-driven generators are shown in Fig. 8.1. These commonly have vertical shafts, whereas the other prime-mover generators are usually horizontal-shaft units.

The scope is to review the application of various protective relays to the generators. An important and vital part of these units is that of the regulating and control systems. These have their own protection, limiters, and safety features. Also, in recent years, sophisticated monitoring and on-line diagnostic systems have become available to monitor the temperatures in various parts of the units, to detect arcs by radio frequency and other methods, to sample the gas or water coolants for pollution, and to provide trends in abnormal situations for early warning of possible problems. A discussion of these is beyond the scope of this book. In many instances they add important additions and/or backup or supplements to the protection discussed here.

The hazards and problems considered are as follows:

1. Internal faults
   a. Phase and/or ground faults in the stator and associated protection zone
   b. Ground faults in the rotor (field windings)

Fig. 8.1 Four 100-MVA, 13.8-kV vertical-shaft hydro generators. The hydraulic turbines (not shown) are located on the floor below. (Courtesy of Seattle City Light.)

2. Abnormal operating and/or system conditions
   a. Loss of excitation (loss of field) or insufficient excitation
   b. Overload
   c. Overvoltage
   d. Under and over frequency
   e. Unbalanced currents — single phasing
   f. Loss of prime mover (motoring)
   g. Nonsynchronized connection of unit
   h. Out-of-step (loss of synchronism)
   i. Subsynchronous oscillations

8.2 GENERATOR CONNECTIONS

The most common connections for generators are:

1. *Directly connected* (one or several) each through a circuit

breaker to a common bus, as illustrated in Fig. 8.2. Usually, the generators are wye connected, but some are delta connected. Generally, these are relatively small kVA or MVA units, frequently at hydro plants and at industrial plants that have their own local machines. They may be connected to grounded power systems directly or through an isolating delta-connected transformer.

2. *Unit connected*, where the generator is connected directly to an associated power transformer without a circuit breaker between, as shown in Fig. 8.3. This is the common connection for the large generators in utilities. Most generators are wye connected, with a few connected delta. These can be a single generator or two separate generators (cross-compound) which are supplied by a common prime-mover system. Cross-compound generators may have the separate units directly connected together to a single transformer, or connected to separate secondary delta windings of a three-winding power transformer. Generators are also connected to the power system through autotransformers.

8.3 STATOR PHASE-FAULT PRIMARY PROTECTION

Phase faults seldom occur, but when they do, very large fault currents flow. Thus differential protection is recommended for all generators except possibly for small units of 1 MVA and less. This provides sensitive protection for phase faults but may or may not provide

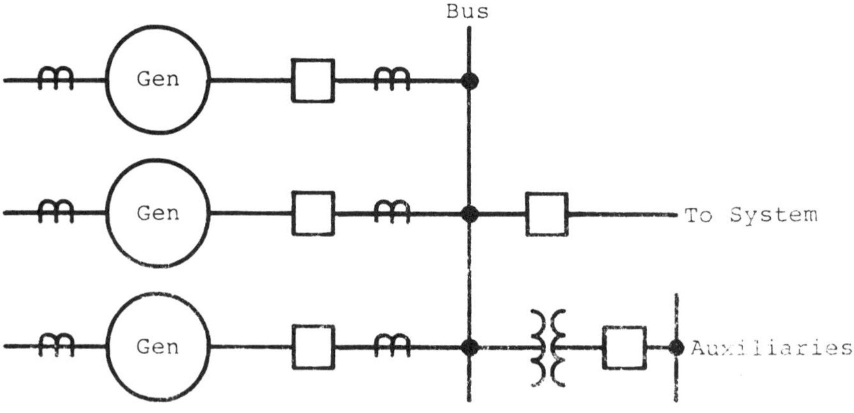

Fig. 8.2 Direct-connected generator units (one or more) to a common system bus.

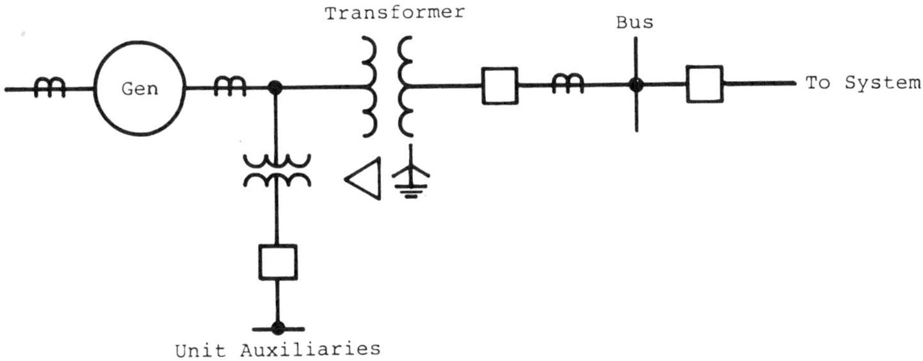

Fig. 8.3 Unit-connected generator.

protection for ground faults, depending on the type of grounding. The fundamentals of differential protection are outlined in Chapter 6.

8.3.1 Differential Protection for Small Generators

The preferred method for small units is shown in Fig. 8.4. The limitation is the ability to pass the two conductors through the window or opening of the CT. Typical opening diameters are about 8 in. maximum. However, where this is possible, high-sensitivity, high-speed protection is obtained and CT performance does not have to be matched, as there is only one involved per phase. The toroidal CT ratio (commonly 50:5) is independent of generator load current. Typical sensitivities on the order of 5 A primary current can be obtained. This provides protection for both phase- and ground-fault currents as long as the fault level for faults within the differential zone is greater than the sensitivity.

This scheme does not provide protection for the connections from the toroidal (ring) CT to the generator breaker unless the CT is on the bus side of the breaker and the generator neutral side leads are carried to that point. This is seldom practical, so other protection must be provided for this area between the toroidal CT and the breaker. In general, this scheme (Fig. 8.4) is more sensitive as long as the generator CT ratio is greater than 150:5 to 200:5. If the toroidal CT is not applicable and differential protection is desired, the scheme of Fig. 8.5 can be used.

8.3.2 Multi-CT Differential Protection for Generators

For medium and large generators the differential (87G) relays of the type discussed and shown in Fig. 6.2 are widely used to provide fast

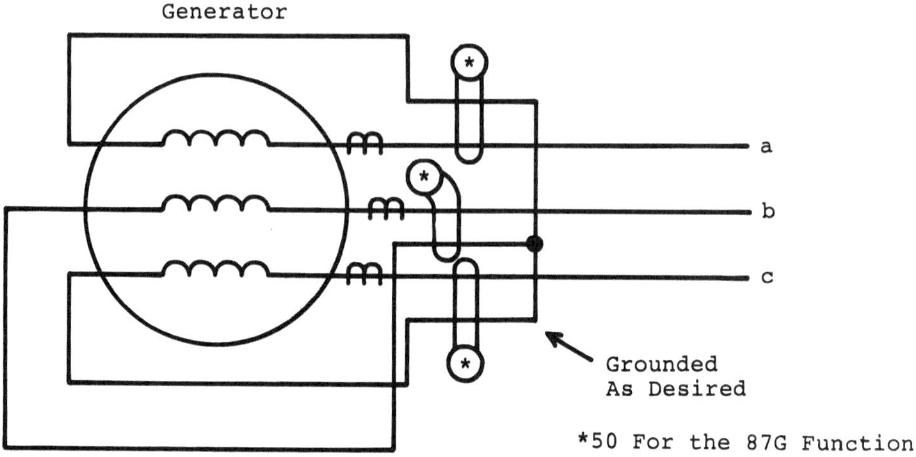

Fig. 8.4 Differential protection for small generator units with toroidal current transformers and an instantaneous overcurrent (50) relay.

and very sensitive protection. This scheme provides primary protection for the generator and associated circuits. It is connected to two sets of current transformers; one set in the neutral leads, the other in the line side. For generators with associated breakers (Fig. 8.2), the line-side CTs are usually associated with the breaker as shown. For unit generators the line-side CTs are usually quite close to the generator, basically at the generator terminals. Typical connections for the three phase are shown in Fig. 8.5 for both wye- and delta-connected generators.

If current transformers are available at each end of the windings for the delta-connected generators of Fig. 8.5b, differential relays can be applied for winding protection. The connections would be similar to those shown in Fig. 8.5a. However, this would not provide protection for the junction points or the phase circuits which are within the protective zone (Fig. 8.5b).

Usually, the differential CTs have the same ratio, and preferably should be of the same type and manufacture, to minimize mismatch errors for external faults. This is possible for the unit generators of Fig. 8.3 but difficult for Fig. 8.2, where the CTs in the neutral are one type and those associated with the breaker are another type. It is preferable not to connect any other equipment in the differential circuit and to keep the lead burden low as possible. Generally, the impedance of the restraint winding of differential relays is low. All this contributes to a total low burden and increased performance margins for the CTs.

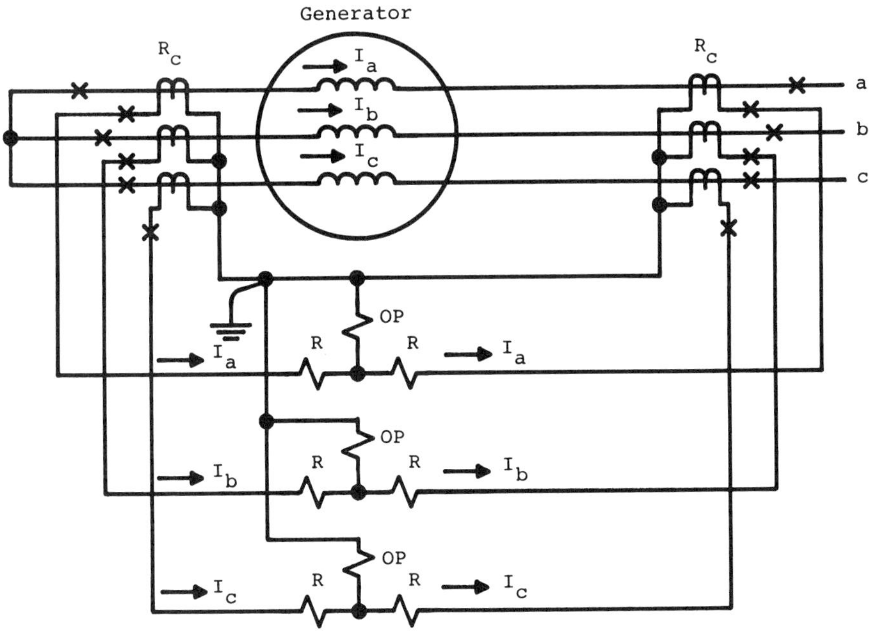

(a)

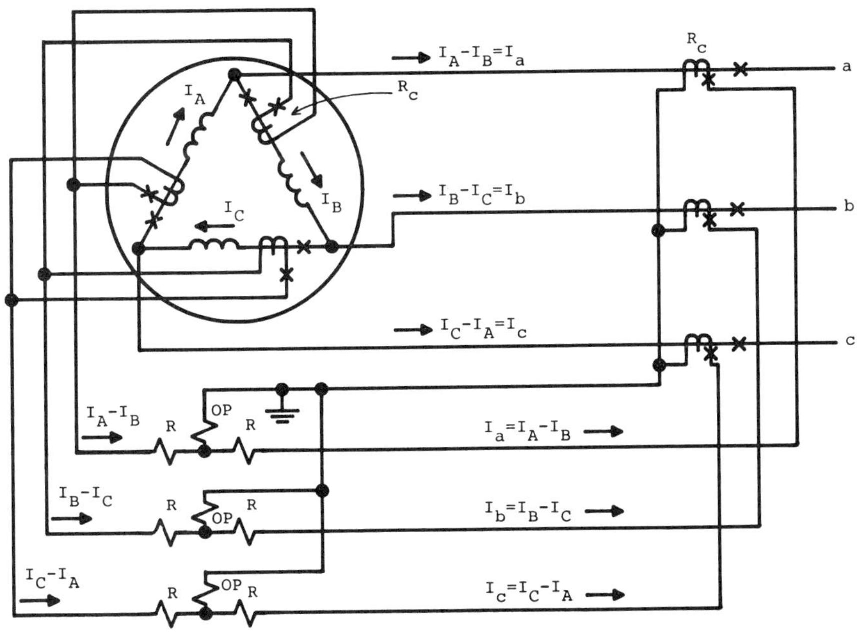

(b)

Fig. 8.5 Typical differential (87G) connections for the protection of wye- and delta-connected generators: (a) wye-connected generator; (b) delta-connected generator.

The application recommendations permit the use of sensitive generator differential relays with low percentage characteristics, typically 10 to 25% for the fixed percentage types and the equivalent or lower for the variable types.

Relay sensitivities (pickup current) are on the order of 0.14 to 0.18 A for the 10% and variable percent types and about 0.50 A for the 25% types. The operating time should be fast to open the breaker(s), remove the field, and initiate reduction of the prime-mover input. Unfortunately, the flux in the machine continues to supply the fault for a number of seconds (on the order to 8 to 16 s), so instantaneous deenergization of generator faults is not possible.

Problems with magnetizing inrush generally are not severe since the voltage on the machine is developed gradually and the generator is carefully synchronized to the power system. However, the differential relays should have good immunity to avoid incorrect operation on an external fault that significantly lowers the voltage, which recovers when the fault is cleared. This can cause a "recovery inrush." This does not apply to those units that are subject to energizing a transformer and/or power system at full voltage (black start).

Cross-compound generators have two units, generally connected to a common power transformer. For these units a separate differential relay is applied for each generator unit connected, as shown in Fig. 8.5a.

Split-winding generators, where the two winding halves are available and with CTs in one of the halves as shown in Fig. 8.6, can be protected with two separate differentials. By comparing one winding half against the total, as shown, protection for shorted turns and open-circuited windings is possible. This is difficult or impossible for conventional differential relaying until the fault develops into the other phases and/or ground. Where a 2:1 CT ratio is not available, auxiliary CTs can be used.

8.3.3 High-Impedance Voltage Differential Protection for Generators

This type of differential protection scheme can be applied as an alternative to the current differential type described. The relays are connected between the phase and neutral leads of the paralleled CTs. For external faults, the voltage across the relay will be low, as the current circulates between the two sets of CTs, as shown in Fig. 8.5. For internal faults the fault currents must pass through each CT's exciting branch and the high-impedance relay, so that the CTs are saturated for most faults, producing a high voltage to operate the relay. This protection scheme is widely used for bus protection and is described further in Chapter 10. The CT requirements are somewhat more critical. They should have identical characteristics, negligible leakage reactance, and have fully distributed secondary windings.

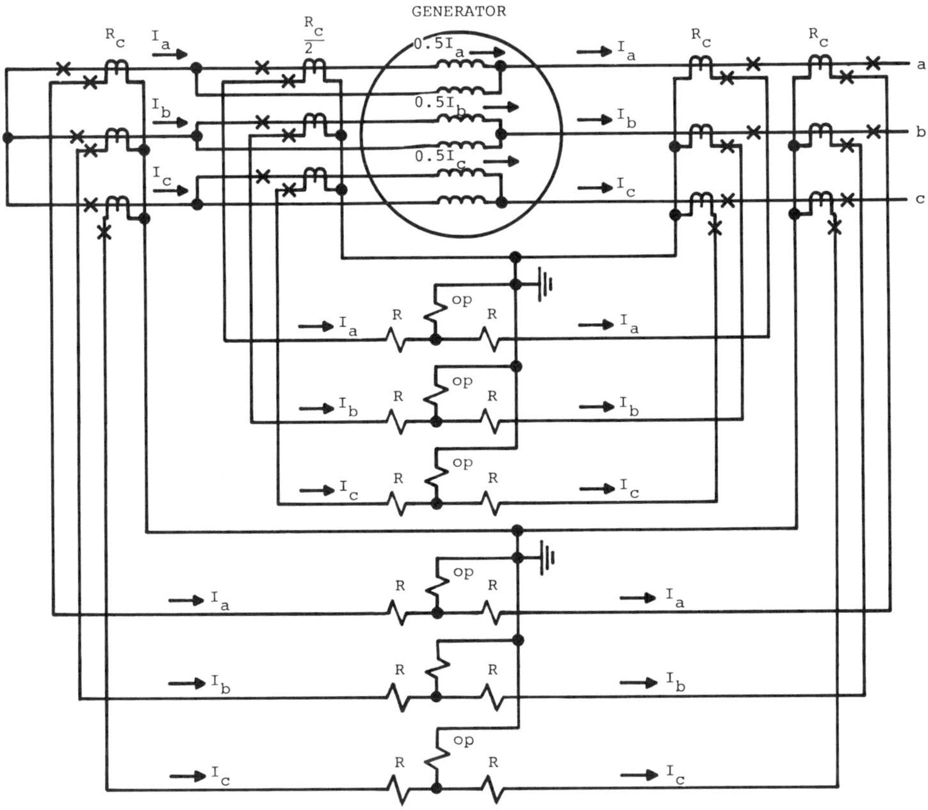

Fig. 8.6 Typical differential (87G) connections for the protection of a split-winding generator.

8.3.4 Generator Current Differential Example

Consider the unit-connected generator tied to a 345-kV power system as shown in Fig. 8.7.

For a three-phase fault on the 18-kV bus at F_1, the positive-sequence network is shown and the total reactance to the fault is calculated as

$$X_{1F1} = \frac{\overset{(0.515)}{0.131} \times \overset{(0.485)}{0.124}}{0.255} = 0.064 \text{ pu} \tag{8.1}$$

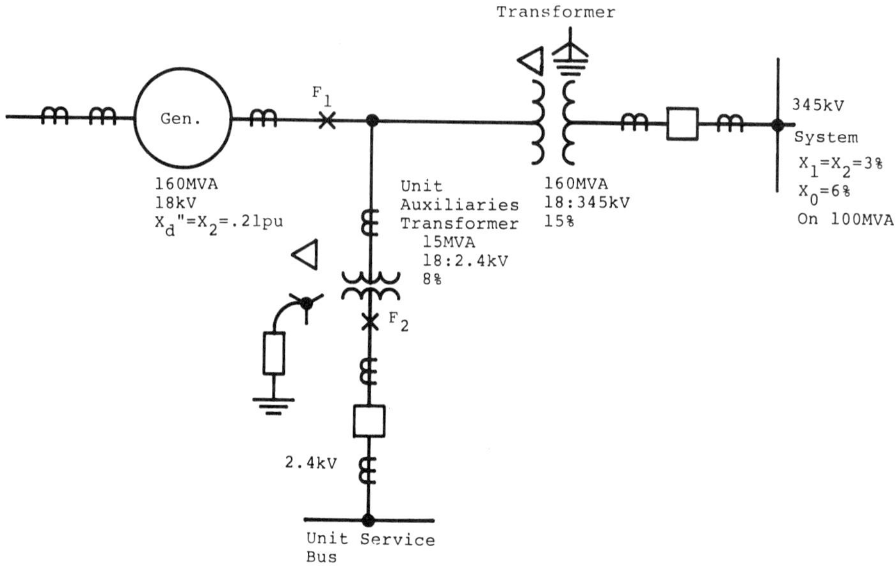

(a)

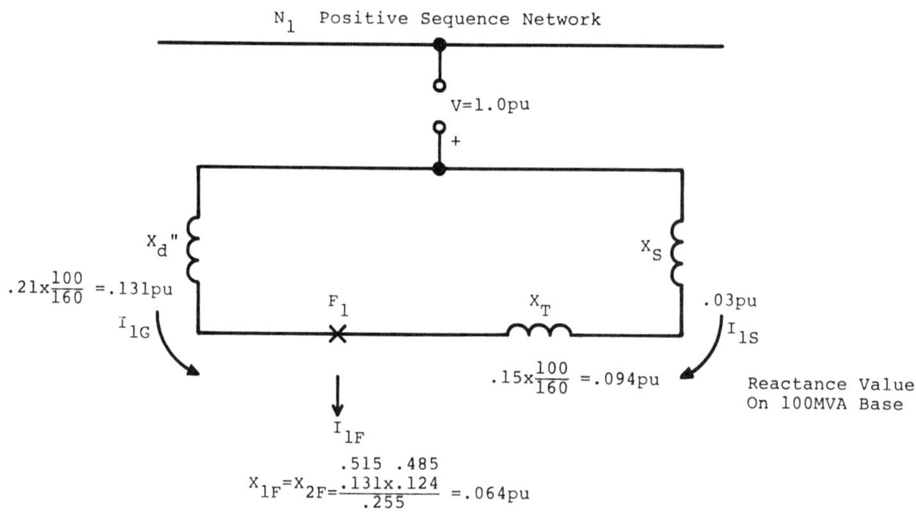

(b)

Fig. 8.7 Typical example of a unit generator (same system as Fig. 7.9 example).

These values are on a 100-MVA base. The values in parentheses define current distribution on either side of the fault. For a solid three-phase fault,

$$I_{1F1} = I_{aF1} = \frac{1}{0.064} = 15.70 \text{ pu} \tag{8.2}$$

$$I_{1\,pu} = \frac{100{,}000}{\sqrt{3} \times 18} = 3207.5 \text{ A at 18 kV} \tag{8.3}$$

$$I_{1F1} = I_{aF1} = 15.7 \times 3207.5 = 50{,}357.3 \text{ A at 18 kV} \tag{8.4}$$

The maximum load on the unit is

$$I_{max\ load} = \frac{160{,}000}{\sqrt{3} \times 18} = 5132 \text{ A at 18 kV} \tag{8.5}$$

From this maximum load a current transformer ratio of either 5500:5 or 6000:5 could be used. The lowest ratio is preferred for increased sensitivity, so suppose that 5500:5 (1100:1) is used. This gives a full-load secondary current of 5132/1100 or 4.67 A.

The three-phase secondary fault current is

$$I_{1F1} = I_{aF1} = \frac{50{,}357}{1100} = 45.78 \text{ A-s} \tag{8.6}$$

If the fault F_1 is inside the 87G differential zone (Fig. 8.7) this 45.78 A would flow through the operating coil. This is many multiples of the differential relay pickup current for positive and fast operation.

For the external fault F_1 as shown, the currents through the restraining windings of the differential relay would be only that contribution from the generator, which is

$$I_{1F1gen} = \frac{50{,}357.3 \times 0.485}{1100} = 22.2 \text{ A-s} \tag{8.7}$$

This would be the internal and total fault current if a three-phase fault occurred before the unit were synchronized to the 345-kV system. For these cases the fault current can be calculated alternatively as

$$I_{1F1gen} = \frac{1.0}{0.131} = 7.62 \text{ pu}$$

$$= 7.62 \times 3207.5 = 24{,}438.11 \text{ A at } 18 \text{ kV}$$

$$= \frac{24{,}438.1}{1100} = 22.2 \text{ A-s} \tag{8.8}$$

The same procedure and technique is applicable to the direct-connected units of Fig. 8.2 for the differential protection shown in Fig. 8.9.

8.4 UNIT TRANSFORMER PHASE-FAULT PRIMARY PROTECTION

Again, differential protection is recommended and is shown as 87TG in Fig. 8.8. Since there is no breaker between the generator and the transformer, this differential is connected to include the generator, since both units must be tripped for either transformer or generator faults. This provides added generator protection. Thus for generator phase faults, 87G and 87TG operate in parallel. In large generator units an additional differential 87T is sometimes connected around the transformer. Thus two primary protective systems are provided both for the generator as shown and for the transformer (only one 87TG shown). The connections and relays for transformers are discussed in Chapter 9.

In Fig. 8.8 the high-side CTs on the unit auxiliaries transformer are included in the 87TG differential, so that faults in this unit auxiliary transformer and 4.16-kV system are external. Alternatively, low-side unit auxiliary transformer CTs could be used so that this transformer is within the 87TG protection zone. The same alternatives exist if the added 87T differential protection is applied.

It is important to recognize that including the unit auxiliary transformer in either the 87TG or 87T zones usually does not provide good or adequate protection for the transformer in many cases. This can be demonstrated by considering a solid three-phase fault on the low 4.16-kV side of this bank or at F_2 in Fig. 8.7. For this fault the total positive-sequence reactance of the system plus the transformer is

$$X_{1F2} = 0.064 + 0.08 \times \frac{100}{15} = 0.064 + 0.533 = 0.597 \text{ pu} \tag{8.9}$$

$$I_{1F2} = I_{aF2} = \frac{1.0}{0.597} = 1.675 \text{ pu}$$

$$= 1.675 \times 3207.5 = 5372.7 \text{ A at } 18 \text{ kV} \tag{8.10}$$

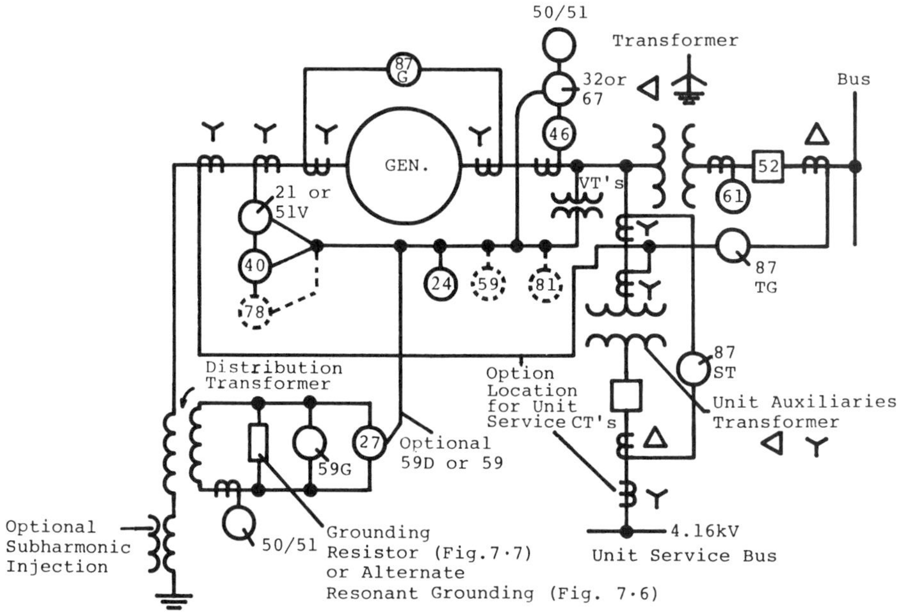

Fig. 8.8 Typical major protection for a unit generator and for large generators in utility systems.

In the overall differential the equivalent CT ratio will be basically around 1100:1, so with this ratio this fault F_2 will be 5372.7/1100 = 4.88 A secondary. The transformer differential relays are less sensitive because they must be applied to different types and ratio CTs, avoid operation of magnetizing inrush, and so on. For most relays this 4.88 A is probably above its minimum pickup, but the low value of fault provides very low multiples of pickup and marginal sensitivity.

This illustrates and emphasizes that a separate transformer differential relay is recommended for the unit auxiliary transformer(s). This is shown as 87ST in Fig. 8.8. For this application, the maximum transformer current would be

$$I_{\text{STmax load}} = \frac{15{,}000}{\sqrt{3} \times 18} = 481.13 \text{ A at } 18 \text{ kV} \tag{8.11}$$

Thus a CT ratio of 500:5 could be used for 87ST rather than the equivalent 1100:1 ratio required for 87TG. With an equivalent 500:5 (100:1), fault F_2 would provide an operating coil current of 5372.7/100 = 53.73, a good margin for fast and sensitive station transformer fault protection. The 87ST relays must also shut down the generator and open the 345-kV breaker.

For generators connected as shown in Fig. 8.2, the unit auxiliaries transformer(s) are connected to the generator bus, in most cases either with or without a breaker to the bus. A separate differential can be used for the transformer, or without a high-side breaker, included in a bus-transformer differential.

Pressure relays, discussed in Chapter 9, are recommended for all the transformers. These supplement differential protection but can operate only for faults inside the transformer tank.

8.5 PHASE-FAULT BACKUP PROTECTION

Backup protection for faults in the generator and transformer, and particularly generator and/or transformer protection for uncleared faults of the generator system bus and connected system, can be provided by the application of a distance relay (21). This relay, connected as shown in Fig. 8.8, is set to reach through the unit transformer into the system, or into the system for units such as those shown in Fig. 8.2. A fixed timer provides the necessary time delay for the distance relay to coordinate with all relays its setting overreaches.

As shown in Fig. 8.8, using CTs at the neutral end of the generator provides backup protection for the generator. This is preferred, but if CTs are not available or suitable, the 21 relay can be connected to CTs on the high side of the generator. With voltage from the VTs on the generator bus as shown, the relay setting is counted from this location independent of the CT location. Thus only the impedance of the unit transformer in Fig. 8.8, plus high-voltage system impedance as desired, are involved.

For generator unit protection only, a distance (21) relay can be connected to CTs and VTs at the generator system bus and set looking into the generator. This provides only generator or unit generator-transformer protection, and no backup for uncleared system faults. No time delay is required for this application, so it provides added fast protection. Both applications of distance (21) relays have been applied to large, important unit generators.

A less expensive alternative and one commonly applied for medium-sized and small generators is a voltage-controlled or voltage-restraint inverse-time-overcurrent relay, 51V. This may be connected to CTs either at the neutral, as in Fig. 8.8, or in the generator leads, as in Fig. 8.9. The function of voltage is to prevent or moderate overcurrent operation until the generator voltage is reduced by a fault. Normal operation is around the knee of the generator iron saturation curve. Thus the synchronous reactance governing load, $X_{d(sat)}$, is a lower value than that for sustained faults, which is $X_{d(unsat)}$. Thus

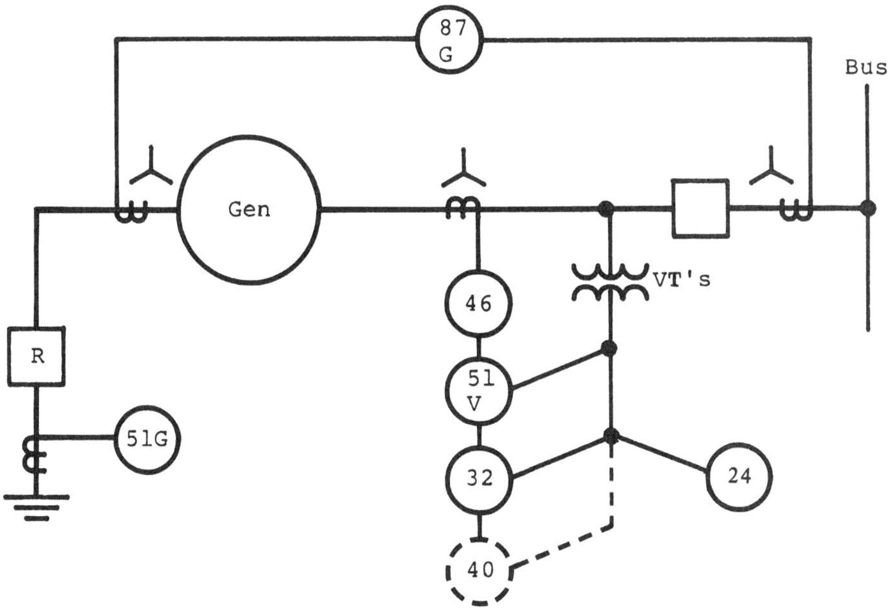

Fig. 8.9 Typical protection for bus-connected generator.

substained three-phase fault current is less than maximum load current. This occurs if the regulator does not increase the voltage for the substained fault reduced voltage. One type prevents the overcurrent unit from operating until the voltage drops below a specific value, typically around 80% of normal. Another type changes the time-overcurrent characteristics with voltage. It will not respond to maximum load if the voltage is normal but will operate on lower currents when the fault reduces the voltage. Generally, this type of relay (51V) is applied to one of the three phases for three-phase fault backup with a negative-sequence relay (46) for unbalanced faults.

8.6 NEGATIVE-SEQUENCE CURRENT BACKUP PROTECTION

Negative sequence in a generator crosses the air gap and appears in the rotor or field as double frequency current. This tends to flow in the surface of the rotor structure, the nonmagnetic wedges, and other lower-impedance areas. Severe overheating and ultimately, the melting of the wedges into the air gap can occur, causing severe damage. ANSI standards establish that limits be expressed as $I_2^2t = K$, where

I_2 is the integrated negative-sequence current flowing for time t in seconds. K is a constant established by the machine design. Typical values for synchronous condensers and older turbine generators were 30 to 40, but for the very large generators K is as low as 5 to 10. Units subject to the specified limit and up to 200% of the limit may be damaged, and early inspection is recommended. For more than 200%, damage can be expected.

Inverse-time-overcurrent units operating from negative-sequence current and with a time characteristic adjustable to $I_2^2t = K$ are available and recommended for all generators. Such a relay (46) is shown in Fig. 8.8 and 8.9. They are set to operate just before the specified machine $I_2^2t = K$ limit is reached. A low-level I_2 auxiliary is available, operating typically about 0.03 to 0.2 pu I_2 for alarm of continued unbalance.

Basically, this protection is backup primarily for system faults that are not adequately cleared, but also backs up the protection for the generator unit and associated equipment for unbalanced faults and conditions.

8.7 STATOR GROUND-FAULT PROTECTION

Insulation failure is the major cause of most faults in a generator. They may start as turn-to-turn faults and develop into ground faults, or start as ground faults initially, so ground-fault protection is very important, although fortunately, such faults are very rare.

Some fault limiting is used except for very small, low-voltage generators. This limits fault damage. In general, the zero-sequence reactance of generators is lower than the positive- and negative-sequence reactance, hence ground-fault current would be greater than three-phase fault current if not grounded through impedance. The various grounding methods have been covered in Chapter 7.

8.7.1 Fault Protection with High-Impedance Grounding

High-impedance grounding is standard for unit generators and is used in industrial systems. The discussion here centers on the common high-resistance grounding, where the fault current is limited to about 1 to 10 A primary (Section 7.5.2). This limits iron burning in the generator, to avoid very costly repairs.

The scheme of Fig. 7.7 is used for the unit generators shown in Fig. 8.8. An inverse-time-overvoltage relay, 59G, is connected across the resistor to respond to the $3V_0$ voltage resulting from a ground fault in the generator and system up to the delta windings of the unit and station service transformers. The example of Section 7.5.3 (Fig. 7.9) provided 138 V for a solid line-to-ground fault to operate 59G. This is quite typical. The pickup of 59G is about 5 to 16 V for good fault sensitivity. It should be detuned or insensitive to third harmonics, which normally flow in the neutral and system in a manner similar to zero sequence.

Where multiple generators are connected to a common bus (Fig. 8.2) or in an industrial system such as that shown in Fig. 7.11, the high-resistance grounding scheme of Fig. 7.8 is recommended. The advantage is that grounding exists independent of which machine(s) are in service at any time. In the example of Section 7.5.4 (Fig. 7.11), 208 V was available to operate relay 59G connected across the grounding resistor to give good fault sensitivity.

For resonant grounding systems (Section 7.5.1, Fig. 7.6), used in a few cases for unit generators, a 59G relay should be applied across the grounding reactor and through a suitable voltage transformer if a primary reactor is used.

With these grounding systems, the sensitive (59G) relays should be coordinated with the voltage transformer primary fuses. If this is not possible or practical, recognize that the unit may be tripped for VT faults. Although these faults are quite possible, the probability generally is very low. Coordination is also important in some applications to avoid misoperation for ground faults on the high-voltage side of the power transformer. This is discussed in a later section.

In many cases the voltage transformers are wye-wye, but open-delta-connected VTs can be used for three-phase voltage. With wye-wye VTs the primary wye should be grounded. Unless the secondary is required for zero-sequence indications, the secondary wye should be ungrounded and isolated. Grounding one of the phases provides a safety ground. Otherwise, the 59G may operate for secondary VT circuit ground faults, and 59G should be coordinated with the secondary VT fuses.

Inverse-time/instantaneous overcurrent (50/51) relays in the resistor secondary circuit (Fig. 8.8) provide alternative and/or backup ground protection. The secondary CT ratio is selected to give approximately the same relay current as that flowing in the generator neutral for a ground fault. Thus using the Fig. 7.9 example, a CT ratio of 400:5 (80:1) would give

$$I_{50/51} = \frac{464.38}{80} = 5.80 \text{ A} \tag{8.12}$$

where the primary fault current in the neutral was 6.19 A, which reflected through the distribution transformer as 464.38 A.

These relays must be set above the maximum unbalance current that normally flows in the neutral circuit. Typical values usually are less than 1 A in the generator neutral. Settings for 51 should be 1.5 to 2 times this unbalance. The 50 relay provides instantaneous protection and it must be set above the normal neutral unbalance as well as above the maximum current resulting from primary system ground faults, whichever is greater. Section 8.7.2 discusses the latter type of fault. Typical 51 settings should be 2 to 3 times the maximum.

In some applications two 51 relays are used, one to provide backup tripping of the unit, the other to initiate breaker failure relaying (when used) associated with the primary breaker, as shown in Fig. 8.8, or breakers if the primary is connected to a ring or breaker and half arrangement.

The 59G relay in the high-impedance grounding systems provides ground protection for about 90 to 95% of the stator winding. Additional protection is required for ground faults that may occur near the neutral end of the stator winding. Several methods exist and are of two basic types: use of third-harmonic voltage or the injection of a subharmonic voltage.

The normal unbalance in the generator neutral may contain several harmonics, with the third harmonic being the major one. This harmonic behaves similarly to zero sequence, and thus appears across the grounding resistor or the broken-delta-connected VTs normally used for $3V_0$ indications. One protection scheme connects an undervoltage (27) relay responsive to third harmonics in parallel with 59G as shown in Fig. 8.8. Its normally closed contact is in series with a voltage-sensing relay connected to the generator VTs. When normal or approximately normal voltage is developed by the generator, the 27 relay contacts are opened by third-harmonic normal voltage, and the supervising voltage relay contacts are closed. When a ground fault occurs near the generator neutral, this shorts out or significantly reduces the third-harmonic voltage, so 27 resets to indicate the fault.

Another scheme applies an overvoltage 59 relay which is responsive to the third harmonics across the broken-delta generator VTs. This must be set above the maximum normal third-harmonic voltage. For the neutral-area ground faults, the third-harmonic voltages will be redistributed and increased at the generator terminals. This scheme may offer limited protection where the normal full-load third-harmonic voltage is high, and time coordination to avoid operation for higher third-harmonic voltage during external faults.

The third-harmonic voltage at both the neutral and generator terminals can vary considerably from maximum to minimum load. Generally, the third-harmonic voltage at maximum load will be at least 50% greater than at minimum load, with variations of about 2:1 to 5:1. In many cases the ratio of the neutral-to-terminal third-harmonic voltage is reasonably constant with load changes, so another scheme operates on the differential voltage between the two ends of the winding. Ground faults near the neutral and near the terminals upset the balance, through the fault reducing the third-harmonic voltage at the faulted end. This provides operating overvoltage for relay 59D (Fig. 8.8).

All these schemes require specific data for the third-harmonic voltages over the operating ranges of the generator, both real and reactive power, before the relays can be set properly. One scheme may be more sensitive for one generator but less sensitive for other generators.

The injection of subharmonic frequency current can be by a separate neutral transformer or through the broken-delta VT connection. Some systems provide coding for increased security. A ground fault decreases the generator capacitance to cause an increase in current for detection. These schemes provide the possibility of 100% winding protection and monitoring capability. This neutral or 100% winding ground-fault protection is a relative new area, so considerable investigation and study are still in progress.

8.7.2 High-Voltage Ground-Fault Coupling to Produce V_0 in High-Impedance Grounding Systems

Ground faults in the primary system produce a voltage in the generator secondary circuit through the primary-to-secondary capacitance of the unit transformer bank. The circuit is shown in Fig. 8.10. With sensitive voltage relays (59G) applied in high-resistance systems, this voltage can cause operation. Thus 59G relays should have a time delay to permit the primary ground relays to clear high-side faults if the coupling voltage is greater than the 59G relay pickup.

This can be illustrated with the example of Fig. 8.7. Assume that the transformer capacitance between the primary and secondary windings (X_{CT}) is 0.012 μF/phase. Using the values of 3R = 5019 Ω from Fig. 7.9, the voltage across the grounding resistor is calculated as shown in Fig. 8.11. Calculation of the currents for a phase-a-to-ground fault on the primary 345-kV bus is shown in Fig. 8.11a. Figure 8.11b shows the calculation for a primary V_0 fault voltage of 81,467 V, which reflects through the interwinding capacitance to produce 1293 V across the grounding distribution transformer primary. With a 18:240 kV ratio, relay 59G receives 17.24 V. This is above the pickup, so the relay would receive operating energy until the 345-kV fault is cleared. These high-voltage faults normally are cleared at high speed, but 59G should be coordinated with the maximum backup time.

If resonant grounding is used for the generator, the primary ground fault may also affect the protection system. For the lower-voltage system, this coupled voltage probably will not be significant, as the V_0 for primary-side faults will be much lower than on the EHV systems.

8.7.3 Ground-Fault Protection with Low-Impedance Grounding

Generators connected as shown in Fig. 8.2 are frequently grounded through low impedance, as described in Chapter 7. Ground protection in these cases may be provided by the 87G differential, depending on the fault level and differential relay sensitivity.

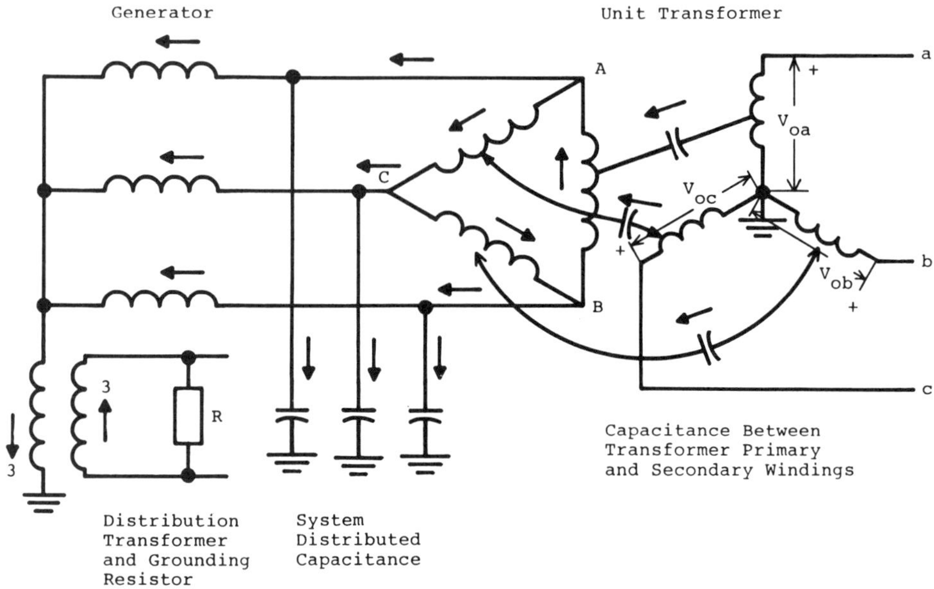

(a)

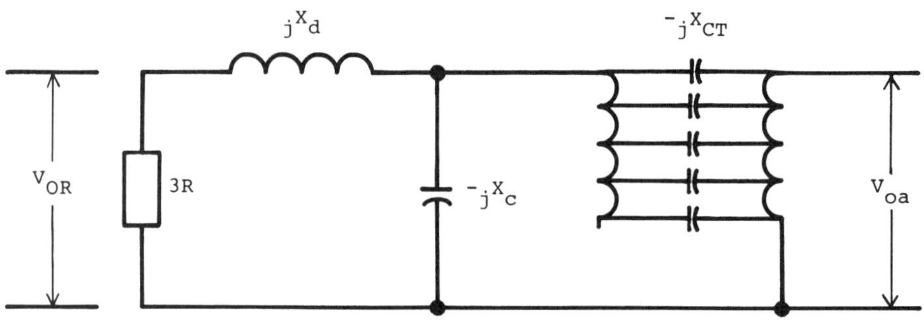

(b)

Fig. 8.10 Capacitive coupling through transformer banks for primary-side ground faults: (a) three-phase system diagram; (b) equivalent circuit.

Higher sensitivity and fast operation for ground faults may be obtained by an additional zero-sequence differential. One type which is relatively independent of the CT ratios and CT performance is shown in Fig. 8.12. A product-type overcurrent relay, 87GD, operates on the product of the two currents. As can be seen from the figure, for an external fault the relay currents are in opposition and the relay does not operate. For internal ground faults, zero sequence from the system reverses to provide operation. One type of relay has a minimum pickup of 0.25 product or 0.5 A in each coil. With maximum operating energy with the two currents in phase, the relay will operate with higher currents out of phase ±90° and with different magnitudes as long as the product times the cosine of the angle between the currents is greater than the tap product. If the system is not grounded, the scheme as shown in Fig. 8.12 will not operate, since the system does not supply zero-sequence current to the internal fault. In this case an auxiliary current transformer can be used to provide internal operating energy with only one zero-sequence source. This is discussed in detail in Section 9.17.

Supplementary and backup protection is provided by 51G, as shown in Fig. 8.9 and 8.12. The CT primary current rating should be about one-half of the maximum ground-fault current with the 51G relay set at approximately 0.5 A. Time coordination is necessary, with other relays protecting the system connected to the generator bus.

When several wye-connected generators are connected to a common bus, such as in Fig. 8.2, it becomes difficult to selectively isolate the fault for minimum system disruption. Whether one or all generator neutrals are grounded similarily, the fault current in the grounded neutrals essentially is the same independent of fault location in the area. That is, for faults in the generator-to-breaker area, on the bus, or in the system just beyond the bus, the fault level is basically the same. This is true for any of the grounding systems.

With low-impedance grounding and sufficient current to operate the differential relays of the generator and bus, proper minimum isolation can be achieved. Then 51G in the grounded neutrals provides backup or "last resort" protection which may be nonselective. Another possibility is to apply sensitive ground directional relays at the terminals of the generators set looking into the generator. These would operate only for ground faults within the machine and not respond to faults in the other machines, bus, or the system. Again 51G in the grounded neutrals provides nonselective backup. This application may be impossible or difficult with high-impedance grounding.

Grounding only one of the generators with an ungrounded system provides the liability of operating ungrounded when that unit is removed from service either manually or through its protection. Thus multiple grounding of each unit seems preferable to avoid ungrounding by fault tripping or a failure to transfer grounding on a planned shutdown of the only grounded unit. Another possibility is to use the

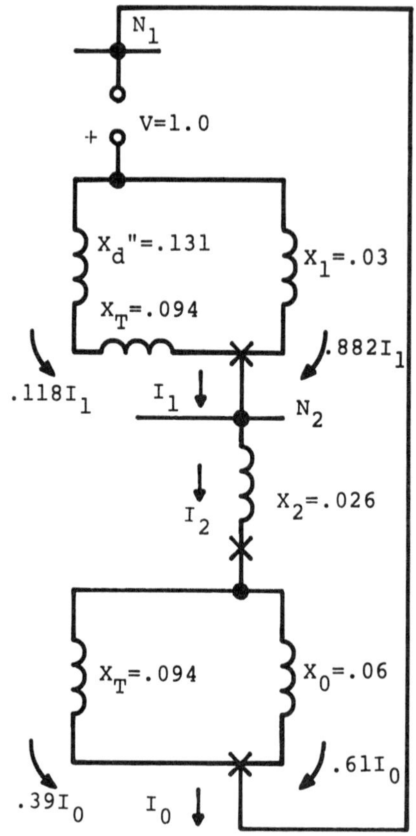

$$X_1 = X_2 = \frac{\overset{(.882)(.118)}{.225 \times .03}}{.255} = .026 \text{ pu}$$

$$X_0 = \frac{\overset{(.61)\ (.39)}{.094 \times .06}}{.154} = .037 \text{ pu}$$

$$I_1 = I_2 = I_0 = \frac{1.0}{.026 + .026 + .037} = 11.17 \text{ pu}$$

Per Unit on 100 MVA

(a)

Fig. 8.11 Example of a primary line-to-ground fault producing voltage across the high-impedance grounding system: (a) sequence networks and fault calculation; (b) grounding network and voltage calculation.

broken-delta distribution grounding system of Section 7.5.4 with a resistor to give a higher primary ground-fault current level.

8.8 REDUCED OR LOST EXCITATION PROTECTION

Protection to avoid unstable operation, potential loss of synchronism, and possible damage is important and is applied for all synchronous

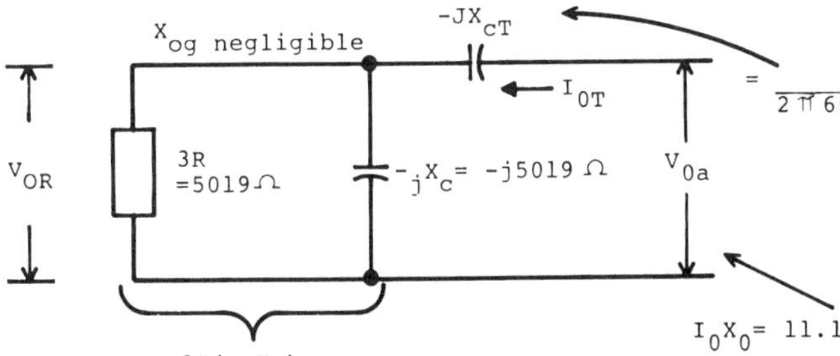

$$= \frac{10^6}{2\pi 60(.012)} = -j221{,}049 \text{ ohms}$$

$$I_0X_0 = 11.17 \text{x} .037 = .409 \text{ pu}$$

$$Z_{OL} = \frac{3R(-jX_c)}{3R - jX_c} = \frac{(5019)(-j5019)}{5019 - j5019} = 3549 \angle{-45^\circ} \text{ ohms}$$

$$V_{0a} = .409\ \frac{345{,}000}{\sqrt{3}} = -j81{,}467 \text{ volts at 345kV}$$

$$I_{OT} = \frac{V_{0a}}{Z_{OL} - jX_{cT}} = \frac{-j81{,}467}{2509.5 - j223{,}559} = .364 \text{ amps at 18kV}$$

$$V_{OR} = I_{OT}Z_{OL} = \frac{V_0 Z_{OL}}{Z_{OL} - jX_{cT}} = 3549 \text{ x } .364 = 1293 \text{ volts at 18kV}$$

$$= \frac{1293}{75} = 17.24 \text{ volts on relay 59G} \quad \frac{18\text{kV}}{240} \text{ ratio}$$

(b)

Fig. 8.11 (continued)

machines. Such protection is included in the excitation system supplied with the machine, but additional protection is recommended to operate independently both as supplemental and backup protection. Distance relays, as outlined in Chapter 6, are applied for this purpose.

Normally, the generator field is adjusted so that slightly lagging power is shipped into the power system. The diagram of Fig. 8.13 provides an overview of synchronous machine operation. With the machine terminal voltage V on the horizontal axis, lagging current for a generator is in the fourth quadrant. This is the normal operating area. If the excitation is reduced or is lost, the current phasor moves into the leading or first-quadrant area. Generators have low or reduced stability in this area. If a total loss of field occurs and the system can supply sufficient reactive power without a larger terminal voltage drop, the unit may run as an induction generator. Otherwise, synchronism is lost. This change is not instantaneous but occurs over a time period depending on the unit and connected system. If the field was accidentally tripped, early alarm may permit the operator to restore it and avoid a costly and time-consuming shutdown and restart. If the field is not promptly restorable, the unit should be shut down.

Generators have characteristics known as capability curves. Typical curves are shown in Fig. 8.14a. Temperature limits are basically zones, so these curves are the designer's thermal limit. As overheating

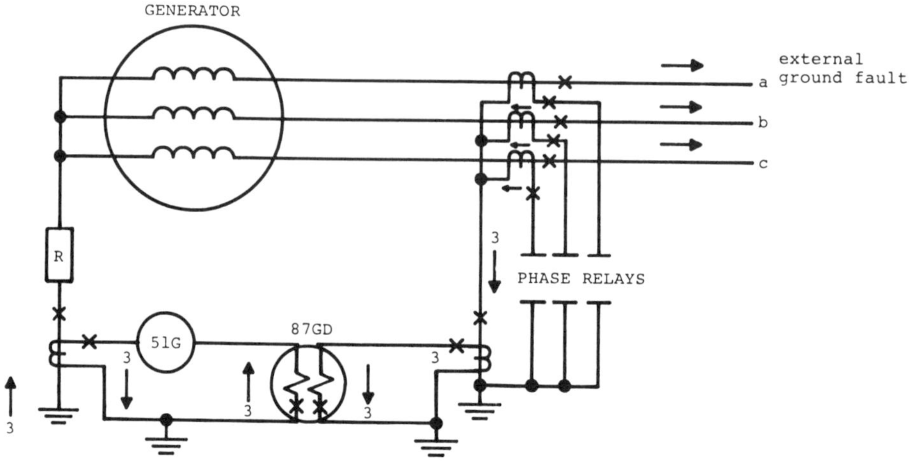

Fig. 8.12 Ground (zero-sequence) differential protection for a generator using a directional ground overcurrent relay.

varies with operation, three arcs of circles define the limits. In one area of operating the limit is the overheating in the rotor windings; in another, in the stator windings; and in the third, in the stator iron.

As indicated earlier, generators should be operated cautiously in the leading or negative reactive zone. The added limit here is the steady-state stability limit (SSSL). This is defined as a circle arc where the offset (center) and radius are:

$$\text{Center offset} = \frac{1}{2}V^2\left(\frac{1}{X_s} - \frac{1}{X_d}\right) \text{ Volt amps} \tag{8.13}$$

$$\text{Radius} = \frac{1}{2}V^2\left(\frac{1}{X_s} + \frac{1}{X_d}\right) \text{ Volt amps} \tag{8.14}$$

where V is the generator line-to-neutral terminal voltage, X_s the total equivalent impedance of the connected system, and X_d the synchronous unsaturated reactance. These are power limits usually expressed in per unit, with X_s and X_d on the generator base.

This steady-state stability curve is typically as shown in Fig. 8.14a. It will vary with the generator and with the system connected, as well as with the voltage. Although the system and voltage will vary during operation, these are usually relatively small for a given system.

The generator excitation system has a minimum excitation limiter (also known as the underexcited reactive ampere limit) to prevent the

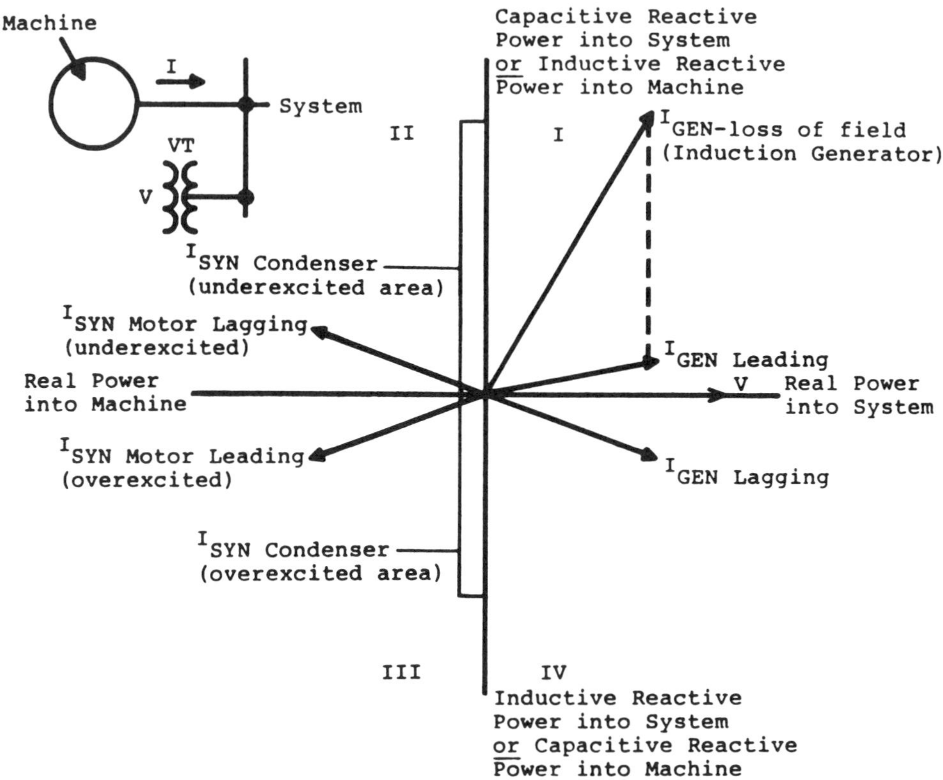

Fig. 8.13 Current and related power diagram for synchronous machines.

exciter regulator from reducing the field below its set point. This is typically set just above the steady-state stability, as shown in Fig. 8.14.

For application of a distance relay, these power curves must be converted to impedances for plotting on the R-X axes. This conversion of Eqs. (2.17) and (6.5) is

$$Z_{relay} = \frac{kV^2 R_c}{MVA\ R_v} Z_{pu\ \Omega} \tag{8.15}$$

R_c and R_v are the respective current and voltage transformer ratios used for the distance relay. If the plot is made in primary ohms, the R_c/R_v factor would not be used. From a value of MVA at the angle

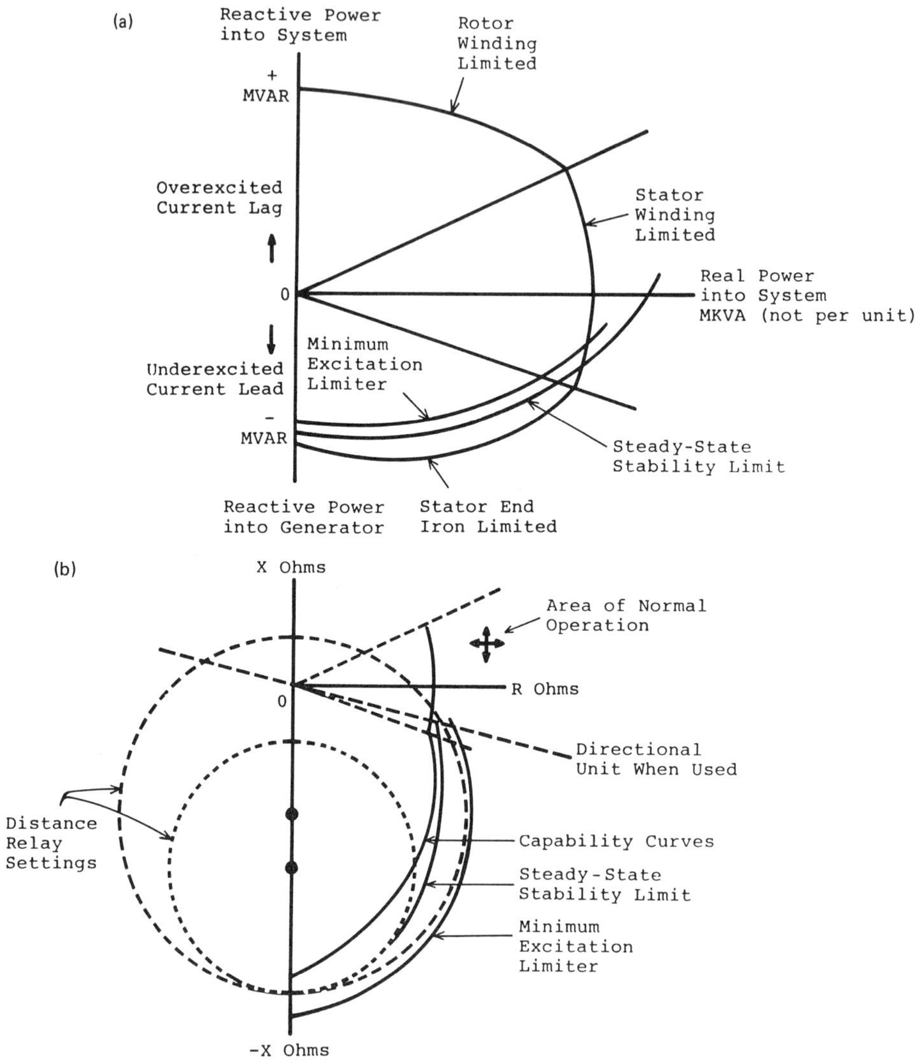

Fig. 8.14 Typical power capability and stability curves for a generator and their conversion to an R-X diagram for (40) relay protection application: (a) capability and stability curves on power axes; (b) power curves transferred to R-X axes with distance-type relay protection.

indicated from the capability curve, Eq. (8.15) converts this to ohms. This converted value is plotted on the R-X diagram at that angle. This conversion is shown in Fig. 8.14b for both the underexcited capability curves and the stability curve. For the stability curve the conversion can be made easier if the values of X_s and X_d are known. Then the stability circle center (offset) from the origin is $1/2(X_d - X_s)$ and the radius is $1/2(X_d + X_s)$. If the plot is in secondary ohms, X_d and X_s must be in secondary ohms per Eq. (8.15).

It should be noted that in the R-X diagram of Fig. 8.13b, the origin is at the generator terminals, with X_d plotted below the origin and X_s plotted above; also, that increasing or higher power is indicated by a longer distance from the origin in Fig. 8.14a, but by the shorter impedance vector in Fig. 8.14b. Thus in the power diagram (Fig. 8.14a), safe operating power is within the capability and stability curves but outside the curves in Fig. 8.14b. The minimum excitation limiter operates on a power level less than the stability limit.

The normal operating area is as shown in Fig. 8.14b. With reduced excitation or loss of excitation, the impedance phasor moves slowly as the flux decreases into the fourth quadrant. A distance relay (40) enclosing this area provides a good means of detecting this condition. Several setting modes are available.

1. For complete loss of field, the distance relay is set as illustrated by the smaller circle in Fig. 8.14b. The diameter is on the order of X_d, with the upper part of the circle 50 to 75% of X'_d below the origin. X'_d is the transient reactance of the generator. The relay operates when the impedance vector moves into this circle. Operating times of about 0.2 to 0.3 s are used with a complete shutdown of the generator.

2. To detect low excitation, partial loss, or complete loss, the diameter is set preferably inside the minimum excitation limiter setting but outside the generator capability and stability limit curves. This is shown by the larger-diameter dotted circle. It is not always easy to make this setting as suggested; good judgment and compromise may be necessary. A directional unit is required to avoid operation of nearby faults and stable transient swings. Relay (40) operation is below the dashed directional line and within the larger dashed operating circle. Where applicable, an undervoltage unit set to drop out between 87 and 80% of normal voltage is used to supervise the relay operation. If the power system can supply reactive power to the generator without a significant drop in voltage, an alarm is sounded for possible corrective action, followed by a shutdown trip after a time delay. Typical delays used vary with machine and system, but are 10 s to 1 min. If the voltage drops below the voltage unit setting, tripping is initiated with operating times about 0.2 to 0.3 s.

3. For large, important generator units, a combination of (1) and (2) is used by applying two (40) loss-of-field relays.

8.9 GENERATOR PROTECTION FOR SYSTEM DISTURBANCES AND OPERATIONAL HAZARDS

In previous sections we have discussed backup protection for uncleared or delayed faults using relays 21, 46, and 51V. In this section we outline problems and protection for other problems that may be encountered. In general, these occur principally in connection with large units in utility power systems. First we give a brief review for small units.

8.9.1 Industrial Generators

Small generators, particularly those in industrial plants or private power sources, may encounter some of the following problems. Since these are invariably connected to a utility, the first action for problems in the tie system is to separate from the tie promptly. In most cases this can be done with an under-frequency relay, unless the tie is floating (i.e., the local generator is supplying all of the local load). This is seldom a normal operating condition. Typical underfrequency settings are 59.5 to 59 Hz in a 60-Hz system with sufficient time delay to override any transient operation but less than any reclose time used by the utility. Power directional relays at the tie may or may not be useful, depending on the system and its operation.

8.9.2 Off-Frequency Operation

For utility generators, faults in the system occasionally can result in a system breakup which leaves an imbalance between available generation and the load. This results in either an excess of or insufficient power for the loads. In the first case, overfrequency with possible overvoltage results as a result of reduced load demand. Operation in this mode will not produce overheating unless rated power and approximately 105% voltage are exceeded. The generator controls should promptly adjust the unit to the load demand.

With insufficient generation for the connected load, underfrequency results, with a heavy load demand. The drop in voltage causes the voltage regulator to increase the excitation. The result is that overheating can occur in both the rotor and the stator. At the same time, more power is being demanded, with the generator less able to supply it at reduced frequency. System load shedding should adjust the load to the available generation before a critical situation develops.

Normal overcurrent protection will protect the rotor, but additional protection is suggested for the stator. This is an extremely inverse-time-overcurrent relay (50/51) set to match and operate just before the generator armature short-time thermal limit is reached. It should be controlled by a high-dropout instantaneous overcurrent unit set at about 115% of rated current to prevent operation below that current level.

The discussion above relates to electrical generator problems. Other problems exist, particularly for steam turbines operating at frequencies different from synchronous. Turbine blades are designed and tuned for efficient operation at rated frequency rotation. Operation with load at different frequencies can result in blade resonanace and fatigue damage in the long blades (18 to 44 in.) in the turbine low-pressure unit. Typical limits for 60-Hz machines with 18- to 25-in. low-pressure unit blades are about 58.5 to 61.5 Hz for continuous operation, but between 56 and 58.5 Hz, only 10 min of operation accumulated over the entire lifetime of the machine is permitted. For turbines with 25- to 44-in. blades, the typical continuous range is between 59.5 and 60.5 Hz, 60 min accumulated lifetime between 58.5 and 59.5, and 10 min accumulated lifetime between 56 and 58.5 Hz. For these units underfrequency relays are suggested. One application uses a three-step unit, one set without time delay at 56 Hz, another at 58.4 Hz with a 2-min delay, and the third at 59.4 Hz with a 6-min delay.

8.9.3 Overvoltage

Generators as well as transformers must not be subject to overvoltage except for short or transient excursions. With normal operation near the knee of the iron saturation curve, small overvoltages result in significant exciting currents in transformers, and excessive flux densities and abnormal flux patterns in generators. These can cause severe and extensive damage.

The field excitation current at rated output is greater than that required at no load, so it is important that the excitation be reduced correspondingly as load is reduced. Normally, this accomplished by the regulating system, but incorrect voltage signals, loss of VT fuses, or other failures in these systems can result in high overvoltage.

A particularly dangerous period is during the time when the generator is disconnected from the system and the speed is being changed. Generator voltage is proportional to frequency and the magnetic flux, so overvoltage protection should have a constant pickup as a function of the ratio of voltage to frequency, a volts/hertz (24) type. Protection supplementary to that in the generator controls is suggested using two-level volts/hertz units. One setting should be around 110% rated voltage to alarm with a subsequent trip within approximately 1 min, the other set around 120% rated voltage to trip on the order of 6 s. A separate VT supply is preferable.

8.9.4 Generator Motoring

If the prime-mover supply is removed while the generator is connected to the power system and the field excited, the power system will drive the unit as a synchronous motor. This is particularly critical for steam and hydro units. For steam turbines it causes overheating and potential damage to the turbine and turbine blades. Low water flow for the hydro units can cause cavitation of the blades of the turbine. This can also occur by closing the steam or water flow valves too rapidly during a load-reduction phase or by tripping the turbine while not correspondingly tripping the generator breaker.

A number of detection means are provided as part of the generator and its control, but a supplementary reverse power relay (32 is recommended. This is shown in Fig. 8.8. It is a power directional relay connected to operate when power flows into the generator. Typical sensitivity is 1/2% of rated power operating on the order of 2 s.

8.9.5 Connecting Generators Out of Synchronism

In recent years several cases of severe damage of generators have resulted from the unit inadvertently being connected improperly to the power system. This can occur by incorrect closing of the circuit breaker while the unit is on turning gear, coasting to a stop, at standstill, or by improper synchronization. The normal protection may operate in many cases but not necessarily for all possibilities. This is an increased danger if some of the protection is not available or useful at low frequencies of startup or shutdown. Since the unit or generator breaker(s) may be the problem, it is important that the adjacent breakers be opened promptly by local breaker failure and/or by remote tripping. Additional protection for this is provided by three directional inverse time-overcurrent units (67), one in each phase, connected to operate for reverse power into the generator. On the basis of current normally flowing out of the generator into the power system, the relay characteristics and connections should provide an operating zone from about 30 to 60° through 180° (reverse power into the generator) to 210 to 240° lead. Typical pickup should be around 0.5 pu current with 2.0 pu current operating times around 0.25 s. This application replaces motoring protection (32) and is more responsive to the conditions outlined above.

8.9.6 Breaker Pole Flashover

Where there is the potential for a pole flashover before synchronizing or after separation from the power system, additional pole disagreement protection is recommended. This condition is likely to occur with no fault and light load, so the current can be very low. In one case a flashover of a 500-kV breaker pole of a 150-mi line which was open at the remote end resulted in sufficient charging current to maintain 70%

rated voltage at rated speed on the generator. I_2 was 12% above the generator rated current. This can be cleared only by prompt reduction of the excitation. Local backup and remote tripping of the far end of the line remove the system, but not the generator. Negative-sequence protection may respond, but with long response time, or may not respond.

Sensitive protection (61) breaker pole failure is available to compare the magnitude levels in the three phases. It operates if one phase current is below a set value while either of the other two are above its set value. Typical sensitivities provide operation if one phase current is less than 20 to 60 mA and the other above 40 to 200 mA. These are in the primary or high-voltage side of the unit transformer. A 3:1 level difference with operating times of around 1/2 s is suggested.

Where two high-side generator breakers are used, as in a ring or breaker and half buses, there can be enough normal unbalance circulating in the bus to operate this sensitive protection. For these applications the unbalance operation should be supervised by a zero-sequence voltage level detector, or a higher setting used.

8.9.7 Subsynchronous Oscillations

The application of series compensation in long EHV transmission lines provides increased power transmission and stability but may result in subsynchronous oscillations. This may also occur for a generator connected to a HVDC transmission line.

The natural frequency of a system (f_n) with X_c capacitive reactance and X_1 inductive reactance

$$f_n = f_s \frac{X_c}{X_1} \qquad (8.16)$$

where f_s is the synchronous or normal system frequency. Since X_c is less than X_1, f_n is subsynchronous. This results in problems in the system, particularly in the generators. The rotor rotating at synchronous frequency is turning faster than the magnetic field produced by the subharmonic frequency. This results in a negative slip and negative resistance effect, with the generator tending to operate as an induction generator. When f_n is close to f_s, the slip is small and the negative resistance large. If this resistance is greater than the system resistance, the circuit becomes self-excited, and the subsynchronous current magnitudes grow. This can cause overheating in the system. One solution is to use low-resistance pole-face damper windings to reduce the rotor resistance.

Another effect of the subfrequency is to produce oscillating torques in the generator with additional transient torques occasioned by switching and faults in the system network. There is a danger that these torques may produce rotor damage.

A number of measures can be taken to mitigate this, and protective relays have been developed to detect subsynchronous resonance. One type measures torsional motion, another senses the level of subsynchronous currents in the armature. This is a complex phenomenon and beyond the scope of this book, particularly as it occurs only in a few systems with series capacitance.

8.10 FIELD GROUND PROTECTION

Ground detection for the exciter and field are important and usually supplied as part of that equipment rather than applied by the user. However, if not supplied, or additional protection desired, protective relays are available.

For units with brushes, a relay (64) with a voltage-divider circuit can be connected across the field and exciter with a sensitive dc-type relay connected between the bridge network and ground. When a ground occurs in the field or exciter circuits, a voltage appears across the relay to produce operation. To avoid no operation for a ground fault at a null point, one branch of the bridge includes a nonlinear resistor that changes this blind spot with voltage variations of the field.

Generators with brushless exciters are supplied with means to drop pilot brushes on slip rings to measure the insulation level of the field on a periodic basis. The insulation of the exciter field is checked continuously. Both these normally provide an alarm for operator action, but can still be used to trip if desired.

8.11 GENERATOR OFF-LINE PROTECTION

All the protection should be examined to see if it is operable during the process of bringing the unit up to rated speed and voltage for synchronizing to the power system. For some units, such as steam turbines, this process can include operation at reduced frequencies for several hours. Thus the protection should be active from about 1/4 to 1/2 rated frequency and above. When the relays and/or instrument transformer's performance is impaired at these low frequencies, supplemental temporary protection should be provided.

8.12 OUT-OF-STEP PROTECTION

For many generators, being out of step is a system problem with the electrical center out in the transmission area. Once the voltages of the separate generation sources swing 180° apart, recovery is not possible and separation must be made. Out-of-step detection and tripping is covered in Chapter 14. No relaying is normally applied at the generator in these cases.

With the larger generators and higher-voltage transmission, the system electrical center can move into the unit transformer and even into the generator. For these systems, out-of-step protection at the generator must be added, as none of the other protection can respond adequately. The large circle setting of the loss-of-field unit (40) will operate for swings passing near and at the bus, and within the unit transformer and generator, but can pass out before the (40) relay times out. This separate protection is discussed in Chapter 14.

8.13 MULTIPLE GENERATOR UNITS CONNECTED DIRECTLY TO A TRANSFORMER: GROUNDING AND PROTECTION

Most often these are large tandem or cross-compound units operating from a common steam prime-mover source. The two units may be paralleled and connected to a common delta-unit transformer winding, or each generator may be connected to separate delta windings on a three-winding unit transformer. For the first or paralleled connection, only one of the generators is grounded, usually with high-impedance grounding. The other unit is left ungrounded. Ground faults in either unit will operate the ground protection as discussed above. The location of the fault will not be indicated, since the fault current level is the same practically for all fault locations. Both units must be shut down, as there is no breaker to isolate the two units.

With the generators connected to separate transformer windings, each unit must be grounded for ground protection. Separate generator protection should be applied to each generator unit. Without breakers between the generators and transformer, the overall transformer differential protection must be of the multirestraint type and connected as discussed in Section 9.8.

Each unit should have loss-of-field protection (40). Only one negative-sequence (46) and one distance (21) or three-phase (51V) relay for backup protection are necessary, and they can be connected to either unit. This is based on both units operating together at all times. If one unit can be operated with the other shut down, each unit should have complete protection, as outlined above and summarized in Fig. 8.8.

8.14 COMBUSTION TURBINE GENERATOR PROTECTION

These units provide an increasing source of power, particularly in areas where coal or hydro is limited or costly and gas or oil is available. Originally, they were relatively small in size and used for peaking, but they have evolved to base-load units with sizes of over 100 MVA. The protection is as shown in Fig. 8.9, supplemented by the additional protection of Fig. 8.8 as the size and connected system warrant.

These units can be motored by the loss of combustion in the turbine. Generally, this is not as critical as for steam or hydro units, except possibly where speed reducers are used. These may not be designed for reverse rotation. Typical settings for motoring protection (32) are on the order of 7 to 8% of the machine power rating.

8.15 SYNCHRONOUS CONDENSER PROTECTION

These units usually operate as an unloaded motor to supply capacitive reactance to the system (Fig. 8.13). Protection for these is as shown in Fig. 8.9, with the additions shown in Fig. 8.8, as indicated by the system and operation requirements. The loss of field protection (40) should be set with its operating circle to enclose an impedance seen at the terminals with zero excitation, or

$$Z = \frac{1}{I_{\text{short circuit}}} \quad \text{pu} \tag{8.17}$$

Operation to supply capacitive reactance will operate the distance unit, but tripping is supervised by the voltage unit. No protection exists for inductive reactance into the system (overexcited), as the directional-sensing unit is open and the distance unit may or may not be operated.

8.16 GENERATOR TRIPPING SYSTEMS

In general, immediate tripping is recommended for all faults in the generator. This means opening the main and field breakers and closing the turbine stop valves or gates. Such action results in sudden and complete shedding of load, which, particularly at full load, can be a severe shock to the mechanical systems. For phase and ground faults, except possibly with high-impedance or resonant grounding systems, immediate tripping is mandatory. With very limited ground-fault current, supervised or delayed tripping is sometimes used. The problem always is that any fault indicates damage and may develop to other phases, and/or a second fault may occur, all with the potential of a high increase in damage and danger. Even with immediate tripping the stored energy in the rotating mass continues the fault and damage for a considerable period.

The alternatives used to preferred complete immediate trip are (1) permissive shutdown where the main and field breakers are tripped after the turbine stop valves or gates are closed, (2) alarm with fixed time-delay trip to permit operator action, and (3) alarm only. Again these are applicable only for ground faults when using high-impedance grounding, and as an individual system preference.

8.17 PROTECTION SUMMARY

Generators, particularly very large units, are extremely critical to system operation and integrity. Thus protection is very important with regard to both dependability and security. While fault incident fortunately is quite low, complete protection with reasonable duplication is well justified. This is the option of the particular system and individual circumstances, so any installation can show variations from others.

As has been indicated, the protection devices of Figs. 8.8 and 8.9 discussed in the text are those applied by the user, in contrast to the protection supplied by the generator and prime-mover manufacturer. They are in common use and represent the general practices of the industry. Figure 8.8 illustrates the maximum protection that would normally be applied to large and major units. Figure 8.9 shows the minimum protection recommended for smaller units, especially those in small and industrial plants. These are relatively isolated from the large system problems in most cases.

BIBLIOGRAPHY

See the Bibliography at the end of Chapter 1 for additional information.

ANSI/IEEE Generator Ground Protection Guide, Standard C37.101-1985, IEEE Service Center.

Griffin, C. H., Relay Protection of Generator Station Service Transformers, *IEEE Trans. Power Appar. Syst.*, PAS 101, 1982 pp. 2780–2789.

Griffin, C. H., and J. W. Pope, Generator Ground Fault Protection Using Overcurrent, Overvoltage and Undervoltage Relays, *IEEE Trans. Power Appar. Syst.*, PAS 101, 1982, pp. 4490–4501.

Gross, E. T. B, and E. M. Gulachenski, Experience on the New England System with Generator Protection by Resonant Neutral Grounding, *IEEE Trans. Power Appar. Syst.*, PAS 92, 1973, pp. 1186–1194.

Guide for Abnormal Frequency Protection of Power Generating Plants, IEEE Standard C37.106 (Draft).

IEEE Guide for AC Generator Protection, IEEE Power System Relaying Committee, publication scheduled in 1986.

Loss-of-Field Relay Operation During System Disturbances, IEEE Power System Relaying Committee, *IEEE Trans. Power Appar. Syst.*, PAS 94, 1975, pp. 1464–1472.

Out-of-Step Relaying for Generators. IEEE Power System Relaying Committee, *IEEE Trans. Power Appar. Syst.*, PAS 96, 1977, pp. 1556–1564.

Pope, J. W., A Comparison of 100% Stator Ground Fault Protection Schemes for Generator Stator Windings, *IEEE Trans. Power Appar. Syst.*, PAS 103, 1984, pp. 832–840.

Potential Transformer Application on Unit-Connected Generators, IEEE Power System Relaying Committee, *IEEE Trans. Power Appar. Syst.*, PAS 91, 1972, pp. 24–28.

Protective Relaying for Pumped Storage Hydro Units, IEEE Power System Relaying Committee, *IEEE Trans. Power Appar. Syst.* PAS 94, 1975, pp. 899–907.

Schlake, R. L., G. W. Buckley, and G. McPherson, Performance of Third Harmonic Ground Fault Protection Schemes for Generator Stator Windings, *IEEE Trans. Power Appar. Syst.*, PAS 100, 1981, pp. 3195–3202.

9

Transformer, Reactor, and Shunt Capacitor Protection

9.1 TRANSFORMERS

Transformers are everywhere — in all parts of the power system, between all voltage levels, and existing in many different sizes, types, and connections. Small transformers of about 3 to 200 kVA can be observed mounted on power distribution poles in many areas. A 325-MVA 230- to 115-kV autotransformer with a 13.8-kV tertiary is shown in Fig. 9.1.

Usually, circuit breakers or other disconnection means are available at or near the winding terminals of the transformer banks. However, economics sometimes dictates omission of a breaker. Thus transformer banks can be connected directly to a bus, line, or generator. The latter is the unit generator arrangement discussed in Chapter 8.

Differential protection, where applicable, provides the best overall protection for both phase and ground faults, except in ungrounded systems or where the fault current is limited by high-impedance grounding. In these latter low-ground-fault current systems, differential provides only phase-fault protection.

Generally, differential protection is applied to transformer banks of 10 MVA and above. The key is the importance of the transformer in the system, so differential may be desirable for smaller units in some cases.

9.2 FACTORS AFFECTING DIFFERENTIAL PROTECTION

In applying differential protection, several factors must be considered:

1. Magnetizing inrush current. This is a normal phenomenon which has the appearance of an internal fault (current into but not out of the transformers).

2. Different voltage levels and hence the current transformers are of different types, ratios, and performance characteristics.
3. Phase shifts in wye-delta-connected banks.
4. Transformer taps for voltage control.
5. Phase shift and/or voltage taps in regulating transformers.

9.3 MAGNETIZING INRUSH

When system voltage is applied to a transformer at a time when normal steady-state flux should be at a different value from that existing in the transformer, a current transient occurs, known as magnetizing inrush current. This phenomenon is illustrated in Fig. 9.2 for a transformer with no residual flux. In the figure the transformer is energized when the system voltage is zero. With the highly reactive circuit involved, the flux ϕ should be at or very near negative maximum, but the transformer has no flux. Thus the flux must start at zero and

Fig. 9.1 Three-phase 325-MVA, 230:115-kV autotransformer bank with a 13.8-kV tertiary. (Courtesy of Puget Sound Power & Light Company.)

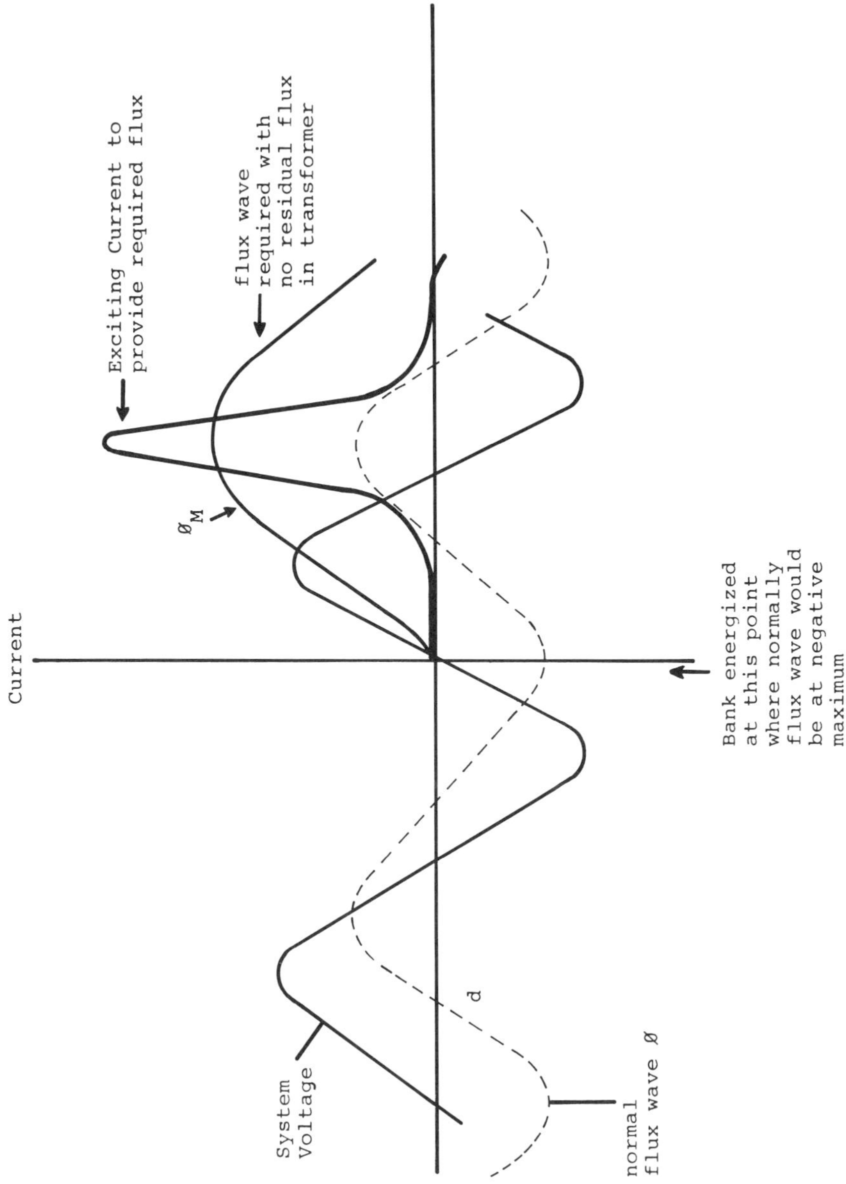

Fig. 9.2 Magnetizing inrush current phenomenon (no residual flux initially in transformer).

reach a value of 2ϕ in the first cycle period. To provide this flux, excursion requires a large exciting current, as shown. Transformers are normally operated near saturation for best efficiency, so values of flux greater than normal ϕ result in severe saturation and a large exciting current.

If a transformer has been energized previously, there is a high possibility that upon deenergization some flux ϕ_R was left in the iron. This could be positive or negative. If in Fig. 9.2, a residual flux of $+\phi_R$ had existed from an earlier energization, the flux maximum required would have been $2\phi + \phi_R$, resulting in a higher maximum magnetizing inrush current. If ϕ_R had been negative, the maximum required flux would be $2\phi - \phi_R$ with less inrush current.

This is a random phenomenon. If the transformer had been energized at or near maximum positive voltage (Fig. 9.2d), the flux requirement at that time is zero. Thus normal exciting current would flow with negligible or no transient inrush. Normal exciting currents for power transformers are on the order of 2 to 5% of full-load current.

The maximum initial magnetizing current may be as high as 8 to 30 times full-load current. Resistance in the supply circuit and transformer and the stray losses in the transformer reduce the peaks of the inrush current so that eventually it decays to the normal exciting current value. The time constant varies from around 10 cycles to as long as 1 min in very high inductive circuits.

The factors involved in the inrush, in addition to the time point of energization with relation to the flux requirements, are the size of the transformer, the size and nature of the power system source, the type of iron in the transformer, prior history, and the L/R ratio of the transformer and system.

In a three-phase circuit some inrush will always occur in one or two and generally all three phases with the voltages 120° apart, although it may or may not be maximum or zero in one of the phases. Figure 9.3 shows a typical magnetizing inrush current trace when a transformer bank is energized from either the wye- or delta-connected terminals. Some years ago studies indicated that the second harmonic component of the inrush wave was 15% or more of the fundamental. In recent years improvements in core steel and design are resulting in transformers where all inrush current harmonics are less, with possibilities of the second harmonic being as low as 7%.

Magnetizing inrush can occur under three conditions and are described as (1) initial, (2) recover, and (3) sympathetic.

9.3.1 Initial Magnetizing Inrush

This may occur when energizing the transformer after a prior period of deenergization. This was described above and has the potential of producing the maximum.

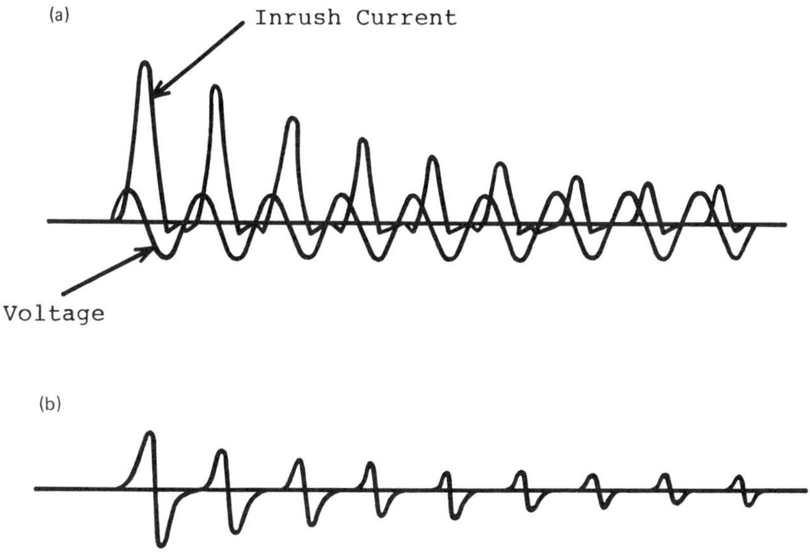

Fig. 9.3 Typical magnetizing inrush current to transformers: (a) A-phase current to wye-connected windings (b) A-phase current to delta-connected windings.

9.3.2 Recovery Magnetizing Inrush

During a fault or momentary dip in voltage, an inrush may occur when the voltage returns to normal. This is called the recovery inrush. The worst case is a solid three-phase external fault near the transformer bank. During the fault the voltage is reduced nearly to zero on the bank; then when the fault is cleared, the voltage suddenly returns to essentially normal value. This may produce a magnetizing inrush, but its maximum will not be as high as the initial inrush since the transformer is partially energized.

9.3.3 Sympathetic Magnetizing Inrush

A magnetizing inrush can occur in an energized transformer when a nearby transformer is energized. A common case is paralleling a second transformer bank with a bank already in operation. The dc component of the inrush current can also saturate the energized transformers, resulting in an apparent inrush current. This transient current, when added to the inrush current of the bank being energized, provides an offset symmetrical total current that is very low in harmonics. This would be the current flowing in the supply circuit to both transformer banks.

9.4 TRANSFORMER DIFFERENTIAL RELAY CHARACTERISTICS

The basic principles of differential relays and protection were outlined in Chapter 6. For applications to transformers the differential relays are less sensitive and with typical percentage characteristics between 20 and 60%. This provides accommodation for the different CT ratios, types, and characteristics, different primary current energization levels, and for transformer taps where they exist.

To avoid undesired operation on the magnetizing inrush current, the relays may (1) be designed with reduced sensitivity to the transient inrush current, (2) use the harmonics of the inrush current to prevent operation, or (3) desensitize or inhibit operation momentarily during the energization time.

For smaller transformer banks, particularly in the lower-voltage subtransmission and distribution systems, the electromechanical induction disk transformer differential relays with typically 50% characteristics and operating about 0.08 to 0.10 s (five to six cycles on a 60-Hz system) are used widely with good immunity to inrush. Generally in these areas the inrush is not too severe and there is sufficient resistance in the system to damp the transient rapidly. Also, the induction disk unit does not operate very efficiently on this high distorted offset wave, and the unit does not operate on dc. The advantage is the relative simple rugged design and low cost. However, it is not possible to assure that the relay will never operate on inrush, although experience can show it to be quite immune. Someone has observed that this type of relay "has a long and enviable record" of good performance, which accounts for its continued use. Typical pickup current is 2.5 to 3.0 A.

For larger banks, banks in the higher-voltage systems and/or where it is important to assure no operation on inrush, harmonic restrained differential relays should be used. For faults the harmonics are of very low values, so the "handle" of harmonics provides an effective means to distinguish the inrush current from fault currents. The second harmonic, which is a major component in the inrush wave, is used to restrain or prevent relay operation. In some designs it is used alone; in others, in combination with the other harmonics. Typical pickup current for these types varies from arount 0.75 to 2.5 A with operating times of 0.015 to 0.03 s.

Momentary desensitizing or disconnecting differential relays is not recommended, as it limits protection. It is used occasionally to solve a problem until a more satisfactory solution can be effected. A slip contact on the breaker control switch can insert a resistor in the operation coil circuit to desensitize the relay or momentarily short the operating coil while the breaker is being closed. A high-set instantaneous trip unit can be used to operate for a heavy current internal fault occurring on energization.

9.5 APPLICATION AND CONNECTION OF TRANSFORMER DIFFERENTIAL RELAYS

The differential protective zone in all cases must account for all circuits into or out of the zone, with one relay per phase for the zone. For two winding transformers with a single set of CTs associated with the windings, a two-restraint relay as shown in Fig. 6.2 is applicable. For multiwinding transformers such as three winding banks, autotransformers with tertiary winding connected to external circuits, or where double breakers and CTs supply a single winding (as in a ring bus or breaker and a half arrangement) a multiple-restraint winding relay should be used. Differential relays are available with two, three, four, and up to six restraint windings with a single operating winding. The characteristics are similar to those described above.

The important fundamentals of application are:

1. Use a restraint winding for *each* fault source circuit.
2. Avoid paralleling the CTs of a feeder (no fault source) with CTs of a fault source.
3. Parallel feeder CTs carefully.

The reasons for and logic of these will be apparent later.

The currents through the differential relay restraint windings should be in phase and there should be a minimum difference (operating) current for load and external faults. Ideally, this difference should be zero, but with different CT ratios on the different voltage levels, this is impossible practically in almost all cases. This suggests two steps for correctly connecting and setting transformer differential relays:

1. *Phasing:* using wye-delta units, to assure that the secondary currents to the differential relay are in phase.
2. *Ratio adjustment:* selecting CT ratios and/or relay taps to minimize the difference current that will flow in the operating circuit.

The recommendations and criteria given above are best explained by typical examples.

9.6 EXAMPLE: DIFFERENTIAL PROTECTION CONNECTIONS FOR A TWO-WINDING WYE-DELTA TRANSFORMER BANK

Consider a delta-wye-connected transformer bank as shown in Fig. 9.4. The ABC delta leads the abc wye by 30°; thus, following the ANSI standard, ABC represents the high-voltage 138-kV side, and abc is the low-voltage of 69-kV side.

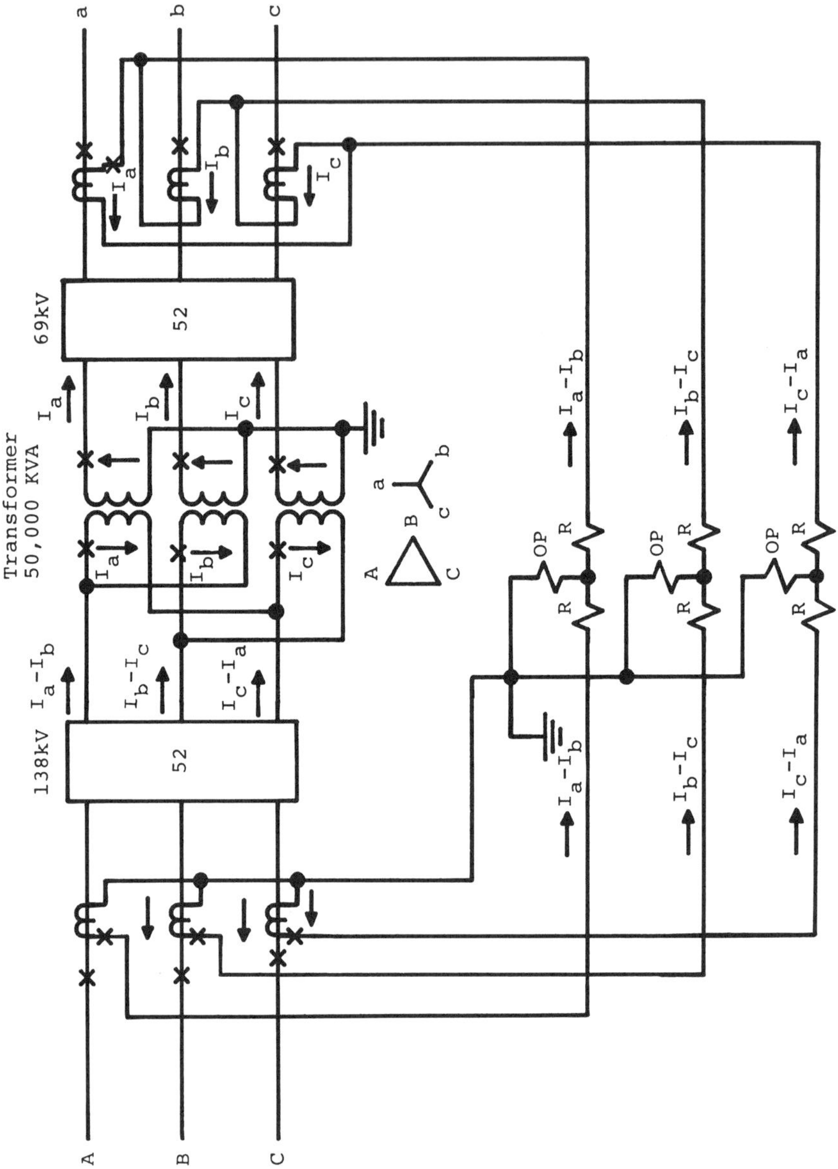

Fig. 9.4 Differential relay connections for the protection of a two-winding transformer bank.

Secondary currents in phase on the differential relay could be provided by connecting the abc set of CTs in wye or in delta with the ABC set of CTs in delta or wye, respectively. However, connecting the abc CTs in wye would result in incorrect operation for external ground faults. Zero-sequence current supplied by the transformer-grounded wye to external faults in the abc system can flow through the wye-connected abc CTs to the relay restraint coil returning through the operating coil. This is because the zero-sequence current circulates in the transformer delta and does not flow in the ABC system to provide proper external fault balancing restraint. Therefore, the CTs on wye transformer windings should be connected in delta. This provides a zero-sequence circulating path within the CT connection so that it cannot flow in the relays.

9.6.1 First Step: Phasing

The two sets of CTs must be connected so that the secondary currents to the relay restraint windings are in phase for through load or any external fault. Assume balanced three-phase currents flowing through the transformer. The direction is not important as long as the currents flow through the bank. It is desirable and easiest to start on the transformer wye side, so in Fig. 9.3 assume that I_a, I_b, and I_c flow in the wye and to the right into the abc system as shown. With transformer polarity as shown, these currents appear in the high-voltage windings, and with the delta connected as shown appear in the high-voltage ABC system as $I_a - I_b$, $I_b - I_c$, and $I_c - I_a$, flowing consistently to the right in the A, B, C phases, respectively.

With the abc CTs to be connected in delta as explained above, the ABC CTs will be connected in wye. With the CT polarity shown, secondary $I_a - I_b$, $I_b - I_c$, and $I_c - I_a$ flow to the differential relay restraint coils as shown. For the external condition, these currents should flow out of the other restraint coils and to the right. Back to the wye abc side, I_a, I_b, and I_c currents flow to the left in the CT secondaries. The last part is to connect these abc CTs in delta to provide the proper restraint secondary currents. This completes the first and phasing step.

In summary, this phasing was done by assuming balanced current flow in the transformer wye circuit, transferring these currents through the transformer to the delta side, connecting the delta-side CTs in wye and to the relay restraint coil, carry these currents through the relays to the other restraint coils, and connect the wye transformer-side CTs in delta to provide these restraint coil currents. If the transformer bank had been connected in delta on both sides, the CTs on both sides could be connected in wye to the differential relays.

For wye grounded-wye grounded transformer banks without a tertiary, or with a tertiary not brought out to terminals, delta-connected

CTs on both sides must be used. It would be possible to use wye-connected CTs if the bank consisted of three independent two-winding transformers connected wye grounded-wye grounded. However, if this type of bank was a three-phase type (all three phases in a common tank), delta-connected CTs are recommended. In these three-phase units there is always the possibility of a phantom tertiary resulting from the interaction of the fluxes because of the construction. The key in all cases is that if wye CT connections are used, the per unit zero-sequence current must be equal on both sides of the bank for all external faults. The delta CT connection is safer, as it eliminates zero sequence from the relays.

At this point a question might arise as to whether the transformer differential relays provide protection for ground faults with delta-connected CTs. The answer is that for ground faults the relays can operate on the positive- and negative-sequence currents involved in these faults. The differential relays operate on the total fault components for internal faults. Thus for a single-line-to-ground fault the total fault current is $I_1 + I_2 + I_0$ and $I_1 = I_2 = I_0$, so the differential relay with delta CTs will receive $I_1 + I_2$, or $2I_1$ for the internal fault.

With reference to Fig. 9.4, an internal single-line-to-ground fault on the 69-kV side would be fed by positive- and negative-sequence currents from the 138-kV source, and positive-, negative-, and zero-sequence current from the 69-kV source. The delta-connected 69-kV CTs eliminate the 69-kV zero-sequence current, but the sum of the positive and negative sequences from the 138 kV and 69 kV both add to flow through the relay operating windings.

For an internal phase-to-ground fault on the 138-kV side, positive- and negative-sequence currents are supplied from the 69-kV source, and all three sequence components from the 138-kV source. Here the differential relays receive the total $I_1 + I_2 + I_0$ fault current with the 138-kV wye-connected CTs.

9.6.2 Second Step: CT Ratio and Tap Selections

It is important to minimize the unbalance current flowing through the operating coils for loads and external faults. Most transformer differential relays have taps available to assist in this process. These provide for differences in the restraint current in the order of 2 or 3 to 1. The percent mismatch (M) can be expressed as

$$M = 100 \times \frac{I_H/I_L - T_H/T_L}{S} \tag{9.1}$$

where I_H and T_H are the secondary current and relay tap associated with the high-voltage (H) winding; I_L and T_L the secondary current and relay tap associated with the low-voltage (L) winding; and S the smaller of the current or tap ratios in the formula. The sign of the subtraction is not significant, so if T_H/T_L is greater than I_H/I_L, the subtraction can be made to give a positive number.

The rated currents in the 50-MVA transformer are

$$I_H = \frac{50{,}000}{\sqrt{3} \times 138} = 209.18 \text{ primary amperes at 138 kV} \tag{9.2}$$

Choose a 250:5 CT ratio which gives

$$I_H = \frac{209.18}{50} = 4.18 \text{ secondary amperes in the left-hand restraint winding of Fig. 9.4} \tag{9.3}$$

$$I_L = \frac{50{,}000}{\sqrt{3} \times 69} = 418.37 \text{ primary amperes at 69 kV} \tag{9.4}$$

Choose a 500:5 CT ratio which gives

$$I_L = \frac{418.37}{100} = 4.18 \text{ secondary amperes in the 69-kV CT secondaries}$$

$$= 4.18\sqrt{3} = 7.25 \text{ secondary amperes in the right-hand restraint windings of Fig. 9.4} \tag{9.5}$$

$$\frac{I_H}{I_L} = \frac{4.18}{7.25} = 0.577 \tag{9.6}$$

Suppose that the particular relay being used has the possibility of $T_H = 5$ and $T_L = 9$, which provides

$$\frac{T_H}{T_L} = \frac{5}{9} = 0.556 \tag{9.7}$$

Then for this application the percent mismatch, Eq. (9.1), is

$$M = 100\,\frac{0.577 - 0.556}{0.556} = 3.78\% \tag{9.8}$$

This is a good match. With transformer differential relays having percentage characteristics between 20 and 60%, 3.78% provides ample safety margin for unanticipated CT and relay differences and performance errors. Theoretically, this mismatch could approach closely the specified percentage of the differential relay, which would then reduce the safety margin correspondingly.

In selecting these CT ratios, it is desirable to keep the ratio as low as possible for higher sensitivity, but (1) not to have the maximum load exceed the CT or relay continuous current ratings as provided by the manufacturers, and (2) the maximum symmetrical external fault should not cause a current transformer ratio error of more than 10%.

For (1) the maximum load should be the highest current, including short-time emergency operation. Transformers often have several ratings: normal, with fans, with forced circulation, and so on. For most transformer differential relays the restraint coils have continuous ratings of 10 a or more.

For (2) the performance of current transformers is covered in Chapter 5 and is applicable here. In general, the burdens of differential relays for external fault conditions are very low. It is accepted practice to use a separate set of CTs for differential protection and not to connect or at least minimize any other relays or equipment in these circuits. This provides a minimum and low total burden to aid the current transformer performance.

Also, the current through the differential for the external fault is only that part of the total fault current that flows through the transformer bank to the fault. Thus it is limited by the transformer bank impedance. In contrast, the internal fault is the total fault current, but here not all of the total flows through any set of CTs except in the case of a single fault source. Some saturation of the CTs may occur for the internal fault. Although this is not desirable, there probably will not be relay operating problems unless the saturation is very severe, as the fault operating current is usually many times greater than the relay pickup current.

9.7 LOAD TAP CHANGING TRANSFORMERS

Usually, these taps provide the possibility of modifying the voltage ratio 10% for voltage and/or var control. Differential relays can be applied as above. The CT ratio and relay tap selection should be made at the midpoint of the tap changing range and with a low value of M. The sum of M plus one-half of the tap range must be within the relay percentage characteristics. Thus in the example of Fig. 9.4, suppose that this bank has 10% taps for changing the 69-kV voltages to a maximum of +10% or a minimum of −10%. With the ratios and taps selected at the 69-kV midpoint of the tap range, a M value of 3.78% was determined [Eq. (9.8)]. With the transformer taps the maximum mismatch is 13.78%, which would occur on the maximum or minimum voltage tap. This value is still within the percentage differential characteristics of applicable relays.

9.8 EXAMPLE: DIFFERENTIAL PROTECTION CONNECTIONS FOR MULTIWINDING TRANSFORMER BANK

Fig. 9.5 shows a three-winding wye-delta-wye transformer bank. This discussion also applies to an autotransformer with a delta tertiary. With three windings connected to external circuits, three restraint transformer differential relays are required. There should be a set of CTs in each circuit, connected to a separate restraint winding. Thus the protective zone is the area between these several CTs.

The two-winding-type differential relay may be applied to multiwinding transformers if:

1. The third winding is a tertiary and not brought out to external circuits.
2. The circuits connected to the third or tertiary winding are considered as part of the protective zone. This might be where there is no adequate breaker, or the winding feeds transformer auxiliaries, and so on.
3. The tertiary winding has a very high reactance such that faults on the associated system would not be large enough in magnitude to operate the transformer differential relays.

The two steps: phasing and CT ratio and tap selection are applicable to the multiwinding applications. It is very important, indeed essential, that these, particularly the second, be done in *pairs*, that is, connect and set the CTs and relays of any two transformer windings, ignoring and assuming zero current in the other winding(s). Then repeat for another winding pair. An example will amplify this.

In Fig. 9.5 there are two grounded wye windings and one delta winding. From previous discussions, the CTs in the wye-connected winding should be connected in delta to avoid operation on external ground faults. The CTs in the delta-connected winding circuit would be connected in a star configuration to accommodate the 30° phase shift. The first step phasing is to choose a pair. Although this is arbitrary, for this example the pair should include the delta and one of the wye windings.

Starting with the left-hand wye winding, assume that balanced currents I_a, I_b, and I_c flow to the right. They pass through the wye-delta winding pair and into the ABC system as $I_a - I_c$, $I_b - I_a$, and $I_c - I_b$. The current in the right-hand wye winding is assumed zero. The wye-connected CTs on the ABC system provide these same currents on a secondary basis in the differential relay restraint windings. These secondary currents must be supplied through the left-hand restraint windings. This is accomplished by connecting the CTs in delta as shown. With this phasing complete, the second pair could be the left-hand star winding just connected and the right-hand wye winding

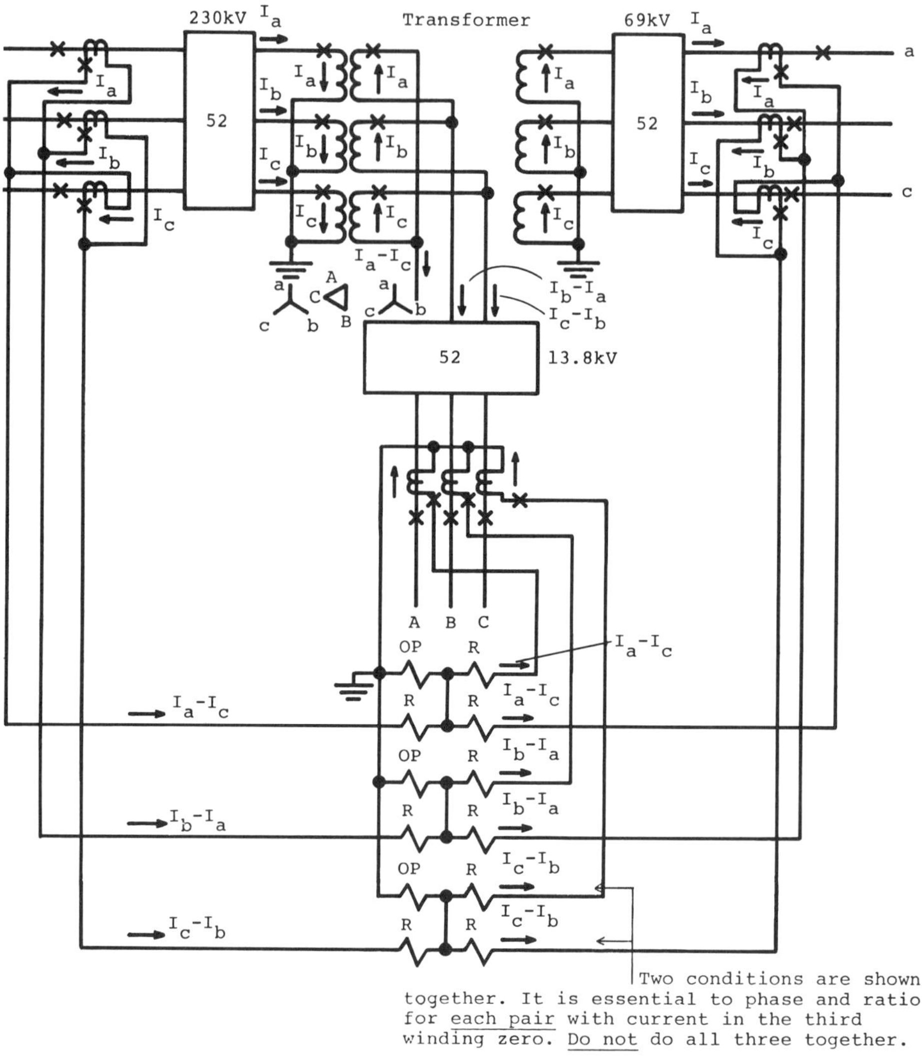

Fig. 9.5 Differential relay connections for the protection of a three-winding transformer bank. Same connections applicable for an auto-transformer with delta tertiary.

with zero in the delta circuit. Now $I_a - I_c$ in the left-hand winding restraint coil must flow out the right-hand and wye winding restraint coil; correspondingly for $I_b - I_a$ and $I_c - I_b$. With I_a, I_b, and I_c flowing to the right out of the right-hand wye winding, the CTs can be connected as required in delta. This completes the first step.

Multiwinding transformer banks usually have different MVA ratings for the several windings, and these are used to determine the CT ratios. Suppose that the Fig. 9.5 bank ratings are 60, 40, and 25 MVA for the 230-, 69-, and 13.8-kV windings, respectively. Then rated current for the windings would be

$$I_H = \frac{60{,}000}{\sqrt{3} \times 230} = 150.61 \text{ A at } 230 \text{ kV} \tag{9.9}$$

Choosing a 150:5 CT ratio, the secondary currents are

$$\begin{aligned} I_H &= \frac{150.61}{30} = 5.02 \text{ A secondary} \\ &= 5.02 \times \sqrt{3} = 8.70 \text{ A in the restraint winding} \end{aligned} \tag{9.10}$$

$$I_M = \frac{40{,}000}{\sqrt{3} \times 69} = 334.70 \text{ A at } 69 \text{ kV} \tag{9.11}$$

Choosing a 400:5 CT ratio, the secondary currents are

$$\begin{aligned} I_M &= \frac{334.70}{80} = 4.18 \text{ A secondary} \\ &= 4.18 \times \sqrt{3} = 7.25 \text{ relay amperes} \end{aligned} \tag{9.12}$$

$$I_L = \frac{25{,}000}{\sqrt{3} \times 13.8} = 1045.92 \text{ A at } 13.8 \text{ kV} \tag{9.13}$$

Choosing a 1200:5 CT ratio, the secondary currents are

$$I_L = \frac{1045.92}{240} = 4.36 \text{ A in CT and relay} \tag{9.14}$$

These rated currents are useful in selecting the CT ratios but are not usable as such in selecting relay taps and calculating mismatch. It is essential for these to select any value of MVA and pass it through the pairs with 0 MVA in the other winding(s). Only if this is done will the differential relay balance correctly for any combination of current division during loads and faults, including one winding out of service. The value chosen is not important for this part, so for example, assume that 40 MVA first flows from the 230- to 69-V systems with zero in the 13.8-kV system. This is convenient and Eq. (9.12) gives the current in the left restraint coil. In the 230-kV right restraint coil, the current to be balanced is

$$I_{230} = \frac{40{,}000}{\sqrt{3} \times 230(30)} \sqrt{3} = 5.80 \text{ A secondary} \tag{9.15}$$

Suppose that the particular relay has taps, including 5 and 6. Using these, the percent mismatch [Eq. (9.1)] is

$$\begin{aligned} M &= \frac{5.8/7.25 - 5/6}{0.8} \times 100 = \frac{0.80 - 0.83}{0.80} \times 100 \\ &= 4.17\% \end{aligned} \tag{9.16}$$

This is well within transformer differential relay characteristics. With tap 5 for the 230-kV restraint winding, and tap 6 for the 69-kV restraint winding established, now pass equal MVA through another pair, arbitrarily select 25 MVA from the 230- to 13.8-kV system, which gives 4.36 A [Eq. (9.14)] in the 13.8-kV restraint. Correspondingly, in the 230-kV restraint,

$$I_{230} = \frac{25{,}000}{\sqrt{3} \times 230(30)} \sqrt{3} = 3.62 \text{ A secondary} \tag{9.17}$$

Choose tap 6 for the 13.8-kV restraint; then the mismatch from Eq. (9.1) is

$$\begin{aligned} M &= \frac{3.62/4.36 - 5/6}{1.67} \times 100 = \frac{0.83 - 0.83}{0.83} \times 100 \\ &= 0\% \end{aligned} \tag{9.18}$$

the CTs are matched. In selecting taps as available on the particular transformer differential relay, a check should be made that the maximum load currents [Eqs. (9.10), (9.12), and (9.14)] do not exceed the continuous current ratings of the manufacturer.

This completes the second step. Again selecting the settings in pairs as described will give proper operation (no trip) for any mix of load or fault currents between the several windings. The final step is to assure current transformers performance for external as well as internal faults, as outlined in Section 9.6.2.

9.9 APPLICATION OF AUXILIARIES FOR CURRENT BALANCING

At times it may not be possible to obtain an acceptable value of mismatch with the available CTs and/or differential relay taps. These cases require the use of auxiliary CTs or current-balance transformers. It is preferable in applying these to reduce the current to the relays if possible. Reducing the secondary current to the relays reduces the relay burden by the square of the current ratio. When the current is increased to the relays, the relay burden is increased by the current ratio squared. This does not include the burden of the auxiliaries, which must be added to the total secondary load on the current transformers.

9.10 PARALLELING CTS IN DIFFERENTIAL CIRCUITS

In Section 9.5 it was recommended that a restraint coil be used for each source and that paralleling a source and feeder should be avoided. This is sometimes considered or done for a multiwinding bank or where two banks are in the same protective zone. The difficulties that may be experienced are illustrated in Fig. 9.6. There is no problem in paralleling two set of CTs such as shown as long as all three winding circuits of the transformer are in service. However, with the possibility of emergency operation where the left-hand breaker is open, restraint is lost and the differential relays operate essentially as sensitive overcurrent units. As shown in Fig. 9.6a, current flow from the right-hand source to the feeder should net zero. This is difficult with different CTs and voltage levels. Even with a perfect match, the difference in the CT exciting currents will flow to the differential relay as explained in Chapter 6. In other words, there is no effective restraint for this condition. This connection might be marginally safe for normal load flow, but is quite subject to misoperation for external faults out on the feeder.

Paralleling the CTs may be done if the transformer would never be operated with the left-hand breaker open. It is not recommended, as there is always the possibility that during an unusual emergency situation, operators may resort to unplanned opening of the left-hand breaker. In contrast, with the recommended restraint for each circuit as in Fig. 9.6b, full restraint exists, as shown.

Feeder circuit CTs may be paralleled in a differential scheme as long as there is negligible fault current supplied through them. Caution should be exercised, recognizing that for external faults out on one of the feeders, some of the secondary current required for balancing the differential is diverted to magnetize the other feeder CTs

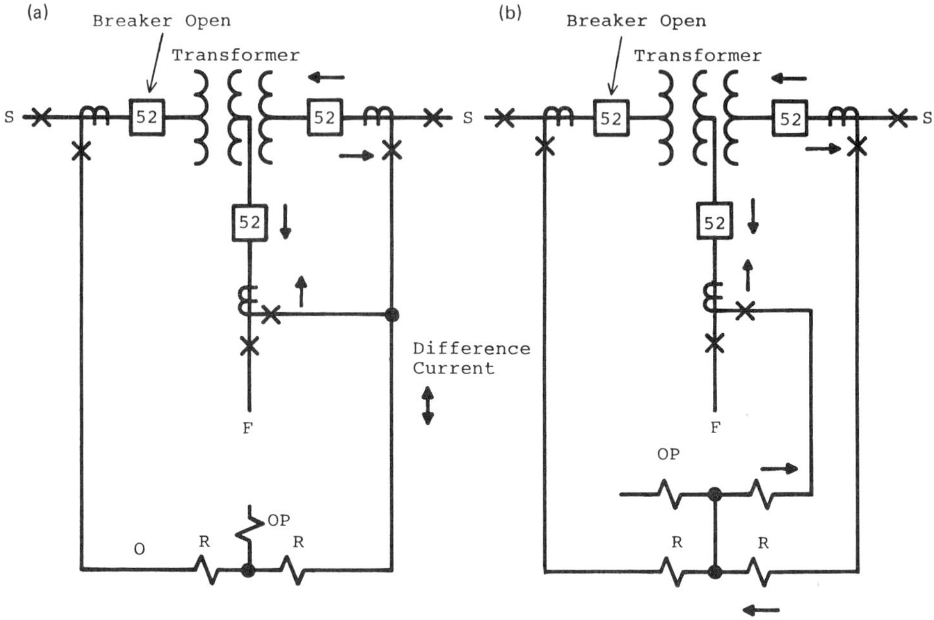

Fig. 9.6 Potential loss of restraint that may result by paralleling circuits in differential protection: (a) no restraint; (b) full restraint (S = source of fault current, F = feeder — no significant fault source).

that are not supplying primary current. This is discussed in Chapter 5.

Typical examples of transformers with multiple circuits are illustrated in Fig. 9.7. As recommended and shown, each circuit should be connected to an individual restraint winding in the differential relay following the procedure outlined earlier. Paralleling of circuits and CTs should be avoided, but when necessary, the problems discussed should be considered carefully and documented to the operating personnel.

9.11 SPECIAL CONNECTIONS FOR TRANSFORMER DIFFERENTIAL RELAYS

Occasionally, it may be necessary to use star-connected CTs on a grounded-wye circuit in a differential scheme intead of the preferred delta CT connection. This can occur in applications where a differential zone is established around several banks. For these the first

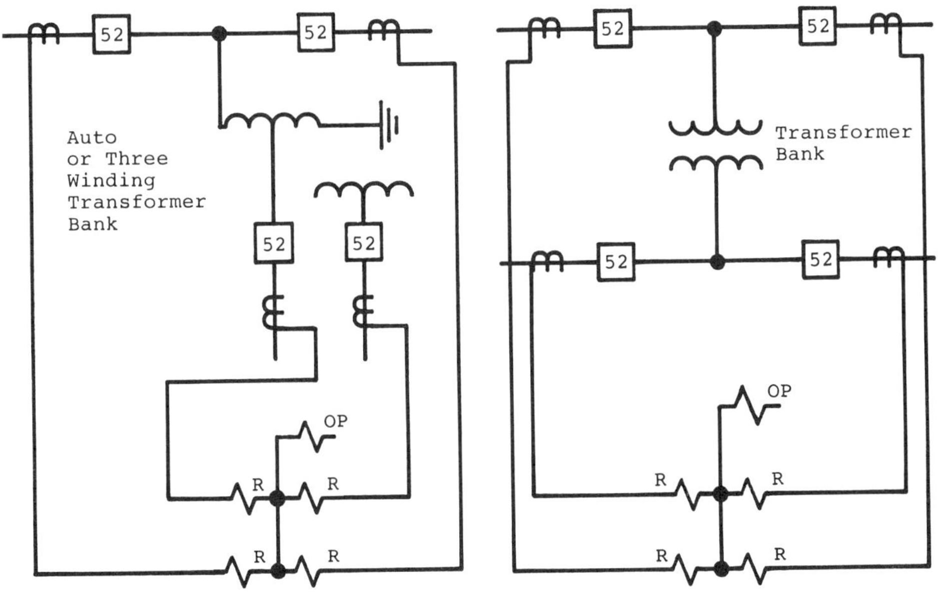

Fig. 9.7 Single-line differential connections for multiple connections to transformers.

consideration is the possibility of a sympathetic inrush. The zone should be examined to see if there is any possible operating condition where one bank can be energized, followed by energizing a second bank. If all banks in the protection zone are always energized together, sympathetic inrush will not occur, and a single differential relay application is possible. Most of these cases are the result of the omission of circuit breaker(s) and associated CTs for cost saving, but at the expense of less flexible system operation.

Figure 9.8 shows the connection of a zero-sequence trap to divert the zero-sequence current from the differential relay with wye-connected CTs on the grounded side of the transformer. It is essential in these types of connections (for any connections) that a path exists for zero-sequence current flowing into or out of a bank in the CT secondaries. This is provided by the trap as shown. If a proper path is not available in the CT secondaries, the equivalent of an open secondary circuit exists with resulting saturation and high and dangerous voltages. This true for positive and negative as well as zero sequences; the absence of a proper path for zero sequence would be apparent only on the occurrence of a ground fault in the otherwise balanced system.

The zero-sequence trap consists of three auxiliary CTs. Their ratio is not important as long as they are the same. Note that the

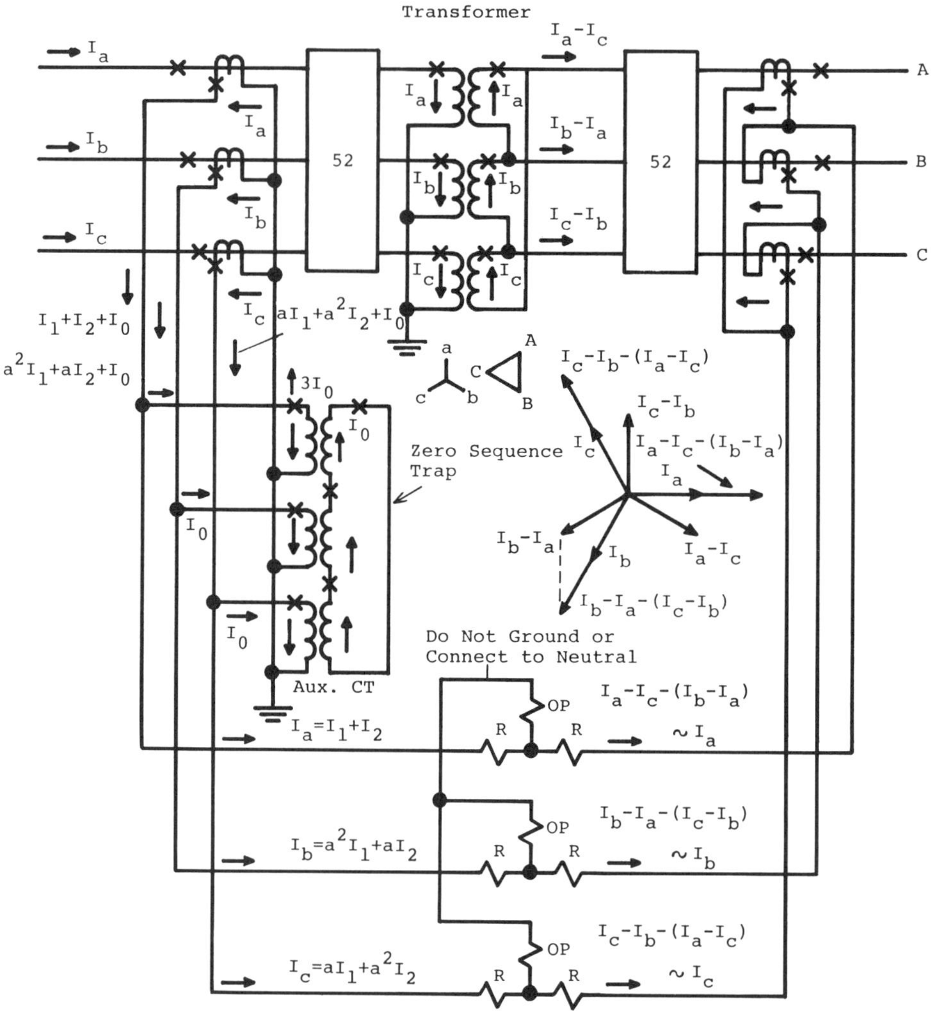

Fig. 9.8 Special differential connections using star-connected CTs on the transformer grounded-wye terminals.

operating coil neutral is *not* connected to the CT neutral or grounded. It is not necessary for zero-sequence current flow for correct relay operation. Connecting this operating coil neutral to the CT neutral places the restraint and operating coil windings in parallel with the primary winding of the trap. This could cause enough diversion of current for an external ground fault and relay misoperation.

With the left-hand CTs connected in wye as shown in Fig. 9.8, it is necessary to connect the right-hand CTs in delta properly to provide currents in phase through the differential relay for external primary current flow. This phasing should be done as above, starting with balanced I_a, I_b, and I_c currents into (or out of, if preferred) the wye transformer windings. Carrying this through it can be seen with a phasor diagram that $I_a - I_c - (I_b - I_a)$ is in phase with I_a, and similarly for the other phases. The per unit magnitudes are three times larger, but this can be adjusted by the CT and relay taps. Ground faults within this differential zone will operate the relays via the positive and negative fault current components. This is shown in Fig. 9.8.

9.12 DIFFERENTIAL PROTECTION FOR THREE-PHASE BANKS OF SINGLE-PHASE TRANSFORMER UNITS

Single-phase transformer units are connected in various three-phase configurations usually wye-delta and with associated circuit breaker(s) and CTs, as illustrated in Fig. 9.4. If it becomes necessary or desirable to use the CTs on the transformer bushings, the normal differential connection cannot be used when the CTs are inside the delta, as in Fig. 9.9. Two sets of CTs connected in parallel are required to provide protection for ground faults in these windings.

Star-connected CTs can be used on both sides. The phasor currents in Fig. 9.9 show the balance for symmetrical currents flowing through the bank. There is a 2:1 ratio difference (per unit) of the CTs on the two sides. This can be adjusted by CT or relay taps or a combination of both. Since each transformer is an individual unit, there will be no possibility of a phantom tertiary effect to plague the differential operation.

For three-phase banks where the three windings and interconnections are all inside the common tank, the standard differential connections of Figs. 9.4, 9.5, 9.7, or 9.8 can be used with CTs located as shown or with CTs in the three-phase transformer bushings on either side.

9.13 GROUND (ZERO-SEQUENCE) DIFFERENTIAL PROTECTION FOR TRANSFORMERS

The ground differential scheme provides compromise protection for delta-wye-grounded transformer banks. It is useful where there are no available or convenient CTs on the delta side. This is a common case for distribution and industrial ties with the delta as the high-voltage side, and possibly protected by fuses.

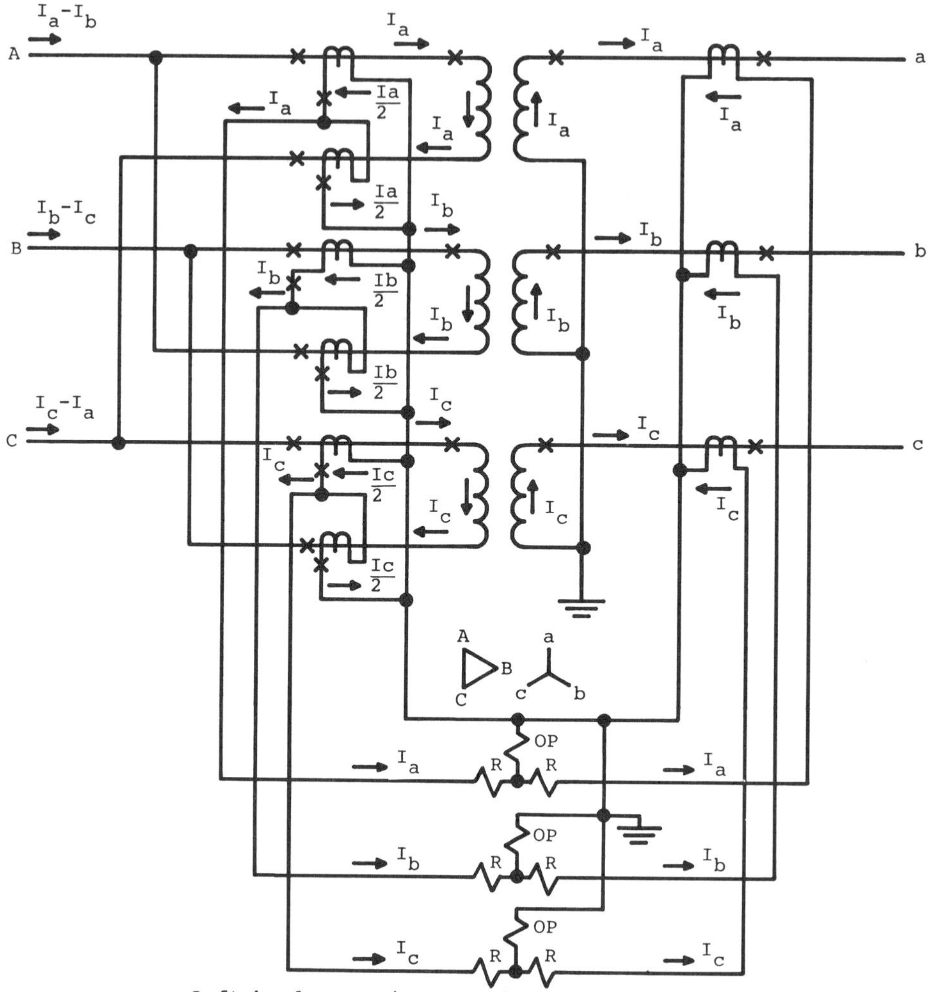

Left-hand per unit CT ratio shown two times the right-hand unit per unit CT ratio.

Fig. 9.9 Differential relay connections for the protection of three single-phase transformers connected delta-wye using CTs in the transformer bushings.

This scheme protects only the wye-grounded windings and associated circuits, and only for ground faults, the most likely and common fault. A typical application is shown in Fig. 9.10 using a conventional differential relay. The differential zone includes the circuits between

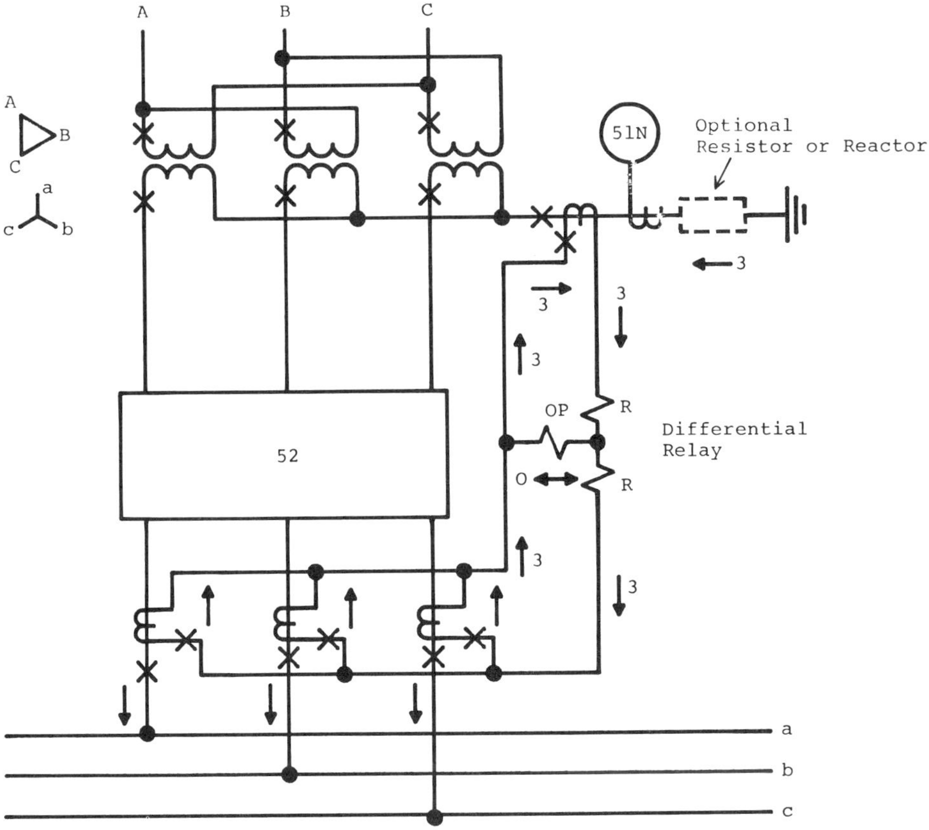

Zero Sequence Current Flow for External Ground Fault.

Fig. 9.10 Ground (zero-sequence) differential protection for delta-grounded-wye transformer bank using a conventional differential relay.

between the two sets of CTs. The delta blocks operation for faults in that area. The techniques of "phasing" and "ratioing" are as outlined previously except using only zero-sequence current flow to an external fault. These currents are shown in Fig. 9.10. The time-overcurrent relay 51N shown connected to a separate CT is recommended for all grounded transformers. It is a "last resort" ground-fault protection and must be set to coordinate with relays it overreaches. This is discussed in more detail later. It may be connected in the differential circuit, which adds burden to the differential circuit and can affect operation.

Where the delta side is protected by fuses, ground faults in the differential zone may not provide sufficient current to clear the faults adequately from the delta-side source. As indicated previously, a 1-pu phase-to-ground fault on the wye side appears as a 0.577-pu phase-to-phase fault on the delta side. This adds to the difficulty of detection. Any fault limiting by neutral impedance and/or fault resistance further reduces the fault magnitudes. Thus in many cases it becomes impossible to clear ground faults by the high-side fuses. Thus the ground differential is useful for the ground faults in its zone. The problem is to clear the fault from the delta-side source without a local breaker, and where the nearest breaker is at a remote station.

9.14 EQUIPMENT FOR TRANSFER TRIP SYSTEMS

Without a fault-interrupting device available at the transformer primary terminals, there are several possibilities for tripping remote breaker(s) as necessary to clear faults. All these methods are in practical use.

9.14.1 Fault Switch

A spring-loaded fault switch is connected to the delta supply side. The transformer protection relays operate to release the switch and thus fault the source system. Protective relays at the remote breaker(s) sense this fault and operate to clear it, thereby removing the source(s).

Most commonly this is a single switch that applies a solid single-line-to-ground fault to operate the remote ground relays. Often, these remote relays can operate instantaneously to provide rapid isolation. There are a few installations of a three-phase-to-ground switch which applies a solid three-phase fault. The advantage is high redundancy; one or two switches can fail and still result in clearing the fault. Also, normally the phase and upon one ground switch failure, the ground relays at the remote location(s) can clear the fault. The disadvantage is higher cost, increased maintenance, and applying a higher-current three-phase fault on the system. This grounding-switch technique is quite practical and relatively simple. However, some shudder at putting any deliberate fault on the system.

9.14.2 Communication Channel

The transformer protective relays initiate a tripping signal that is transmitted over a transfer trip channel to operate remote breaker(s). This channel may be via power-line carrier (radio frequency over the power lines), audio tones over a telephone circuit or microwave channel, or a direct-wire pilot pair. High security is very important to avoid undesired trip operation from extraneous signals on the channel. This and the usually higher cost are the major disadvantages. It is most important that these equipment transfer trip systems be carefull and thoroughly engineered. It is not possible to enhance their security with fault detectors at the breaker tripping terminal(s), as the low-side faults generally will be too low a level.

9.14.3 Limited Fault Interruption Device

A circuit switcher or breaker with limited fault interrupting capability is installed at the transformer on the delta-source side. The transformer protective relays initiate the tripping of this device directly or with a slight time delay. The scheme is based on the good probability that the remote relays will operate and clear any high current fault at high speed and before the circuit switcher or breaker can open. If the remote relays do not sense the fault or operate slowly, the level of fault current is within the interrupting capabilities of the local device.

9.15 MECHANICAL FAULT DETECTION FOR TRANSFORMERS

The accumulation of gas or changes in pressure inside the transformer tank are good indicators of internal faults or trouble. These devices are recommended, wherever they can be applied, as excellent supplementary protection. In many cases they are more sensitive, so will operate on light internal faults that are not detected by differential or other relays. However, it is important to recognize that their operation is limited to problems inside the transformer tank. They will not operate for faults in the transformer bushings and the connections to the external CTs. Thus their protective zone is only within the tank, in contrast to the differential protective zone of Fig. 9.4.

9.15.1 Gas Detection

Gas detection devices can be applied only to transformer units built with conservator tanks. This type of transformer is common in Europe but not widely used in the United States. For these units with no gas space inside the transformer tank, a gas accumulator device commonly known as the Buchholz relay is connected between the main and conservator tanks. It collects any gas rising through the oil. One part of the relay accumulates gas over a period of time to provide sensitive indication of low-energy arcs. It is used generally to alarm, since gas may be generated by tolerable operating conditions. The other part responds to heavy faults, forcing the relay open at high velocity. This is used to trip in parallel with the other transformer protection.

9.15.2 Sudden Pressure

These are applicable to oil-immersed transformers. One type operates on sudden changes in the gas above the oil, another on sudden changes in the oil itself. Both have equalizing means for slow changes in pressure, which occur with loading and temperature changes. They are sensitive to both low- and high-energy arcs within the transformer and have inverse-time characteristics: fast for heavy faults and slower for light faults. Generally, they are used to trip with the contacts in parallel with the differential and other relay trip contacts, although they can be used for alarm if preferred.

9.16 GROUNDING TRANSFORMER PROTECTION

To provide system grounding on the delta side of power system transformer banks, a shunt-connected grounded wye-delta or zigzag transformer bank is applied. With wye-delta units, the delta operates as an unloaded tertiary to circulate zero-sequence current. If this tertiary has a CT available inside the delta, it can be connected to a time overcurrent 51N relay. This relay receives I_0 for ground faults out on the system, so it must be coordinated with other ground relays.

The ground differential protection, as shown in Fig. 9.10, is applicable to the grounded wye-delta units, and as shown in Fig. 9.11 for zigzag units. An alternative is to use three-phase time-overcurrent 51 relays, each connected to one of the line-side CTs. Since load does not pass through these units, the CT ratios and relay taps can be based on ground-fault current.

The zigzag bank essentially consists of three 1:1 ratio transformers interconnected as shown in Fig. 9.11. The zero-sequence current flow is illustrated. Positive- and negative-sequence currents cannot flow, since they are 120° out of phase.

Sudden pressure- or gas-actuated trip relays are highly recommended where applicable for light internal fault protection. Turn-to-turn faults are always difficult to detect. In a zigzag unit they can be limited by the magnetizing impedance of an unfaulted phase.

Grounding transformers frequently are directly connected to the bus or associated power transformer without a fault-interruption device. A typical example would be to connect the grounding unit between the power transformer delta and the right-hand circuit breaker in Fig. 9.8. This application requires the use of a zero-sequence trap, as shown, since there is a zero-sequence source now on both sides of the wye-delta bank with a 30° shift across the bank.

With the grounding bank within the transformer differential zone as indicated, an alternative connection to Fig. 9.8 would be to connect

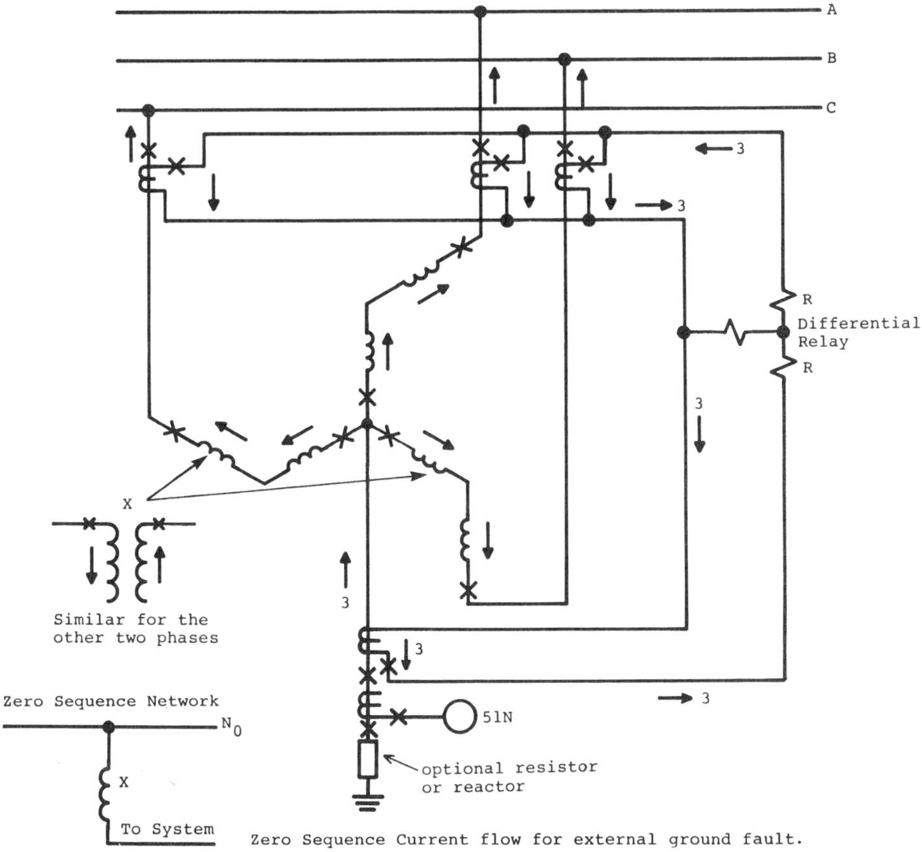

Fig. 9.11 Ground (zero-sequence) differential protection for a zigzag transformer bank using a conventional differential relay.

the right-hand CTs in wye with a zero-sequence trap, and the left-hand CTs in delta.

A set of auxiliary CTs could be used instead of the zero-sequence trap to provide the necessary zero-sequence isolation and the 30° shift. The trap is preferred, as it is involved only during ground faults. Several arrangements are possible. In working these out, do not connect the main CTs in wye and to the delta-connected auxiliary, as there will be no zero-sequence path for primary fault currents.

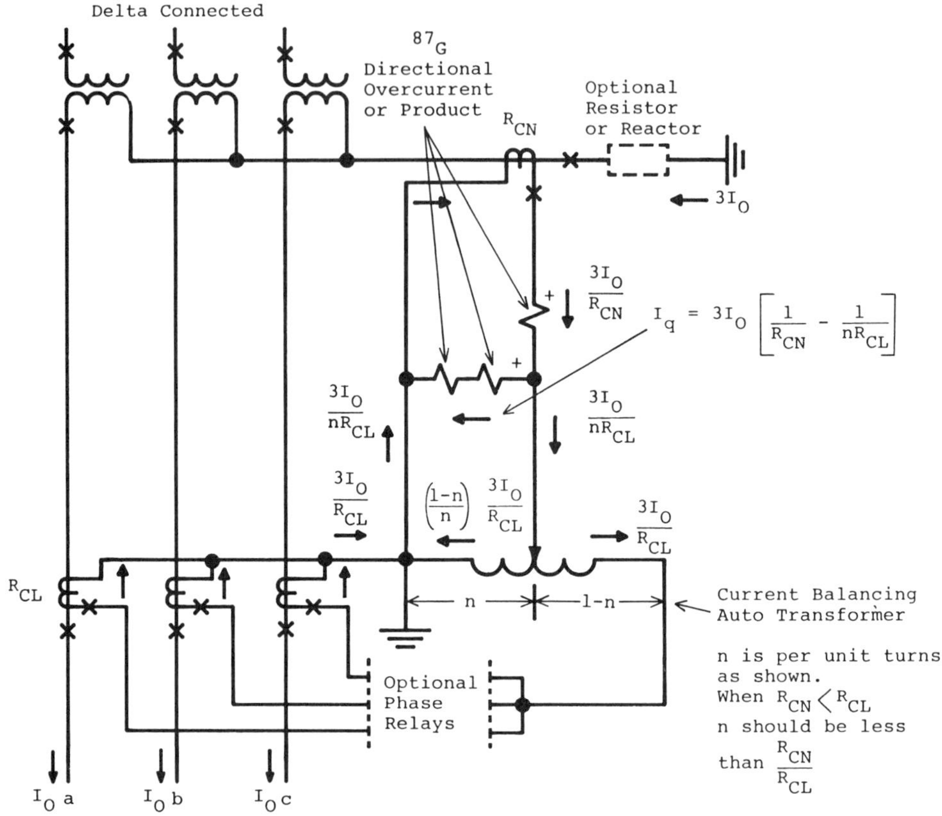

(a)

Fig. 9.12 Ground (zero-sequence) differential protection for a delta-grounded-wye transformer bank using a directional overcurrent relay with current-balancing autotransformer. The currents shown are for external faults: (a) zero-sequence current flow for external ground faults; (b) example of zero-sequence current flow for an external ground fault (directional sensing relay does not operate).

9.17 GROUND DIFFERENTIAL PROTECTION WITH DIRECTIONAL RELAYS

Where the CT ratios and/or characteristics are not suitable for the application of conventional differential relays, a directional overcurrent relay differentially connected can be used. This is particularly

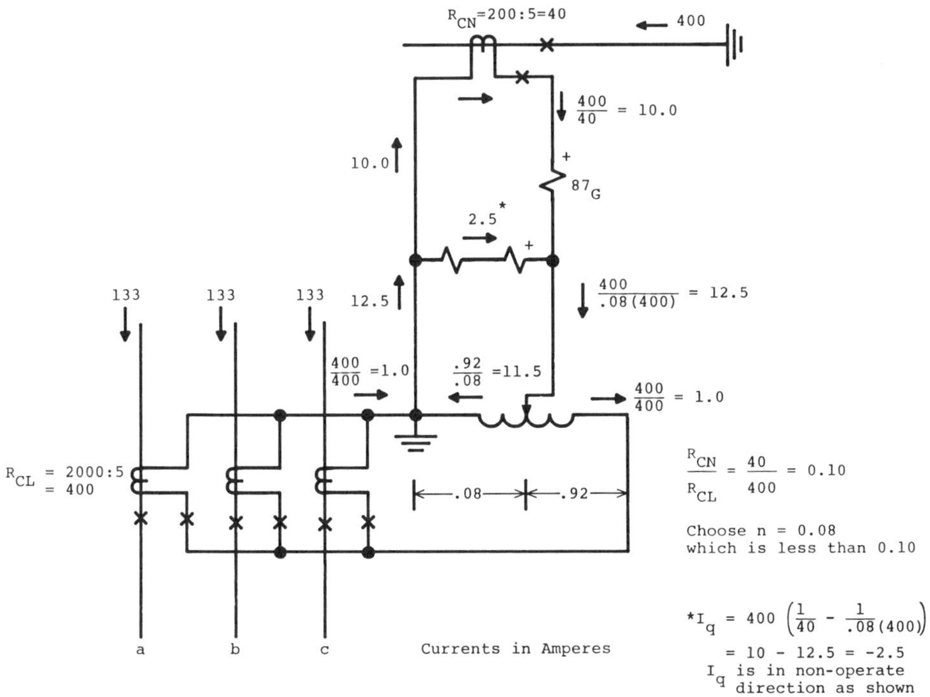

(b)

Fig. 9.12 (continued)

applicable where ground-fault current is limited by neutral impedance or where a low neutral CT is used for higher sensitivity for remote faults on distribution feeders.

Two applications are shown. Figures 9.12 and 9.13 use an auxiliary current-balancing autotransformer. Figures 9.14 and 9.15 use an auxiliary 1:N transformer. One of the two diagrams shows the operation for external faults, the other for internal faults. The two types are equivalent where $(1 - n)/n = N$.

Two types of relays can be used in either scheme: a product overcurrent or a directional overcurrent. The product relay can be an induction disk unit, where the torque control or lag circuit is available as one circuit and the main coil as the other. These are shown in the diagrams as the two coils with + polarity marks. The operation is the product of the currents in the two circuits times the cosine of the angle between them. When currents in phase are flowing into the polarity marks ($\cos 0° = 1$), the relay has maximum operating torque to

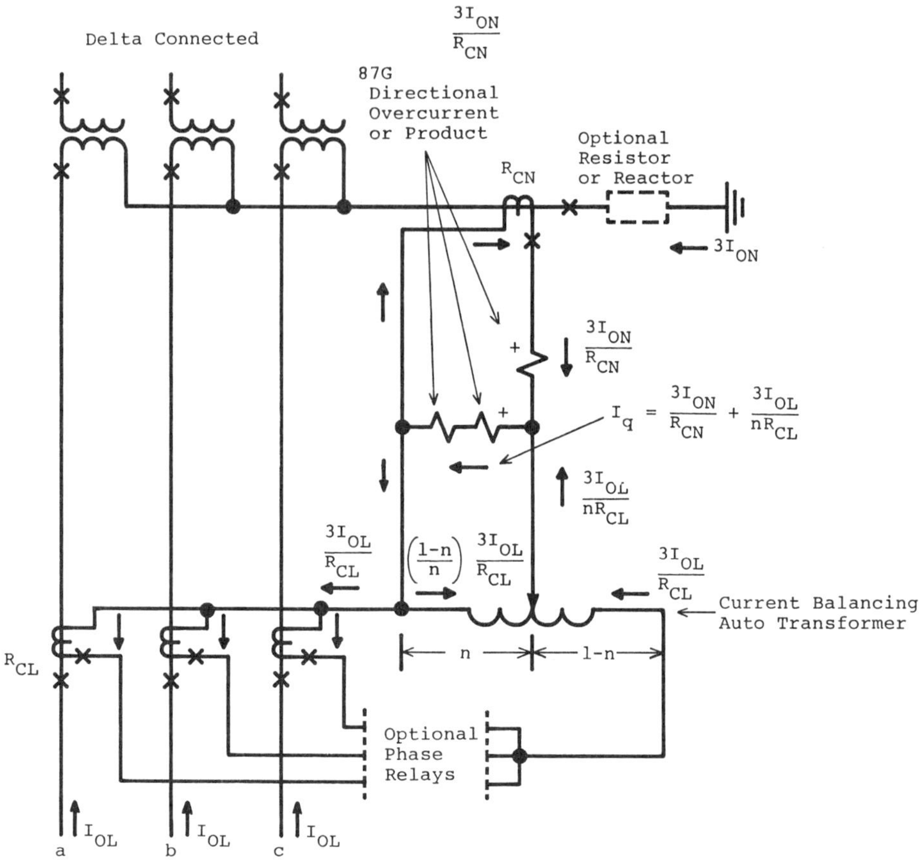

(a)

Fig. 9.13 Operation of the differential system of Fig. 9.12 for internal faults; (a) zero-sequence current flow for internal ground fault; (b) example of zero-sequence current flow for an internal ground fault with no current from line (currents in amperes); (c) example of zero-sequence current flow for an internal ground fault with current from line (currents in amperes).

close the contacts. If one current is in polarity in one coil and out of polarity on the other ($\cos 180° = -1$), the relay has maximum nonoperating or restraint torque. Zero torque occurs when the two currents are $\pm 90°$. This relay can operate with wide differences of current magnitudes as long as the product is greater than the pickup minimum, and with relative independence of the phase-angle variations within

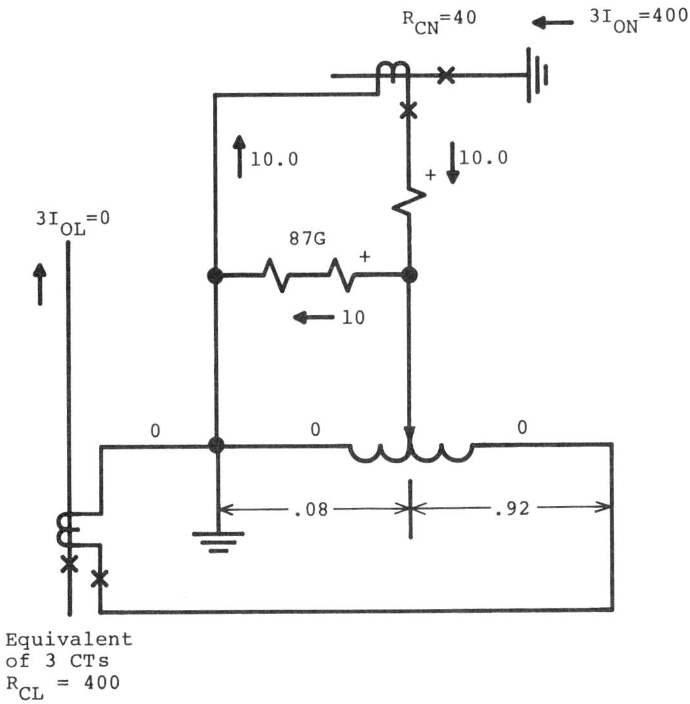

(b)

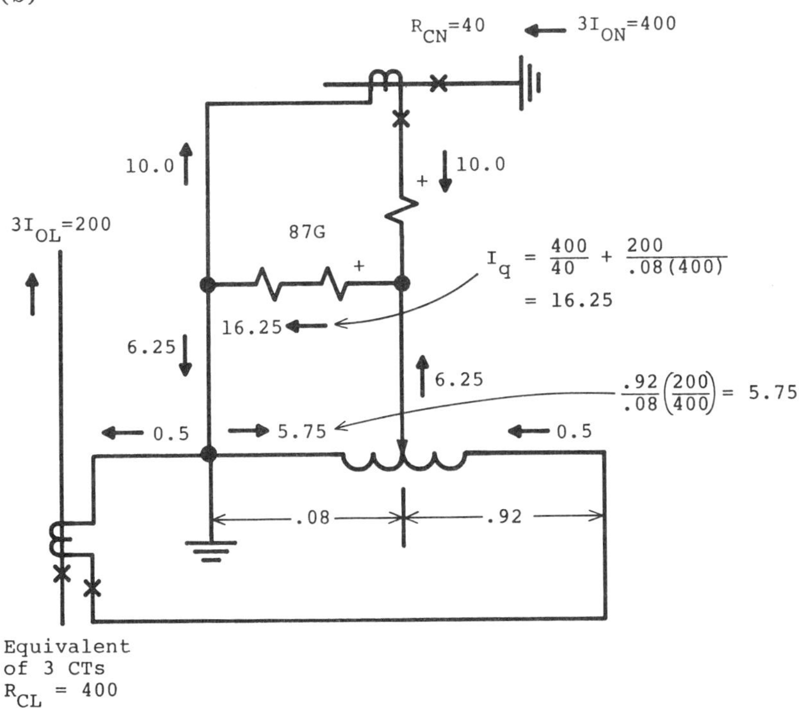

(c)

Fig. 9.13 (continued)

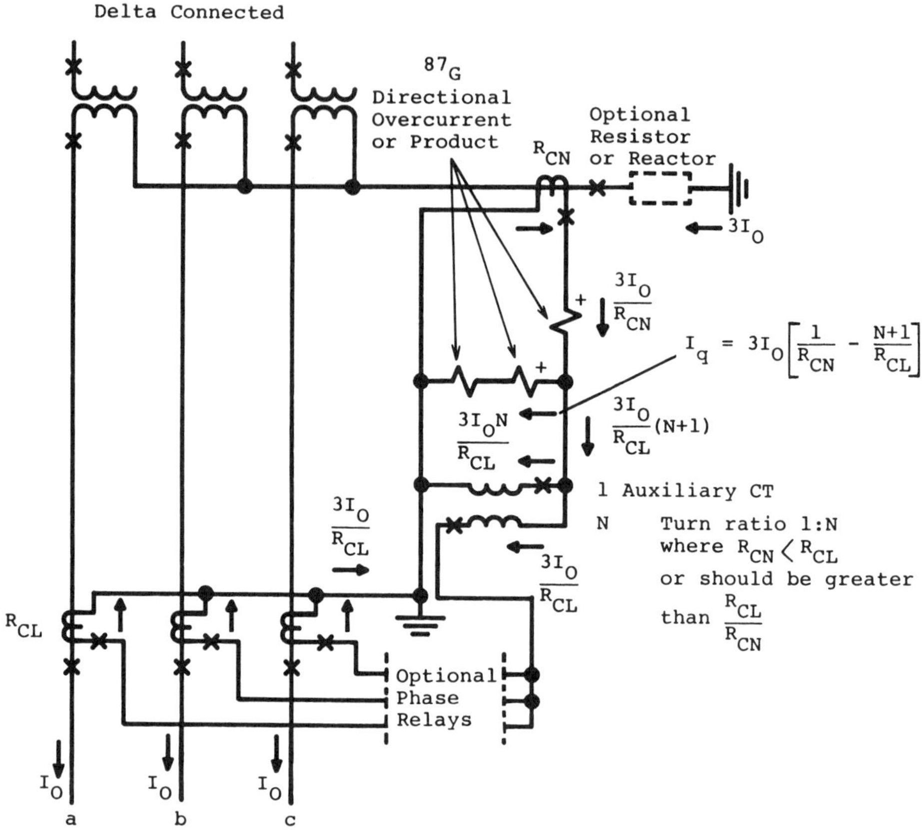

(a)

Fig. 9.14 Ground (zero-sequence) differential protection for a delta-grounded-wye transformer bank using a directional overcurrent relay with auxiliary CT (currents shown are for external faults): (a) zero-sequence current flow for external ground fault; (b) example of zero-sequence current flow for an external ground fault directional (currents in amperes).

±90°. It has inverse-time characteristics, operating very fast for large current internal faults.

The directional overcurrent relay has a separate directional unit that operates as the product unit. The overcurrent unit of the relay is nondirectional and operates on the magnitude of current in its main coil. In the figures it is the coil without polarity and connected basically across the CTs. Again it has inverse-time characteristics,

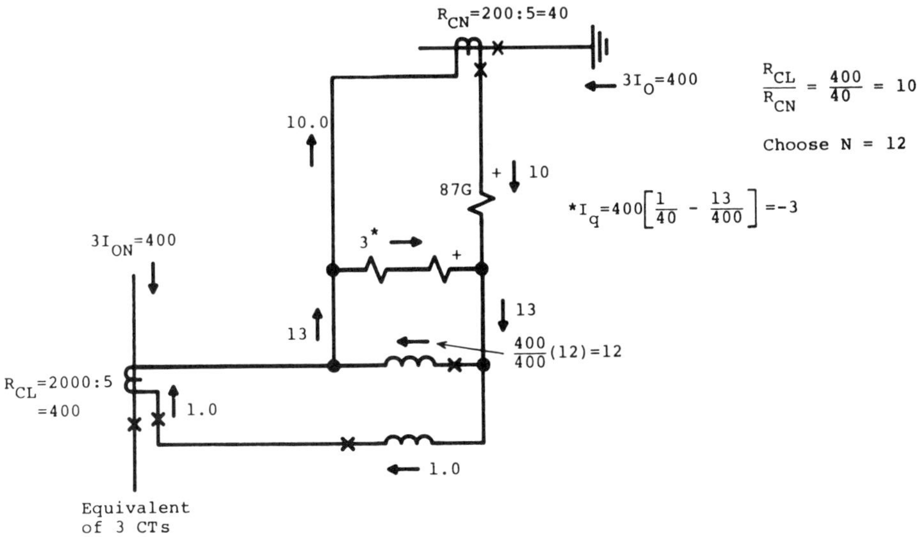

(b)

Fig. 9.14 (continued)

operating fast for high current, but only if the directional unit has operated. It is torque controlled.

The auxiliary transformers are required to provide correct differential operation: not to operate for external faults outside the differential zone, operate for the ground faults inside the differential zone. It is shown that these schemes operate for cases where no ground current is supplied from the system to the internal faults. This would be common for many distribution and industrial applications. With no current, as in Figs. 9.13b and 9.15b, it must be recognized that some of the neutral secondary current will be diverted to excite the auxiliaries and line CTs, as discussed in Chapter 5. This should be small with good-quality CTs and is neglected in the figures.

If ground current always is supplied from the system for internal faults, the product relay could be applied with one coil connected across the line CTs and the other coil across the neutral CT without any auxiliaries. As indicated, it can operate with wide differences in currents and CT performances. The directional overcurrent relay could be connected only to the line CTs, to operate for faults in the bank without any connection to the neutral CT. In many cases the latter connection would not be very sensitive, as the remote ground sources generally are quite weak.

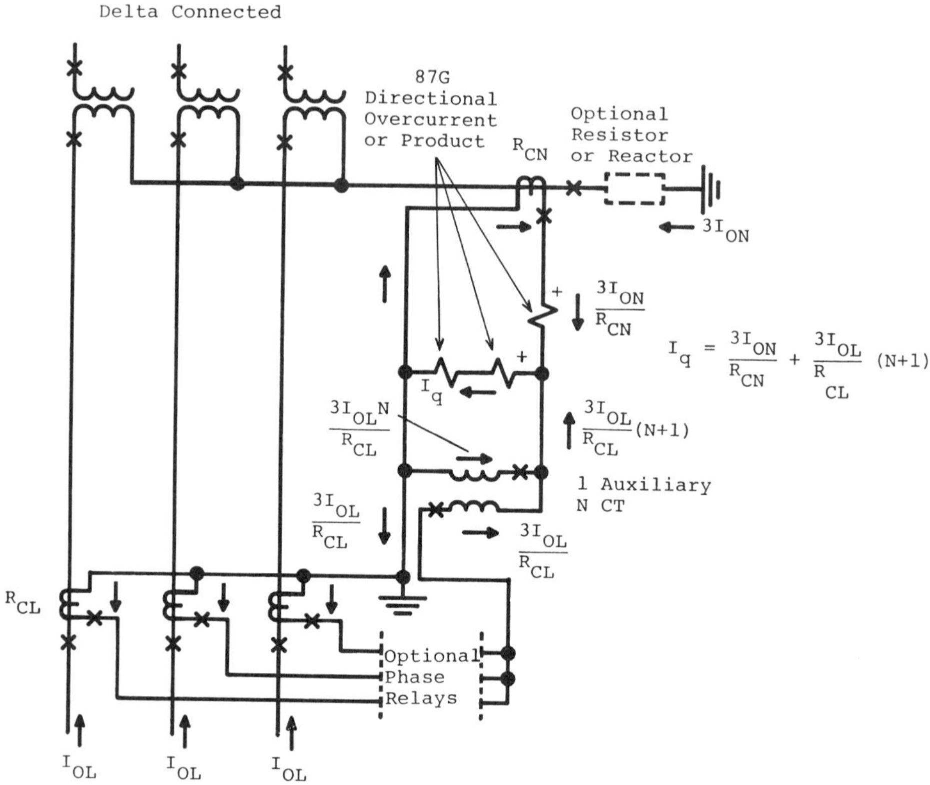

(a)

Fig. 9.15 Operation of the differential system of Fig. 9.14: (a) zero-sequence current flow for internal ground faults; (b) example of zero-sequence current flow for an internal ground fault with no current from line (currents in amperes); (c) example of zero-sequence current flow for an internal ground fault with current from the line (currents in amperes).

9.18 PROTECTION OF TRANSFORMER BANKS WITH ADJUSTABLE PHASE SHIFTS

Phase-shifting regulating transformers are designed for specific problems involved in the interchange of power between two systems. Their design is complex and specialized for the specific application. Differential protection is difficult or perhaps impossible, so that sudden pressure- or gas-actuated tripping relays serve as good first-line

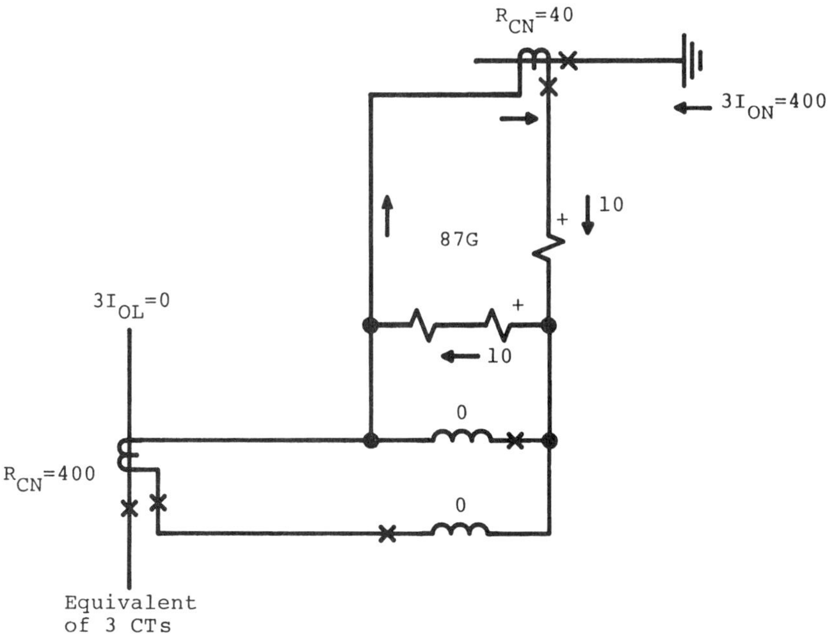

(b)

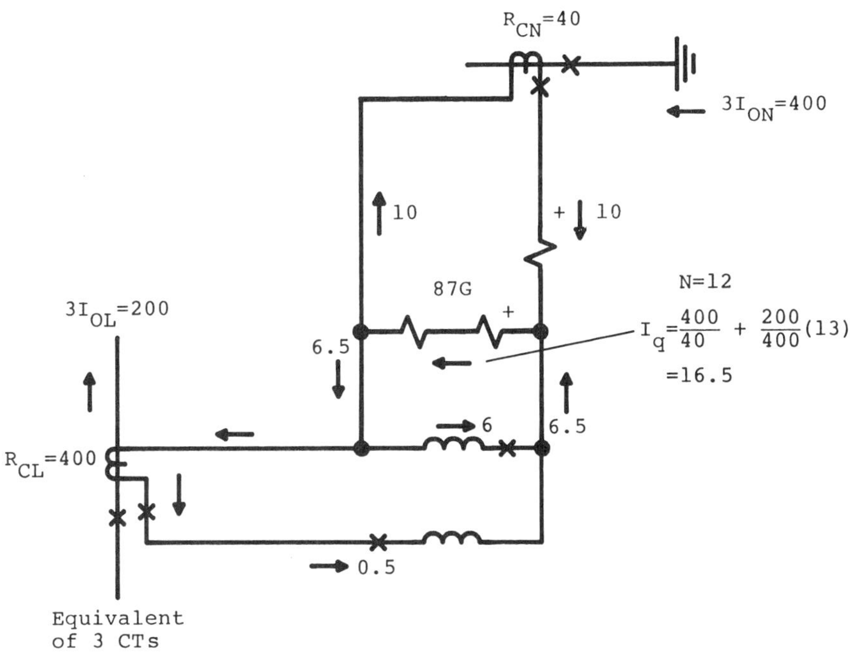

(c)

Fig. 9.15 (continued)

protection, with overcurrent backup in many cases. If differential protection is applicable, it probably requires CTs installed inside the unit, so that the protection is best planned and worked out with the manufacturer before delivery.

9.19 TRANSFORMER OVERCURRENT PROTECTION

Phase- and/or ground-fault overcurrent protection is common for transformers. This is either as the primary protection for smaller units or for any unit without differential protection, or as backup protection on larger units protected by differential relays. Around 10 MVA and below, primary fuses may be used. Otherwise, inverse-time-overcurrent relays and at higher voltages, distance relays provide protection for the transformer and associated circuits. Since these devices can operate well outside the transformer protection zone, their application and setting is a combination of transformer and associated system protection. The emphasis here is on transformer protection and the subject will be expanded later in connection with the protection of other equipment, primarily feeder and line protection.

It is desirable to set the protective devices as sensitive as possible, but the fuses and phase overcurrent relays must not operate on any tolerable condition, such as magnetizing inrush, the maximum short-time overloads such as may occur when reenergizing circuits after an outage (cold load pickup) or any emergency operating condition. The ground relays must be set above the maximum zero-sequence unbalance that can exist principally as the result of single-phase loading.

Instantaneous overcurrent relays may be applied to supplemental differential or overcurrent protection and provide protection for heavy primary transformer faults. They must be set not to operate on magnetizing inrush (unless a harmonic restraint is used), on the maximum short-time load (cold load) or on the maximum secondary three-phase fault. Typical setting would be 150 to 200% of the greatest of these currents. This may limit their operation on primary faults.

On the other hand, the relays and/or fuses should protect the transformers against damage from through faults. High fault current passing through the transformer can cause thermal as well as mechanical damage. High temperatures can accelerate insulation deterioration. The physical forces from high currents can cause insulation compression, insulation wear, and friction-induced displacement in the windings. ANSI/IEEE define the limits for these faults.

9.20 TRANSFORMER OVERLOAD-THROUGH-FAULT WITHSTAND STANDARDS

The ANSI/IEEE standards for distribution and power transformers specifying their overload-through-fault capabilities were changed in about 1977 (see IEEE C57.12-1980 and the article by Griffin in the Bibliography). Comparison of the change is given in Table 9.1. The multiples of rated current represents the maximum possible and are derived assuming an infinite source (zero impedance). Thus the maximum symmetrical fault through the transformer bank with 4% impedance would be 1/0.04 = 25 pu or 25 times the bank rated current, as shown.

The source impedance is never zero, but it can be very small relative to the transformer bank impedance, particularly for an industrial plant or small distribution substation connected to a large utility power system. Hence the multiples represent a maximum limit.

Soon after this change to all 2-s limits, protection engineers discovered that they had been using a thermal overload curve (Fig. 9.16a) published in ANSI Standard C37.91-1972, Guide for Protective Relay Applications to Power Transformers, for overcurrent transformer protection, and that the new 2-s standard would severely limit power transformer protection for through faults. This dilemma has lead to further changes, which were approved late in 1982 and are summarized in Table 9.2 and Fig. 9.16. These changes are being incorporated in the several applicable standards. The conflict existed because C37.91 was

Table 9.1 ANSI/IEEE Transformer Overcurrent Capability Standards until Modified in 1982[a]

Multiples of rated current	Transformer impedance (% of rated kVA)	Maximum current withstand (s) 1977 (1980)	1973
25	4	2	2
20	5	2	3
16	6	2	4
14 or less	8 or greater	2	5

[a]ANSI/IEEE C57.12.00. Old standard dated 1973, approved 1977 but not officially published until 1980. To be revised as indicated in the text.

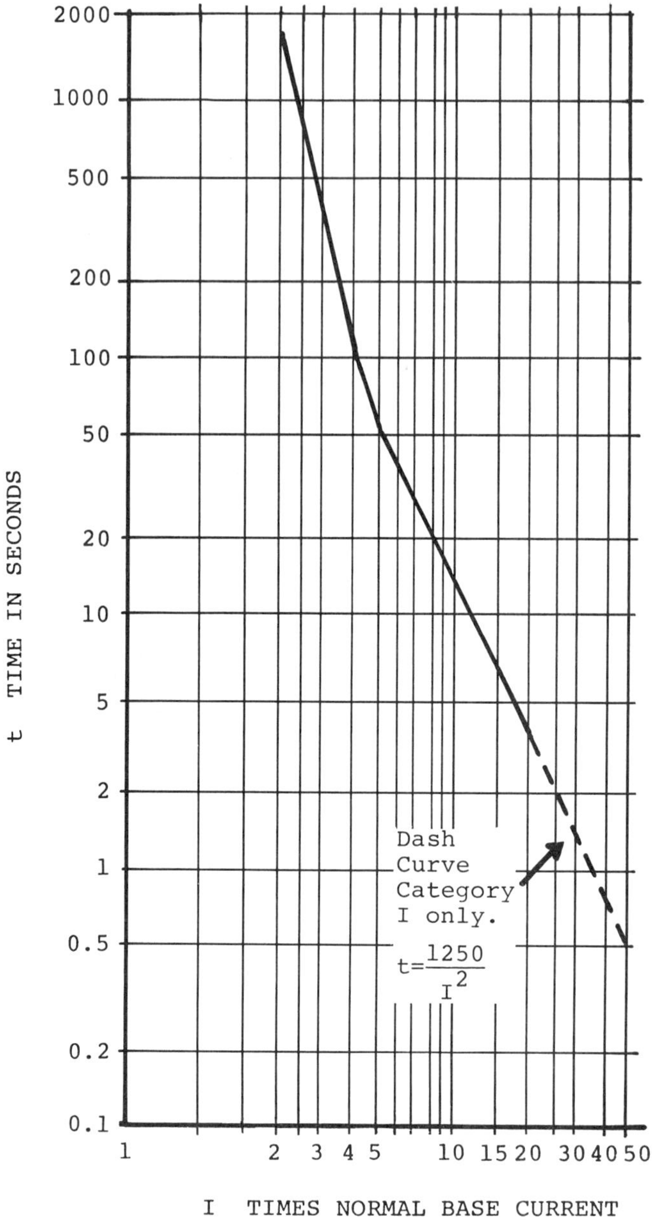

(a)

Fig. 9.16 Through-fault protection curves for transformers: (a) for Category I frequent or infrequent faults and for infrequent faults with Categories II and III; (b) for frequent faults with Category II transformers; (c) for frequent faults Category III and frequent or infrequent Category IV transformers.

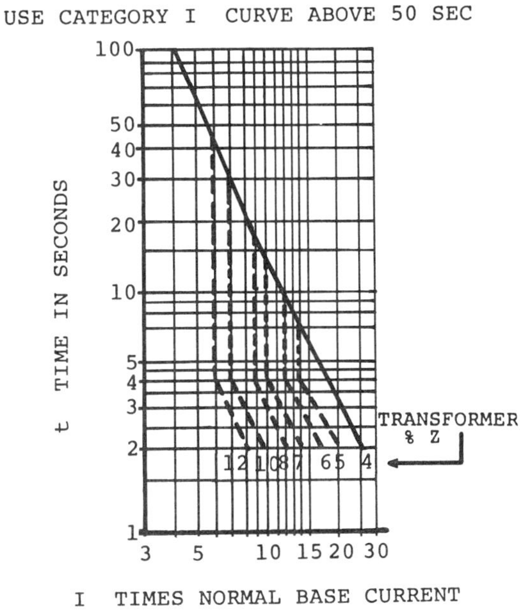

(b)

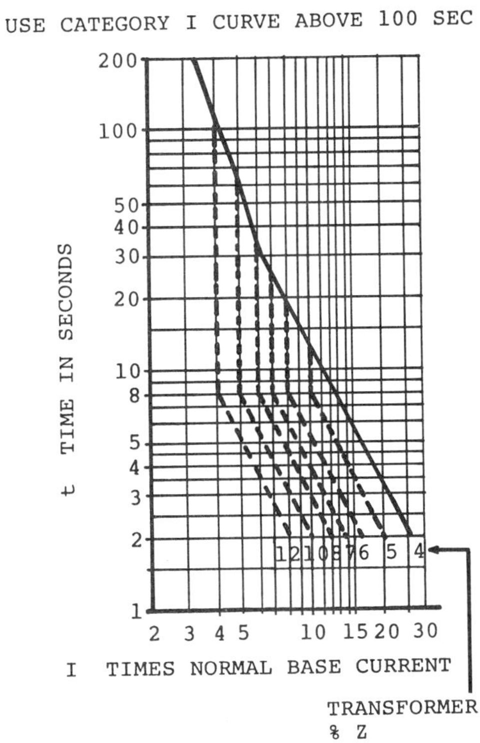

(c)

Fig. 9.16 (continued)

Table 9.2 ANSI/IEEE Transformer Overload-Through-Fault Standard Categories

Category	kVA transformer ratings Single-phase	Three-phase	Use curve Fig. 9.16	Frequent faults[a]	Dotted curves apply from[b]
I	5–500	15–500	a	—	25 to 50I, where $t = \frac{1250f}{60I^2} = \frac{1250}{I^2}$ at 60 Hz
II	501–1667	501–5000	a or a + b	10	70 to 100% of maximum possible fault where $I^2t = K$; K is determined at maximum I where t = 2
III	1668–10,000	5001–30,000	a or a + c	5	50 to 100% of maximum possible fault where $I^2t = K$; K is determined at maximum I where t = 2
IV	Above 10 MVA	Above 30 MVA	a + c	—	Same as category III

[a] Faults that occur frequently are typically more than the number shown and are for the lifetime of the transformer. Infrequent faults are less than that shown for the life of the transformer. Category II and III infrequent-fault curve (Fig. 9.15a) may be used for backup protection where the transformer is exposed to frequent faults but is protected by high-speed primary relays.

[b] I, = Symmetrical short-circuit current in per unit of normal base current using minimum nameplate kVA; t, = time in seconds; f, = frequency in hertz. Per unit currents of 3.5 or less probably result from overloads, in which case the transformer loading guides should be followed.

a thermal damage curve, while C57.12.00 was concerned primarily with mechanical damage from the through faults. The latest changes, as shown in Fig. 9.16, cover both thermal and mechanical limits.

The new standard has six curves; one each for categories I and IV, and two each for categories II and III. The basic curve shown by the solid line in Fig. 9.16 for all categories is the same and as such applies to all 4% impedance transformers. These can withstand through currents of 25 times normal base current for 2 s. Modifying curves shown dotted apply for transformers with other than 4% impedance, depending in the fault frequency over the entire life of the unit.

The frequency of faults that might occur during the entire life of a transformer is an estimation based on past experience and judgment. A guide to aid in this determination is outlined in Fig. 9.17. In cases of doubt, the frequent-fault curves should be used.

The application of transformer protection to meet this standard can be outlined as follows: For a given transformer:

1. Determine the category from Table 9.2.
2. If category II or III, determine whether the service is subject to frequent or infrequent faults (Fig. 9.17).
3. Select the appropriate curve from Fig. 9.16 (or the ANSI standard).
4. Replot this curve on suitable log-log paper using amperes related to the specific transformer (per unit could be used, but since relays and fuses operate on current, amperes is preferable). Either system primary or system secondary current can be used, properly translated as required through the transformer bank. Generally, secondary side currents are preferred for coordinating with other secondary downstream devices.
5. Select proper fuses and/or relays with tap and time dials, and so on, to protect the transformer and coordinate with all other devices in the area.

The details of this are best illustrated by examples.

9.21 EXAMPLES: TRANSFORMER OVERCURRENT PROTECTION

The several examples are from IEEE draft revisions C37.91, Guide for Protective Relay Applications to Power Transformers. These are good and typical applications.

9.21.1 An Industrial Plant or Similar Facility Served by a 2500-kVA 12-kV:480 Volt Transformer with 5.75% Impedance

The protection consists of power fuses on the primary and low-voltage direct-acting circuit breakers with series overcurrent trip units on the secondary side and associated feeders. From Table 9.2, this transformer is category II, and with metal-clad or metal enclosed secondary switchgear, the fault frequency can be considered infrequent. Thus Fig. 9.16a applies.

This curve is replotted on Fig. 9.18, where the abcissa is secondary amperes. This translation is

$$I_{\text{per unit}} = I_{\text{rated}} = \frac{2500}{\sqrt{3} \times 0.48} = 3007 \text{ A at } 480 \text{ V} \qquad (9.19)$$

and so for various times:

Time (s) from Fig. 9.16a	Per unit I from Fig. 9.16a	Equivalent I amperes at 480 V (pu x 3007)
1,000	2.3	6,916
300	3.0	9,021
100	4.0	12,028
50	5.0	15,035
12.5	10.0	30,070
4.13	17.39	52,296

For 50 s and less,

$$t = \frac{1250}{I^2} \quad \text{such as} \quad \frac{1250}{5^2} = 50 \text{ s} \qquad (9.20)$$

as shown. The maximum possible current with an infinite source is

$$I = \frac{1}{0.0575} = 17.39 \text{ pu} \qquad (9.21)$$

where

$$t = \frac{1250}{17.39^2} = 4.13 \text{ s} \qquad (9.22)$$

so this is the termination of the transformer through-fault protection curve.

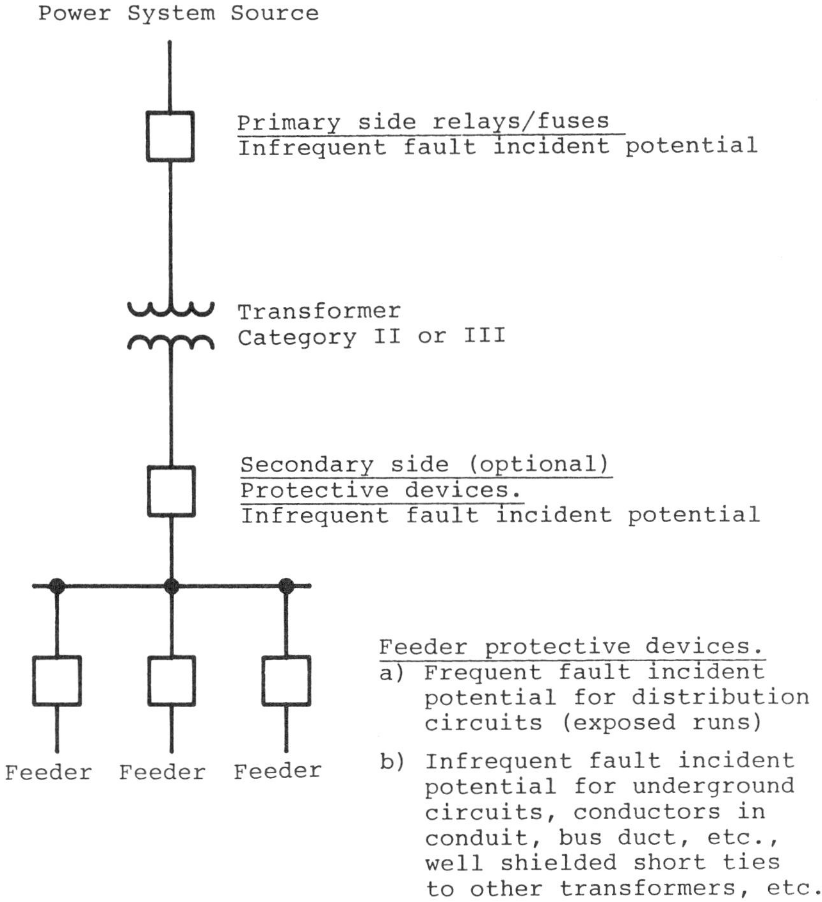

The primary protection zone for each of the protection devices is between that device and the next downstream device.

Fig. 9.17 Suggested guide for determining zones of infrequent and frequent incident potential.

On the primary side, rated current is

$$I_{\text{per unit}} = I_{\text{rated}} = \frac{2500}{\sqrt{3} \times 12} = 120.3 \text{ A at } 12 \text{ kV} \tag{9.23}$$

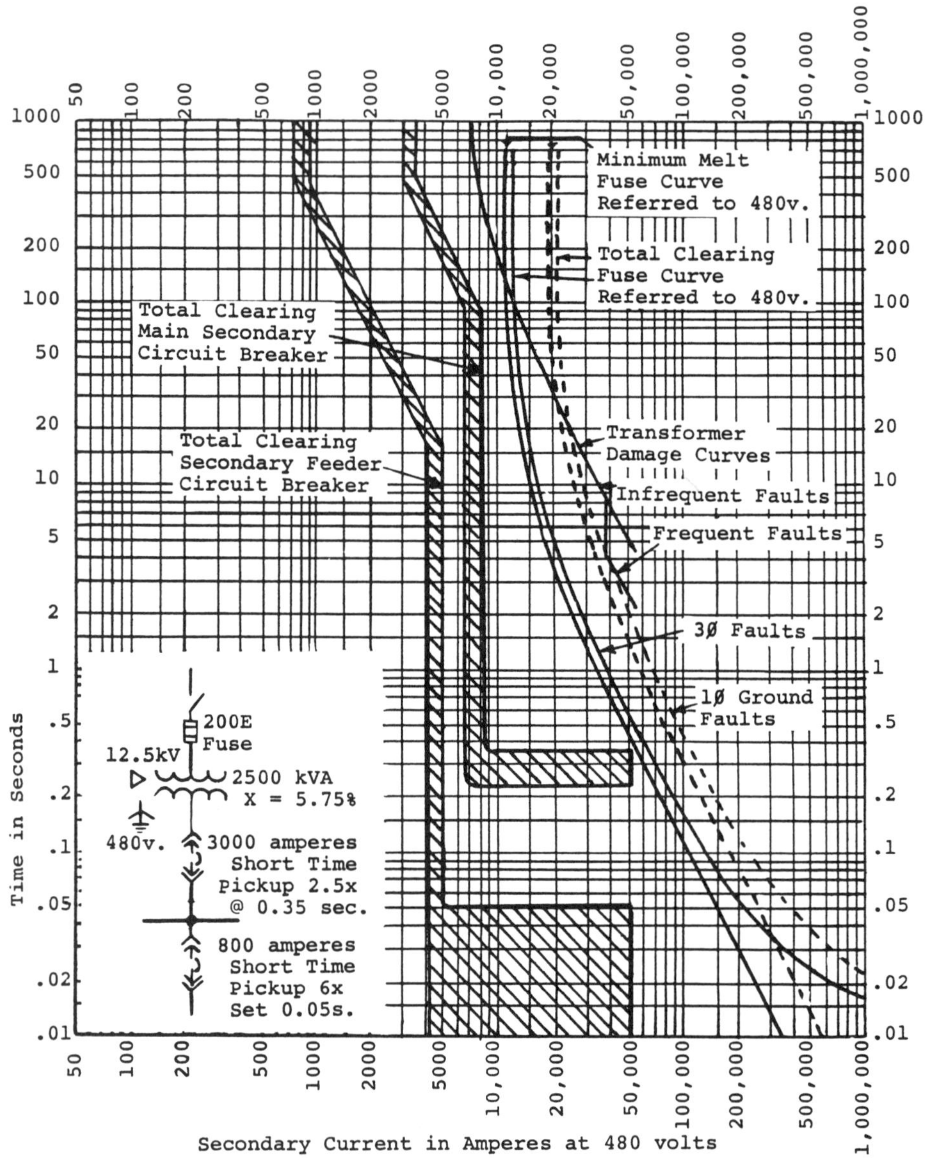

Fig. 9.18 Overcurrent protection for a category II transformer serving an infrequent-fault secondary system.

To avoid operation on magnetizing inrush, short-time transient load, and so on, yet provide protection for secondary faults, typical fuse ratings are selected around 150% of rated current. Thus 1.5 × 120.3 = 180.4 A, so 200-A fuses were selected and their time characteristics plotted. Both total clearing and minimum melt curves are shown. Since the current magnitudes in the primary are only 57.7% of the secondary current for phase-to-ground faults, the fuse curves are moved to the right by multiplying all values by 1.73 ($\sqrt{3}$) in addition to the transformer winding ratio. This is shown in Fig. 9.18. The three-phase fault magnitudes are the same in per unit. This is reviewed in Fig. 9.19. The fuse curves extend to higher currents, as these are possible for high side faults (in primary amperes) between the fuses and the transformer.

The transformer secondary and the feeders have low-voltage circuit breakers with direct-acting overcurrent units. These have long-time and short-time/instantaneous elements. The characteristics are a band between the total clearing times and the reset times, as illustrated in Fig. 9.18. For this example, the transformer breaker long-time unit is set to pick up at $1.2I_{rated}$ = 1.2 x 3007 = 3608 A at 480 V, where the time is 450 s. The short-time pickup is set at $2.5I_{rated}$ or 2.5 × 3007 = 7518 A at 480 V. A time delay of 0.35 s is used to provide coordination with the feeders. The feeder circuit breakers are set with the long time unit at 1.2 × I_{rated} = 1.2 × 800 = 960 A and the short time set instantaneously (0.05 s) at 6 times 800, or 4800 A at 480 V. More details on these setting criteria are in Chapter 12.

The protection plot of Fig. 9.18 shows good protection and coordination except for light secondary faults. The primary fuse curves cross the transformer through the fault protection curve at about 13,000 a for three-phase faults at about 23,000 A for phase-to-ground faults. This means that the transformer is not protected according to the standard for faults of these magnitudes or less by the fuses. Such faults are possible. If they are in the transformer, damage has already occurred and must become heavier before the source can be removed by the fuses. If between the transformer and the secondary circuit breaker, it must also develop to a heavier fault, which means more damage and time; however, the probability of faults in this area will usually be small.

Typical industry data indicate that at 480 V arcing phase-to-ground faults may be as low as 19% of the rated fault value. Thus for a secondary fault maximum of 52,296 A, the primary current for the fuses would be 52,296 × 0.19 × 0.577 × 0.48/12 = 229 A at 12 kV, just above the 200-a rating, and it is doubtful that the fuses will give any protection until severe burning increases the fault current.

Secondary faults on the bus should be cleared by the secondary transformer circuit breaker, and faults on the feeders by their circuit

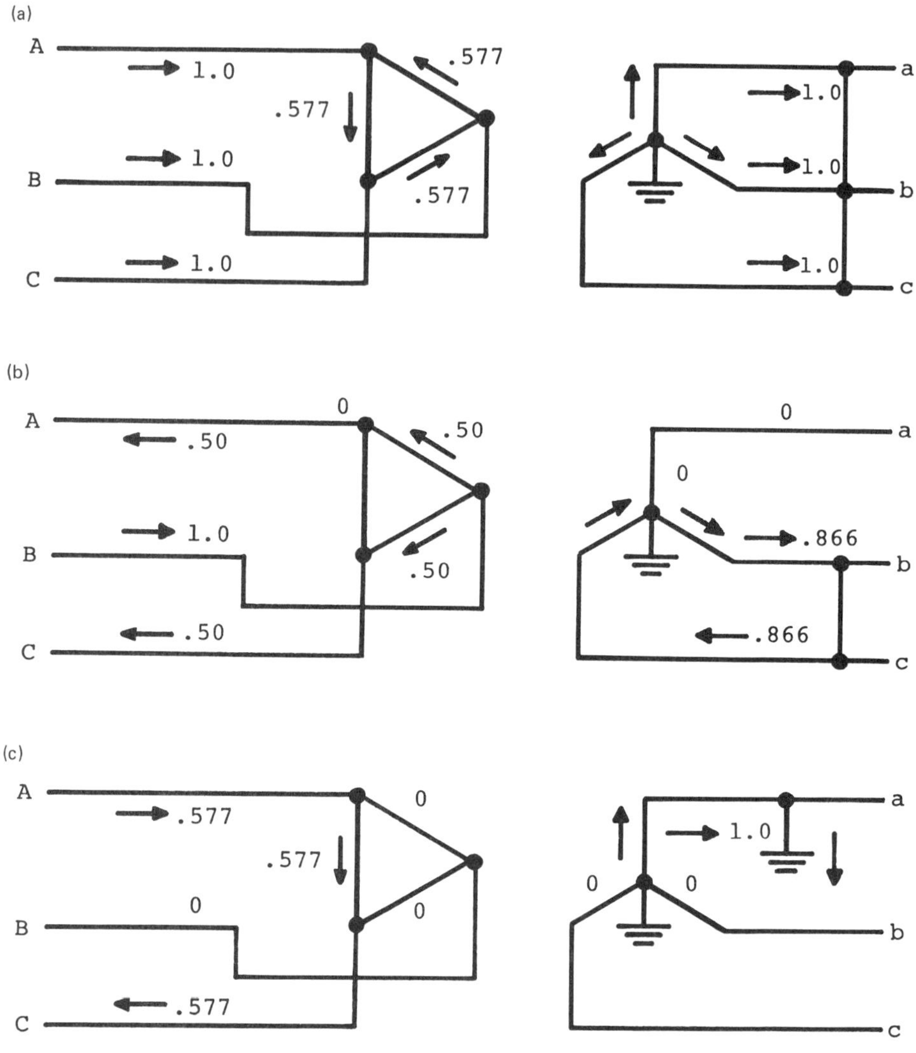

Fig. 9.19 Review of faults through delta-wye transformer banks (currents shown in per unit): (a) three-phase faults; (b) phase-to-phase faults; (c) phase-to-ground faults where $X_1 = X_2 = X_0$.

breakers backed up by the transformer breaker. Thus the primary fuses are backup for these faults if they can "see" these secondary faults.

The frequent fault curve modification from Fig. 9.16b has been shown on Fig. 9.18 for comparison. As can be observed, the primary fuse protection for through faults or secondary phase-to-ground faults is very marginal.

9.21.2 A Distribution or Similar Facility Served by a 7500-kVA, 115:12 kV Transformer with 7.8% Impedance

The primary protection is power fuses. A transformer secondary breaker is not used and the feeders have circuit reclosers.

This transformer is category III and the secondary is subject to frequent faults. Similar to the example in Section 9.21.1, the transformer through-fault protection curve is translated from Fig. 9.16a and c to secondary amperes for Fig. 9.20.

$$I_{\text{per unit}} = I_{\text{rated}} = \frac{7500}{\sqrt{3} \times 12} = 360.84 \text{ A at } 12 \text{ kV} \qquad (9.24)$$

The maximum secondary fault assuming that the source is very large relative to the transformer, and thus is infinite (X = 0), is

$$I_{3\phi} = I_{\phi G} = \frac{1}{0.078} = 12.82 \text{ pu} = 4626 \text{ A at } 12 \text{ kV} \qquad (9.25)$$

Points on the transformer through-protection curve are:

Time (s) from Fig. 9.16 a and c	Per unit I from Fig. 9.16 a and c	Equivalent I amperes at 12 kV (pu x 360.84)
1000	2.3	830
300	3.0	1082.5
100	4.0	1443.4
50	5.0[a]	1804.2
30.42	6.41[a]	2313
8	6.41[b]	2313
3.29	10[b]	3608.4
2	12.82[b]	4626

[a]For infrequent fault incident, K = 1250 from 5.0 to 12.82 pu; for frequent fault incident, K = 1250 from 5.0 to 6.41 pu.
[b]For frequent fault incident, K = 328.73 from 6.41 to 12.82 pu.

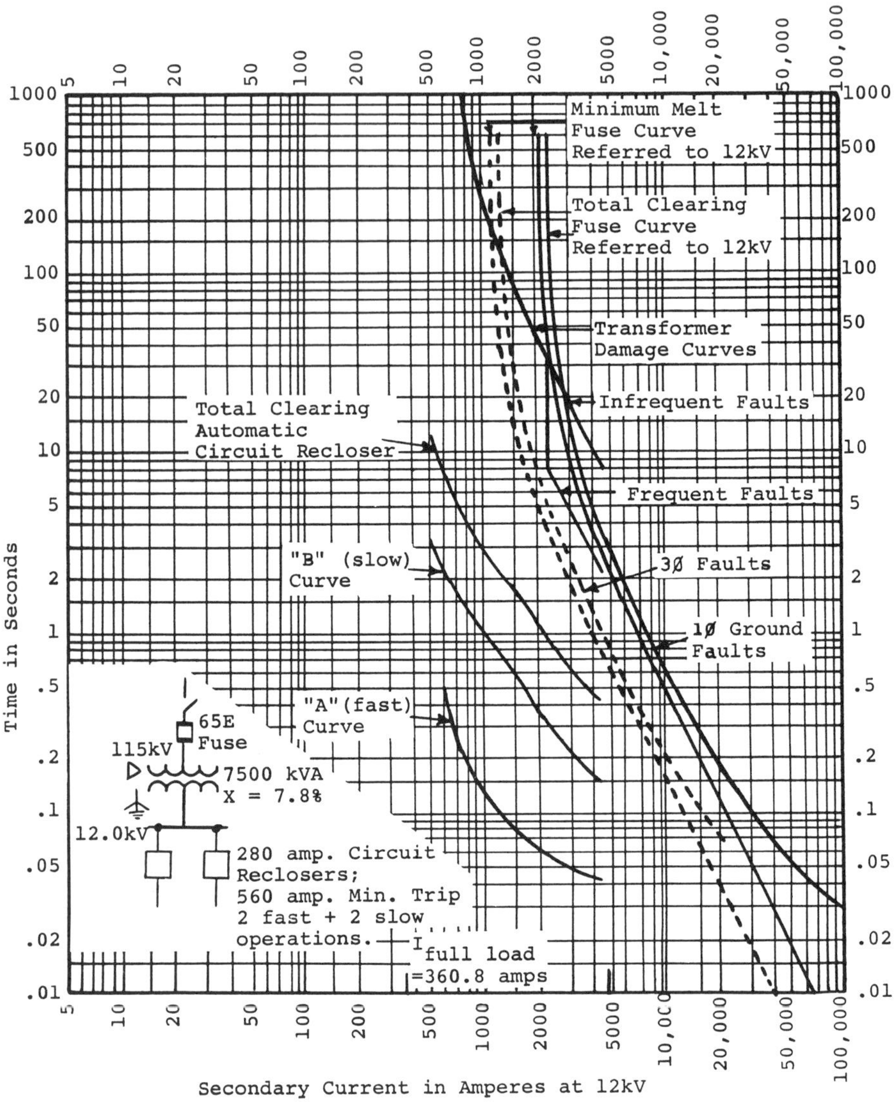

Fig. 9.20 Overcurrent protection for a category III transformer serving a frequent-fault secondary system. Bank has primary fuses.

On the primary side,

$$I_{\text{per unit}} = I_{\text{rated}} = \frac{7500}{\sqrt{3} \times 115} = 37.65 \text{ A at 115 kV} \tag{9.26}$$

and 65E ampere fuses were used (1.73 × 37.65). Their characteristics are plotted in terms of secondary side amperes for three-phase faults and secondary side phase-to-ground faults.

The two feeder circuits are rated 280 A and automatic circuit reclosers are applied with a 560-A minimum setting. Their characteristic curves for fast and slow operations are shown on Fig. 9.20.

While there is good coordination between the primary fuses and secondary circuit reclosers, the transformer is not protected for seconary phase-to-ground faults with the frequent-fault curve. Also, with the infrequent-fault curve there is inadequate protection for ground faults of about 3000 A and less. Again the phase shift through the bank provides 57.7% less ground-fault current (Fig. 9.19).

For faults out on the feeders the primary side fuses serve as backup, so hopefully they will seldom be called upon to operate. However, the fuses are primary protection for faults on the secondary side up to the circuit reclosers. Hopefully, these ground faults will be infrequent and greater than 3000 A to avoid potential damage to the transformer. These are risks that must be evaluated by the power system personnel.

9.21.3 A Substation Served by a 12/16/20-MVA 115-:12.5 kV Transformer with 10% Impedance

The primary has a relayed breaker, but without a transformer secondary breaker. This is category III transformer, so that the curves of Fig. 9.16a and c apply, which are replotted on Fig. 9.21 using secondary amperes. The maximum fault current with an infinite source (X = 0) is

$$I_{3\phi} = I_{OG} = \frac{1.0}{0.1} = 10 \text{ pu} = 5542.6 \text{ A at } 12.5 \text{ kV} \qquad (9.27)$$

where;

$$I_{\text{per unit}} = I_{\text{rated}} = \frac{12{,}000}{\sqrt{3} \times 12.5} = 554.26 \text{ A at } 12.5 \text{ kV} \qquad (9.28)$$

K in Table 9.2 is 1250 for infrequent faults from 5.0 pu to the maximum of 10 pu, and K = 200 for infrequent faults over the same range.

The transformer has three MVA ratings; the first is a self-cooled rating; the second a rating with forced oil, and the third a rating with both forced oil and air. Thus the maximum load is

$$\frac{20{,}000}{\sqrt{3} \times 115} = 100.4 \text{ A at } 115 \text{ kV} \qquad (9.29)$$

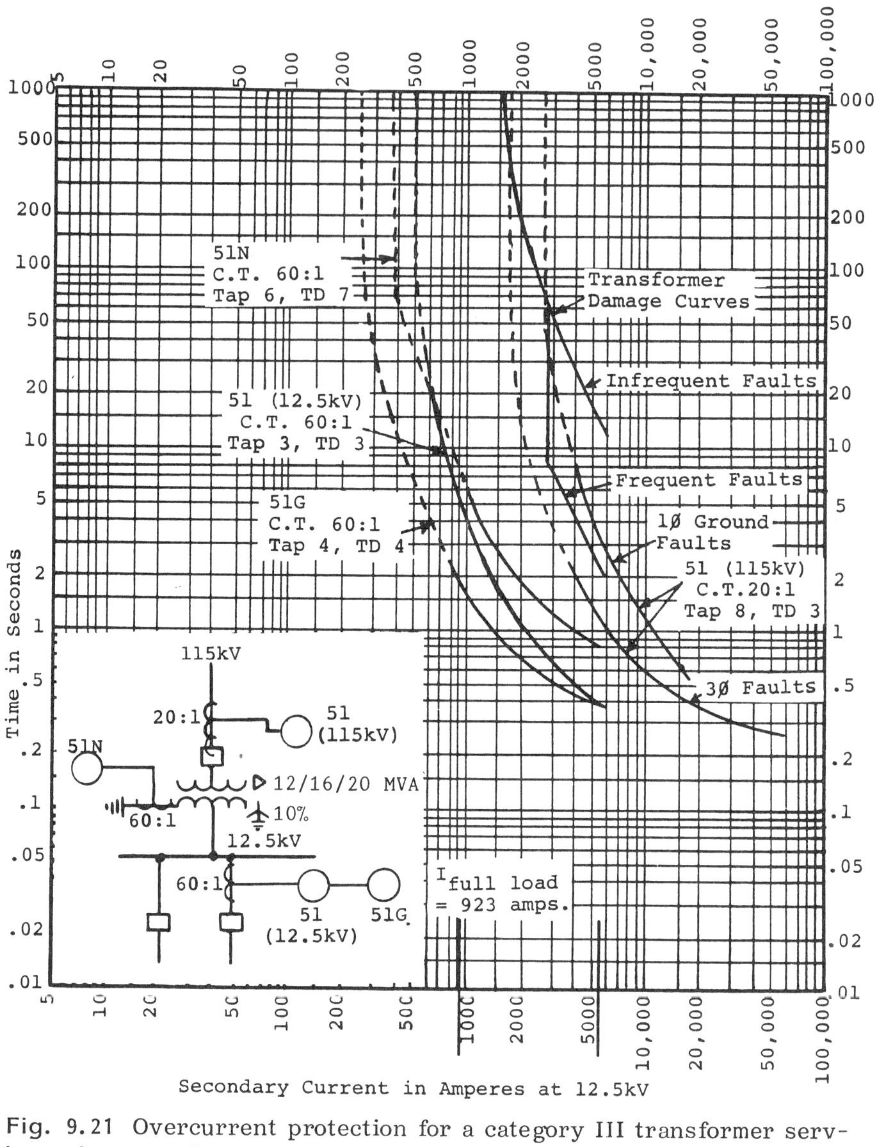

Fig. 9.21 Overcurrent protection for a category III transformer serving a frequent-fault secondary system with primary circuit breaker and relays.

and 100:5 (20:1) primary CTs are selected. Inverse-time-overcurrent phase relays (51) are set on tap 8, so pick up on 8 × 20 = 160 primary amperes, equivalent to 1472 A secondary amperes. Their pickup for secondary phase-to-ground faults is 1472/0.577 = 2551 A. The relay operating curves are plotted on Fig. 9.21 for specified time dials.

The 12.5-kV feeders are 6000 kVA with 300:5 (60:1) CTs, relay pickup phase 480 secondary amperes (51), 240 secondary amperes ground (51G). These curves are plotted in Fig. 9.21 for specified time dials.

This transformer bank is protected for through-three-phase secondary faults but not for secondary phase-to-ground faults. In this application differential protection should be applied and connected around the bank and secondary bus together with gas/pressure relays for primary protection. Then this service might be considered as infrequent service, and the overcurrent primary relays provide reasonable backup protection.

9.22 TRANSFORMER THERMAL PROTECTION

Thermal protection is usually supplied as a part of the transformer. Generally, it is used for monitoring and alarm, but may be used for tripping. Various types of thermal indicators are used to detect overheating of the oil, tank, tank terminals, failures of the cooling system if used, hot spots, and so on. These devices may initiate forced-cooling equipment. This protection is beyond the scope of this book.

9.23 OVERVOLTAGE ON TRANSFORMERS

Transformers must not be subject to prolonged overvoltage. For maximum efficiency they are operated near the knee of their saturation curve, so at voltages above around 110% of rated, the exciting current becomes very high. Just a few percent increase in voltage results in a very large increase in current. These large currents can destroy the unit if not reduced promptly.

Protection against overvoltage seldom is applied directly but is included in the regulating and control devices for the power system. An exception exists for unit generator-transformers, where overvoltage of these units is more likely to occur as they are brought onto or removed from service. This protection is discussed in Chapter 8.

9.24 SUMMARY: TYPICAL PROTECTION FOR TRANSFORMERS

The protection recommended and commonly applied for transformers is summarized in the following figures. The application for the various devices has been discussed in previous sections. It should be noted and recognized that these are general recommendations. More or less protection may be applied for any specific case, and depend on local circumstances and individual preferences.

9.24.1 Individual Transformer Units

The protection is summarized in Fig. 9.22 for banks where fuses are used on the primary. For larger or important banks in this category, differential protection may be applied; overall using CTs in the transformer primary bushings, or a ground differential as discussed in Section 9.13. Both require primary source tripping as discussed in Section 9.14.

For transformer banks with primary breakers, the protection is summarized in Fig. 9.23. Relay 151G provides backup protection for secondary bus and feeder faults and must be time coordinated, with other ground relays protecting the various feeder circuits on the secondary bus. Similarly, phase relays 51 must be coordinated with the phase relays on the feeders. Relay 51G is set with a longer time and to coordinate with 151G.

9.24.2 Parallel Transformer Units

The protection for transformer banks where the secondaries are connected together by a bus tie breaker is summarized in Fig. 9.24. The arrangement shown is typical for large or critical load substations, especially for industrial plants. The loads are supplied from separate buses connected together by a bus tie breaker (52T) which may be operated either normally closed (NC) or normally open (NO). If operated normally open, the protection of Fig. 9.22 or 9.23 applies.

If operated with 52T normally closed, the protection of Figs. 9.22 and 9.23 are applicable with the secondary side modified as shown in Fig. 9.24 b or c.

With the bus tie breaker closed, there is a possibility for the interchange of power between the two surces. For this case, current flows from one source through its transformer, the secondary buses, and back through the other transformer to the second source. Generally, this is not desirable or permitted. To prevent this operation, directional time overcurrent relays (67, 67N) are applied to each transformer. The single line connections are shown in Fig. 9.24 b and c, with complete three-line connections in Fig. 9.25. They operate only for

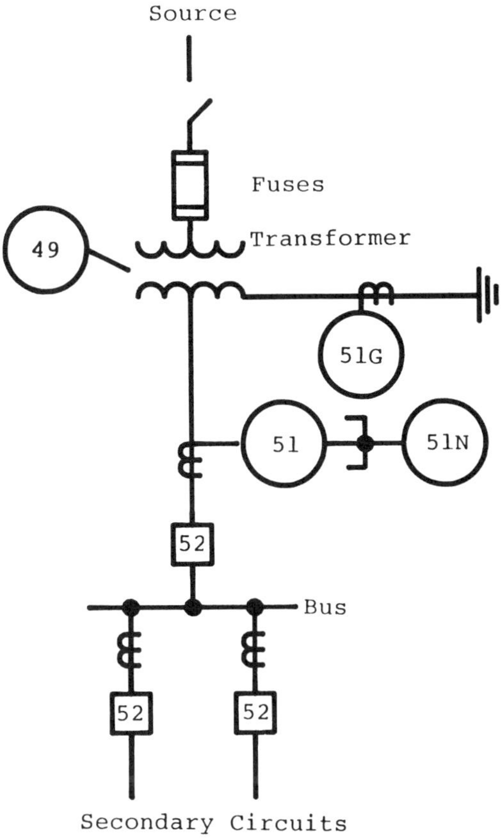

Fig. 9.22 Transformer protection without primary-side circuit breaker. Common connection shown with delta on the source (primary) side and wye-grounded on secondary side. Other possible connections: delta-delta, wye-wye, or primary wye secondary-delta. Secondary circuits should have 51 and 51N relays so transformer secondary breaker and relays may be omitted unless another source connects to the secondary bus. 51N relay can be omitted with 51G available.

fault current that flows into the transformer and trip the secondary breaker (52-1 or 52-2). This is also important in removing a secondary fault source for faults in the transformer bank. The phase relays (67) can be set on a low or minimum tap. Load current does flow through the relay but normally not in the operating direction. The low tap continuous rating must not be exceeded by maximum load current. The 67 time setting must coordinate with the protection on the transformer

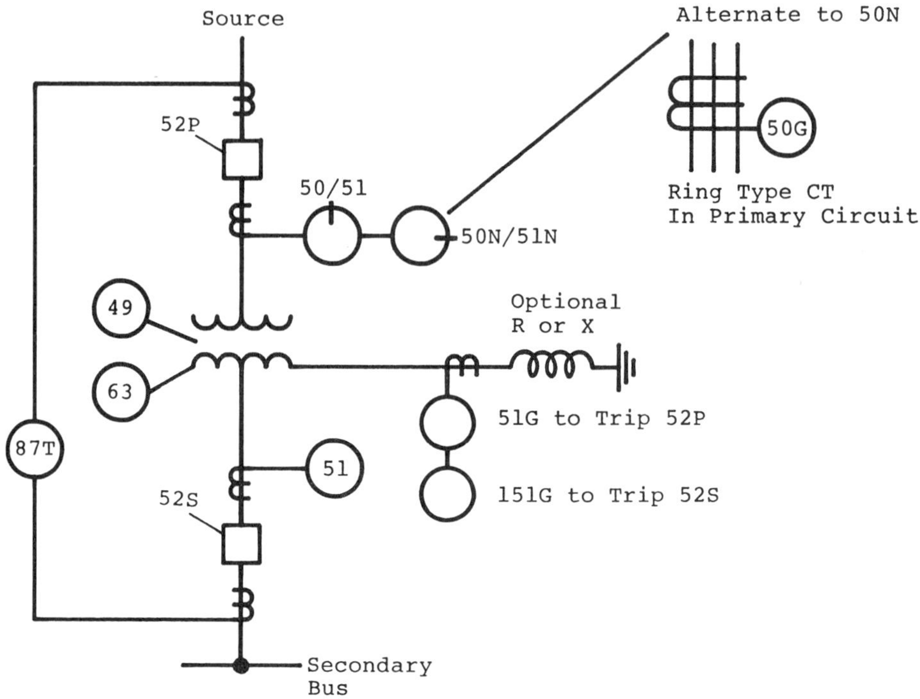

Fig. 9.23 Transformer protection with primary-side circuit breaker. Common connection shown with delta on the source (primary) side and wye-grounded on the secondary side. Other possible connections: delta-delta, wye-wye, primary-wye secondary-delta, three-winding, or autotransformer. 52S may be omitted in some applications requiring 151G to coordinate with and trip the secondary circuit devices, if used.

primary. When used, the ground relay can be set on minimum setting and time, as coordination is not necessary.

The inverse-time overcurrent relays (51, 51N) provide bus protection and backup protection for the feeder circuits. These relays trip both 52-1 (or 52-2) and 52T. This is a partial differential connection which is discussed in Section 10.11.4 and shown in Figure 10.10. These units must be time coordinated with the protection on the several feeders connected to the bus. Only two phase relays are required, but the third relay (shown optional in Fig. 9.25), provides additional redundancy. When a ground differential is used, as illustrated in Fig. 9.24c, 67N and 51N are omitted. The connections shown are compatible with Figs. 9.12 through 9.15.

Ground-fault backup is provided by 51G, 151G, and 251G inverse-time overcurrent relays (Fig. 9.24c). Relay 251G provides bus ground-fault protection and backup for the feeder circuit ground relays. It must be time coordinated with these. It trips the bis tie 52T, as the fault could be either on the bus or on the associated feeders. If the fault still exists with the bus tie open, relay 151G trips breaker 52-1 (or 52-2). Thus 151G must coordinate with 251G. If the fault is still present, it is between the secondary breaker, in the transformer winding, or in the grounding impedance. Relay 51G set to coordinate with 151G is the last resort. It trips the high-side or primary breaker to remove the transformer from service.

9.25 REACTORS

Reactors are used in power systems primarily (1) in grounded neutrals to limit fault current (2) in series in the phases to reduce phase fault current magnitudes, and (3) in shunt to compensate for the capacitive reactance of long transmission lines and pipe-type cable circuits. Other uses are as harmonic filter banks and to suppress secondary arc currents in single-pole relaying.

Application (1) is covered in Chapter 7 and in this chapter. Phase-fault limiting reactors (2) are used between buses each of which have high short-circuit levels, and in feeder circuits connected to similar high-short-circuit-capacity buses. In this section we briefly review shunt reactors.

9.25.1 Types of Reactors

Shunt reactors are either dry or oil-immersed types. The dry type, available to voltages of about 34.5 kV, usually are applied to transformer tertiaries. They are single-phase air-core construction with the winding exposed for natural convection for either indoor or outdoor mounting. The location should be in an area where the high-intensity magnetic field is not a problem or hazard.

As single-phase units, phase-type faults are not common, but may occur by simultaneous faults in more than one reactor or by a fault spreading to involve the bus. Thus the principal hazards are ground and turn-to-turn faults. A common arrangement of tertiary-connected reactors is in ungrounded wye with the system grounded by a broken-delta resistor method (Fig. 7.8).

Oil-immersed reactors can be either single-phase or three-phase units in a tank, with an appearance similar to that of a transformer bank. Voltage is not a limitation, so this type is used for line connections. Solid grounding is normal. The principal hazards are phase

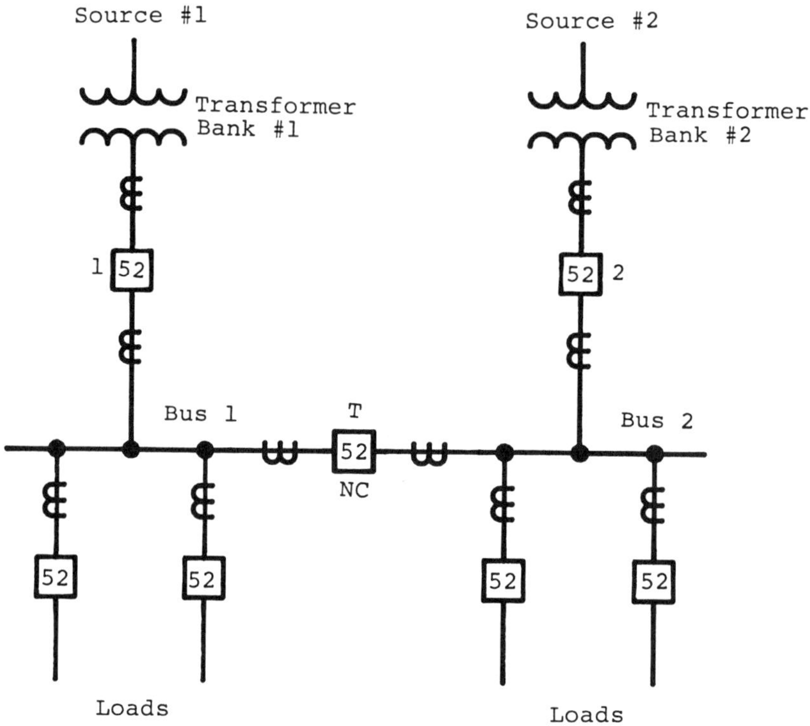

(a)

Fig. 9.24 Transformer and secondary bus protection for a typical double source supply with secondary tie and breaker. For 67 and 51 connections, see Fig. 9.25. For 87G connections, see Figs. 9.12 through 9.15. (a) Single-line diagram; (b) secondary protection with high-side fuses (see Fig. 9.21); (c) secondary protection with high-side breaker (see Fig. 9.22).

and/or ground faults from insulation failure, bushing failure, turn-to-turn faults, low oil, and loss of cooling.

9.25.2 General Application of Shunt Reactors

Generally, shunt reactors are connected directly (without breakers) to the line, or when the line is terminated without a breaker to a transformer bank with tertiary, the reactors may be connected to the tertiary. Reactors may also be connected to a bus. With direct line

(b)

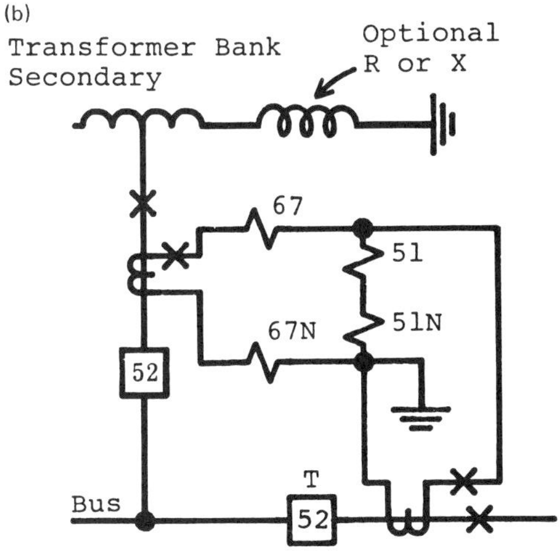

(c)

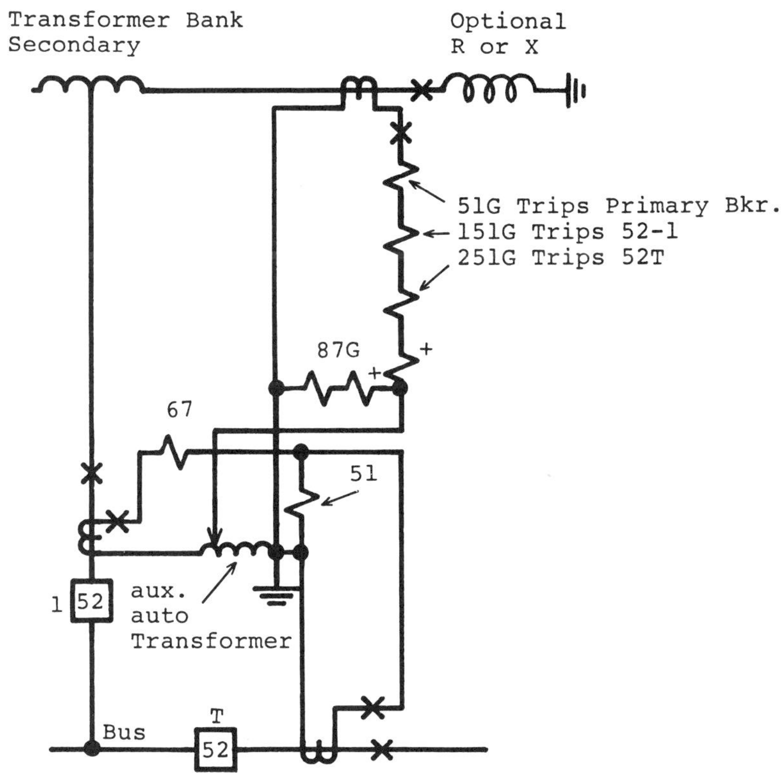

Fig. 9.24 (continued)

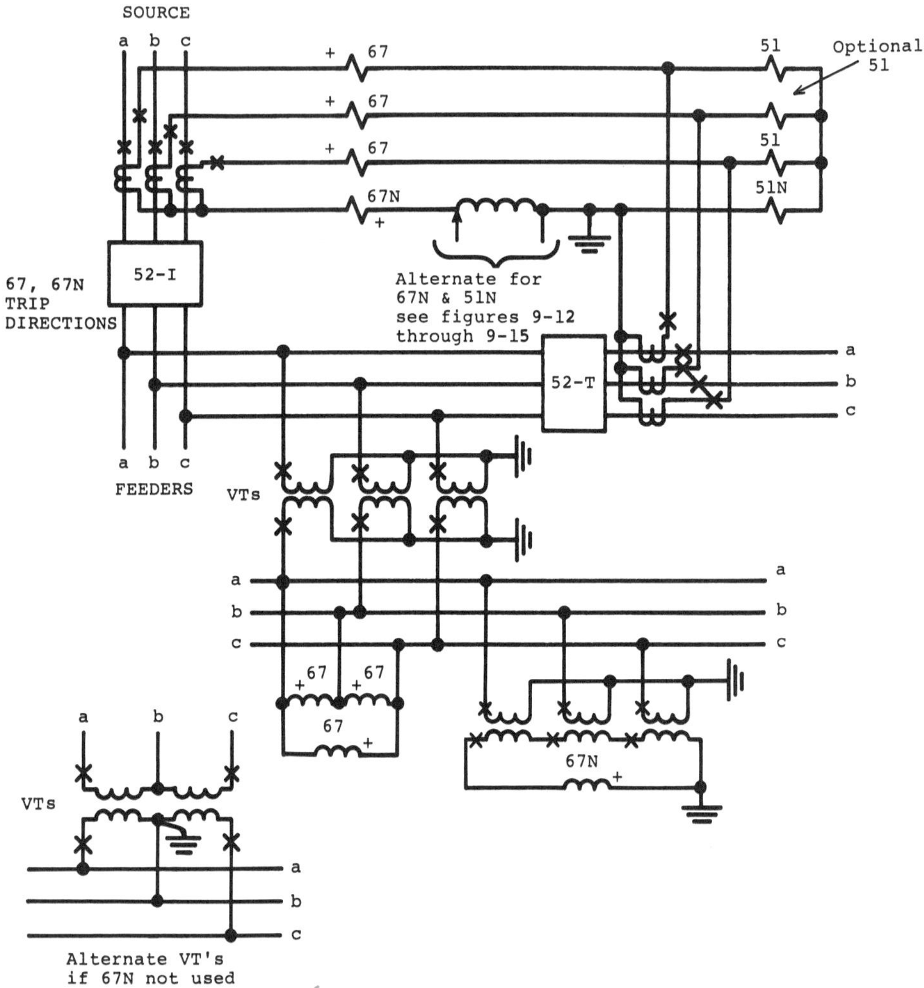

Fig. 9.25 Three-line connections for reverse-phase and partial differential backup protection.

connection, problems in the reactor units require deenergizing the line by tripping the local breaker and transfer-tripping all remote breakers that can supply fault current. Disconnect switches provide means of manual isolation, or circuit switchers may be used to disconnect the reactors automatically when the line has been opened. This may be necessary for automatic reclosing of the line; however, operation of the circuit without reactors can produce overvoltage. Deenergizing a circuit with shunt reactors produces a transient oscillation between the reactive and capacitive elements with a frequency generally less than 60 Hz. This can be a problem for some of the protection. Also, if there is an energized parallel circuit, this coupling can produce overvoltage and/or ferroresonance. An example was cited in Chapter 7.

Reactors connected to transformer tertiary circuits usually result in very low fault levels in the higher-voltage circuits for reactor faults because of the relative high impedance through the transformer windings to the tertiary. Breakers usually are available for isolation of these tertiary-connected units. Again, removing the reactor can result in overvoltages on the line and associated system.

9.25.3 Reactor Protection

Basically, the protection of reactors is the same as for transformers with the size, importance to the system, and "personality" providing many variations and different approaches.

In all applications, differential (87) protection is the most widely used, with phase and, where applicable, ground overcurrent as backup or sometimes the primary protection. Differential connections are as shown in Fig. 8.5a or for ground differential, in Fig. 9.10. Instantaneous-overcurrent (50) devices set above the inrush and transient currents, and inverse-time-overcurrent (51) devices set to coordinate with all other protection equipment overreached are both used. Impedance relays set to "look into" the reactor are used for both primary and backup. They must be set to avoid operation on the energizing inrush current and not to operate on the natural frequency oscillations that can occur when the compensated line is deenergized. Negative-sequence current relays are also applied in some cases. Connection of the protection is covered in this chapter and elsewhere.

Turn-to-turn faults are a concern in reactors as well as in generators, motors, and transformers. They can produce considerable current and damage in the area of the problem but reflect very few indications ("handles") at the terminals for detection until they develop and involve other phases and/or ground. For dry reactors a voltage balance using an overvoltage relay (59) compares the neutral-to-ground voltage measured from the ungrounded reactor wye neutral with the grounded broken-delta voltage. Normal unbalance which can exist is balanced out so that the scheme can detect between this unbalance and that caused by a fault. Negative sequence has been used but is relatively insensitive, as the amount of negative sequence measurable for turn-to-turn faults is very low for light faults. Impedance relays can provide protection by the change in reactor impedance. Probably the most sensitive protection for these very difficult-to-detect faults are mechanical relays (Section 9.15), but gas accumulator or pressure relays are applicable only to the oil-immersed units.

Wherever possible, these relays should always be applied in addition to the other protection.

9.26 CAPACITORS

Both series and shunt capacitors are utilized in power systems. Series capacitor banks in long HV transmission lines reduce the total impedance between large power sources. This increases the power transfer capabilities and enhances stability. This is discussed further in Chapter 14. The protection of these banks can be similar to transformers. Often they are part of the transmission line and so are protected as part of the line, as outlined in Chapter 12.

9.26.1 Shunt Capacitor Applications

Shunt-connected capacitor banks are used for reduction in the amount of vars required by reactive loads and to aid in the regulation and control of power system voltages. They may be connected as required throughout the system, but generally they are located at distribution stations and out on the distribution feeders. The maximum benefit to the power system is obtained by locating the capacitors as near as practicable to the reactive loads.

Shunt capacitors are var generators. They supply lagging vars, generally kilovars, to the system, which also can be supplied by overexcited rotating machines, such as synchronous condensers. This effect is illustrated in Fig. 8.13.

Synchronous condensers provide variable control, but at a relative high cost. Static capacitor banks are fixed or switched. The fixed banks are connected to the system and remain connected for long periods. The switched banks are added and removed, usually in response to the voltage levels, and so are essentially voltage regulators.

9.26.2 Shunt Capacitor Protection

The primary protection of a capacitor bank is by the individual capacitor fuses, which are part of the capacitor design and supplied as part of the bank. This protects the individual units, but if several fuses open, damage can result to other sound capacitors. Hence additional and/or backup protection is applied. Phase fuses or phase and ground overcurrent (51, 51N) between the system and the bank provide protection for major bank faults. Unbalance or neutral ground protection is used, with the scheme depending on the grounding and bank connections.

Capacitor groups are connected in single-wye banks either grounded or ungrounded, double-wye banks either grounded or ungrounded, or delta in most applications. For single ungrounded wye banks, a 59G relay across the broken-delta secondary of VTs connected phase to ground is used. For double ungrounded wye banks a VT or CT between the two neutrals with a secondary 59 or 51 relay, respectively, provides good balance protection.

For grounded-wye banks, a neutral CT with a resistor and 59G overvoltage relay provides unbalanced protection. Other schemes and a more complete discussion are provided in IEEE Standard C37.99-1980. Relays connected in the neutral circuits should be insensitive to third harmonics, or a filter used as normal harmonic quantities can be in the same order of magnitude as a light fault quantity.

Energizing and deenergizing capacitor banks can produce severe transients and possible overvoltages, so attention to these is important in applying circuit breakers or other interrupting devices and in relays and their setting. Transients of high magnitude and high frequency can occur when switching banks back to back. This where a capacitor bank is connected to the system near a capacitor bank that is already energized.

BIBLIOGRAPHY

See the Bibliography at the end of Chapter 1 for additional information.

General Requirements for Liquid-Immersed Distribution, Power and Regulating Transformers, ANSI/IEEE Standard C57.12-1980, IEEE Service Center.

Griffin, C. H., Development and Application of Fault-Withstand Standards for Power Transformers, *IEEE Trans. Power Appar. Syst.*, PAS 104, 1985, pp. 2177–2188.

Guide for Protection of Shunt Capacitors, ANSI/IEEE Standard C37.99-1980, IEEE Service Center.

Guide for Protective Relaying Applications to Power Transformers, ANSI/IEEE Standard C37.91-1984, IEEE Service Center.

Shunt Reactor Protection Practices, IEEE Power System Relaying Committee, *IEEE Trans. Power Appar. Syst.*, PAS 103, 1984, pp. 1970–1976.

Sonnemann, W. K., C. L. Wagner, and G. D. Rockefeller, Magnetizing Inrush Phenomenon in Transformer Banks, *AIEE Trans.*, 77, P. III, 1958, pp. 884–892.

Transformer Through Fault Current Duration Guide, ANSI/IEEE Standard C57.109-1984, IEEE Service Center.

10

Bus Protection

10.1 INTRODUCTION: TYPICAL BUS ARRANGEMENTS

Buses exist throughout the power system and wherever two or more circuits are interconnected. The number of circuits that are connected to a bus varies widely. Bus faults can result in severe system disturbances, as all circuits supplying fault current must be opened to isolate the problem. Thus buses are often split when there are more than six to eight circuits involved by a circuit breaker (bus tie), or a bus arrangement is used that minimizes the number of circuits that must be opened for a bus fault. There are many bus arrangements in service dictated by the above and by the economics and flexibility of system operation. The major types are illustrated in Figs. 10.1 through 10.8. Four circuits for each bus have been chosen arbitrarily for convenience and comparison. The bus circuit breakers usually have disconnect switches on either side as shown, to provide means of isolating them from the system after trouble or for maintenance. Generally, these switches are operated manually at no load. The circuits shown connecting to the buses can be generators, transformers, lines, motors, and so on.

The buses typically illustrated are:

Single bus—single breaker	(Fig. 10.1)
Double bus with bus tie—single breaker	(Fig. 10.2)
Main and Transfer bus—single breaker	(Fig. 10.3)
Double bus—single breaker	(Fig. 10.4)
Double bus—double breaker	(Fig. 10.5)
Ring bus	(Fig. 10.6)
Breaker-and-a-half bus	(Fig. 10.7)
Bus and Transformer—single breaker	(Fig. 10.8)

Other arrangements exist and can be considered as combinations or variations of the above.

Fortunately, bus faults are not too common, but are serious since they can result in considerable loss of service via the circuits that must be opened to isolate the fault. The most common causes of bus faults are equipment failures, small-animal contacts, broken insulators, wind-driven objects, and contamination.

Differential provides sensitive and fast phase and ground-fault protection and is generally recommended for all buses. In the figures the dotted box or boxes outline the bus differential protection zone. This is the primary protection zone. Backup is usually provided by the protection associated with the connecting circuits. A second differential scheme is sometimes used for very important buses.

10.2 SINGLE BREAKER—SINGLE BUS

This type, shown in Fig. 10.1, is the most basic, simple, and economical bus design, and is used widely particularly at distribution and lower transmission voltages. For this type of bus, differential is easy to apply as long as suitable CTs are available, with the protective zone enclosing the entire bus as shown.

This bus arrangement provides no operating flexibility. All bus faults require opening all circuits connected to the bus. Breaker problems or maintenance require that the circuit be removed from service. However, maintenance may not be too much of a problem if maintenance on the entire circuit and the protection can be scheduled together.

One set of voltage transformers on the bus can supply voltage for the protection on all the circuits.

10.3 SINGLE BUSES CONNECTED WITH BUS TIE

This is an extension of the single bus--single breaker arrangement and is shown in Fig. 10.2. It is used where a large number of circuits exist, and especially at lower voltages, such as for distribution and industrial substations. It provides flexibility where the substation is fed from two separate power supplies. One supply connected to each bus permits operation with the bus tie (52T) either open or closed. If one supply is lost, all circuits can be fed by the other with 52T closed. Separate differential zones for each bus are applied. A fault in one bus zone still permits partial service to the station via the other bus.

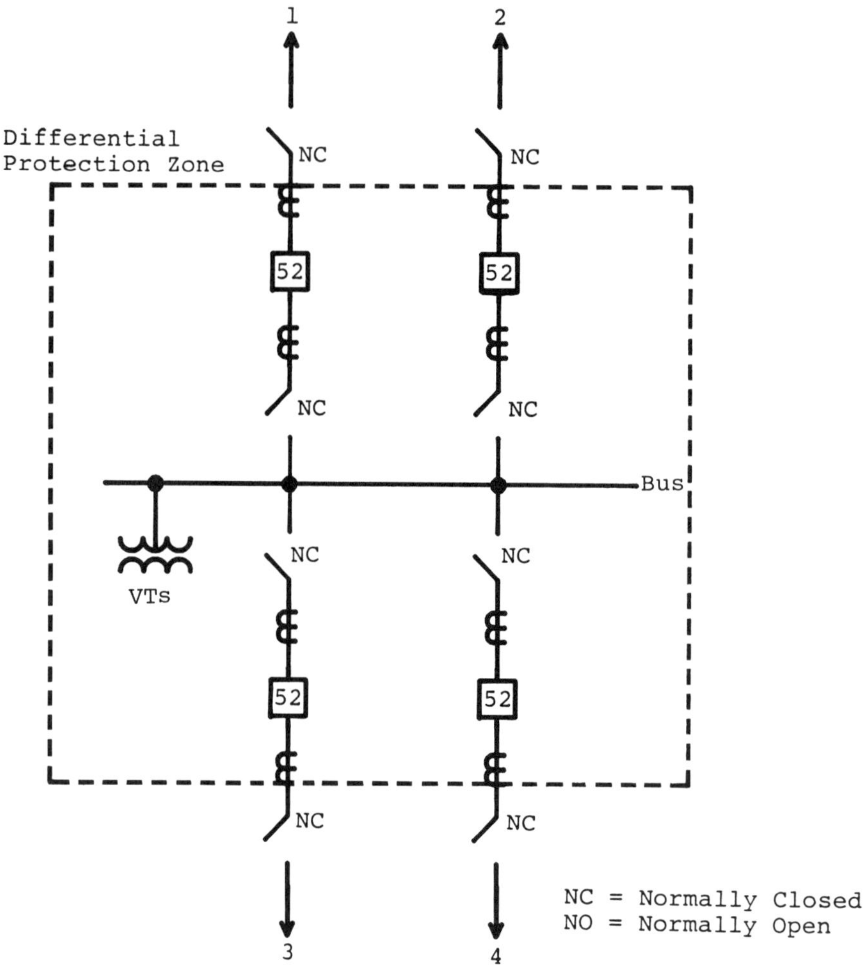

Fig. 10.1 Typical four-circuit single breaker–single bus and the bus differential protection zone.

10.4 MAIN AND TRANSFER BUSES WITH SINGLE BREAKER

Increased operating flexibility is provided by the addition of a transfer bus, as shown in Fig. 10.3. Normal operation is similar to Fig. 10.1, with all circuits supplied from the main bus. This bus is protected by a single differential zone, shown dotted. A bus fault requires tripping all breakers, thus interrupting all service connected to the bus.

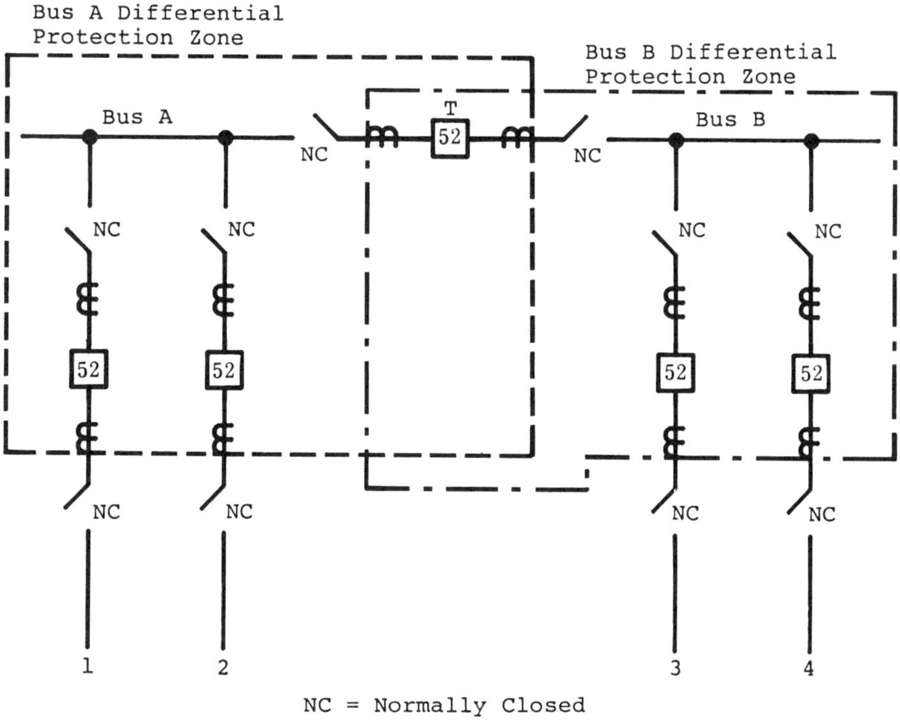

Fig. 10.2 Typical four-circuit single breaker–double bus with bus tie, and the bus differential protection zones.

Normally, the transfer bus is not energized. For any breaker trouble or maintenance, that circuit is connected to the transfer bus by closing its normally open (NO) disconnect switch and closing the bus tie (52T) breaker to continue service. Only one circuit is thus connected to the transfer bus at any one time. The protection associated with the bus tie breaker must be suitable and adaptable for the protection of any of the circuits of the main bus. this can require diferent settings, which must be made for each circuit transferred or operating with compromise protection for the period of transfer bus operation. This is a disadvantage from a protection standpoint. As a generality it is not desirable to switch or modify protection systems because the potential for error that can result is no protection or misoperation. One set of voltage transformers on the bus can supply voltage to all the protection for the several circuits.

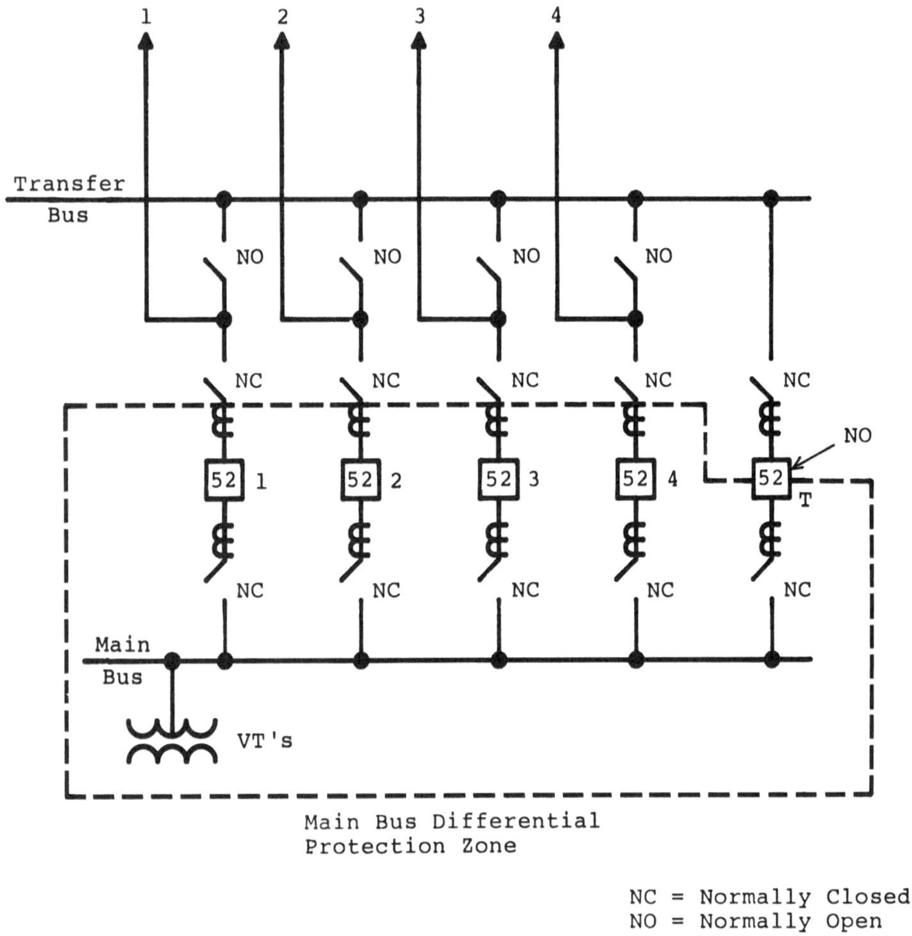

Fig. 10.3 Typical four-circuit single breaker—main bus with transfer bus, and the bus differential protection zone.

10.5 SINGLE BREAKER—DOUBLE BUS

This arrangement, shown in Fig. 10.4, provides high flexibility for system operation. Any line can be operated from either bus, the buses can be operated together as shown or independently, and one bus can be used as a transfer bus if a line breaker is out of service. The disadvantage is that it requires complicated switching of the protection: both the bus differential and line protection. Two differential zones for the buses are required. In the Fig. 10.4, lines 1 and 2 are shown

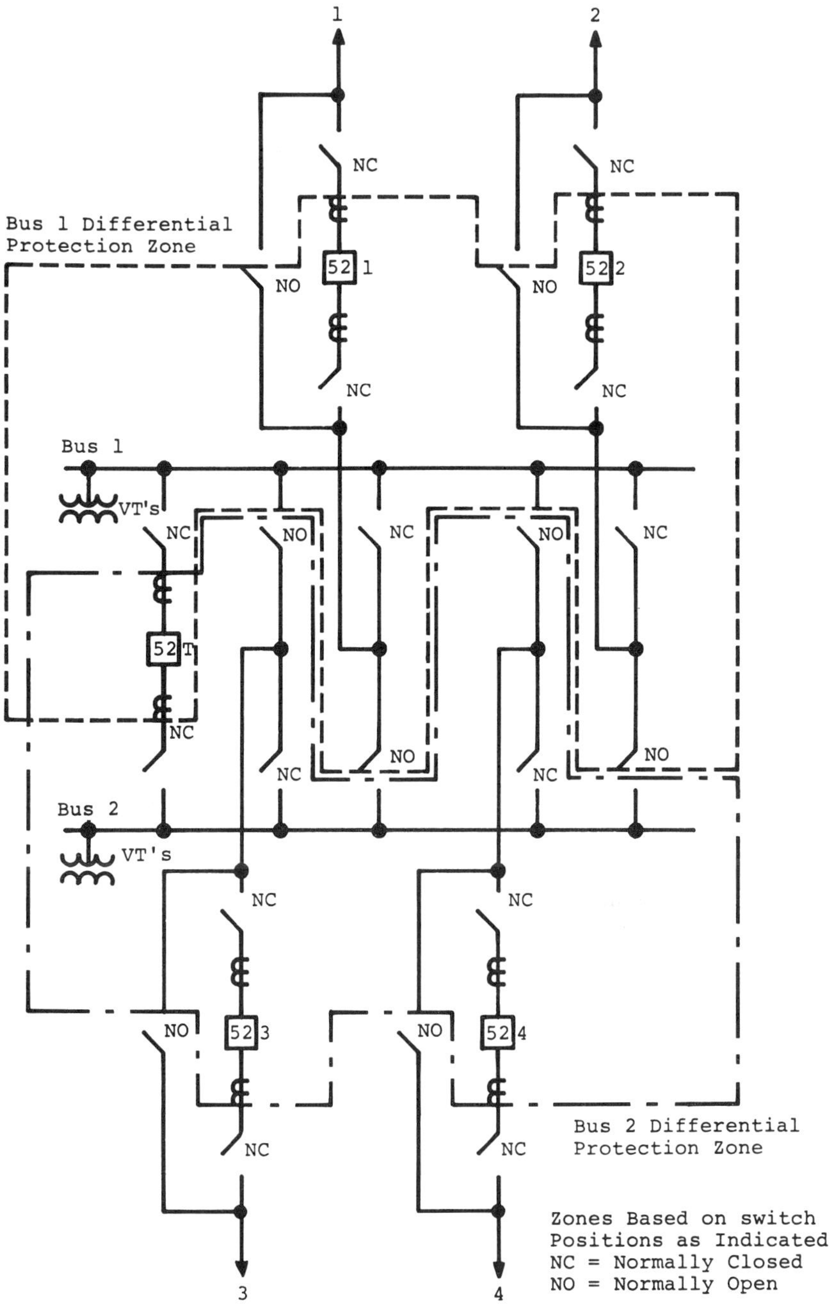

Fig. 10.4 Typical four-circuit single breaker–double bus and the bus differential protection zones.

connected to bus 1, with lines 3 and 4 connected to bus 2. For this operation the differential zones are outlined: dotted for bus 1, and dash-dot for bus 2.

As for the previous bus arrangement (Fig. 10.3), the bus tie protection must be adaptable for the protection of any of the lines when 52T is substituted for any of the line circuit breakers. When a line breaker is bypassed and the bus tie (52T) breaker substituted using one bus as a transfer bus, the differential protection on that bus must be removed from service.

Faults on either bus or associated circuits require tripping of all circuits connected to the bus at that time. Faults in the bus tie breaker (52T) must trip both buses and all circuits.

Voltage transformers for protection are required for each bus as shown. However, line-side VTs are preferable to avoid switching if voltage is required for line protection.

This bus arrangement is not in wide use in the United States principally because of the protection complications, and is not recommended from that standpoint.

10.6 DOUBLE BREAKER—DOUBLE BUS

This is a very flexible arrangement that requires two circuit breakers per circuit, as shown in Fig. 10.5. Each bus is protected by a separate differential, with zones as illustrated. The line protection operates from paralleled CTs, and this provides protection for the bus area between the two zones overlapping the two breakers. Line protection operates to trip both breakers.

With all disconnect switches normally closed (NC) as shown, a fault on either bus does not interrupt service on the lines. All switching is done with breakers and either bus can be removed for maintenance.

Line-side voltage, either VTs or CCVTs, is necessary if required by the line protection.

10.7 RING BUS

This arrangement (fig. 10.6) has become quite common, particularly for higher voltages. High flexibility with a minimum of breakers is obtained. Each breaker serves two lines and must be opened for faults on either line. The bus section between the breakers becomes part of the line, so that bus protection is not applicable or required. The interconnection of the CTs for protection of each line is shown dotted in Fig. 10.6, and line faults must trip two breakers. If the ring is

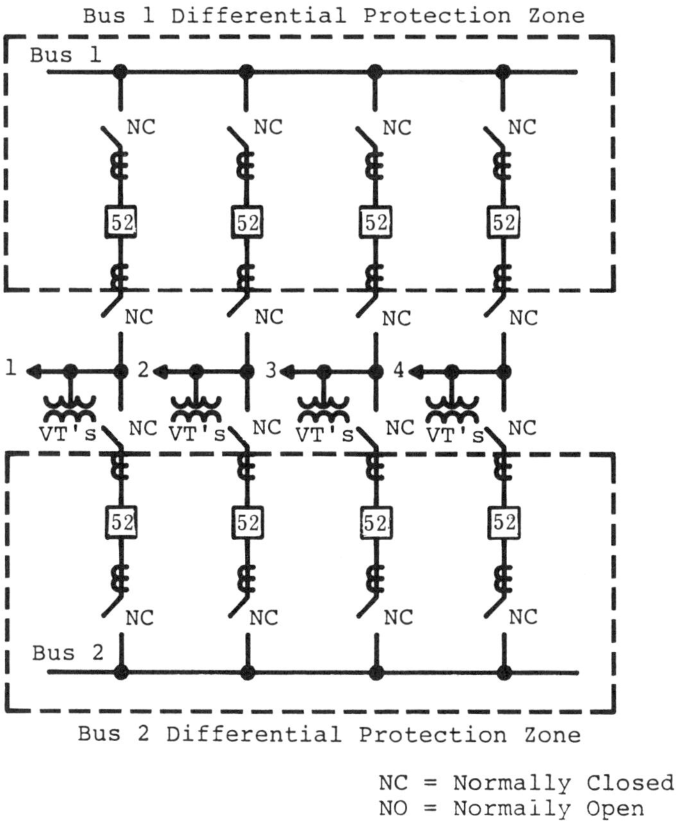

Fig. 10.5 Typical four-circuit double breaker–double bus and the bus differential protection zones.

open for any reason a fault on a line may separate the other lines and the bus.

Line protection voltage, if required, is obtained from VTs or more commonly at the higher voltages by CCVTs connected to each line.

10.8 BREAKER-AND-A-HALF BUS

This arrangement, shown in Fig. 10.7, provides more operating flexibility but requires more circuit breakers than the ring bus. It, too, is widely used, especially for larger multiple-circuit higher-voltage systems. Two operating buses each have separate differential protection. Each line section is supplied by both buses through two circuit

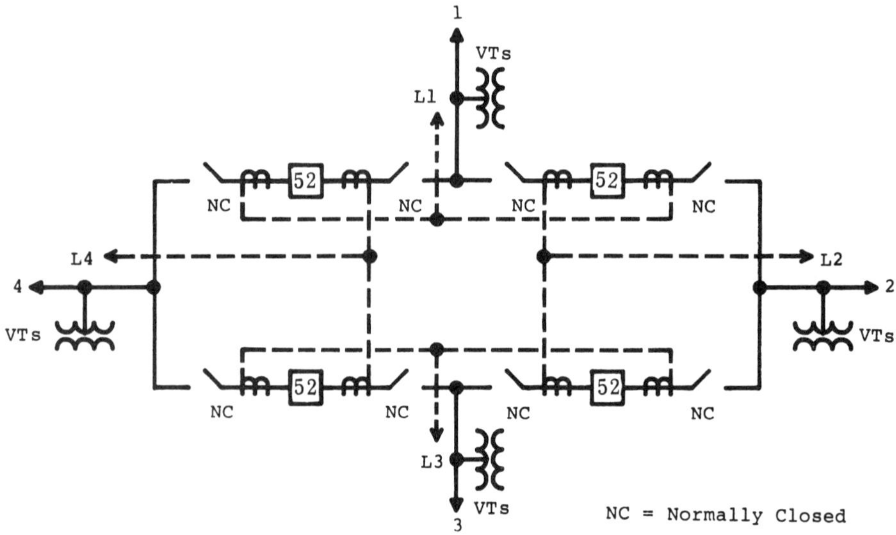

Fig. 10.6 Typical four-circuit ring bus. Differential protection not applicable. Bus sections are protected as part of the lines or connected equipment, as shown dotted.

breakers. The center circuit breaker serves both lines, hence the half designation.

The CT interconnections are shown for each line section dotted in Fig. 10.7. Voltage for line relays must use line-side CCVTs or VTs. Line faults trip two breakers but do not cause loss of service to the other lines if all breakers are normally closed as shown.

10.9 TRANSFORMER-BUS COMBINATION

This is the single breaker—single bus of Fig. 10.1 with a transformer bank directly connected to the bus as shown in Fig. 10.8. The advantage is the cost saving of the circuit breaker between the transformer and the bus. It is practical for small stations such as distribution, where there is only one transformer to supply several circuits. Here a fault in either the transformer or on the bus requires that all service be interrupted with or without the intervening breaker.

The differential zone includes both the bus and the transformer, as shown dotted. In these applications transformer differential relays must be used. The application and setting for these is presented in Chapter 9.

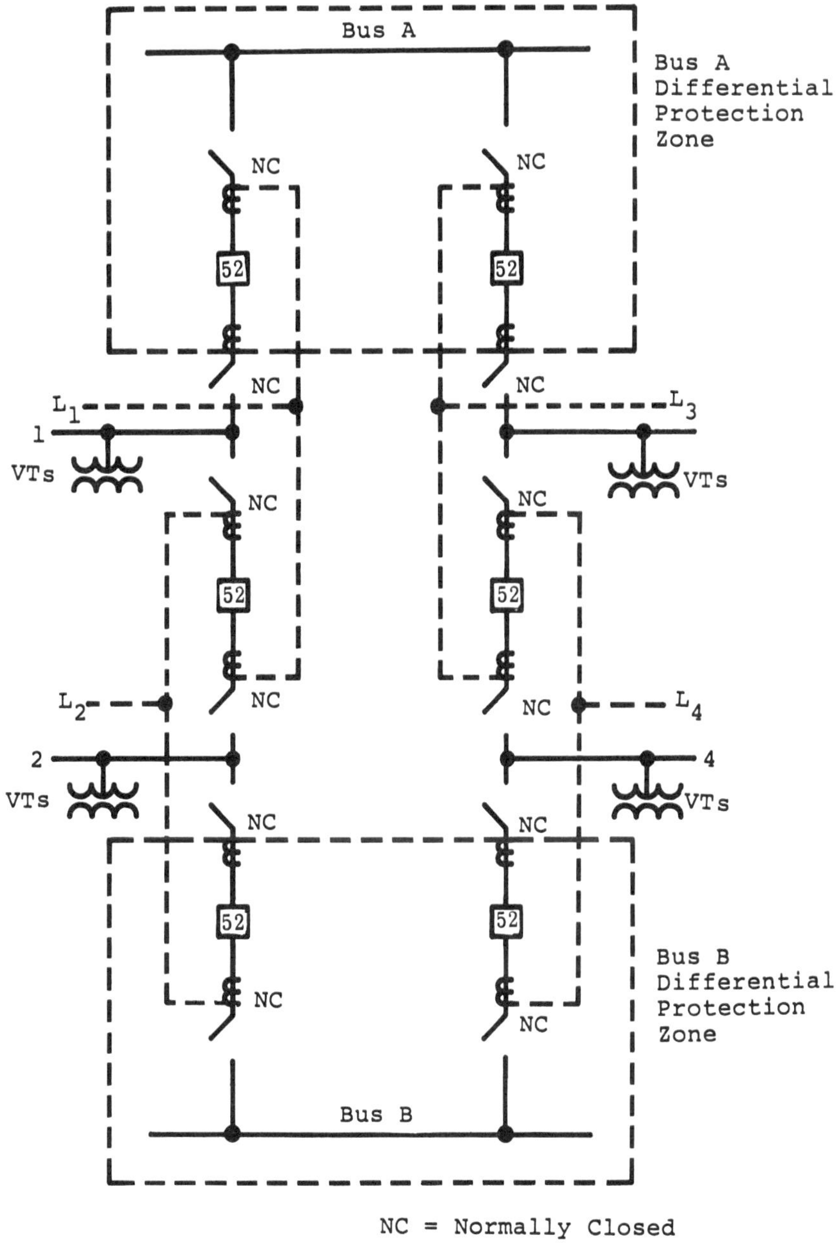

Fig. 10.7 Typical four-circuit breaker-and-a-half bus and the bus differential protection zones. The mid-bus sections are protected as part of the lines or connected equipment, as shown dotted.

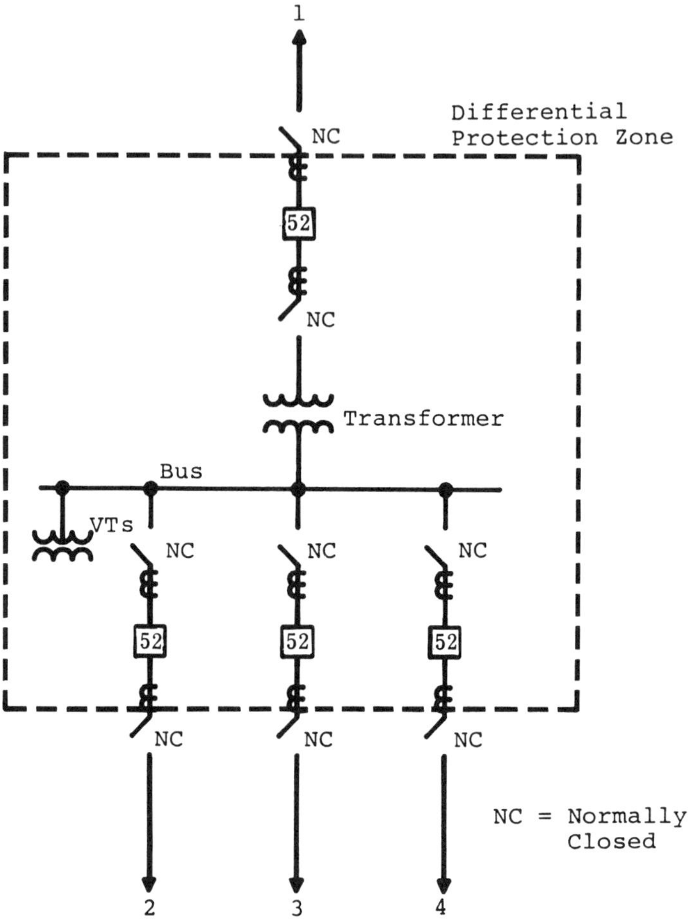

Fig. 10.8 Typical four-circuit single breaker bus and transformer with combined bus-transformer differential protection zone.

10.10 GENERAL SUMMARY OF BUSES

Table 10.1 provides a brief summary of the several bus arrangements outlined above. The far-right column indicates the tripping requirements in the event that one of the breakers fails to clear a fault on a circuit connecting to the bus. This could result from malfunction of the protective system, or the breaker failing to open properly. Local breaker failure-backup protection is assumed, and this is discussed in Chapter 12.

Table 10.1 General Summary of System Buses

Bus type	Fig. No.	Bus sect.	Required for one line Bkrs.	Required for one line Disc.	Required for each added line Bkrs.	Required for each added line Disc.	Total four lines Bkrs.	Total four lines Disc.	Three-phase voltage supply for line relays	Breaker failure: local backup requires tripping[a]
Single bus—single breaker	10.1	1	1	2	1	2	4	8	1 on bus	All breakers on bus
Single buses with bus tie	10.2	2	For two-line minimum 3	6	1	2	5	10	On each bus	All breakers on bus[b]
Main and transfer bus	10.3	2	2	5	1	3	5	14	1 on main bus	All breakers on bus
Double bus—single breaker	10.4	2	2	7	1	5	5	22	1 on each bus[c]	All breakers on bus[b]
Double bus—double breaker	10.5	2	2	4	2	4	8	16	On each line	All breakers on bus
Ring bus	10.6	0	For two-line minimum 2	4	1	2	4	8	On each line	Adjust breaker and remote line terminal
Breaker and a half	10.7	2	For two-line minimum 3	6	Each two lines 3	6	6	12	On each line	All breakers on bus; for center breaker, adjacent breaker and remote line terminal

[a]Also, the remote terminal of the stuck breaker must be opened by relay action at that terminal and/or by transfer tripping.

[b]Failure of the bus tie breaker requires tripping all circuits on both buses.

[c]Line voltage preferred to avoid voltage transfer as lines are switched to a different bus.

10.11 DIFFERENTIAL PROTECTION FOR BUSES

Complete differential protection requires that all circuits connected to the bus be involved, since it compares the total current entering the zone with the total current leaving the zone. Except for a two-circuit bus, this means comparisons between a number of CTs operating at different energy levels and often with different characteristics. The most critical condition is the external fault just outside the differential zone. The CTs on this faulted circuit receive the sum of all the current from the other circuits. Thus it must reproduce a potential high current magnitude with sufficient accuracy to match the other CT secondary currents and avoid misoperation. Thus CT performance is important. The relays and CTs are both important members of a "team" to provide fast, sensitive tripping for all internal faults, yet restrain all faults outside the differential zone. Two major techniques are in use to avoid possible unequal CT performance problems: (1) multirestraint current and (2) high-impedance voltage. A third system employs air-core transformers to avoid the iron-core excitation and saturation problems. All are in practical service. They exist with various features, depending on the design, and each has specific application rules. These should be followed carefully, as they have been developed to overcome the inherent deficiencies of conventional CTs on both symmetrical and asymmetrical fault currents.

10.11.1 Multirestraint Current Differential

This is the most versatile method for general application using conventional current transformers, but in general is more difficult to apply. However, each manufacturer has developed application connections and criteria that simplify the process. Multirestraint relays are used with a restraint winding connected to each circuit that is a major source of fault current. Feeders and circuits with low fault current contribution may be paralleled. The fundamentals discussed in Section 9.5 apply here. All CTs are connected in wye and to the restraint windings since there are no phase-shift problems with buses except for the case shown in Fig. 10.8.

These schemes are designed to restrain correctly for heavy faults just outside the differential zone with maximum offset current as long as the CTs do not saturate for the maximum symmetrical current. This can be accomplished by CT ratio selection and by keeping the secondary burden low. Thus it is important and recommended that no other devices be connected in the differential circuits. The restraint windings of the differential relays normally have quite low impedance, so that often the major burden encountered is that of the leads connecting the CTs to the relays. This can be kept low by large wire, which is also desirable to minimize physical damage. Accidental breakage or an open in the differential circuit can result in incorrect operation and loss of a critical part of the power system.

Multirestraint bus differential relays do not have ratio taps. These are not required in most applications, as a common CT ratio normally can be obtained among the several bus CTs. Otherwise, auxiliary CTs are required for the CTs that do not match. When these are used it is desirable that they step down the current if possible. This reduces their secondary burden, as discussed in Chapter 5.

The relays exist with up to six restraint circuits, and can have either fixed or variable restraint characteristics. Typical sensitivities for internal faults are on the order of 0.15 A, with operating times of 50 to 100 ms.

10.11.2 High-Impedance Voltage Differential

This scheme loads the CTs with a high impedance to force the error differential current through the CTs instead of the relay operating coil. The basic principles are illustrated in Fig. 10.9. For an external fault, the maximum voltage V_R across the differential relay Z_R will occur if the CT on the faulted circuit (1) is completely saturated and the other CTs (2 and 3) do not saturate. This is a worst case, as in practice all CTs may not saturate on light external faults or will have varying degrees of saturation for the heavy faults. An empirical margin with a safety factor is provided by the manufacturer to modify this maximum voltage calculation for setting the relay. This calculation is made for both the maximum symmetrical three-phase and phase-to-ground faults. The fault currents are different and the lead resistance R_L (maximum for the various circuits) is R_L for three-phase faults and $2R_L$ for phase-to-ground faults.

For internal bus faults as shown in Fig. 10.9, the high-impedance Z_R of the bus differential relay forces much of the secondary current through the CT exciting impedances. Thus V_R will be high to operate the relay, and is essentially the open-circuit voltage of the CTs. A varistor or similar protective device across Z_R provides circuit protection by limiting the voltages to a safe level. A tuned circuit provides maximum sensitivity at rated system frequency, and filters out the dc transient components. The impedance between the junction and the relay R_{LR} is negligible compared to the high value of the relay Z_R.

The scheme requires that the total resistance of the CTs and leads to the junction point ($R_S + R_L$) be kept low. Thus bushing or toroidal wound CTs, where their secondary impedance is very low, can be used, and they should be interconnected together as near the CT locations as possible, preferably equidistant, so that the several R_L values are essentially equal and low.

All CTs should have the same ratio and operate on the full winding. Operating at CT taps is not recommended, but if necessary, the windings between the taps must be completely distributed, and the

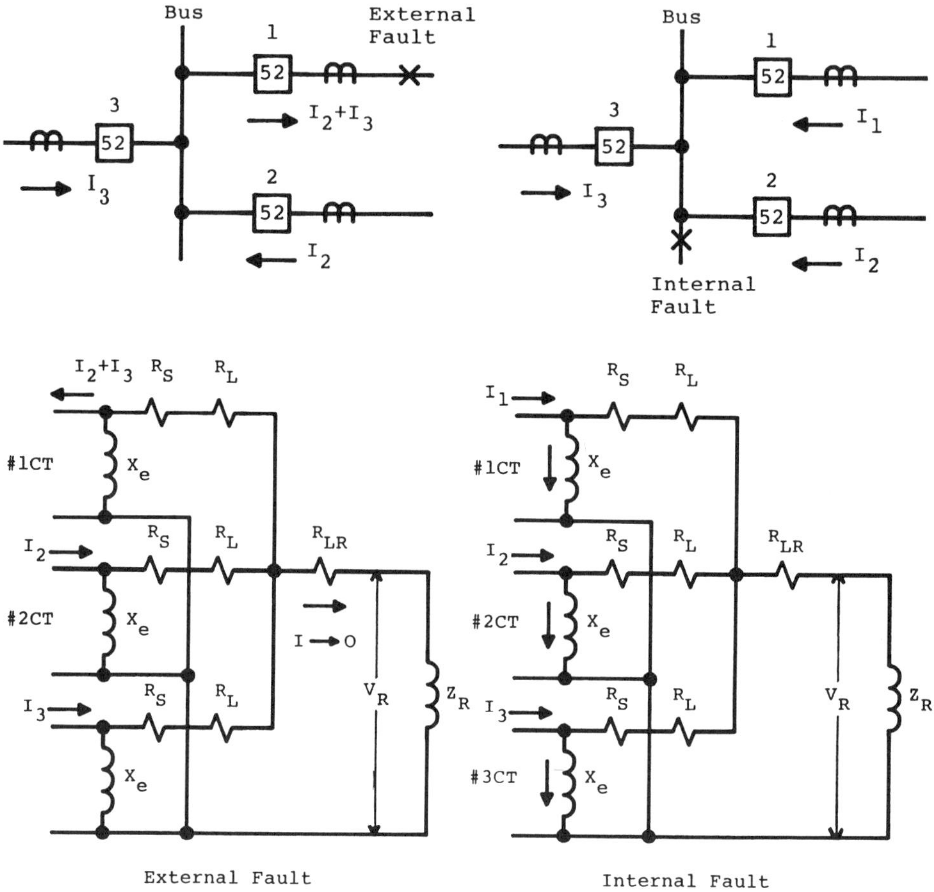

Fig. 10.9 Operating principles of the high-impedance voltage bus differential system.

unused end well insulated to avoid high-voltage breakdown from the autotransformer effect. Auxiliary CTs are not recommended. If they are required, a detailed analysis and/or special relays may be applicable. One type of relay employs added restraint circuits for applications with widely diverse CTs.

The several limitations outlined are not too difficult to meet with modern CTs and proper bus design, so this is a very effective and widely used bus protection system.

Typical operating times are on the order of 20 to 30 ms, and if a supplementary instantaneous unit is used for high-current internal faults, times of 8 to 16 ms are available.

10.11.3 Air-Core Transformer Differential

As has been pointed out, the major problem in differential schemes results because of the CT iron, which requires exciting current and saturates at high fault currents. By elimination of the CT iron, this problem does not exist, and a simple, fast, reliable bus differential system results. This is known as the linear coupler differential scheme, and a number of these are in service. It has not become popular primarily because existing and conventional CTs cannot be used, nor can linear coupler CTs be used for other applications.

The linear coupler in appearance is the same as conventional iron-core CTs and can be mounted on a bushing or connected as a wound-type CT in the primary circuit. It operates as an air-core mutual reactor where

$$V_{Sec} = I_{Pri} M \quad V \tag{10.1}$$

where M has been designed to be 0.005 Ω at 60 Hz. Thus a secondary voltage of 5 V is induced for 1000 primary amperes. The linear coupler secondaries for each circuit on the bus are all connected in series and to a sensitive relay unit. For an external fault or load, the sum of the voltage for all current flowing into the bus is equal and opposite to the voltage developed by the current flowing out of the bus. Thus the voltage across the relay essentially is zero for no operation.

For an internal fault, with all current flowing into the bus, the linear coupler secondary voltages add to produce an operating voltage. Thus the relay operating current (I_R) is

$$I_R = \frac{V_{sec}}{Z_R + Z_C} \tag{10.2}$$

where Z_R is the relay coil impedance and Z_C is the linear coupler secondary impedance. Typical values of Z_C are about 2 to 20 Ω for Z_R. Lead impedance is not significant with these values. The relays operate from 2 to 50 mA for high sensitivity. Typical times are 16 ms and less.

The system is quite flexible, as the linear coupler secondaries do not have to be shorted if open circuited, and circuits can be added or subtracted with minimum problems. Changing the number of circuits affects the value of the Z_C sum, which is basically offset by a corresponding change in the V_{Sec} sum.

If the primary circuit is subject to high frequencies such as occur with large capacitors or back-to-back switching of capacitors, secondary lightning arrestors may be required. Linear couplers very efficiently transform all frequencies.

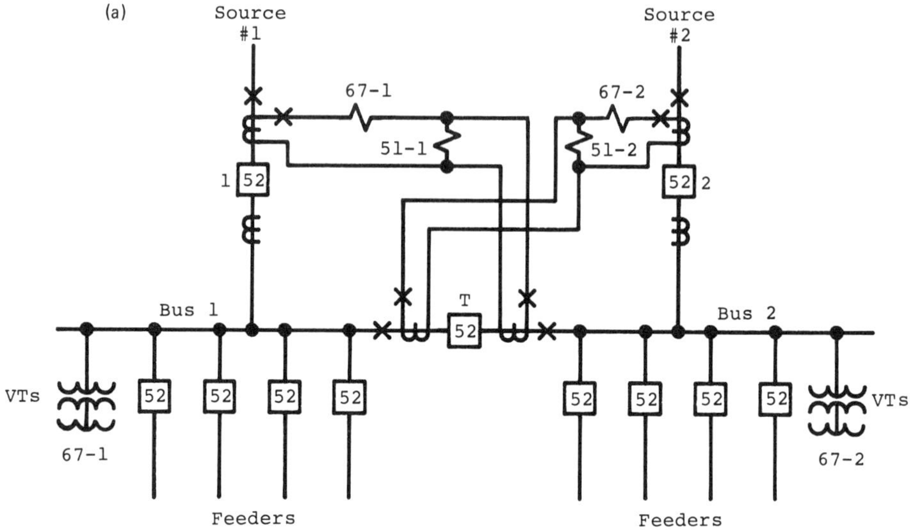

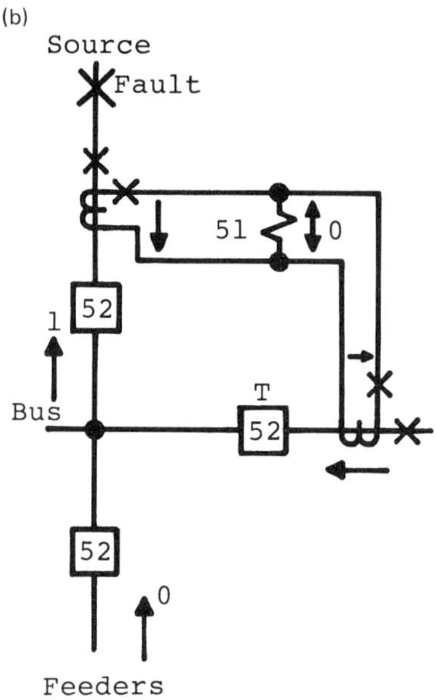

Fig. 10.10 Partial differential bus and feeder backup protection: (a) single-line diagram for a typical bus arrangement; (b) operation for source-side faults; (c) operation for bus and feeder faults; (d) operation for adjacent bus faults.

(c)

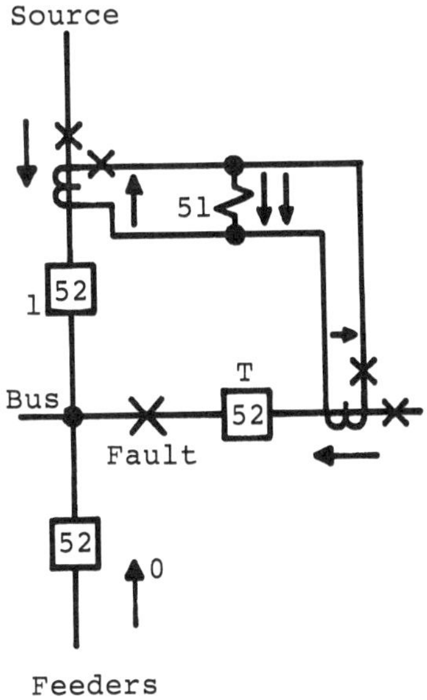

(d)

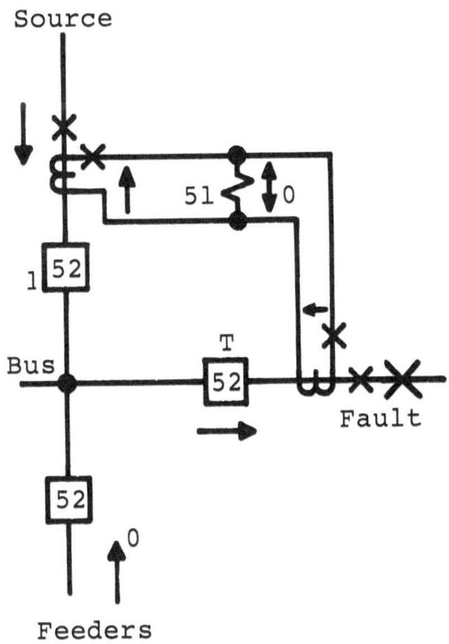

Fig. 10.10 (continued)

10.12 OTHER BUS DIFFERENTIAL SYSTEMS

Several other schemes exist but are in limited use. These are the time-overcurrent, directional comparison, and partial differential systems.

Time-Overcurrent Differential

In this the secondaries of all the current transformers are paralleled and connected to an inverse-time-overcurrent (51) relay. There is no restraint, so the relay must be set above the maximum CT magnetizing difference current for the external fault.

The inverse characteristic, which provides long times for low current magnitudes, is an advantage to override unequal CT saturation, particularly on the dc component. Thus the dc time constant should be short for these applications. Typical operating times for these schemes are 15 to 20 cycles (60-Hz base) for internal faults where the total fault current adds for relay operation. This provides a relatively inexpensive but slow differential, and one that is difficult to set with security except through long experience. It is used only for small, low-voltage buses.

Directional Comparison Differential

A directional sensing unit as described in Chapter 3 is connected to each bus circuit "looking into" the bus, with the trip contacts in series. For normal operation one or more contacts are opened by the load passing through the bus. For internal faults all contacts should close to trip the bus. Normally closed contacts are necessary for feeder circuits with no feed to the fault.

The major advantage is its almost complete independence of CT performance, characteristic, and ratio. The disadvantages are relative high cost (relays required for each circuit), that a voltage or reference source is required, and that contacts in series are difficult to coordinate. It has not been in general use in recent years.

Partial Differential

This scheme is used frequently to provide protection for buses in industrial and lower-voltage distribution substations. It is applicable where there are feeder circuits that (1) supply negligible current to bus faults, and (2) do not have adequate or suitable CTs for a complete differential application. Good examples are feeders supplying only static or induction motor loads and where there is a great diversity in their power ratings. A typical bus of this type is shown in Fig. 10.10. The CTs of the fault source circuits are paralleled and connected to inverse-time-overcurrent (51) relays. As illustrated in Fig. 10.10 b and d, negligible current flows in the 51 relays for external faults in the sources. The total fault current is available to operate the relays for bus and for faults out on the feeders. This requires that the 51 relays be time coordinated, with the protection on all the feeders not included in the differential. The use of an inverse relay and this scheme provides a good compromise for low-voltage substation buses. Three-phase connections are included in Fig. 9.25.

If current-limiting reactors are used in the feeder circuits, distance-type relays (21) can be substituted for the 51 relays. The 21 relays are set into but not through the lowest reactor impedance. This does not require selective settings with the feeder protection, and so avoids the time delay necessary with the 51 relays to provide fast and sensitive bus protection. It does require bus voltage for the distance relays.

10.13 GROUND-FAULT BUS

The bus supports and the substation equipment are insulated from ground and are all connected together to be grounded at one point through an overcurrent relay. A ground fault that involves this interconnection passes current through the relay to trip the respective protected area. A separate fault detector is used to supervise tripping for added security. This relay operates on system zero-sequence current or voltage. This scheme basically is not used in the United States because of relatively higher cost, difficulties of construction, and difficulties in protecting personnel.

10.14 PROTECTION SUMMARY

Differential protection should be considered and/or applied wherever possible for all buses as the primary protection. Although bus faults are relatively infrequent, they can be very disruptive to a power system, so fast, sensitive protection as available through the various differential systems is recommended. The most probable causes of bus faults are animals, wind-blown objects, and insulation failures (natural or gunshot). Lightning may result in bus faults, but stations and substations generally are well shielded and protected against lightning. The protection applied to the several circuits connected to the bus generally provides the backup protection for the bus, or the primary protection where bus differentials are not or cannot be applied.

BIBLIOGRAPHY

See the Bibliography at the end of Chapter 1 for additional information.

Guide for Protective Relay Applications to Power buses, ANSI/IEEE Standard C37.97-1979, IEEE Service Center.

11

Motor Protection

11.1 INTRODUCTION

The protection of motors varies considerably and is less standardized in general than the protection of the other apparatus or parts of the power system. This results from the very wide variety of sizes, types, and applications of motors. The protection is based principally on the importance of the motor, which usually is closely related to size. This chapter deals with motors and protection applied directly to them. Backup often associated with the connecting circuits is discussed in Chapter 9. The present discussion is for motors that are switched by a circuit breaker, contactors, or starters, and where the protection is separate from these devices and from the motor. Essentially, this covers motors at the 480- to 600-V level and above. Not covered specifically or directly are motors where the protection is built into the motor or starter, or where fuses are the only protection.

11.2 POTENTIAL MOTOR HAZARDS

The potential hazards normally considered are:

1. Faults — phase and/or ground
2. Thermal damage from
    a. Overload (continuous or intermittent)
    b. Locked rotor (failure to start, or jamming)
3. Abnormal conditions
    a. Unbalanced operation
    b. Undervoltage and overvoltage
    c. Reversed phases
    d. High-speed reclosing (reenergizing while still running)
    e. Unusual ambient and/or environment (cold, hot, damp)
    f. Incomplete starting sequence

These are for induction motors, which represent the large majority of all motors in service. For synchronous motors, additional hazards are:

4. Loss of excitation (loss of field)
5. Out-of-step operation (operation out of synchronism)
6. Synchronizing out of phase

These can be reclassified as to their origins:

A. Motor-induced
    1. Insulation failure (within motor and associated wiring)
    2. Bearing failure
    3. Mechanical failures
    4. Synchronous motors — loss of field
B. Load-induced
    1. Overload (and underload)
    2. Jamming
    3. High inertia (Wk^2)
C. Environment-induced
    1. High ambient temperature
    2. High contaminant level — blocked ventilation
    3. Cold, damp ambient temperature
D. Source- or system-induced
    1. Phase failure (open phase or phases)
    2. Overvoltage
    3. Undervoltage
    4. Phase reversal
    5. Out-of-step condition resulting from system disturbance
E. Operation- and application-induced
    1. Synchronizing, closing or reclosing out of phase
    2. High duty cycle
    3. Jogging
    4. Rapid or plug reversing

11.3 MOTOR CHARACTERISTICS INVOLVED IN PROTECTION

The primary motor characteristics available and involved in the protection are:

1. Starting-current curves
2. Thermal capability curve, which should include the permissible locked rotor thermal limit
3. The K constant (R_{r2}/R_{r1})

These characteristics normally are obtained from the motor manufacturer and are basic for the application of protection. Typical curves are illustrated in Fig. 11.1.

The maximum starting current curve is at rated voltage. Currents for lower voltages exist to the left, with the knee at a higher time level.

The thermal limits are three different curves, which in many cases approximately blend together to a general curve such as that shown. Thermal limits are relative indeterminate zones which engineers desire to have represented by a specific curve.

1. The higher current portion indicates the permissible locked rotor times . This is the time the rotor can remain stalled after the motor has been energized before thermal damage occurs in the rotor bars, rotor end rings, or in the stator, whichever is the limit for a particular design.

   In very large motors this locked rotor thermal limit can be less than the starting time, so these motors must start rotating instantly to avoid thermal damage. This curve is from locked rotor current at full voltage to that current at the minimum permissible starting voltage.
2. The accelerating thermal limit curve from locked rotor current to the motor breakdown torque current, which is around 75% speed.
3. The operating or running thermal limit curve, which represents the motor overload capacity during emergency operation.

11.4 INDUCTION MOTOR EQUIVALENT CIRCUIT

As an aid in the protection and for protection analysis, the equivalent motor diagram can be reduced as shown in Fig. 11.2. Typical values for induction motors in per unit on the motor kVA or kV as shown in Fig. 11.2b are R_s and R_r = 0.01 pu, jX_m = j3.0 pu, and $jX = jX_d''$ = 0.15 pu, and from these, the typical locked rotor or starting is

$$I_{starting} = \frac{1}{jX_d''} = \frac{1}{0.15} = 6.67 \text{ pu} \tag{11.1}$$

This is the symmetrical value; the asymmetrical current is higher, as shown in Fig. 11.1.

Since the shunt jX_M is high relative to the other impedances, the equivalent at the motor input reduces with the typical values above to

$$Z_{M1} = Z_{M2} = 0.144 \angle 82.39° \tag{11.2}$$

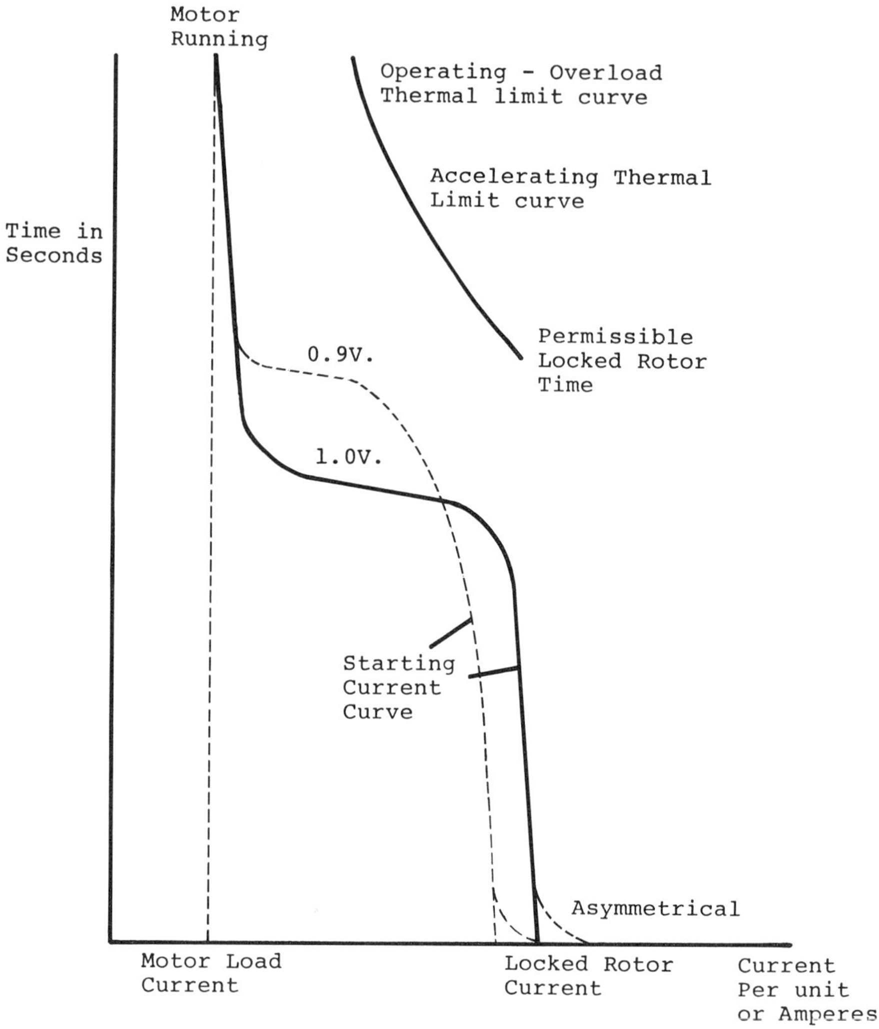

Fig. 11.1 Typical induction motor characteristics.

or practically equal to $jX''_d = 0.15$ pu as commonly used for a stalled motor (S = 1.0). If the motor is running (S = 0.01), the values above give

$$Z_{M1} = 0.927 \underline{/25.87°} \quad \text{and} \quad Z_{M2} = 0.144 \underline{/84.19°} \text{ pu} \qquad (11.3)$$

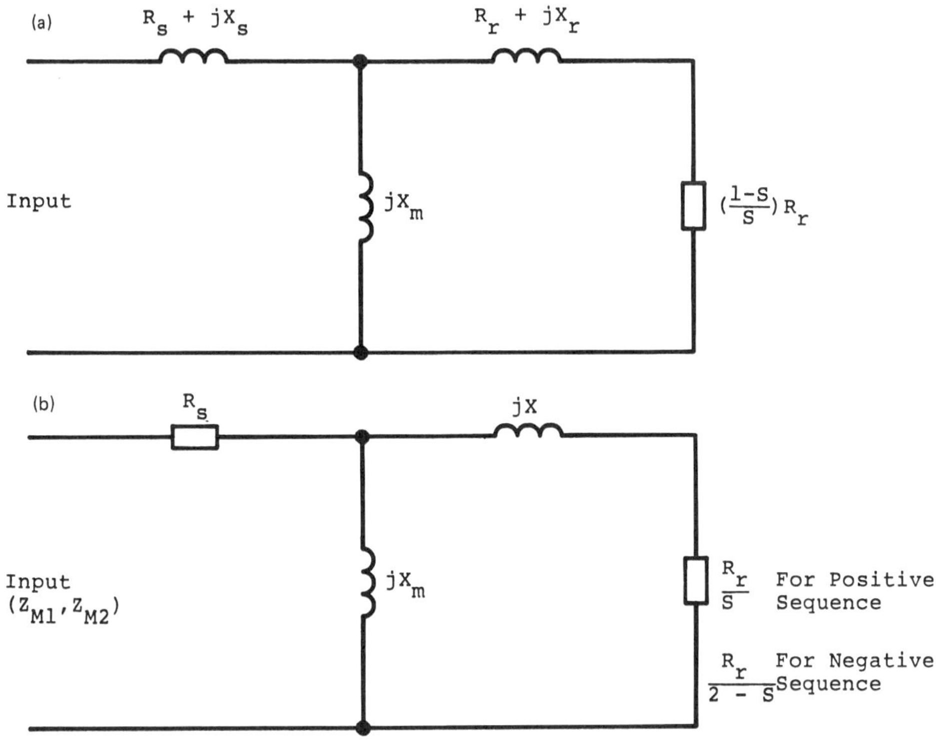

R_s = Stator Resistance

jX_s = Stator Leakage reactance at rated frequency

R_r = Rotor Resistance

jX_r = Rotor Leakage reactance at rated frequency

jZ_m = Shunt exciting impedance

$jX \cong jX_s + jX_r \cong jX_d''$

$$S = \frac{\text{Synchronous RPM - Rotor RPM}}{\text{Synchronous RPM}} \begin{array}{l} = 1.0 \text{ stalled} \\ = 0+ \text{ running} \end{array}$$

Fig. 11.2 Equivalent-circuit diagrams for induction motors: (a) equivalent diagram for an induction motor; (b) simplified equivalent induction motor diagram.

Thus, practically,

$$Z_{M1} = 0.9 \text{ to } 1.0 \text{ pu} \qquad \text{and} \qquad Z_{M2} = 0.15 \text{ pu} \tag{11.4}$$

From stalled to running the positive-sequence impedance changes from approximately 0.15 to 0.9 or 1.0 pu, while the negative-sequence impedance remains essentially the same, at approximately 0.15 pu. Again these are on the base of the rated motor kVA, which is roughly equal to the motor horsepower (hp). These values will vary with each individual motor, but these typical values are close and quite useful if specific data are not available.

11.5 GENERAL MOTOR PROTECTION

Protection for motors exists in many forms, a variety of designs, and either packaged individually or in different combinations. Each has its features which will not be restated or evaluated here. The fundamentals and basic aim should be to permit the motor to operate up to but not to exceed its thermal and mechanical limits for overloads and abnormal operation conditions and provide maximum sensitivity for faults. In most cases these can be achieved in general terms as follows:

11.6 PHASE-FAULT PROTECTION

Instantaneous nondirectional overcurrent relays can be used to protect induction motors. Faults generally provide current greater than the locked rotor starting current except for turn-to-turn faults. Considerable current can flow between turns, but unfortunately very little evidence of this is available at the motor terminals until it develops into other types, phase to ground and/or phase to phase.

The motor is the end device in the electrical system, so instantaneous relays can be used. There is no coordination problem. The induction motor backfeed to system faults is relatively small ($1/X''_d$ + offset) and decays rapidly in a few cycles, so that nondirectional relays can be applied. The CT ratios supplying these relay should be selected so that the maximum motor current provides between 4 and 5 A secondary current.

Phase instantaneous relays should be set well above the asymmetrical locked rotor and well below the minimum fault current. This can be equated where I_{LR}, the locked rotor symmetrical is

$$I_{LR} = \frac{1}{X_{1S} + X''_d} \tag{11.5}$$

where X_{1S} is the total reactance (impedance) of the power system or source to the motor. This equation is similar to Eq. (11.1), where the maximum starting or locked rotor current is with a very large or infinite source, so X_{1S} approaches zero. A fault at the motor is

$$I_{3\phi} = \frac{1}{X_{1S}} \tag{11.6}$$

and for a phase-to-phase fault with $X_{1S} = X_{2S}$,

$$I_{\phi\phi} = 0.866 I_{3\phi} = \frac{0.866}{X_{1S}} \tag{11.7}$$

If P_R is the ratio of the relay pickup (I_{PU}) to locked rotor current,

$$P_R = \frac{I_{PU}}{I_{LR}} \tag{11.8}$$

Typically, P_R should be 1.6 to 2.0 or greater.

If P_F is the ratio of the minimum fault-to-relay pickup current,

$$P_F = \frac{I_{\phi\phi\,min}}{I_{PU}} \tag{11.9}$$

Desirably, P_F should be 2 to 3 or greater. From Eq. (11.9) and (11.8),

$$I_{\phi\phi} = P_F I_{PU} = P_F P_R I_{LR}$$

and

$$\frac{I_{\phi\phi}}{I_{LR}} = P_F P_R \quad \text{or} \quad \frac{I_{3\phi}}{I_{LR}} = 1.155 P_F P_R \tag{11.10}$$

or the three-phase solid fault at the motor should be $1.155P_FP_R$ or larger for good instantaneous overcurrent protection. If the minimum recommended values of $P_R = 1.6$ and $P_F = 2$, the three-phase fault should be 3.7 times the locked rotor current. If $P_R = 2$, $P_F = 3$, the three-phase fault should be at least 6.9 times larger than the locked rotor current.

Equating Eq. (11.7) and (11.10) and substituting Eq. (11.5) yields

$$I_{\phi\phi} = \frac{0.866}{X_{1S}} = \frac{P_F P_R}{X_{1S} + X_d''}$$

$$X_{1S} = \frac{0.866 X_d''}{P_F P_R - 0.866} \tag{11.11}$$

Thus with $P_R = 1.6$, $P_F = 2$,

$$X_{1S} = \frac{0.866 X_d''}{2 \times 1.6 - 0.866} = 0.371 X_d''$$

and with the typical $X_d'' = 0.15$, $X_{1S} = 0.056$ pu, or with $P_R = 2$, $P_F = 3$, and $X_d'' = 0.15$, $X_{1S} = 0.025$ pu. This defines the source impedance, which should be as indicated or less for instantaneous overcurrent protection. The per unit values in the examples are on the motor kVA, kV base, where

$$kVA_{Rated} = \frac{(\text{horsepower})(0.746)}{(\text{efficiency})(\text{power factor})} \tag{11.12}$$

In many applications the source X_{1S} for all practical purposes is the reactance of the supply transformer, which is connected on its primary to a large utility, which is a relatively infinite source. Further, generally the supply transformer is supplying other loads and so is much larger than any specific motor, so its reactance on the motor base will tend to be small. This can be illustrated with the example of Fig. 9.18, where a 2500-kVA transformer with 5.75% X is shown supplying an 800-A feeder. Suppose that this is a motor, and 800 A at 480 V is 665 kVA. At 665 kVA, the transformer reactance is 5.75(665)/2500 = 1.53%, or 0.0153 pu. This is well below the limits above to provide good instantaneous relay protection.

Where the recommended setting criteria of P_R and P_F above cannot be met, or where more sensitive protection might be desired, the instantaneous relay (or a second relay) can be set more sensitively if delayed by a timer. This permits the asymmetrical starting component to decay out. A typical setting recommendation for this is a P_R of 1.1 to 1.2 with a time delay of 0.10 s (six cycles at 60 Hz).

If on loss of voltage, running motors are transferred from a bus to a hot bus, or if high-speed system reclosing reenergizes motors before their residual voltage has dropped to around 33% rated, very high transients can occur. These currents are very hard on motors unless specifically designed for them. Careful attention must be given to set the relays above this transient if the condition is tolerable.

When the criteria above indicate that there is not sufficient margin between locked rotor and fault current, differential protection is indicated.

11.7 DIFFERENTIAL PROTECTION

Differential protection is preferred. However, for some motors the two ends of the windings may not be available, and differential protection cannot be applied. If both ends of the windings are available, the best differential in terms of sensitivity, speed, and security is to pass the conductors of the windings through a toroidal (ring) CT, as shown in Fig. 11.3a. These CTs are described in Chapter 5 and also applied for small generator protection in Chapter 8. Typical maximum opening or window in these CTs are around 8 in. in diameter. With a fixed ratio of 50:5 and a sensitive instantaneous overcurrent relay (50), the combination can provide a pickup of around 5 A primary current. This is a flux-balancing differential independent of load and starting current magnitudes and with only one CT per phase, so that matching CT performance does not exist. Internal phase and ground protection is provided within the motor and up to the CT location. Other protection is required for the connections to the circuit breaker, starter, and so on. The limitation is the conductor size relative to the CT opening.

Conventional differential with CTs in the neutral and output leads should be applied where the toroidal type cannot be used. Normally, the two sets of CTs would be of the same type and ratio, so conventional 87 two-restraint differential relays are applicable, as illustrated in Fig. 11.3b. With equal CT ratios the secondard currents through the relay restraint windings (R) would be essentially the same for all external faults and load, and the operating current (OP) very small to zero. For motor faults between the two sets of CTs, all of the fault current flows through the (OP) operating winding for high sensitivity to both phase and ground faults. The line-side CTs should be as shown, so that the differential zone includes the circuit breaker and connecting leads as well as the motor.

11.8 GROUND-FAULT PROTECTION

As for phase protection, instantaneous overcurrent relays are applied for ground-fault protection. Where applicable, the preferred method is to use a ring-type current transformer, with the three motor conductors passed through the CT opening. This provides a magnetic summation of the three-phase currents so that the secondary output to the

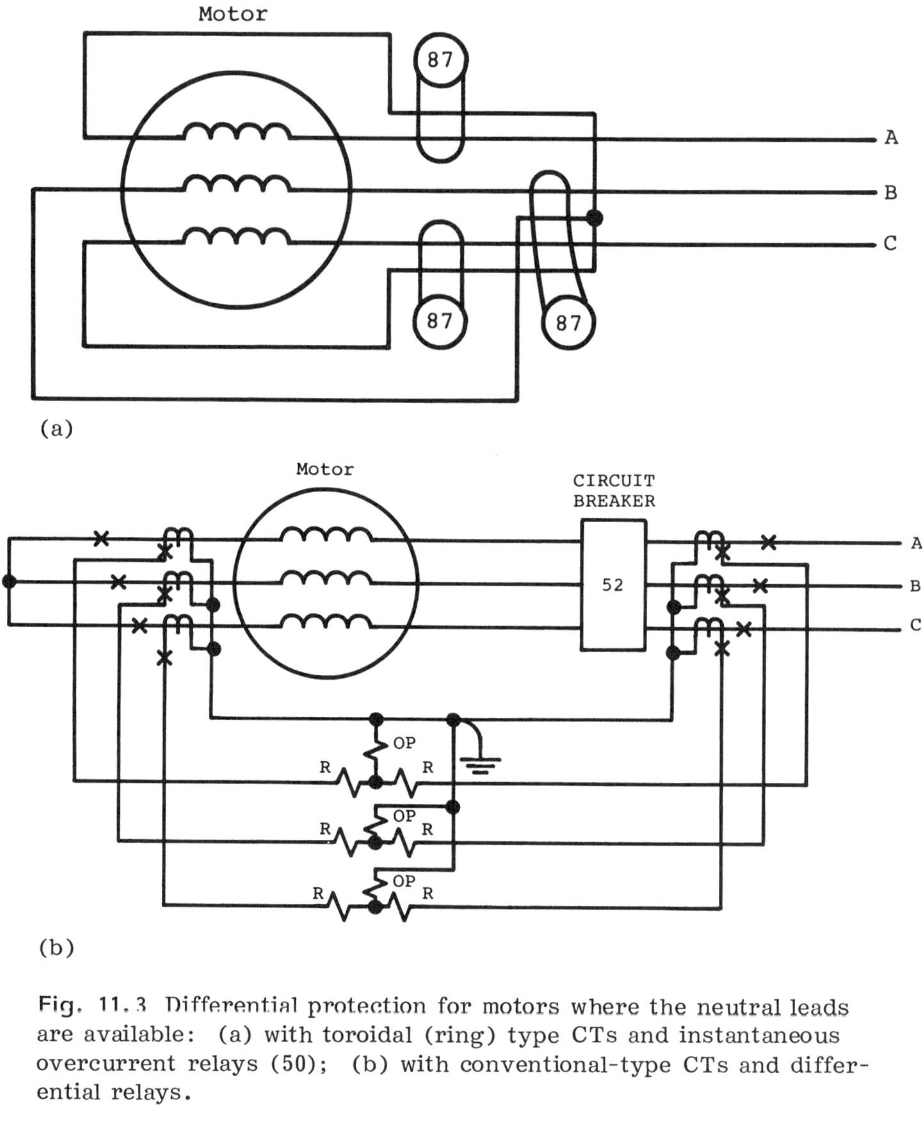

Fig. 11.3 Differential protection for motors where the neutral leads are available: (a) with toroidal (ring) type CTs and instantaneous overcurrent relays (50); (b) with conventional-type CTs and differential relays.

relay is zero-sequence ($3I_0$) current. This is shown in Fig. 11.4a. The CT ratio, commonly 50:5, is independent of motor size, while the conventional CTs in the phases must be sized to the motor load. The advantage is high sensitivity with good security, but is limited by conductor size which can be passed through the CT opening. As

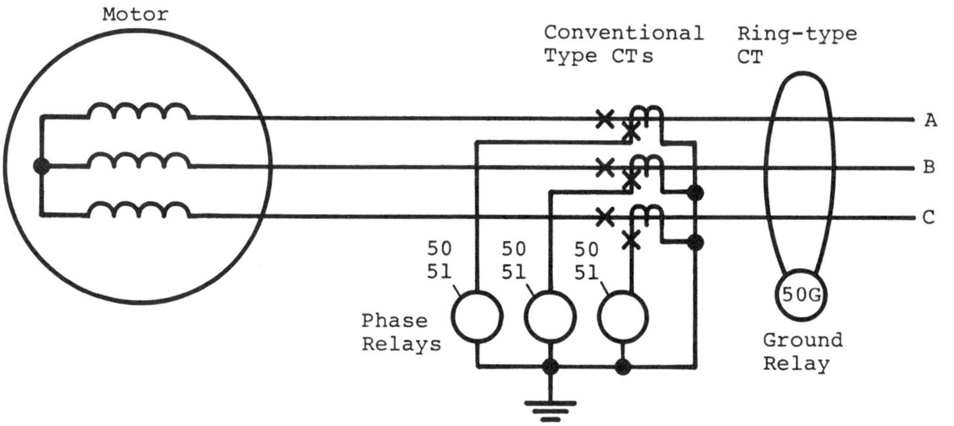

(a)

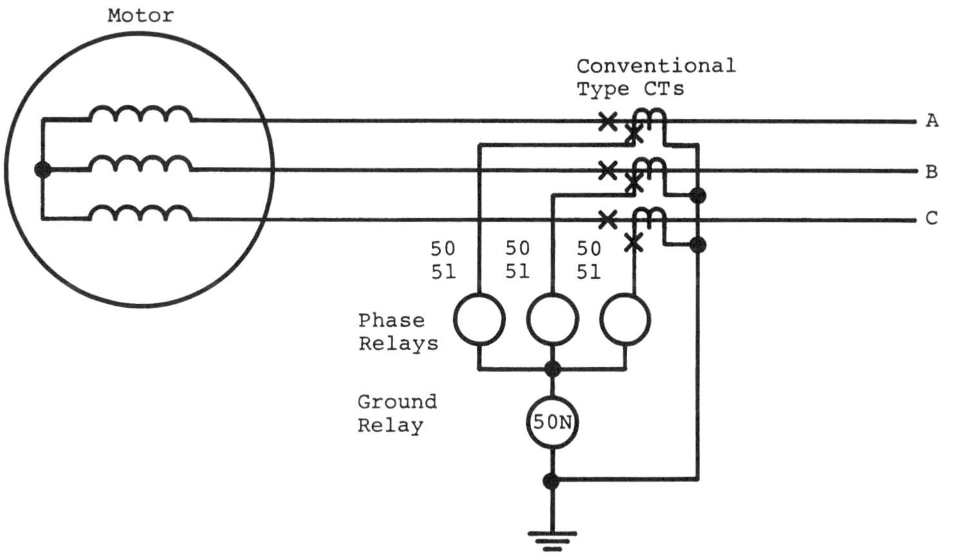

(b)

Fig. 11.4 Ground overcurrent protection for motors: (a) with the three conductors passed through a ring-type current transformer; (b) with conventional-type current transformers.

indicated in the preceding section, typical sensitivity is 5A primary current.

For larger motors and conductors, ground relay in the neutral must be used as in Fig. 11.4b. Although load influences the CT ratios, the ground relay can be set sensitively and well below motor load. 50N must be set above any "false" residual current that can result from the unequal performance of the three CTs on high unequal offset starting currents. This is difficult to predetermine, but the probability of a problem is very low if the phase burdens are balanced and the CT voltage developed by the maximum starting current is not more than 75% of the CT accuracy class voltage. A lower 50N relay tap and consequent higher burden may help by forcing all three CTs to saturate more evenly. Resistance in the neutral circuit may also help. This increased burden, however, should not be great enough to significantly inhibit the relay sensitivity. These later "fixes" are used generally after trouble is encountered during startup. Time delay could be used until the offset has decayed, but this delays tripping for actual faults.

With ground-fault limiting, as is common in supply systems for motors, the ground-fault current will be smaller than for phase faults. If high-resistance grounding is used (Chapter 7), ground-fault currents will be in the order of 1 to 10 A primary. The protection of Fig. 11.4a can provide reasonable sensitivity for these systems if the ground-fault current is greater than 5 A. Considerably more sensitive protection can be obtained by the application of a product overcurrent relay (32N). This general type was described in Section 9.17. For this application a relay with a current coil and a voltage coil is used. It operates on the product of voltage times current and for use in high-resistance grounded systems, the maximum torque occurs when the current leads the voltage by 45°. The current coil is connected in the neutral of the CTs in place of 50N of Fig. 11.4b, and the voltage coil across the grounding resistor and in parallel with 59G in Fig. 7.7 or 7.8. The polarity is that the relay will operate when zero-sequence current flows into the motor. As indicated in Chapter 7, the current in high-resistance grounding is low, but the zero-sequence voltage is high. Typical pickup for this product overcurrent relay is about 7 to 8 mA with 69.5 V. This is well below the ground-fault levels of 1 to 10 A.

11.9 THERMAL AND LOCKED-ROTOR PROTECTION

This protection involves the application of relays to closely match the thermal and locked-rotor curves of Fig. 11.1. Again it should be remembered that these motor thermal curves are approximate representations of thermal damage zones for general or normal operation. The relays should operate just before the limits are reached or exceeded.

Over many years this is accomplished by thermal relays to match the thermal limit curves, and inverse-time-overcurrent relays for locked-rotor protection. This protection, designed and packaged in various ways, provides very good protection for most motors. Typical applications to the motor characteristics shown in Fig. 11.1 are illustrated in Fig. 11.5.

Thermal relays are available in several forms:

1. A "replica" type, where the motor heating characteristics are approximated closely with a bimetallic element within a current heater unit. These operate on current only.
2. Relays operating from exploring coils, usually resistance temperature detectors (RTD), embedded in the motor windings. These operate on winding temperature only and the detectors are placed in the motor by the designer at the most probable hot spot or danger area. These are common in motors from about 250 hp and above but may not be installed in some motors unless specified.
3. Relays that operate on a combination of current and temperature. Care should be exercised with combinations where both are required for relay operation to assure that there are not operating conditions that may not be covered. High current and high temperature normally indicate problems, but high current without measurable high temperature might exist for overheating in the rotor, bearings, drive machine problems, and in the controllers or connections. For these the combination may provide limited to no protection.

The comparison of motor starting and inverse-time-overcurrent relay curves on the same plot, as is commonly done and shown in Fig. 11.5, can provide false information. This may occur where the space between the starting current and locked-rotor limit is very narrow, which it usually is for very large motors. Often in these cases it appears possible to set the overcurrent relay so that its characteristic is above the motor start curve and below the locked-rotor limit, only to discover in service that the overcurrent relays operate under normal start.

Actually, the curves for motor starting and relay operation are two quite different characteristics. The motor starting curve is a plot of the changing current with time from locked rotor or starting condition to the motor load operating current. The relay characteristic represents the operating times for different constant-current values. With the overcurrent relay typically set at about one-half locked-rotor current or less, it begins operation at the moment the motor is energized. Unless the starting current drops below the relay pickup before the relay times out, it will initiate tripping undesirably. Relay operating times for variable current are not directly available from its time

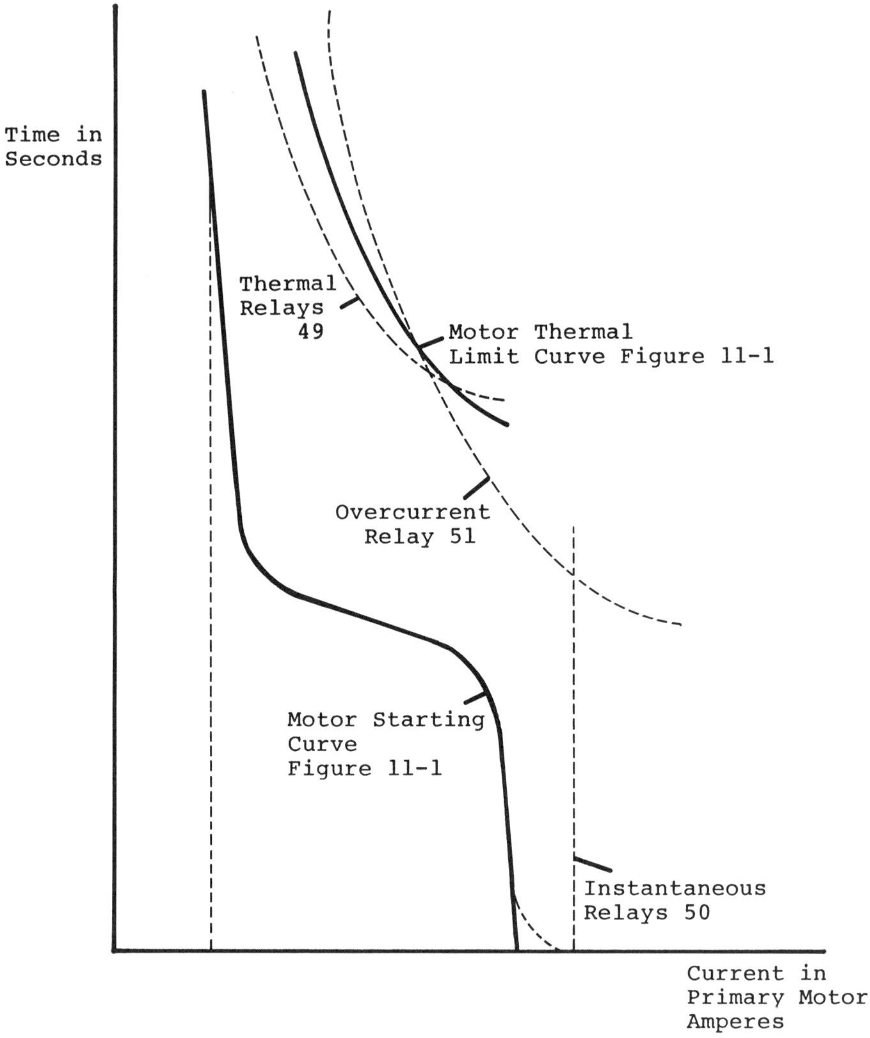

Fig. 11.5 Typical overload, locked-rotor, and fault protection for a motor.

characteristic. This is a complex calculation, but manufacturers have developed criteria for individual relays.

11.10 LOCKED-ROTOR PROTECTION FOR LARGE MOTORS

As indicated, the permissible locked-rotor current can be very close or less than the starting current. In general, this occurs for very large modern motors. The protection for this can be a zero-speed switch built into the motor. If the motor does not accelerate upon energization to open or operate the switch in a prescribed manner, the supply circuit is opened. The concern with using this type of protection is that the motor could start and lock up at less than full-load speed, and the difficulty of testing and maintenance.

Protection for locked rotors can be obtained by applying a distance relay, as described in Chapter 6. The relay is set looking into the motor, as shown in Fig. 11.6. The ratio of system voltage and starting current is an impedance, which can be determined and plotted as a vector on the R-X diagram. From a specific value at start it increases in magnitude and changes phase angle as the motor accelerates. The distance 21 relay is set so that its mho operating circle encloses the locked-rotor impedance vector. When the motor is energized by closing breaker 52, the distance 21 relay operates and the timer 62 is energized. By using an ac-operated timer, variable time with voltage is obtained to match the longer permissible locked-rotor times at lower voltage. The heavy starting current can cause the voltage to drop momentarily during the starting period. If the start is successful, the impedance phasor moves out of the 21 operating circle before the 62 timer contact closes. If the start is not successful, the impedance vector stays in the circle and when timer 62 operates, the trip is initiated. The timer is set as determined by the permissible locked-rotor time curve from full voltage to about 75 or 80% voltage. This protection does not cover failure to accelerate to full speed, or to pull out with rotation continuing.

11.11 SYSTEM UNBALANCE AND MOTORS

The most common cause of unbalance for three-phase motors is the loss of phase resulting from an open fuse, connector, or conductor. Unbalances in other connected loads can also effect the motor. A voltage unbalance of 3.5% can produce a 25% or greater increase in motor temperature rise. This results primarily from negative sequence produced by the unbalance. This current produces flux in the motor air gap rotating in the opposite direction to the actual motor direction. The relative effect is essentially double frequency current in the rotor. Skin effect results in higher resistance, and as outlined in Section 11.4, negative-sequence impedance remains essentially at locked-rotor value. Thus high current and high resistance compound the heating effect.

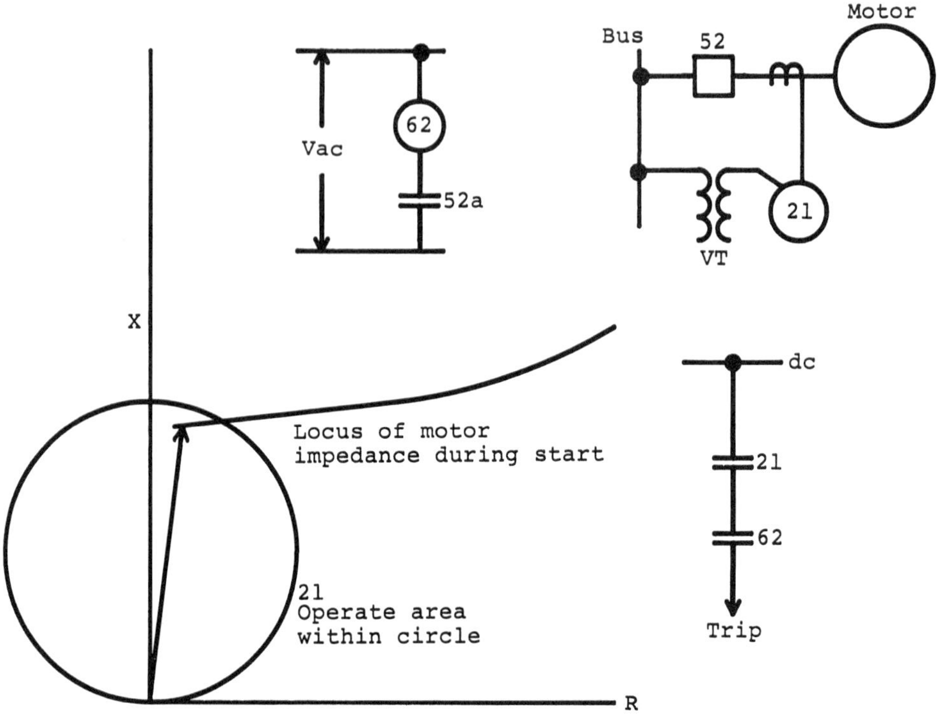

Fig. 11.6 Locked-rotor protection with a distance (21) relay and timer.

The total heating in a motor is proportional to

$$I_1^2 t + K I_2^2 t \tag{11.13}$$

where I_1 and I_2 are the positive- and negative-sequence currents, respectively, in the motor and K is

$$K = \frac{R_{r2}}{R_{r1}} = \text{conservative estimate as } \frac{175}{I_{LR}^2} \tag{11.14}$$

where R_{r1} and R_{r2} are the motor rotor positive- and negative-sequence resistances, respectively, I_{LR} the locked-rotor current in per unit. Equation (11.13) shows that there is a high increase in heating from the negative-sequence component.

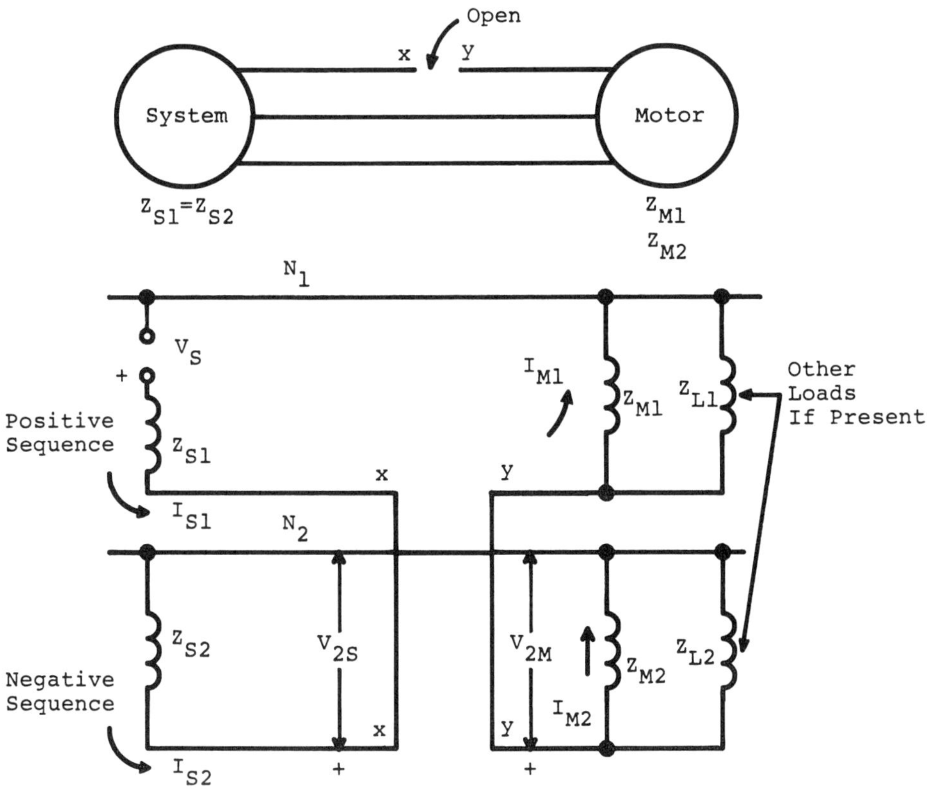

Fig. 11.7 Simplified symmetrical component representation for an open phase.

The symmetrical components network for an open phase is shown in Fig. 11.7. This is a simplified circuit showing the total source system as lumped impedances $Z_{S1} = Z_{S2}$. For any specific case, this circuit can be expanded to show more detail of the source or other loads. The supply transformer, for example, can be represented by its reactance (impedance) X_T. For an open phase between the transformer and motor, X_T would be added in series with the source impedances for the equivalent values of Z_{S1} and Z_{S2}. When the open phase is between the system and the transformer, X_T would not be included in the source equivalents but added in series with the motor impedance. This circuit is for an ungrounded motor, as is the common practice. The zero-sequence network is not involved for one phase open unless the systems on either side of the open are both grounded.

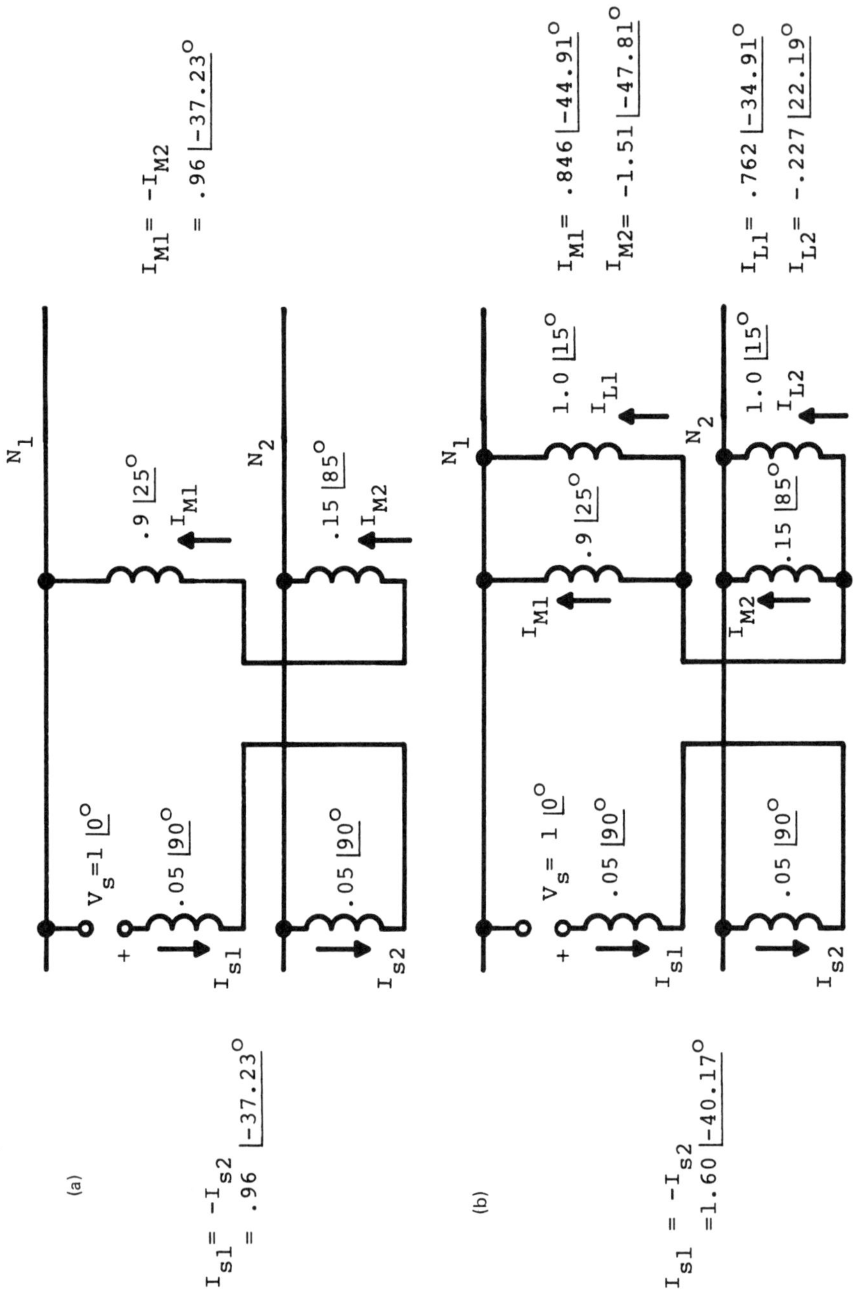

(a)
N1
N2
V_s = 1 ∠0°
.05 ∠90°
.9 ∠25°
.15 ∠85°
I_M1
I_M2
I_s1
I_s2
I_M1 = -I_M2 = .96 ∠-37.23°
I_s1 = -I_s2 = .96 ∠-37.23°
(b)
N1
N2
V_s = 1 ∠0°
.05 ∠90°
.9 ∠25°
.15 ∠85°
1.0 ∠15°
I_L1
I_L2
I_M1
I_M2
I_s1
I_s2
I_M1 = .846 ∠-44.91°
I_M2 = -1.51 ∠-47.81°
I_L1 = .762 ∠-34.91°
I_L2 = -.227 ∠22.19°
I_s1 = -I_s2 = 1.60 ∠-40.17°

(c)

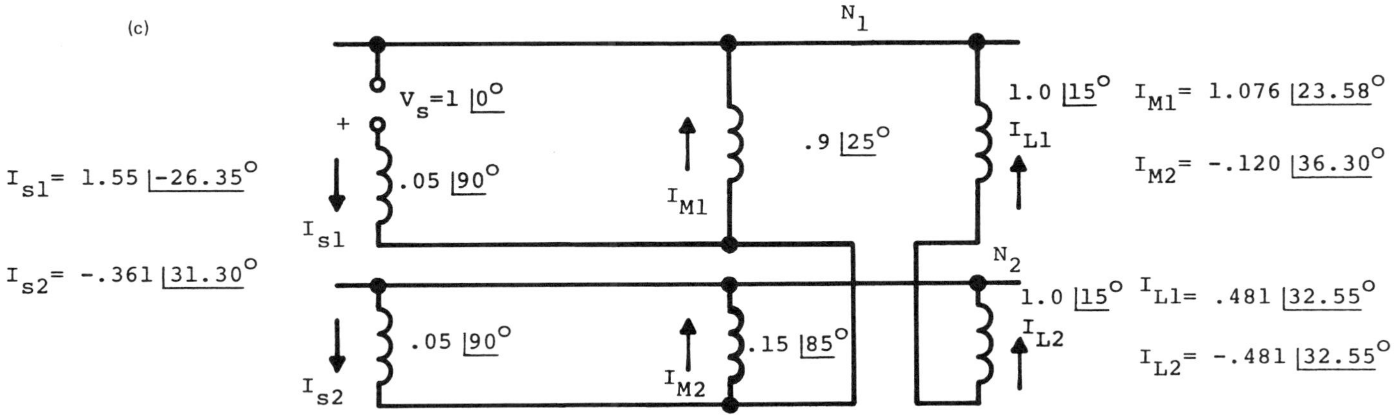

Fig. 11.8 Positive- and negative-sequence currents for an open phase supply to a motor with and without associated static load: (a) open-phase sequence currents with motor only; (b) open-phase sequence currents with static load on the motor bus, phase open on system side of motor and load; (c) open-phase sequence currents with static load on the motor bus, phase open between motor and the load.

The distribution of currents for an open phase using the network of Fig. 11.7 is shown in Fig. 11.8 for several cases. Typical per unit values of impedances shown are all on the motor kVA base, and are:

$$\begin{aligned} Z_{S1} &= Z_{S2} = 0.05 \angle 90^\circ \text{ pu} \\ Z_{L1} &= Z_{L2} = 1.0 \angle 15^\circ \text{ pu for static loads at the motor} \\ Z_{M1} &= 0.9 \angle 25^\circ \\ Z_{M2} &= 0.15 \angle 85^\circ \end{aligned} \tag{11.15}$$

These angles were included in the calculations, but the simplification of assuming all impedances at the same angle gives close approximations and does not change the trends shown. With all values at 90°, for example, $I_{S1} = 0.87$ pu instead of the 0.96 pu in Fig. 11.8a.

From these sequence currents it is observed that on either side of the open, $I_a = I_1 + (-I_2) = 0$ correctly. The sound phase currents are

$$\left.\begin{aligned} I_b &= a^2 I_1 + a I_2 = -j\sqrt{3}\, I_1 \\ I_c &= a I_1 + a^2 I_2 = +j\sqrt{3}\, I_1 \end{aligned}\right\} \text{ when } I_1 = -I_2 \tag{11.16}$$

so in Fig. 11.8a, I_b and I_c currents are 1.66 pu. Thus it is seen that an open phase provides very low phase currents relative to the normal motor load current of around 1 pu, so that overcurrent relays are not adequate to detect an open phase.

When static load is connected in parallel with the motor as shown in Fig. 11.7 and calculated in the examples of Fig. 11.8b, the continued rotation of the motor generates a voltage on the opened phase. This continues to energize the load connected to this phase. The power is transferred across the motor air gap and reduces the motor shaft power so that pullout may occur. One example indicated that the motor would pull out at 20% of rated load with static load three times larger than the motor load, or at 50% of rated load with static load equal to the motor load. In addition, the low value of the motor negative-sequence impedance means that a larger portion of negative-sequence current flows in the motor to increase heating. This distribution is illustrated in Fig. 11.8b. The motor negative-sequence current can be low, as shown in Fig. 11.8c, when only the static load is single phased.

The fundamental for an open phase is that the positive- and negative-sequence currents are equal and opposite as long as zero sequence is not involved. This is useful to develop the unbalance currents through wye-delta transformer banks. This, as well as the calculations in Fig. 11.8, are for the conditions just after the phase opens

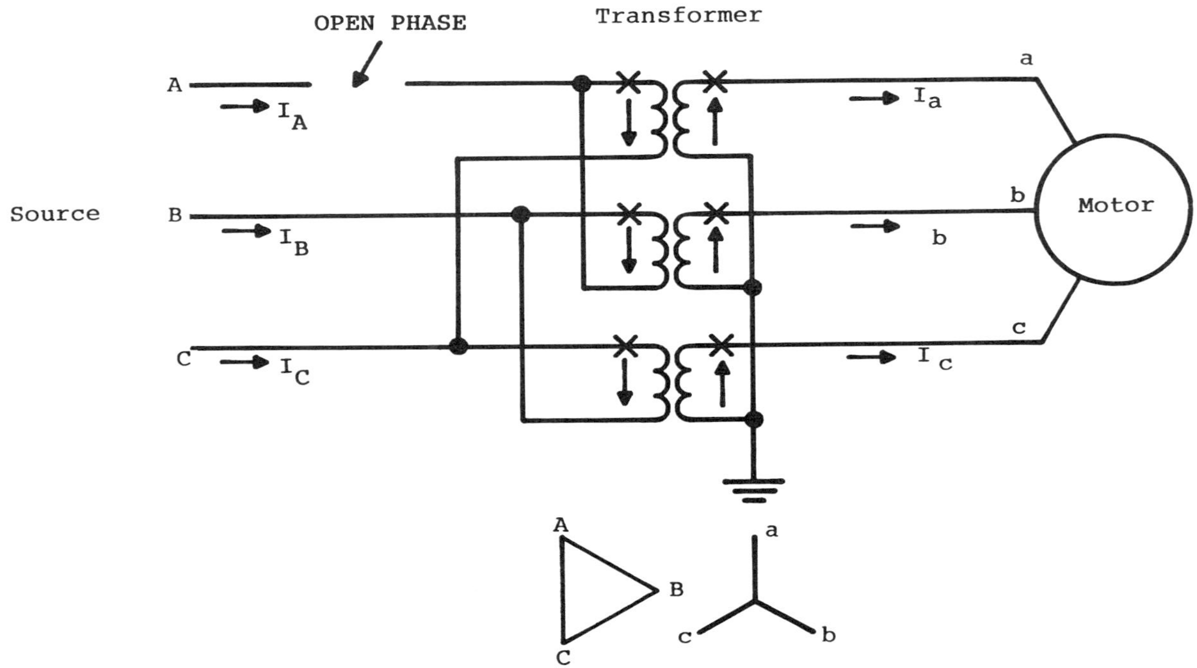

Fig. 11.9 Unbalanced current flow through a delta-wye transformer bank to a motor for phase A open on the source side.

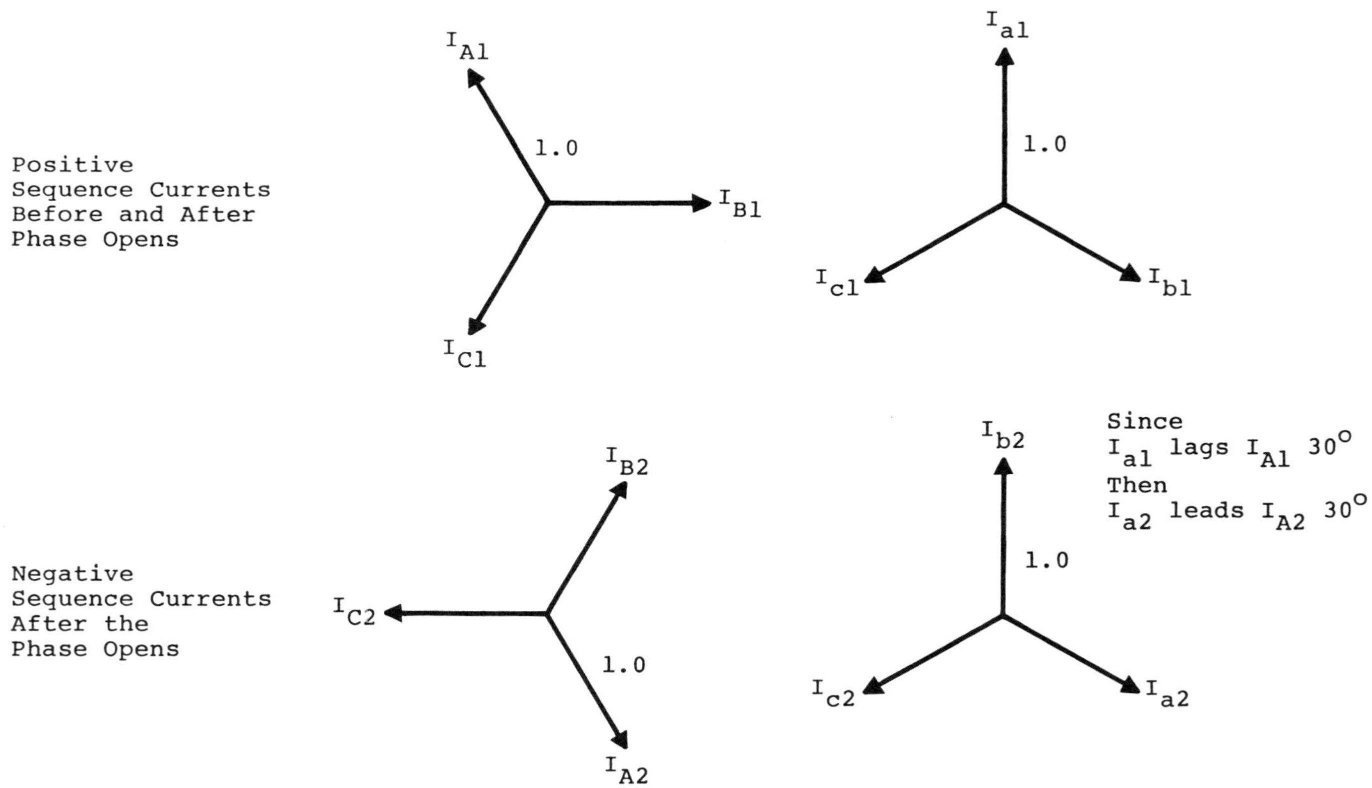

Fig. 11.9 (Continued)

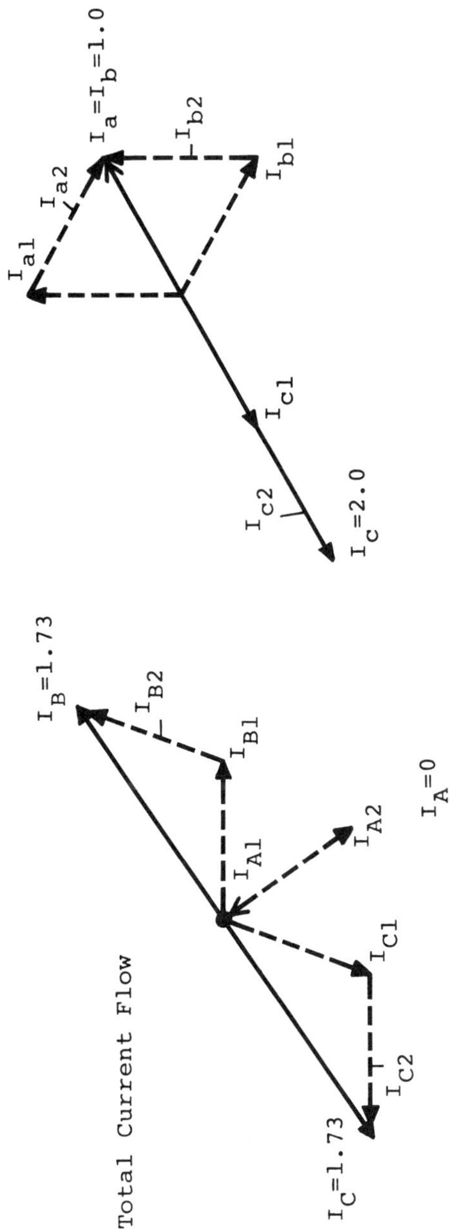

Fig. 11.9 (Continued)

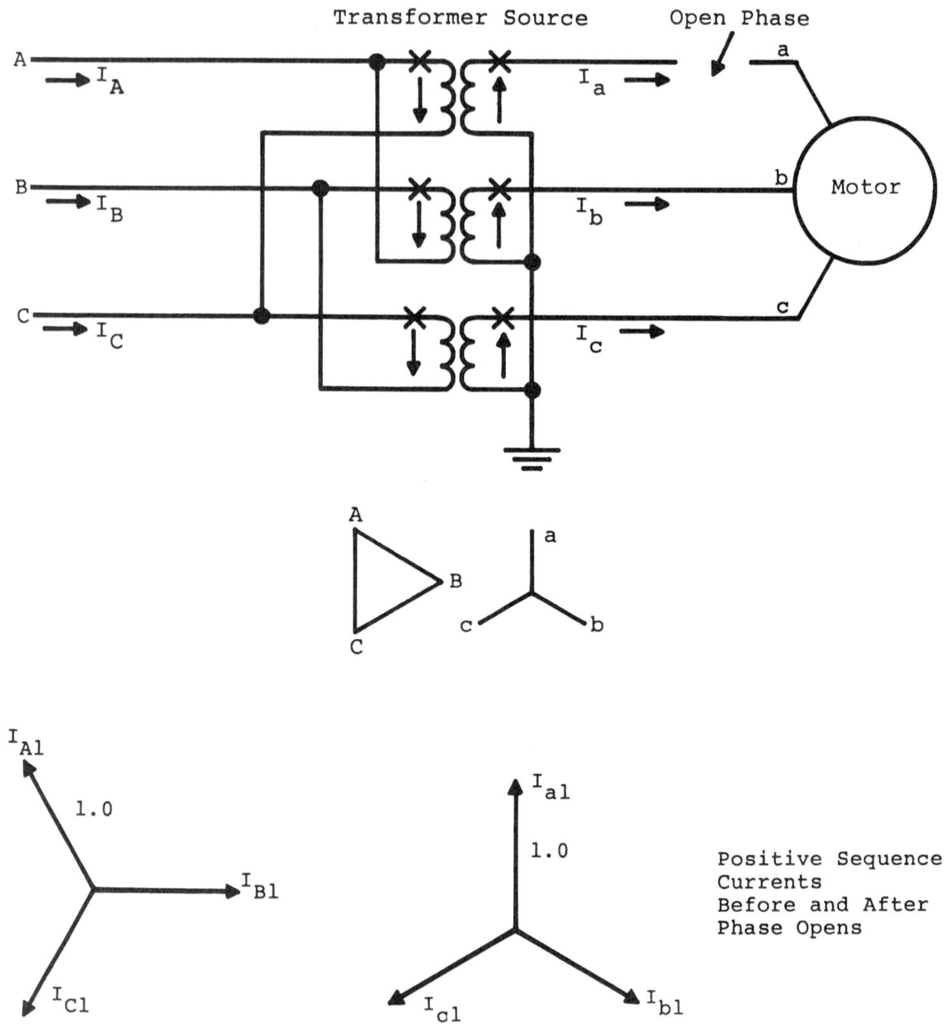

Fig. 11.10 Unbalanced current flow through a delta-wye transformer bank to a motor for phase A open on the motor or secondary side.

and before the motor slows down, stalls, or its internal impedances change, and so on.

The currents for an open phase on the primary side of a delta-wye transformer supplying a motor are shown in Fig. 11.9, and in Fig. 11.10 for the open phase on the secondary motor side. As developed in Chapter 3, if positive-sequence current is shifted 30° in one direc-

Since
I_{A1} leads I_{a1}
by 30° then
I_{A2} lags I_{a2}
by 30°

Negative Sequence
Currents
After the Phase Opens

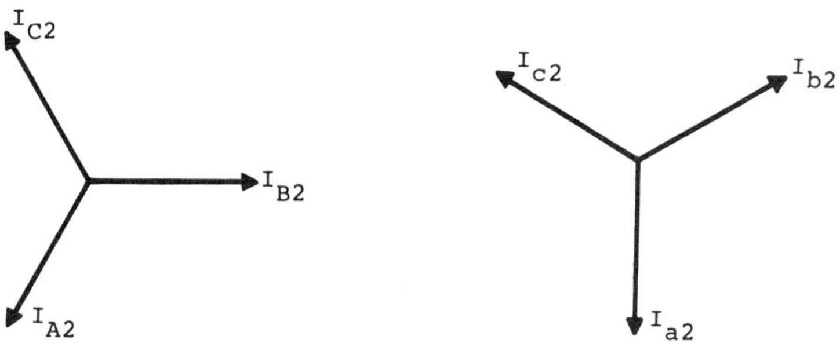

Fig. 11.10 (continued)

tion through the bank, the negative-sequence current shifts 30° in the opposite direction.

The current directions shown in the circuit diagram for these two figures are correct for the phasor diagrams. Without these specific phasor diagrams, the I_B in Fig. 11.9 could be shown into the motor as indicated, but at $\sqrt{3}$ magnitude with I_C at $\sqrt{3}$ flowing to the source. This is actually the flow, and is indicated in Fig. 11.9 by the phasor diagram, showing that I_B flows as indicated, but I_C is 180° from I_B, both at $\sqrt{3}$ magnitude. If I_B and I_C were shown in opposite directions in the circuit diagram, the correct phasor diagram would be to show I_B and I_C in phase. These currents can be traced through the transformer with the unbalances shown, remembering that 1.0-pu current in the wye winding appears as 0.577-pu current in the delta winding.

Negative-sequence voltage is useful in motor protection. By fundamental definition $V_2 = -I_2Z_2$. With reference to the examples of Fig. 11.8, for case a, $V_2 = 0.96 \times 0.15 = 0.144$ pu; for b, $V_2 = 1.51 \times 0.15 = 0.227$ pu, and for c, $V_2 = 0.12 \times 0.15 = 0.018$ pu. This indicates that V_2 is usable for loss of phase protection if the open phase is between the source and the point where V_2 is measured, but generally too low if the open phase is between V_2 measurement and the motor. This generality is based on the relatively low source impedance with respect to the motor impedances, as is usually the case. V_2 for Fig. 11.8a, when measured on the source side for a downstream open

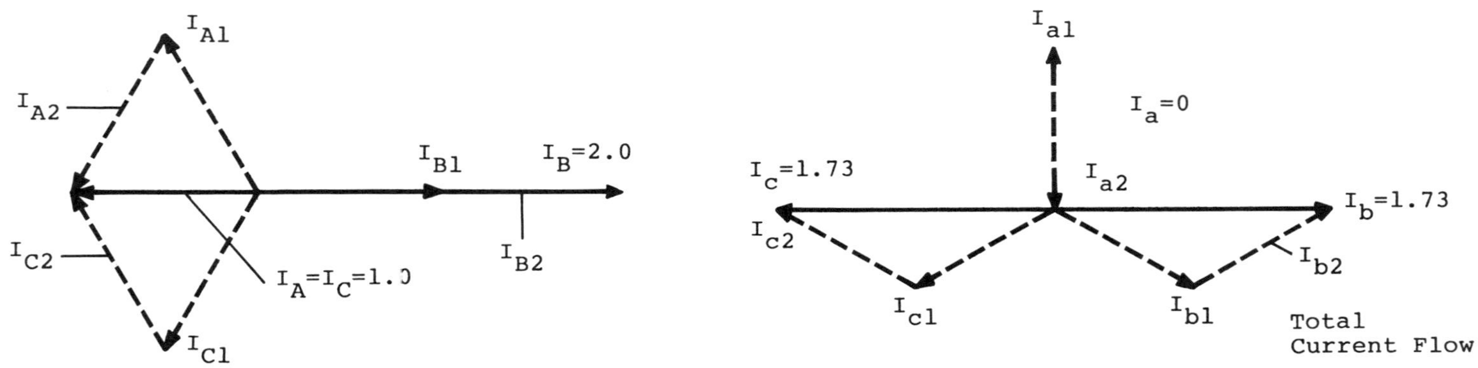

Fig. 11.10 (continued)

phase, is $0.96 \times 0.05 = 0.048$ pu, compared to 0.144 pu for the upstream open phase.

11.12 UNBALANCE AND PHASE ROTATION PROTECTION

As suggested in Section 11.11, there are several "handles" available for unbalance detection; (1) magnitude differences between the three phase currents, (2) the presence of negative sequence current, and (3) the presence of negative-sequence voltage. All three of these are used for protection.

The current balance type (46) compare the phase current magnitudes and operate when one phase current is significantly different in magnitude from either of the other two phase currents. This is very effective protection for individual motor feeders to detect opens or unbalances in that circuit. If other loads are supplied by the circuit to which this protection is connected, care should be taken to assure that any open phase or unbalance will not be camouflaged by the balanced current to the sound load. One relay should be applied for each load or feeder. The typical minimum sensitivity of these relays is about 1 A in one phase with zero current in the other, or 1.5 pu in one phase and 1 pu in the other.

Another type (46) responds to the negative-sequence current, either instantaneously with a fixed time delay added, or following the $I_2^2t = K$ characteristic as used for generator protection. These types of relays are not widely applied for motor protection.

The negative-sequence voltage type (47) is recommended to detect phase unbalance and phase reversal in the supply or source circuits. Typical operating sensitivities are around 0.05 pu V_2. One such relay should be connected through VTs (either wye-wye or open delta VTs) to each secondary supply bus. As shown in Section 11.11, sufficient V_2 voltage generally is available for open phase in the source or upstream system. They should not be applied for open phase downstream or between the relay and motor, for as shown, this V_2 voltage may be quite low.

When the phases are reversed, 1 pu V_1 becomes 1 pu V_2, so the negative-sequence relay positively responds to phase reversals. Phase-reversal relays also are available equivalent to a small motor. Normal phase rotation produces restraint or contact opening torque, while phase reversal causes operation or contact closing torque.

11.13 UNDERVOLTAGE PROTECTION

Low voltage on a motor results in high current and either failure to start, to reach rated speed, or to lose speed and perhaps pull out.

Very often, protection for undervoltage is included as part of the motor starter, but an inverse time undervoltage relay (27) is recommended to trip when prolonged undervoltage exists and as backup.

11.14 BUS TRANSFER AND RECLOSING

When motors, either induction or synchronous, are reenergized before they have stopped rotating, high transient torques can result, with possible damage and/or destruction. This can occur when a rapid transfer of motors is made from a bus that has lost voltage to a live auxiliary bus. Such transfers are necessary to maintain vital services to the auxiliaries supplying large generating plants or to critical industrial processes.

Another example are industrial plants supplied by a single utility tie. Problems in the utility will require opening the tie. Most experience indicates that many faults in the utility are transient in nature, resulting from lightning induction, wind, and/or tree contacts. As the utility is anxious to restore service promptly to its customers, they frequently use high-speed reclosing (around 0.20 to 0.60 s), and thus reenergize the motors, with possible damage.

The safe limits for reconnection of motors are complex and beyond this scope. If rapid transfer is mandatory, special attention should be given to this to it during the design stage. Otherwise, the best policy is either to delay any reclosure or reenergization of induction motors, or to assure that the motors are quickly disconnected from the system. For induction motors reenergization should not occur until the motor voltage has dropped to 33% or less of normal.

For synchronous motors, reclosing or reenergizing must not be permitted until proper resynchronizing can be effected. This means opening the motor supply promptly on the loss of supply.

An effective means to open the supply breaker under these conditions is the application of an underfrequency relay. Typical underfrequency relay (81) settings would be 98 to 97% of rated, with time to override the momentary voltage dip effects but before reenergization can take place. If the plant has local generation, or there are other ties with generation of the supply feeder, care should be taken to assure that the frequency will decline on loss of the utility. Generation sufficient to maintain load, particularly at lightload periods, will result in negligible frequency change.

11.15 REPETITIVE STARTS AND JOGGING PROTECTION

Starting motors repetitively with insufficient time between or operating with extreme load variations (jogging) can result in high motor temperatures. It is possible for a high temperature to follow a short-time load peak with subsequent low current load in normal operation and not exceed the motor limits. Thermistors on smaller motors and an integrating thermal overload unit responding to total heating [Eq. (11.13)] for larger motors provide a means of protection. Relays (49) operating on both overcurrent and temperature have been used. They operate with high current and high temperature. A high temperature without overload or a high overload without high temperature may not cause operation. Application of these requires careful analysis of the motor.

11.16 SYNCHRONOUS MOTOR PROTECTION

The protection discussed above for induction motor protection is applicable to synchronous motors, with additional protection for the field and for asynchronous operation. These motors usually include control and protection for starting, field application, and synchronizing. In addition, the following protection should be applied or considered.

As illustrated in Fig. 8.13, reduction or loss of excitation requires reactive power from the power system, and thus lagging current flows into the motor. For small units a power factor relay (55) is recommended. This relay operates when current into the motor lags more than 30°. Typical maximum current sensitivity occurs when the current lags 120°, so a fairly large current is required for typical loss of field lagging currents of 30 to 90°. For large synchronous motors, distance loss-of-field relays as described for generators are recommended. This provides improved protection for partial loss of field in addition to complete loss.

Out-of-step protection may or may not be required, depending on the system and motor. Faults in the system that momentarily reduce the motor voltage may cause the voltage angles between the system and motor to swing apart sufficiently that on fault clearing, the motor cannot recover and so will go out of step.

If the motor is running synchronously, only dc voltage exists in the field or exciter windings. When it falls out of step from system disturbances or loss of field, ac voltage appears which can be used to detect these conditions if the machine has brushes. With modern brushless exciters, the power factor relay (55) will provide both pull-out and loss-of-field protection.

11.17 SUMMARY: TYPICAL PROTECTION FOR MOTORS

Typical protection recommended and commonly applied for the protection of motors is summarized in Fig. 11.11. The application of the

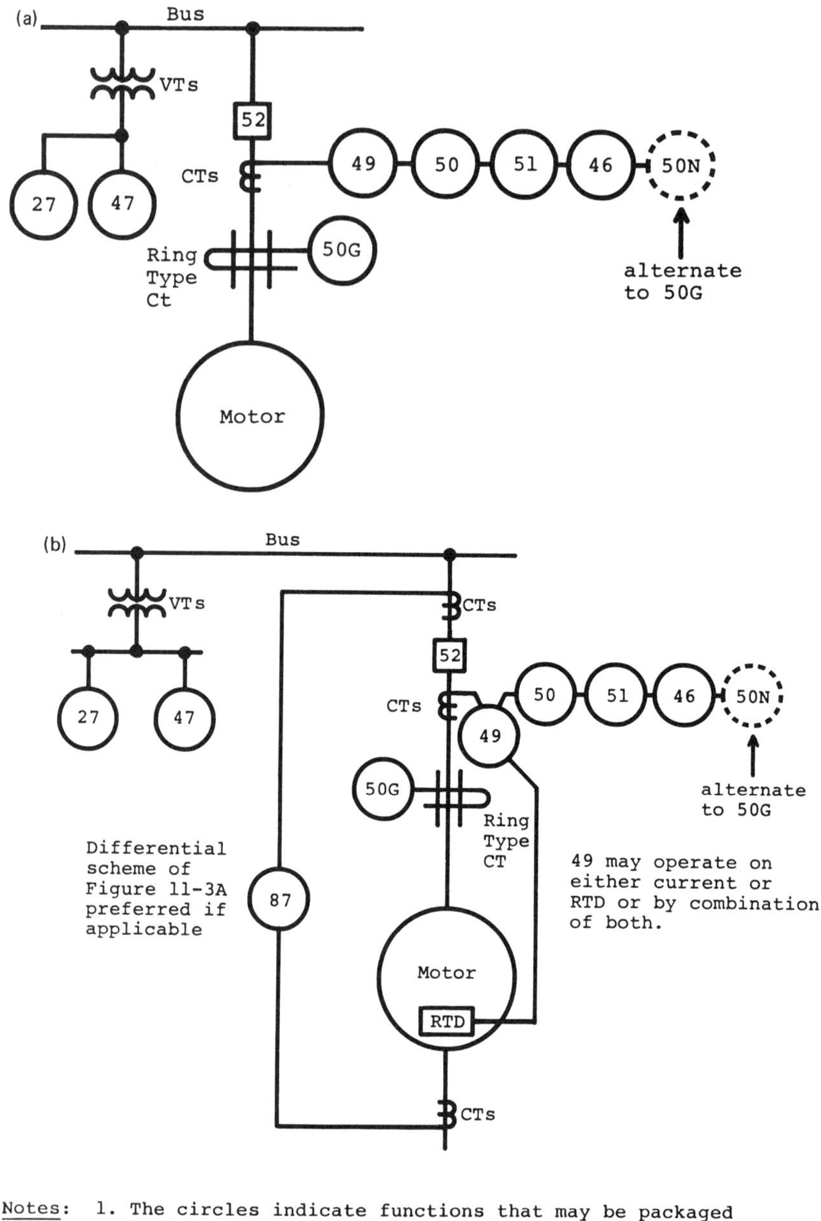

Notes:
1. The circles indicate functions that may be packaged individually or in various different combinstions.
2. Additional protection for Synchronous motors covered in Section 11.16.
3. Protection to avoid re-energization after short-time interruptions covered in Section 11.14.

Fig. 11.11 Typical recommendations for motor protection: (a) for motors without neutral leads and RTD's available; (b) for motors with neutral leads and RTD's available.

various relays is covered above. These are general recommendations recognizing that any specific application may use more or less depending on local circumstances, economics, and individual preferences.

BIBLIOGRAPHY

See the Bibliography at the end of Chapter 1 for additional information.

Brighton, R. J., Jr., and P. N. Ranade, Why Overload Relays Do Not Always Protect Motors, *IEEE Trans. Ind. Appl.*, IA 18, November-December 1982, pp. 691–697.

Dunki-Jacobs, J. R., and R. H. Kerr, A Quantitative Analysis of Grouped Single-Phased Induction Motors, *IEEE Trans. Ind. Appl.*, IA 17, March-April 1981, pp. 125–132.

Gill, J. D., Transfer of Motor Loads between Out-of-Phase Sources, *IEEE Trans. Ind. Appl.*, IA 15, July-August, 1979, pp. 376–381; discussion in IA 16, January–February 1980, pp. 161–162.

Guide for AC Motor Protection, ANSI/IEEE Standard C37.96-1981, IEEE Service Center.

Lazar, I., Protective Relaying for Motors, *Power Eng.*, 82, No. 9, 1978, pp. 66–69.

Linders, J. R., Effects of Power Supply Variations on AC Motor Characteristics, *IEEE Trans. Ind. Appl.*, IA 8, July–August 1972, pp. 383–400.

Regotti, A. A., and J. D. Russell, Jr., Thermal Limit Curves for Large Squirrel Cage Induction Motors, IEEE Joint Power Generation Conference, September 1976.

Shulman, J. M., W. A. Elmore, and K. D. Bailey, Motor Starting Protection by Impedance Sensing, *IEEE Trans. Power Appar. Syst.*, PS 97, 1978, pp. 1689–1695.

Woll, R. E., Effect of Unbalanced Voltage on the Operation of Polyphase Induction Motors, *IEEE Trans. Ind. Appl.*, January–February 1975, pp. 38–42.

12

Line Protection

12.1 CLASSIFICATION OF LINES AND FEEDERS

Lines provide the links — the connections — between the various parts of the power system and the associated equipment. Power generated at low voltages is stepped up to higher voltages for transmission to various stations, where it is stepped down for distribution to industrial, commercial, and residential users.

Most power systems have from two to many voltage levels. Over the 100-year history of ac systems, the actual three-phase voltages utilized have varied considerably, with no standards internationally. In the United States, the IEEE Standards Board adopted a set of standards in 1975 proposed by the Industrial Applications Society. These are summarized in Table 12.1. The class designations are not completely or uniformly agreed upon between the IEEE Power Engineering Society and the Industrial Application Society, both of whom are involved in power systems. Within the power utility area there has been and continues to be a general understanding of voltage classes as follows:

Industrial—distribution	34.5 kV and below
Subtransmission	34.5 to 138 kV
Transmission	115 kV and up

with the last class generally divided into:

High voltage (HV)	115 to 230 kV
Extra high voltage (EHV)	345 to 765 kV
Ultra high voltage (UHV)	1000 kV and above

Table 12.1 ANSI/IEEE Classifications of Three-Phase Power System Voltages. (From IEEE Standard Dictionary, p. 1006)

		Nominal rms line-to-line[a] system voltage				
		Three-wire systems		Four-wire systems		
IAS[b]	PES[c]	Preferred	Optional/ Existing	Preferred	Optional/ Existing	Maximum voltage
Low (LV)	No voltage class stated (ANSI C84.1-1977)			208Y/120		220Y/127
			240	240/120Tap		245/127
		480		480Y/277		508Y/293
			600			635
Medium (MV)			2400			2540
		4160		4160Y/2400		4400Y/2540
			4800			5080
			6900			7260
					8320Y/4800	8800Y/5080
					12000Y/6930	12700Y/7330
				12470Y/7200		13200Y/7620
				13200Y/7620		13970Y/8070
		13800			13800Y/7970	14520Y/8380
					20780Y/12000	22000Y/12700
					22860Y/13200	24200Y/13970
			23000			24340
				24940Y/14400		26400Y/15240
			34500	34500Y/19920		36510Y/21080
	Higher voltage ANSI C84.1-1977		46 kV			48.3 kV
		69 kV				72.5 kV
High (HV)		115 kV				121 kV
		138 kV				145 kV
			161 kV			169 kV
		230 kV				242 kV
Extra high (EHV)	ANSI C92.2-1978	345 kV				362 kV
		500 kV				550 kV
		765 kV				800 kV
Ultra high (VHV)		1100 kV				1200 kV

[a]The second voltage for the three-phase four-wire systems is line to neutral or a voltage tap, as indicated.

[b]IAS: IEEE Industrial Applications Society. Their voltage class designations adapted by IEEE Standards Board (LB 100A of April 23, 1975).

[c]PES: IEEE Power Engineering Society. Standard references indicated in the table.

As indicated, these are not well defined, so the range can vary somewhat in any specific power system. With time, more conformity with the standards will take place, but there exist many different and nonstandard voltages in use that were adopted many years ago and will not be changed for a long time. Some of these, but not all, are indicated in the table as optional/existing.

The voltage values indicated represent the nominal and typical rms system voltages (line to line unless so indicated) commonly designated in technical information and communication. As power flows through a system the actual voltage varies from point to point from the generators to the ultimate terminals from the drops, transformer ratios, taps, regulators, capacitor banks, and so on, so that arbitrary nominal or typical values are selected as indicated.

The terminals of the lines and feeders and the location of equipment such as transformers, generators, and circuit breakers, are known as stations, plants, and substations. The specific designation is not well defined or standardized. A generating station or plant is obvious. A substation is smaller and less important than a station. In one system relative to another a station in one might be designated as a substation in the other, and vice versa. So common designations are generating stations, switching stations, power stations, distribution stations, substations, distribution stations, and so on.

Many lines are two terminal, but there are a number of multiterminal lines with three or more terminals. These may interconnect stations or substations, but frequently are taps to supply loads. Thus lines of all voltage levels can become distribution circuits. When an industrial or commercial complex with a large load requirement is located near a transmission line, it often becomes economical to tap the line. Distribution lines usually have many taps, either three-phase or single-phase, as they supply loads along their routes.

For protection purposes, lines may be classified as (1) radial lines or feeders, or (2) loop types. Feeders or radial lines supply power principally to nonsynchronous loads. A fault on these is supplied from the source end, with negligible or no fault current from all other terminals or taps. As indicated in Section 4.8, in most cases induction motors are not considered as sources.

Loop circuits are those where for line faults, fault current is supplied from both or all terminals. A line may be a feeder type for positive (and negative) sequence but a loop type for zero sequence. This circuit would have a power source at one end but be grounded at both terminals. Similarly, a line could be a loop type for positive (and negative) but a feeder type for ground faults. This line would have a ground source only at one end.

12.2 TECHNIQUES APPLICABLE FOR LINE PROTECTION

The protection available for line protection includes:

1. Nondirectional instantaneous overcurrent
2. Nondirectional inverse-time overcurrent
3. Directional instantaneous overcurrent
4. Directional inverse-time overcurrent
5. Current balance
6. Directional distance — instantaneous and/or step or inverse time
7. Pilot with a communication channel between terminals

All of these are utilized individually or in various combinations for both phase and ground protection of lines. Several of these relay types have been discussed briefly in Chapter 6 and their application are covered in the present chapter. The current-balance type compares the currents in parallel lines to detect an unbalance resulting from a fault in one line. It is not in general use in the United States. The reasons are that it is not applicable to single lines; must be disabled for single-line operation; requires interconnections between the controls for the two lines, which is not desirable for operation and testing; and can experience difficulties for a fault involving both paralleled lines.

12.3 FUNDAMENTALS OF PROTECTION COORDINATION

Except where differential protection (7, for lines) is used, the protection for the lines will extend into adjacent lines and equipment (buses, transformers, motors, etc.) Similarly, equipment protection overlaps into the lines. As a result, the settings must assure that the protection does not operate for faults in this overlapping (backup) area until the primary relays assigned to that area have had the opportunity to clear the fault. This process of setting is called coordination or selectivity.

Both phase-fault and ground-fault relays must be selectively set or coordinated. The process is basically the same for both. Phase relays are set using three-phase fault data and maximum short-time load or transient inrush. Ground relays are set using single-phase-to-ground fault data and the maximum zero-sequence load unbalance.

At the distribution-utilization levels, fuses are used and must be included in the coordination process. Fuses are often associated with equipment protection as well as for line protection. Fuses provide both phase- and ground-fault protection, but they receive only the phase or line currents, while ground relays operate on $3I_0$ currents. Fortunately, fuses generally are applied only to radial or feeder circuits, where the line current equals the $3I_0$ current (i.e., for a phase-a-to-ground fault, $I_a = 3I_0$). This is not true for loop circuits, where the faulted phase current and the $3I_0$ current can be quite different. This is discussed and illustrated in Chapter 4 (see Fig. 4.25 for an example.

As a result, the coordination of relays and fuses, even on feeder circuits, can be difficult or impossible where the ground relays are set more sensitively. This is normal where only ground relays without fuses are coordinated. Thus with fuses, the situation often requires raising the ground relay setting to that essentially equivalent to the phase relays. This suggests the elimination of separate ground relays, since little to no increased sensitivity can be obtained. This is not recommended, as the absence of separate ground relays, commonly used, can suggest inadequate protection in liability cases.

The protection problem was introduced in Chapter 6. Applying this to line protection, Fig. 12.1 is an extension of Fig. 6.4 and shows the fault data required with typical time coordination curves. The objective is to set the protection to operate as fast as possible for faults in the primary zone, yet delay sufficiently for faults in the backup area. This requires setting well below the minimum fault current for which they should operate, but not operate on all normal and tolerable conditions. In some cases these requirements provide very narrow margins or no margin. This is especially true where there is a large variation in fault current magnitude with system operation. Fault currents can be high at peak-load periods with all the generation and lines in service, but quite low as equipment is removed during the light-load periods. Where this is significant, the fault study should document the extremes. Where coordination is not possible, either a compromise must be made or pilot protection applied. The latter area is discussed later.

Thus coordination is often a "cut and try" process. For radial lines and feeders, the coordination is done conveniently by overlaying time-current characteristics drawn on transparent log-log paper and shifting them until the objectives are obtained. This was the method used earlier for transformer protection (Figs. 9.18, 9.20, and 9.21). In these applications the currents in the several protective devices is the same or a fixed difference, as for the currents through wye-delta transformer banks.

In loop systems with multiple fault current sources, the current in the relays being coordinated will be different for the fault, so that the curve overlay method can become difficult. A coordination chart or table is then suggested. A later example will amplify this.

With fault current flowing in both directions through the relays, all time-overcurrent relays should be directional. This limits their operation, and hence coordination, only to fault current flowing in the "trip direction," which normally is into the line being protected.

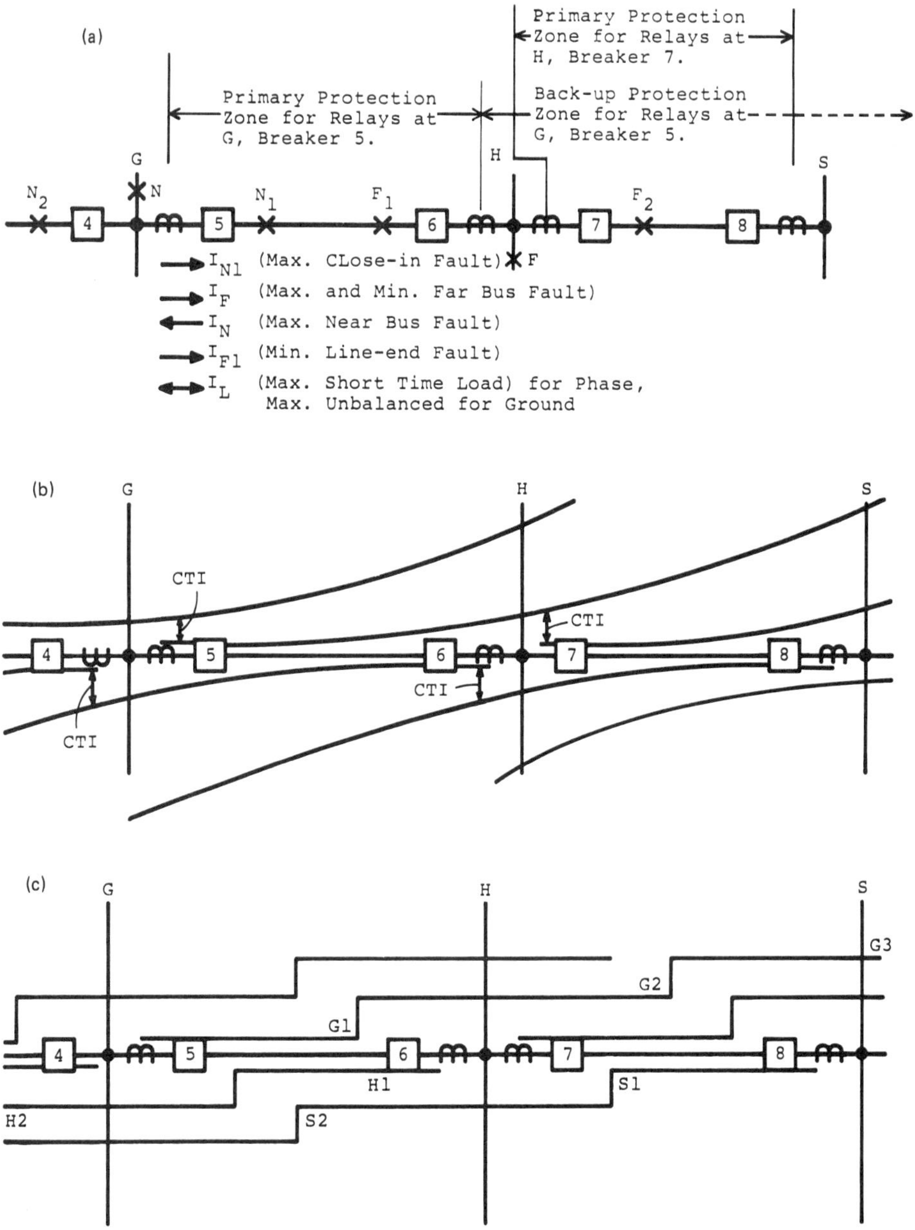

Fig. 12.1 Fault data requirements and typical time coordination settings: (a) the key currents for setting relays at G breaker 5 to protect line GH; (b) coordination with instantaneous overcurrent and inverse time units; (c) coordination with directional distance units.

12.4 GENERAL SETTING AND COORDINATION CRITERIA

The basic requirements and limits for overcurrent relays and/or fuses can be defined as follows:

12.4.1 Phase Overcurrent Relay Current Setting

For lines there is rarely a thermal limit as there was for transformers. The minimum operating current (relay pickup) must be set so that operation will not occur on the largest transient or short-time current that can be tolerated by the system. The key factors to be considered are:

1. Short-time maximum load (I_{STM}). This is the current that the circuit may be required to carry during emergencies or unusual operating conditions for intervals which can be from about 1 hour up. Practically, it is often the maximum capability of the transformer or maximum limit of the load.
2. Magnetizing inrush of the transformers supplied by the circuit. Where overcurrent type relays are applied, the inrush often is not too severe and/or decays rapidly because of the appreciable amount of resistance in the circuit.
3. Cold load is a short-time increase in load current that occurs when a distribution feeder is reenergized after an outage. Normal feeder load is based on diversity, since not all customers require maximum load at the same time. After an outage this diversity is lost momentarily as all of the load is energized at the same time. The amount and duration is quite variable, depending on the circuit and the length of the outage, so experience history is usually necessary to document this.

A phase overcurrent relay pickup of 1.25 to 1.5 times the maximum short-time load or greater will be required to avoid operation on short-time transients with inverse relay characteristics. The lower multiples can be used with the extremely or very inverse types, as the operating times are very long just above pickup. The transient overcurrents may energize the relays but subside below relay pickup before the operating time is reached. Generally, the extremely inverse characteristics more nearly match the fuse characteristics and motor starting curves, so are preferable for protection in the load areas. Moving back toward the source, the less inverse types are applicable.

12.4.2 Ground Overcurrent Relay Current Setting

The miniumum operating current (pickup) must be set above the maximum zero-sequence current unbalance that may exist and can be tolerated by the system. This unbalance usually is the result of unequal loading of single-phase taps between the three phases. Monitoring the unbalance and changing taps are used to keep the unbalance minimum. With this, and except for problems in coordinating with fuses, ground relays can be set much lower than the phase relays for increased fault sensitivity.

Typically, 0.5- and 1.0-A taps are used where the unbalance is low, especially at the higher voltage levels.

12.4.3 Fuses

The continuous rated current of the fuse must be equal to or greater than the maximum short-time load that will pass through it. Also, the symmetrical interrupting rating of the fuse should be equal to or greater than the maximum fault current. Attention must be given to system voltage and insulation level and the system X/R ratio. Fuses operate in a time-current band between maximum clearing times and minimum melt (or damage) times. The difference is the arcing time within the fuse.

The minimum melt time is important when the fuse backs up or overreaches other devices. The latter devices must clear the fault before thermal damage can occur to the fuse, as is indicated by the minimum melt time.

12.4.4 Automatic Circuit Reclosers

Commonly known as reclosers, these are a type of circuit interrupters with self-contained controls to sense overcurrent and open on faults either instantaneously or with time delay. They can be programmed to initiate automatic reenergization of the circuit (reclosing) at variable intervals if the fault persists, and eventually lock out. They are used frequently in distribution systems, as their cost is less than that of conventional circuit breakers and separate relays. The setting criteria for these devices is similar to those outlined for fuses.

12.4.5 Coordinating Time Interval

The coordinating time interval (CTI) is the time interval between the primary and remote backup protective devices. It is illustrated in Fig. 12.1. For example, it represents the delay required for the relays at breaker 5, station G, to permit breaker 7 relays at station H to clear faults in line HS, such as fault F_2. It consists of:

1. Breaker fault interruption or clearing time, typically two to eight cycles (0.033 to 0.133 s).
2. Relay overtravel (impulse) time. The energy stored in the electromechanical induction disk or solid-state circuitry will continue

operation after the initiating energy is removed. Typically, this is not more than 0.03 to 0.06 s for electromechanical units, less but not zero for solid-state units.
3. Safety margin for errors or differences in equipment operating time, fault current magnitudes, CT ratios, and so on.

CTI values frequently used in relay coordination range between 0.2 to 0.5 s, depending on the degree of confidence or the conservatism of the protection engineer.

12.5 EXAMPLE: COORDINATION FOR A RADIAL DISTRIBUTION FEEDER

A typical 13-kV feeder of several fed from a 115-kV power system through a 15/20/25-MVA transformer is shown in Fig. 12.2. There are four load taps protected by fuses. Three taps have an automatic circuit recloser ahead of them which is set to operate and reclose for faults to and beyond the three fuses on the right. The sequence can be one or more instantaneous trips and reclosures to clear temporary faults and restore service. If the fault persists, the fuses will clear when the fault is to their right (downstream), or the recloser on timed trip will clear faults downstream to the fuse location. Approximately 80 to 95% of the faults on open-wire circuits of this type are temporary and can be cleared by momentarily deenergizing the circuit. They are usually caused by tree contact from wind or are lightning induced. Thus tripping-reclosing cycles provide a high service continuity for these type faults, which is highly desirable for the customers of power. Reclosing can provide a potential liability where the circuit is physically contacted by people, such as downed lines at or near ground level. The particular sequence is a judgment based on many local factors and experience.

Separate overcurrent phase and ground relays are applied at the feeder breaker. These may or may not use reclosing relays coordinated with the downstream devices. One practice is to set an instantaneous unit to overreach the first fused tap with reclosing, after which the operation of the instantaneous unit is locked out. This permits the fuse or timed units to clear the permanent faults.

For this example, coordination only will be covered. The maximum load for 25-MVA tap is

$$\frac{25,000}{\sqrt{3} \times 115} = 125.5\text{A} \quad \text{at } 115\text{kV}$$

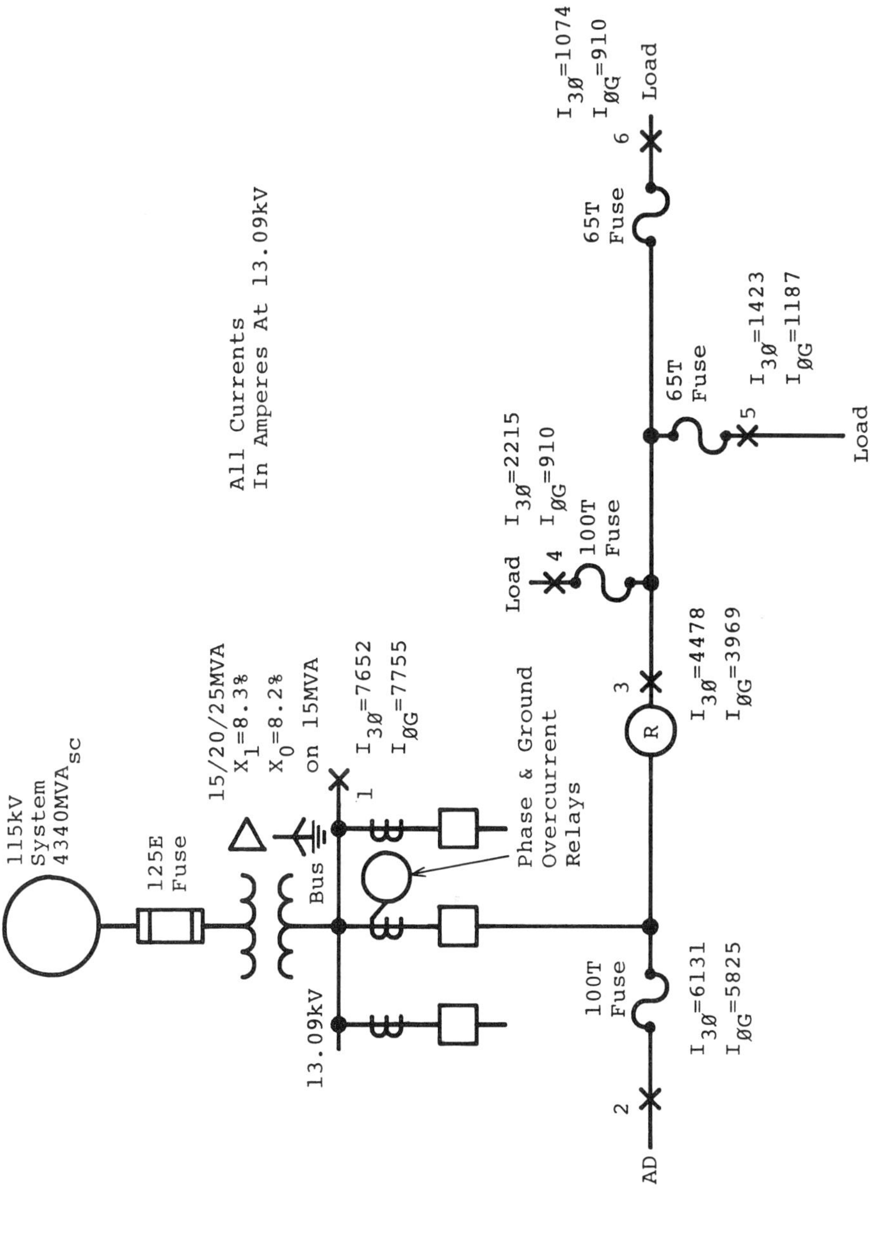

Fig. 12.2 Typical distribution feeder serving multiple load centers.

A 125E fuse was selected for the transformer bank primary. Its operating time at around 250 A is 600-s, which should override cold load and magnetizing inrush transients.

The characteristics are plotted on log-log fuse coordinate paper, as shown in Fig. 12.3. The abscissa is amperes at 13 kV, so the 125E fuse in the 115-kV circuit is plotted at 115/13.09 = 8.79 times the manufacturer's curves. Thus the 600-s minimum melt current of 250 A becomes 250 × 8.79 = 2196 A for balanced currents. The dotted minimum melt curve shown to the left reflects the effect of load current preheating the fuse.

Whereas phase-to-phase faults on the 13-kV side are 0.866 of the three-phase fault value, as shown in Fig. 9.19, the current in one phase on the primary is the same as the three-phase fault value. However, the primary fuse sees only 0.577 of the secondary one per unit current for 13-kV phase-to-ground faults (Fig. 9.19). The dotted curve to the right is the primary fuse minimum melt characteristic for secondary ground faults. For the 600-s operating time, $2196 \times \sqrt{3} = 3804$ A is equivalent to 2196 A for a phase-to-ground fault.

The transformer through-fault overcurrent limit curve is plotted as shown. This area is discussed in Chapter 9. As shown, the transformer is protected satisfactorily against thermal damage.

The 65T and 100T fuses selected on the basis of the loads served from the taps are shown plotted on Fig. 12.3 from manufacturers' curves. The left curve is minimum melt, the right maximum clearing.

The maximum load through the recloser is 230 A. A recloser was selected with a minimum trip rating of 560 A phase, slightly more than twice the load needed to override cold load with a safety factor. The ground unit is set at 280 A by choice. The time characteristics for both units are plotted for the timed and the instantaneous operations from the manufacturers's data.

The maximum load through the breaker and relays at the 13-kV bus is 330 A. Thus the CT ratio of 400:5 will give a secondary current of 330/80 = 4.13 A.

Extremely inverse time-overcurrent relays provide good coordination with the fuses and the recloser. Selecting tap 9 provides a phase relay pickup of 9 × 80 = 720 A, just over twice the maximum load needed to override cold load. The ground relay is set on tap 4 by choice. This provides a primary 13-kV pickup of 4 × 80 = 320 A. Time dial 2 for the phase relays and dial 1 for the ground relay provides a CTI of at least 0.2 s above the recloser. This is satisfactory where the recloser time curves include fault interruption time. The fault currents in fig. 12.2 and 12.3 represent the maximum values. If these values are less as a result of system operating changes or from fault impedance, it is seen that coordination still exists.

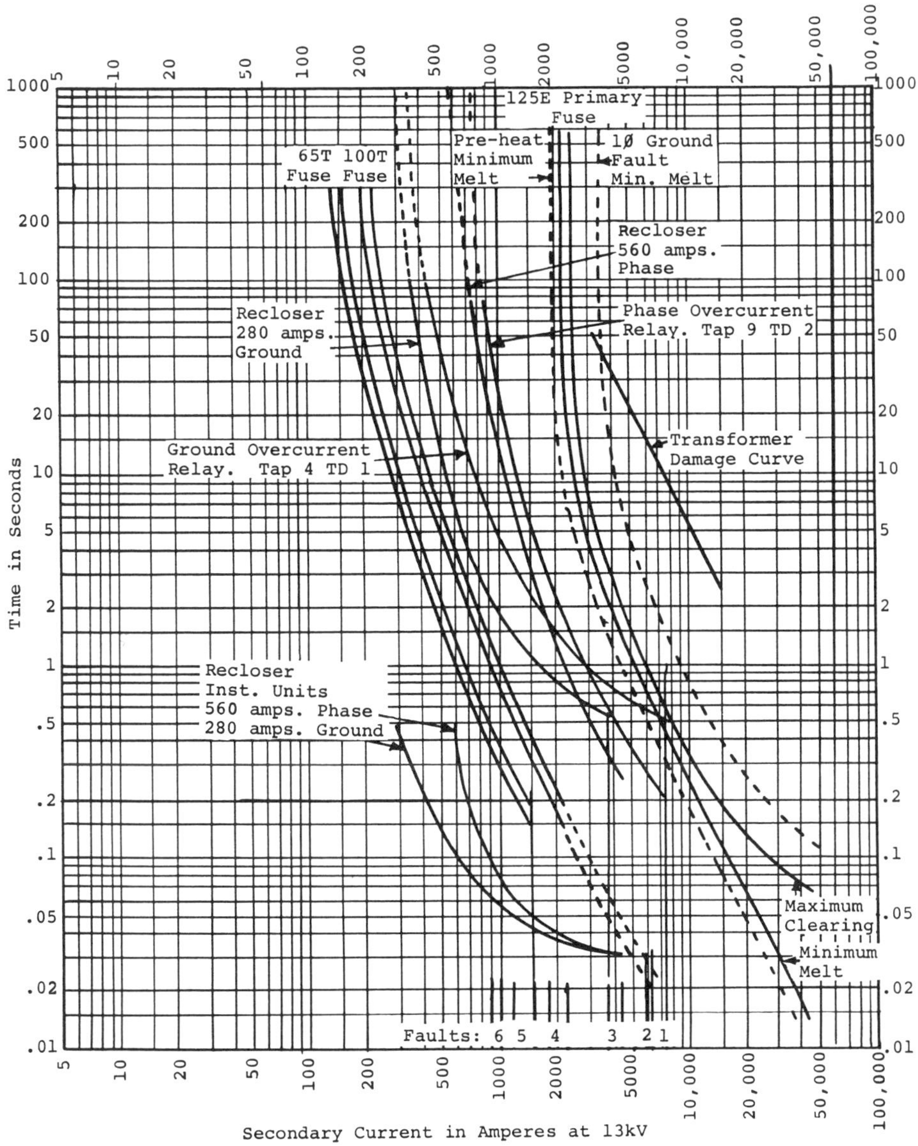

Fig. 12.3 Typical time coordination curves for the distribution system feeder of Fig. 12.2.

12.6 INSTANTANEOUS OVERCURRENT UNIT APPLICATION

Instantaneous overcurrent (IT) units operate with no intentional time delay and generally on the order of 0.015 to 0.05 s. This requires that they be set not to overreach any other protective device. An exception is for "fuse saving," such as discussed for the recloser in the preceding example. At the breaker of Fig. 12.2 feeder, IT units might be used overreaching the first 100T fuse tap with a fast reclose of the breaker to restore service on the feeder for transient faults on the load tap and the feeder circuit up to the recloser. Then the IT units are locked out so that if the fault is permanent, it is cleared by either the fuses or the time-overcurrent relay units. If the instantaneous trip first reclose is successful (transient fault cleared), the reset of the relays and recloser restores the instantaneous unit for the next fault occurrence. A recent industry survey by the IEEE showed that 81% use this for phase faults and 61% for ground faults.

The fundamental principles for setting the instantaneous units with reference to Fig. 12.1, breaker 5 at bus G, are:

1. Set at $kI_{Far\ Bus\ Max}$ ($I_{F\ Max}$ in the figure). If there is a tap or a recloser before the far bus, use the maximum current at that device or point. k typically is 1.1 to 1.3. The value depends on the responses of the IT unit to a possible fully offset current and the degree of conservatism of the protection engineer.
2. If this setting value current is greater than the maximum near-bus fault (I_N in Fig. 12.1), a nondirectional instantaneous unit can be used.
3. If this setting value current is less than the maximum near-bus fault (I_N in Fig. 12.1), a directional instantaneous overcurrent unit is required, or the IT setting must be increased to avoid operation with a nondirectional type.

For the example of Fig. 12.2, a nondirectional instantaneous trip unit could be used at the feeder breaker. Using k = 1.2, a setting of 7357 for the phase type and 6990 A for a ground type would be required for setting not to operate for fault 2, or with fuse saving, a setting of 5374 A, phase unit; 4763 A, ground units not to operate for fault 3. The coverage of the IT units might be considered quite limited, especially if a lower fault current level can exist for some operating conditions. Generally, these IT units are relatively inexpensive and are recommended even if the coverage is limited or may not exist for light faults. At least they provide fast operation for the large current close-in faults.

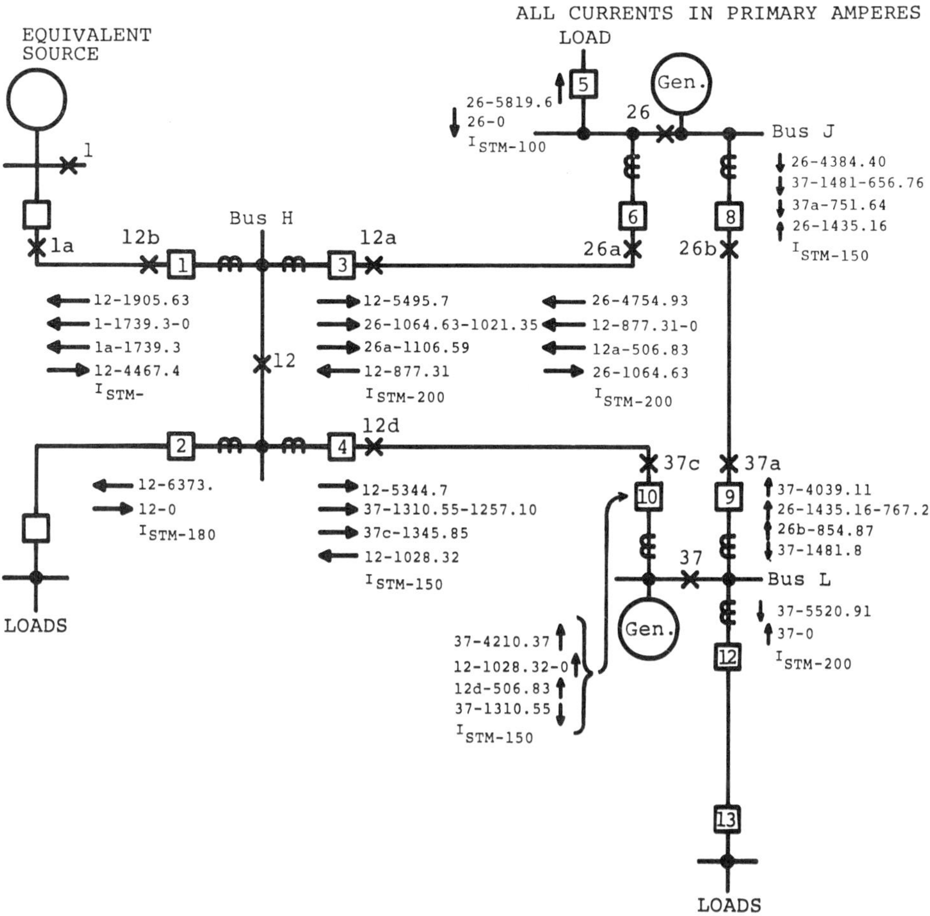

Fig. 12.4 Typical loop system with multiple sources. Three-phase fault currents at 34.5 kV. First value is the maximum current, second value is the minimum current. The minimum operating condition considered was the generators at buses J and L out of service for light load. The generator-transformer units have fast differential protection.

12.7 EXAMPLE: COORDINATION FOR A LOOP SYSTEM

Coordination for a loop system is much more complex and difficult. This can be illustrated with the typical system of Fig. 12.4. The key faults are documented for the several breakers at the three buses of the loop for three-phase faults. Typical settings will be made for the phase relays. Setting ground relays for the system is similar using phase-to-ground fault data and relay pickup values, as outlined earlier. In general, these taps will be one half or less of the phase relay taps for most systems.

The directional time-overcurrent relays are applied at breakers 1, 3, 4, 6, 8, 9, and 10, with each directional unit "looking into the line" or operating when current is flowing into the line section. Around the loop clockwise:

Relays at 3 must coordinate with relays at 5 and 8,
Relays at 8 must coordinate with relays at 10 and 12,
Relays at 10 must coordinate with relays at 1, 2, and 3.

Around the loop counterclockwise:

Relays at 4 must coordinate with relays at 9 and 12,
Relays at 9 must coordinate with relays at 5 and 6,
Relays at 6 must coordinate with relays at 1, 2, and 4.

Thus it is seen that the loops are not completely independent. The settings in both are dependent on the settings of the relays on other circuits (and loops) from the several buses. In the example these other circuits are the relays at breakers 1, 2, 5, and 12 and the generators at buses J and L. In setting relays around the loop, the first step is to determine the settings and operating times for these relays. To simplify the example, assume that the settings for these are:

Phase relays breaker 1: Pilot relays used on this short line with operating time not exceeding 0.06 s.
Phase relays breaker 5: Maximum operating time for fault 26 on the line is 0.24 s.
Phase relays breaker 12: Maximum operating time for fault 37 on the line is 0.18 s.
Phase relays breaker 2: Maximum operating time for fault 12 on the line is 0.21 s.

In setting relays around a loop, a good general rule is to attempt to set each relay to operate in less than 0.20 s for the close-in fault and at least 0.20 plus the CTI interval for the far-bus fault. Where the relays protecting the lines extending from the remote bus have operating times greater than 0.20 s, the setting should be that maximum time plus the CTI interval. For this example, a CTI of 0.30 s will be used.

The relay coordination information for setting the relays around the loop in the clockwise direction, starting arbitrarily at breaker 3;

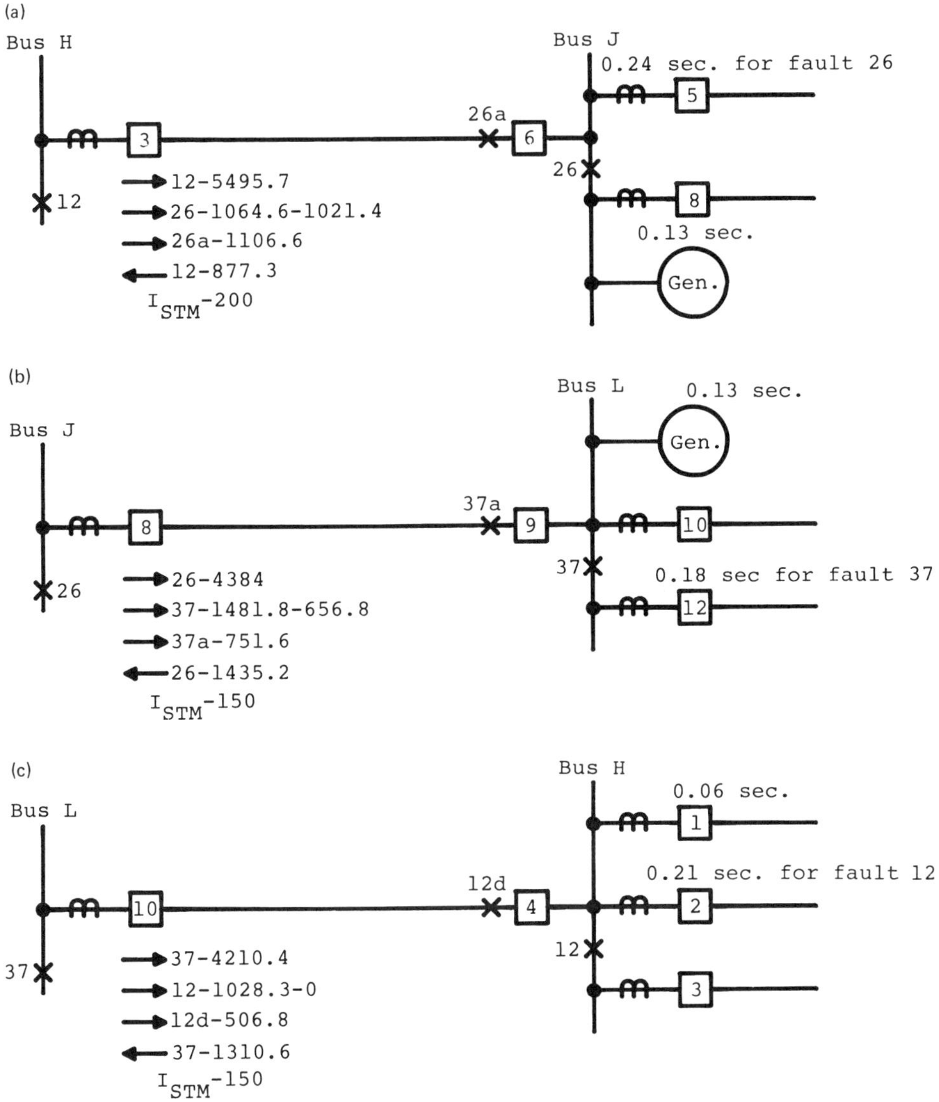

Fig. 12.5 Information for setting relays for phase-fault protection clockwise around the loop of Fig. 12.4: (a) data for setting breaker 3 phase relay; (b) data for setting breaker 8 phase relays; (c) data for setting breaker 10 phase relays.

is documented for convenience in Fig. 12.5a. With the short-time maximum load of 200 A, 250:5 CTs can be used. The maximum load is then 200/50 = 4 A secondary. Select relay tap 6, which is 1.5 times this maximum load and gives a primary fault current pickup of 6 × 50 = 300 A.

Typical time overcurrent relay curves are illustrated in Fig. 12.6 for determining the time dial setting for coordination. In Fig. 12.5a, relay 3 operating times for fault 26 at the far bus must be at least 0.24 + 0.30 = 0.54 s, assuming that relays at breaker 8 can eventually be set to operate for close-in fault 26 at not more than 0.24 s. For this maximum fault (26), relay 3 receives 1064.6 A or 1064.6/300 = 3.55 multiple of its pickup current, From Fig. 12.6, a time dial of 1 provides an operating time of 0.58 s at this multiple and thus coordination. Relay 3 operating times: for minimum fault 26, 0.61 s (1021.4/300 = 3.4 mult.); maximum close-in fault 12, 0.18 s (5495.7/300 = 18.32 mult.); and minimum line-end fault 26a, 0.54 s (1106.6/300 = 3.69 mult.). This line-end fault is not a coordination concern, as for it, breaker 6 is open. With directional relays at 3, bus fault 12 is not involved.

Now move to bus J to set relays at breaker 8. The data are shown in Fig. 12.5b. With a 150-A load, 200:5 ratio CTs are suggested. With these the secondary load is 150/40 = 3.75 A. Tap 5 provides a margin of 1.33 times maximum load and a primary fault current pickup of 5 × 40, or 200 A. Now with relay 3 operating at 0.61 s minimum for fault 26 from above, relay 8 should not operate more than 0.61 − 0.3 = 0.31 s for fault 26. The time of relays at 10 are unknown, but for the rest at bus L, relay 8 for faults 37 must be at least 0.18 + 0.3 = 0.48 s. Maximum close-in fault 26 is 4384 A, to provide a multiple of 4384/200 = 21.9. For the far-bus fault 37, the multiple is 1481.8/200 = 7.41. From the time curves (Fig. 12.6) time dial 2 provides 0.35 s. for the close-in fault and 0.56 s for the far-bus maximum fault. This does not coordinate. Going back to relay 3 and increasing its time dial to 1.5 changes the operating times to 0.25 for the close-in fault and 0.85 s for the far-bus maximum fault. This is 0.5 s above relay 8.

Continuing around the loop to relays at breaker 10, the 150-A load suggests 200:5 CTs, giving a secondary load current of 150/40 = 3.75 A. Tap 5 provides a margin of 1.33 above the maximum load, and a primary current pickup of 5 × 40 = 200 A. For the close-in fault 37, the relay multiple is 4210.4/200 = 21. For the maximum far-bus fault 12, the multiple is 1028.3/200 = 5.14. The limits for relay 10 are less than 0.26 s (0.56 − 0.30) for the close-in fault and greater than 0.55 s (0.25 + 0.30) for the far-bus fault. Time dial 1.5 just meets this, to provide coordination.

Numbers are confusing, so the coordination around the loop is summarized in Fig. 12.7. The relays at bus H are repeated to show coordination. The times in parentheses are the operating times for the far-bus minimum fault and the line-end fault. With the generators out

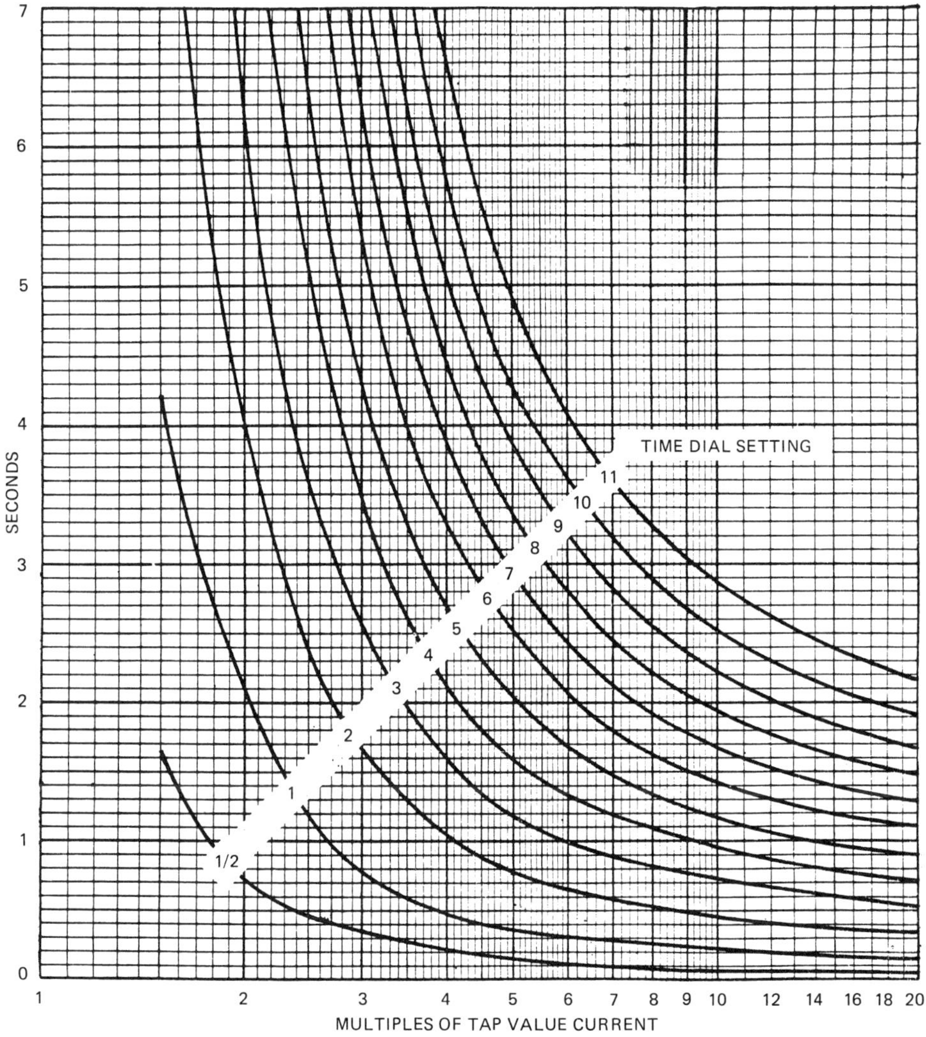

Fig. 12.6 Typical inverse-time-overcurrent relay curves. (Courtesy of Westinghouse Electric Corp.)

of service at both buses J and L for the minimum condition, no current flows through breaker 10 and 6 for far-bus faults. This changes after the far-bus relays 4 or 3 open, which provides fault current per 12d and 12a, respectively, the line-end faults. It is important to assure that the relays can respond to these line-end faults; otherwise, they cannot be cleared.

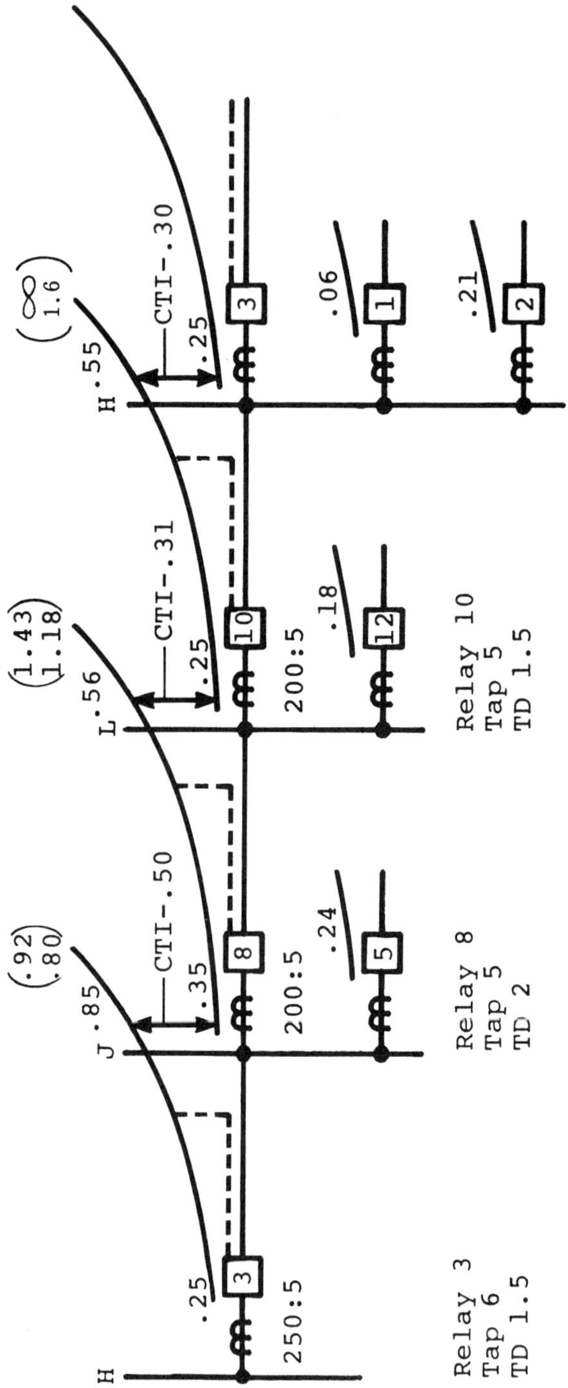

Fig. 12.7 Summary of the phase relay settings around the clockwise loop of Fig. 12.4 system.

The loop for this minimum operating condition becomes a single source loop. Although most distribution systems are radial, some are of this single-source loop type. It is also utilized in industrial plant complexes where there are several separate load areas. The advantage is that any one line circuit can be removed with service available to all the loads. In this type of system, where the source is at bus H only (Fig. 12.4), the relays at breaker 6 do not need to coordinate with the relays at 1, 2, and 4 since they do not have current for fault 12. Similarly, relays at 10 do not need to coordinate with relays at 1, 2, and 3. Faults on the lines such as at 12a and 12d can be detected only after breaker 3 or 4, respectively, have opened when the fault current of the line-end fault exists. Thus these faults are "sequentially cleared." Load current into these lines also would be zero unless there are other line taps. As a result, directional instantaneous relays can be applied at breakers 6 and 10 and set very sensitively and below the line-end faults. This provides "high-speed sequential operation" for these terminals. The next phase in the coordination process for the loop example is to set relays 4, 9, and 6 counterclockwise around the loop. This will not be continued, as the basic principles have been covered.

In most actual systems, attention must be given to the possibilities of various lines out of service and other operating conditions that may occur. It would be desirable to set all the relays to provide complete backup protection over all the adjacent remote lines. In the example, this would be to have the relays at breaker 3 provide protection for faults to bus L and out on load line 5 to any sectionalizing point. This may or may not be possible. The "infeed" of fault current by the source at bus J tends to reduce the fault current through 3 for faults on line JL and at bus L.

The computers that provide fault data for many variables and operating conditions also provide an excellent tool for setting and coordinating relays. A number of such programs exists with varying degrees of capability and sophistication, with others being developed. These can be of great value in reducing the time and drudgery of hand coordination, and often consider more alternatives and conditions than would otherwise be convenient.

12.8 INSTANTANEOUS TRIP APPLICATION FOR A LOOP SYSTEM

Where a reasonable difference in the fault current exists between the close-in and far-bus faults, instantaneous units can be utilized to provide fast protection for faults out on the line. The fundamentals were outlined in Section 12.6. For the example of Fig. 12.4 and using Fig. 12.5, instantaneous unit at relays 3 must be set at k times the maximum far-bus fault current of 1064.6. Using k = 1.2, the setting would be 1277.5 or 1278 A. This gives good coverage for the line compared to the close-in fault of 5495.7 A. The percent coverage is not linear on a loop system as it would be on a radial line, where the coverage would be

$$\%\ \text{Instantaneous coverage} = 100\left(\frac{I_{CI} - I_{IT}}{I_{CI} - I_{FB}}\right) \tag{12.1}$$

where I_{CI} is the close-in fault current, I_{FB} the far-bus fault current, and I_{IT} the instantaneous unit setting current. Thus for the loop the coverage will be less than this value of

$$100\left(\frac{5495.7 - 1278}{5495.7 - 1064.6}\right) = 95.2\%$$

possibly about 85 to 90%. Since the setting of 1278 is greater than the reverse or near-bus current of 877.31, a nondirectional instantaneous unit can be used. However, a directional unit is required at the opposite end, at breaker 6, unless IT is set above 1.2 × 1065 = 1278 A.

For breaker 8 at bus J, set the instantaneous unit at 1.2 × 1481.8, or 1778 A, and a nondirectional type is applicable.

For breaker 9, set the instantaneous unit at 1.2 × 1435.2, or 1722 A. Since the near-bus current is 1481.8 A and slightly larger than the far-bus current, a nondirectional type might be used since the value is less than the setting. However, the margin for transients and errors is less, so a directional type should be considered.

For breaker 10, set the instantaneous unit 1.2 × 1028.3, or 1234 A. This must be a directional type since the near-bus current is 1311 A. A nondirectional type with a setting of 1.2 × 1310.6, or 1573 A could be applied with good coverage of the line.

These instantaneous trip units supplement the time-overcurrent protection to provide fast operation over part of the line sections. They are shown dotted on Fig. 12.7.

Often, these units do not reach as far as was possible in the example, and in some cases offer limited protection for the maximum fault condition and none for minimum faults. Thus their application becomes marginal from a protection coverage standpoint. Still, they can provide fast clearing for the heaviest close-in faults and for the nondirectional type add very little to the total cost.

The operating times of the time-overcurrent relays can be reduced with instantaneous units by coordinating at their pickup point rather than at the far bus. In other words, the CTI for relay 3, Fig. 12.7, would be used at the instantaneous pickup point, the vertical dotted line to the right of 8. This becomes more difficult where the instantaneous unit reach varies considerably with fault-level changes from system changes. This can be programmed into the computer relay setting programs.

12.9 SHORT-LINE APPLICATIONS

Short lines are low in miles or kilometers, but are better defined electrically for overcurrent protection purposes by considering the relative values of the close-in (CI) and far-bus (FB) fault currents. Figure 12.8 shows these two faults. A short line is when Z_L is low compared with the source impedance Z_S. Then I_{CI} almost equals I_{FB} and current magnitude differences between the two faults do not provide a good indication for fault location. For a long line Z_L is large relative to Z_S, to provide a significant and measurable difference between the two faults.

Thus for the short lines, essentially fixed time discrimination must be used to coordinate between the relays at bus G and those at bus H. On a radial feeder this means that the operating time is largest nearest the source end and minimum at the load end. On loop systems, coordination is possible only where a significant magnitude difference exists between the close-in and far-bus faults around the loop. If the difference does not provide the means of fast clearing for close-in faults and delayed tripping for far-bus faults, coordination around a loop is impossible. In these cases, pilot protection, quite often pilot wire protection, must be used for the primary protection. This type is discussed in Chapter 13.

Time-overcurrent relays can be used for back-up protection, recognizing that they will not coordinate somewhere around the loop. This point of miscoordination should be selected at (1) a point of least probable fault incident, and/or (2) a point in the system where the minimum system disturbance would result. With highly reliable pilot protection, the backup would seldom be called on for protection.

12.10 NETWORK AND SPOT NETWORK SYSTEMS

High-density load areas such as exist in major metropolitan areas, commercial buildings, and shopping centers are served by a low-voltage grid network supplied from at least two relatively independent supply sources. Spot networks are unit substations with two step-down transformers, with primaries connected to separate power sources. In both types the secondaries are connected through circuit breakers or network protectors into the secondary grid. The protection is built into the protectors for automatically (1) disconnecting the supply transformer or circuit from the secondary network for faults, (2) opening

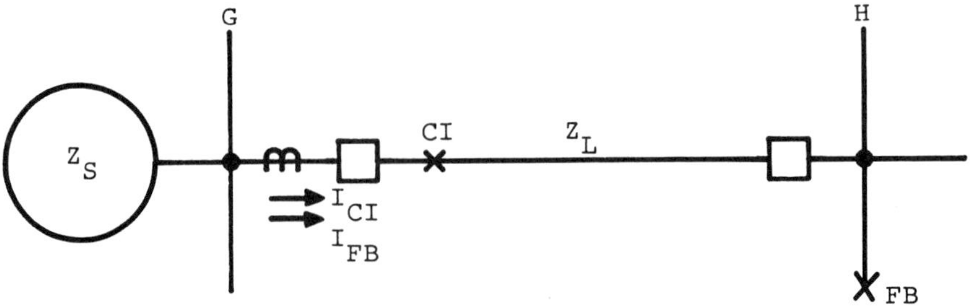

For three-phase faults:

$$I_{CI} = \frac{1}{Z_S}$$

$$I_{FB} = \frac{1}{Z_S+Z_L}$$

Fig. 12.8 Close-in and far-bus faults on line Z_L for defining short versus long lines.

the circuit when the supply or source voltage is lost, and (3) closing when the source is restored and the system source and network voltages are within designated limits. These are highly specialized areas of design and protection and beyond the scope of this book.

12.11 DISTANCE PROTECTION FOR PHASE FAULTS

The basic characteristics and fundamentals of distance relays have been covered in Chapter 6. This type of protection is applied almost universally for phase protection of lines at 69 kV and above. The "mho" characteristic of Fig. 6.11b is used most commonly. The major advantages are: (1) fixed reach as a function of the protected line impedance and so independent of system operating and fault levels over a very wide range, (2) ability to operate for fault currents near or less than maximum load current, and (3) minimum to no transient overreach. They are more complex and costly than overcurrent relays. They are applicable at the lower voltages but not widely used except for special problems, such as load and fault current magnitudes close together.

A minimum of two zones are necessary for primary protection because of the impossibility of determining if the far-bus fault is within or outside the line section. Both zones operate instantaneously, but one (zone 2) is delayed by the CTI, to provide coordination. A fixed timer T2 is used. It has been customary over many years to apply a third zone toward backup protection of the remote line(s).

The common practice in the United States has been to use separate distance units for the several protection zones. This is in contrast to distance relays that use a single distance measuring unit initially set for zone 1 reach. If the fault persists, the reach is extended by switching to zone 2 ohms after T2 time delay, then after T3 to zone 3 ohms. Both designs provide good protection. Separate units provide the comfort of redundancy since for faults in the zone 1 primary reach area, all three distance units will operate. Thus zones 2 and 3 are backup for failure of the zone 1 unit. The switched types do not provide this backup but are more economical.

These zones and typical settings are illustrated in Fig. 12.9. Part a shows the zones at several locations. Typically, zone 1 is set for 90% (range 85 to 95%) of the positive-sequence line impedance, zone 2 approximately 50% into the next adjacent line, and zone 3 approximately 25% into the adjacent line beyond. Where this is possible, zones 2 and 3 provide backup for all the adjacent line at operating times of T2 and T3.

Figure 12.9b shows the operating circles for the three zones at bus G, breaker 1 (solid) and at bus H, breaker 2 (dotted), plotted on the R-X diagram. The several lines are shown at their respective r + jx positions. The relays operate when the ratio of fault voltage to current falls within the circles. Load impedance [Eq. (6.2)] normally falls in the general areas shown.

On long lines, where large mho operating circles can include the load areas, the restrictive characteristics illustrated in Fig. 6.12c, d, or e are used. They provide a long reach in the fault area with quite restricted reach in the load areas.

The operating circles must be set so as not to operate on any system swings from which the system can recover. Such swings occur after a system disturbance, such as faults, sudden loss of generation or load, or from switching operations. This is discussed later. These swings may also require application of the restricted operating characteristics.

Zone 1 at each end of the line provides the most desirable protection: simultaneous high-speed operation for the middle 80% of the line section. This can be increased to 100% only with pilot relaying.

Backup protection as suggested in Fig. 12.9 is ideal and seldom obtainable. In practice, most buses have multiple lines of different lengths and with power sources at their remote ends. A typical example is illustrated in Fig. 12.10. The relays at breaker 1, bus G protecting line GH look into lines HR and HS extending from bus H. Where line HR is short and line HS is long, zone 2 set for 50% of line HR will cover only a small percentage of line HS. Setting for 50% of line HS would result in possibly overreaching and miscoordinating

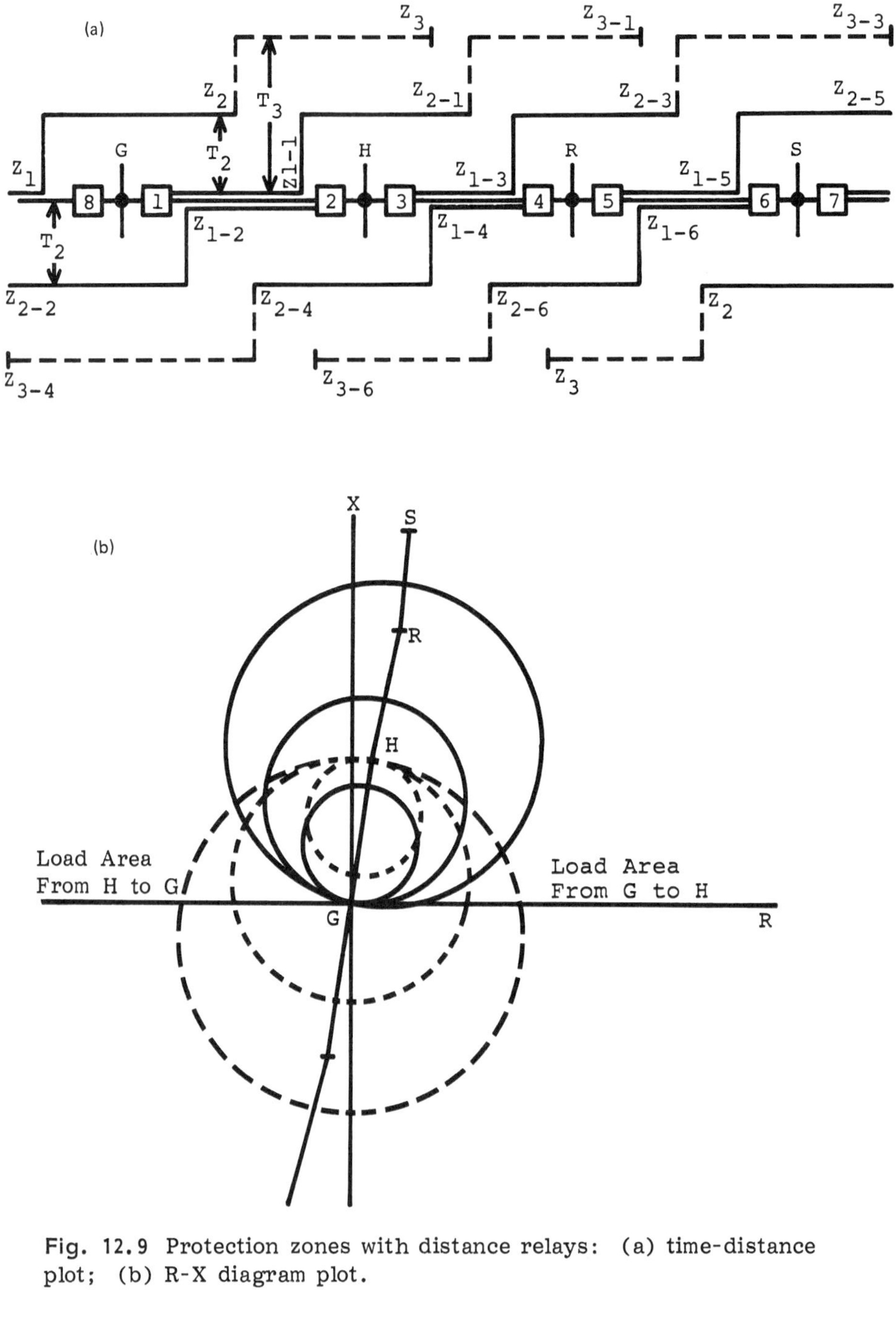

Fig. 12.9 Protection zones with distance relays: (a) time-distance plot; (b) R-X diagram plot.

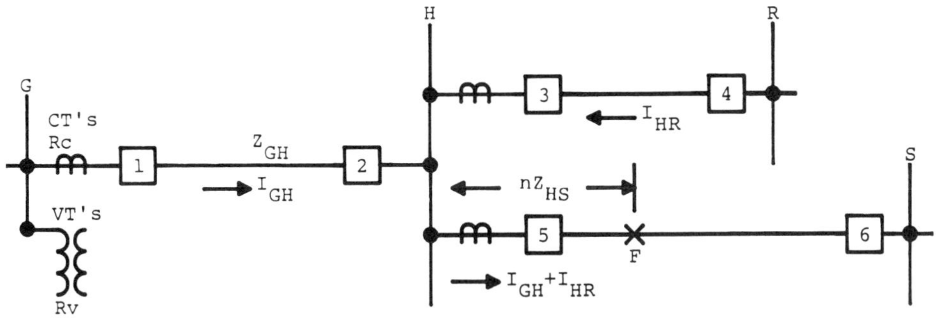

Fig. 12.10 Protection for multiple lines and infeed at a remote bus.

with Z2 of line HR unless T2 time was increased. This problem is multiplied with other lines of different lengths extending from bus H. However, the reach will not be as far as indicated because of the "infeed effect." Fault current from other lines will cause relays at 1 to under reach. This effect can be seen by considering a solid three-phase fault at F. With $V_F = 0$, the relays at 1 receive current I_{GH}, but the bus G voltage is the drop $Z_{GH}I_{GH} + nZ_{HS}(I_{GH} + I_{HR})$. Thus the relays at 1 "sees" an apparent impedance of

$$Z_{Apparent} = \frac{Z_{GH}I_{GH} + nZ_{HS}(I_{GH} + I_{HR})}{I_{GH}}$$

$$= Z_{GH} + nZ_{HS} + \frac{I_{HR}}{I_{GH}}(nZ_{HS}) \qquad (12.2)$$

This is a larger value by the third term, $(I_{HR}/I_{GH}(nZ_{HS}))$, than the actual impedance, which is

$$Z_{GH} + nZ_{HS} \qquad (12.3)$$

As a result, relay 1 set to a value of the actual impedance of Eq. (12.3) would not see fault F; in other words, relay 1 "under reaches" as a result of the fault contributions from other lines connected to bus H. Setting the relays for the apparent impedance value has the danger of overreaching and miscoordination when the infeeds are removed or changed by system operation.

The infeed term can be quite large and also variable when a large part of the current fault is supplied via the other circuits. The relay underreach will approach the remote bus but not quite reach it, so primary protection is not inhibited. This is true for a two-terminal line without infeed taps, such as that shown in Fig. 12.10.

12.12 DISTANCE RELAY APPLICATIONS FOR TAPPED AND MULTITERMINAL LINES

A typical line of this type is shown in Fig. 12.11. The tap at T may be a transformer at or near the line, so that Z_{TR} would be the impedance from the tap plus the transformer bank impedance. Sometimes the tap ties through Z_{TR} to a bus, as shown in the figure. The tap may serve a load, so that negligible fault current is supplied through it to line faults, or it may tie into a fault source at R, as shown dotted. Another variation is illustrated in Fig. 12.12.

The fundamentals for setting distance relays on these types of lines for primary phase fault protection are as follows:

1. Set zone 1 for k times the lowest actual impedance to any remote terminal for Fig. 12.11-type circuits, or for k times the lowest apparent impedance to any remote terminal for the special case of Fig. 12.12. k is less than 1, normal 0.9.
2. Set zone 2 for a value greater than the largest impedance, actual or apparent, to the remote terminals.
3. The zone 2 timer (T2) must be set so as not to cause misoperation when any terminal is out of service to cause the distance unit to overreach.

For some arrangements of circuits, these requirements can make primary protection quite difficult and/or limited. In Fig. 12.11, consider that tap T is very near bus G, so nZ_{GH} is small and $(1 - n)Z_{GH}$ large with Z_{TR} very small. Then zone 1 at breaker 1, bus G must be set at 90% of $(nZ_{GH} + Z_{TR})$, which is a very small value compared with $(1 - n)Z_{GH}$. Thus high-speed coverage of the line is almost negligible.

On the other hand, if the tap is a load transformer where Z_{TR} is high relative to Z_{GH}, zone 1 at breakers 1 and 2 can be set for 90% of the line to provide good high-speed line protection.

If R is a load tap in Fig. 12.11, with negligible current to line faults, distance relays (and overcurrent) are not applicable at breaker 3, and basically not necessary, as opening breakers 1 and 2 terminates the line fault. The worst case is a small generator or source connected to R, large enough to maintain a fault on the line, but not large enough to support fault-detecting relays. In other words, the impedance to a line fault from bus R is very large and approaches infinity.

For the case of Fig. 12.12, current can flow out R terminal for an internal line fault near the H bus. Thus distance or directional relays

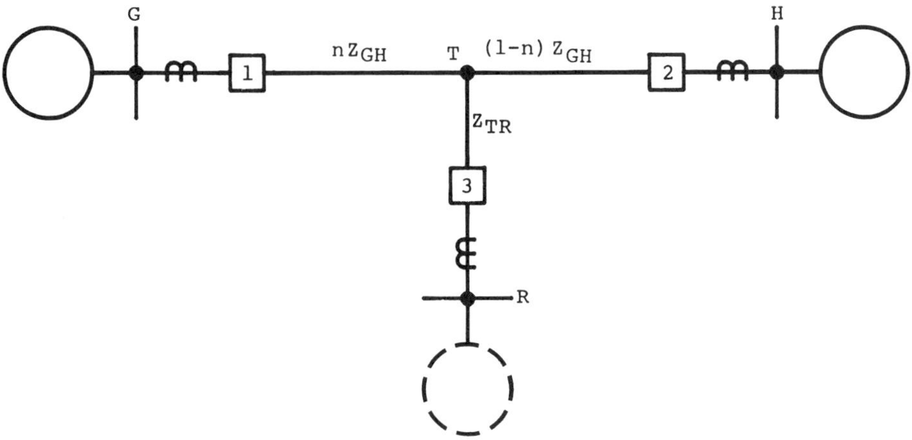

Fig. 12.11 Typical tapped or multiterminal line.

at breaker 3 see the internal fault as external for no operation until after breaker 2 has opened.

Thus protection of tapped and multiterminal lines is more complex and requires specific data on the line impedances, location and type of tap or terminal, and fault data with current distributions for the various system and operating conditions. Most often, except for small transformer load taps, these types of lines are protected best by pilot relaying.

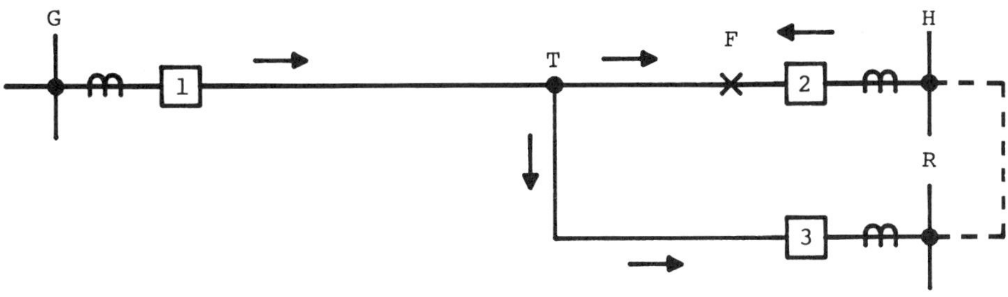

Fig. 12.12 Multiterminal line where fault current can flow out at one terminal for internal faults.

12.13 VOLTAGE SOURCES FOR DISTANCE RELAYS

Three-phase voltage is required and provides reference quantities to which the currents are compared. For phase distance relays, either open-delta or wye-wye voltage transformers (VTs) or coupling capacitor voltage devices (CCVTs) can be utilized and connected either to the bus or to the line being protected. Both are widely used and the decision is economic as well as involving use of line-side CCVTs for radio-frequency coupling for pilot and/or transfer trip relaying.

These voltage sources involve fuses/ primary and secondary for VTs and secondary for CCVTs. These fuses should be generously sized, carefully installed, and well maintained, as a loss of one or more phase voltages may result in an undesired, unwanted relay operation. Where this is of great convern, overcurrent fault detectors can be added to supervise the trip circuit of the distance relays. For loss of voltage in the absence of an actual fault, the overcurrent units would not operate. The disadvantages are the need for additional equipment and the loss of the feature of distance relays operating for fault levels less than maximum load.

Recently, detectors measuring V_0 but no I_0 have been used to supervise the relays. This requires wye grounded-wye grounded voltage sources.

12.14 DISTANCE RELAY APPLICATIONS IN SYSTEMS PROTECTED BY INVERSE-TIME-OVERCURRENT RELAYS

Zone 1 distance relays can be applied to provide increased instantaneous protection over that obtainable by instantaneous overcurrent relays. This is especially helpful where there is a large variation in fault level because of system and operating changes. Distance relays can provide around 90% instantaneous coverage of the line section independent of system and fault-level changes.

To coordinate with existing inverse-time characteristics, a zone 2 distance relay can be set into or through the adjacent line section, with an inverse-time-overcurrent relay as a timer. The distance relay torque controls the overcurrent units; that is, the overcurrent relay cannot operate until the distance relay operates. This permits setting the overcurrent relays below maximum load. Hence this application is valuable where fault and load currents are close together or the maximum possible load is greater than minimum fault current.

12.15 GROUND-FAULT PROTECTION FOR LINES

The setting and coordination procedure for ground relays, both inverse time and instantaneous overcurrent, is the same as discussed above for phase relays. Ground relay taps for the inverse-time units must be set above the tolerable zero-sequence unbalance on the line, and single-phase-to-ground fault data are used. These relays are connected to operate on zero-sequence current from three paralleled CTs, or from a CT in the grounded neutral.

In distribution, industrial, and commercial systems where line fuses are utilized, the ground relays receiving $3I_0$ must coordinate with fuses receiving line I_a, I_b, and I_c currents. While the line current is equal to $3I_0$ in radial feeder circuits, fuses must be set above load and short-time inrush. This may require a high tap setting on the ground relays. The example of Section 12.5 discussed ground relay applications in these types of circuits.

For subtransmission and transmission lines, generally 34.5 kV and above, the systems are usually solidly grounded at many of the stations and so are multigrounded systems. The system unbalance at these levels tends to be quite low. Fuses are not used in the lines. Thus separate ground relays can be set very sensitively relative to phase relays. In this area either ground distance or ground overcurrent relays are utilized.

12.16 DISTANCE VERSUS OVERCURRENT FOR GROUND FAULTS

Ground distance relays are not in wide use in the United States. This may seem unusual in view of the common application of distance relays for phase faults, and that most line faults (80 to 90%) involve ground. However, ground distance relays are more complex, and the directional overcurrent relays in most cases can provide quite satisfactory ground-fault protection in multigrounded systems at less cost.

Before exploring this further, first consider ground distance relays. As discussed in Chapter 4, zero-sequence voltage is maximum at the fault and decreases toward the grounding point. As a result, the ratio of system voltage to current, such as V_0/I_0, does not provide a suitable indication of fault location and so cannot be used. As a result several methods are in use. Either the voltage or the current can be modified to provide a relay whose setting is proportional to the positive-sequence line impedance and the ratio of the line Z_0/Z_1. Another type operates on the principle that at the fault $V_{0F} + V_{1F} + V_{2F} = 0$. This relation is reproduced at the relay location by compensators. The modified V_0 is used as an operating quantity, and modified $V_1 + V_2$ as restraint. For single-phase-to-ground faults within the set reach, V_0 operating is greater than $V_1 + V_2$ restraint to trip. Faults outside the set zone provide restraint greater than operating quantity.

While ground distance relays have the advantage of a relatively fixed reach, the setting for zone 1 is generally 80 to 85%, since Z_0 cannot be determined as accurately as can Z_1. The reach is also affected by mutual impedance and arc resistance. These effects are covered in a later section.

The advantages for overcurrent relays are: (1) relative independence of load, (2) generally a larger margin between the close-in and far-bus faults since the X_0 value of lines is around 3 times X_1, (3) system unbalance low, (4) fuses not used in the lines, and (5) current level for ground faults tends to be more constant than for phase faults, as ground source transformers are seldom switched.

Perhaps the most important reason for the general use of overcurrent relays for ground protection is the general practice of utilizing pilot protection on lines of 115 kV and above. As has been indicated, the ideal protection is high-speed simultaneous operation at all terminals for 100% of the line faults. Pilot relaying provides this possibility. Ground distance relays provide high-speed simultaneous protection for only the middle 60 to 70% of the line. Ground overcurrent relays are usable and provide quite sensitive protection in the most widely applied pilot systems.

12.17 DIRECTIONAL SENSING FOR GROUND OVERCURRENT RELAYS

For multigrounded loop systems, inverse-time-overcurrent units are directional. Instantaneous units may or may not need to be directional, depending on the relative magnitude of the maximum near-bus and maximum far-bus fault currents. The criteria are given in Section 12.6. Generally, separate directional units are used to torque control the fault-detecting overcurrent units. A reference quantity is required to which the various line zero-sequence currents can be compared to determine if the line current is flowing into the line (tripping direction) or out of the line into the bus (nontripping direction). This reference is known as the polarizing quantity. Zero-sequence current or voltage, or negative-sequence voltage and current, are used for polarizing and directional sensing.

The most common type is current polarizing utilizing $3I_0$ from a CT connected in the grounded neutral of a two-winding wye-delta power transformer. The transformers are as shown in Fig. A4.2-1a and b. Voltage polarization uses $3V_0$ across a grounded wye-delta voltage transformer as shown in Fig. 4.4.

None of the other two-winding transformer bank connections in Fig. A4.2-1 can be used for current polarization, as zero-sequence current cannot pass from the N_0 bus to any system fault via the transformers.

With the three-winding transformer banks of Fig. A4.2-3, a single CT in the grounded wye neutral of diagrams a and c can provide polarizing similar to the grounded wye-delta two-winding banks. For type b banks, CTs with ratios inversely proportional to the voltage ratio and connected in parallel can be used. The connections are shown in Fig. 12.13. This provides polarization for the ground relays on both the high (H) and low (M) systems. For ground faults in the high-voltage (H) system, zero-sequence current flows up the left-hand neutral and is equal to the sum of the current from the low-voltage (M) system, which flows down the right-hand neutral, and that circulating in the tertiary (L winding). For ground faults in the low-voltage (M) system, zero-sequence current flows up the right-hand neutral and is the sum of the high-voltage (H) current down the left-hand neutral and the current circulating in the tertiary (L winding). On a per unit basis the net current from the paralleled CTs is always in the same direction and equivalent to that circulating in the tertiary. An alternative to the two paralleled CTs is to use a single CT inside the tertiary, as shown. Usually, this tertiary CT is not available unless specified.

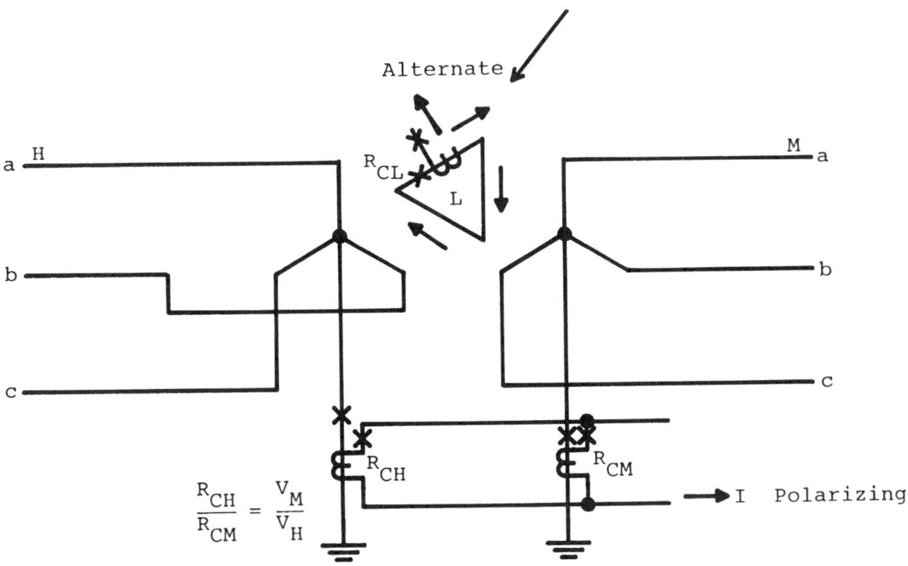

Fig. 12.13 Ground relaying directional polarizing from a three-winding transformer bank.

If a load or a fault source is connected to the tertiary three CTs, one in each winding, paralleled CTs may be required to cancel out positive and negative sequence to provide only $3I_0$ current for polarizing. The validity of the foregoing current flows are easily checked from the zero-sequence network given in Fig. A 4.2-3b.

12.18 POLARIZING PROBLEMS WITH AUTOTRANSFORMERS

The neutral of the autotransformer is seldom usable for polarizing. For a grounded neutral autotransformer without tertiary, zero-sequence current passes through the autotransformer and the current in the neutral reverses for faults on the two sides. Thus the neutral is not suitable for polarizing.

For ungrounded autotransformers with a tertiary, zero sequence passes through the bank, and the current flowing in the tertiary reverses for faults on the two sides. Thus the tertiary is not suitable for polarizing.

For the more common grounded autotransformers with tertiary, an example was given in Chapter 4 (Section 4.14). Expanding this in general terms will provide the criterion for use of these autotransformers for ground relay polarizing. The transformer and zero-sequencing networks are shown in Fig. 12.14. For a phase-to-ground fault on the high (H) side (Fig. 12-14b), the currents shown are the per unit zero-sequence values. With I_{0H} in amperes at kV_H, the currents through the autotransformer are:

H winding: kI_{0H} amperes at kV_H

M winding: $pI_{0H}\dfrac{kV_H}{kV_M}$ amperes at kV_M

L winding: $(1 - p)\, I_{0H}\dfrac{kV_H}{\sqrt{3}\, kV_L}$ amperes at kV_L

This is the current inside the delta tertiary winding. The amperes follow Kirchhoff's laws, so the current up the neutral is

$$I_N = 3\left(kI_{0H} - pI_{0H}\frac{kV_H}{kV_M}\right)$$

$$= 3I_{0H}\left(k - p\frac{kV_H}{kV_M}\right) \text{ A} \tag{12.4}$$

Examining Eq. (12.4), the distribution factors k and p are less than 1, normally k is greater than p, and the voltage ratio kV_H/kV_M is greater than 1. Thus the product p kV_H/kV_M can be greater than k, equal to k, or less than k for a specific autotransformer and connected system. In most cases the product is larger than k, so that I_N is negative and the current flows down the neutral.

Consider now a phase-to-ground fault on the low-voltage (M) winding. This is shown in Fig. 12.14c. With I_{0L} in amperes at kV_M, the currents through the auto are:

H winding: $tI_{0L}\dfrac{kV_M}{kV_H}$ amperes at kV_H

M winding: sI_{0L} amperes at kV_M

L winding: $(1 - t)I_{0L}\dfrac{kV_M}{\sqrt{3}\,kV_L}$ amperes at kV_L

This is the current inside the delta tertiary winding. Again by Kirchhoff's laws, the current up the neutral is

$$I_N' = 3\left(sI_{0L} - tI_{0L}\frac{kV_M}{kV_H}\right) = 3I_{0L}\left(s - t\frac{kV_M}{kV_H}\right) \text{ A} \qquad (12.5)$$

In this equation the distribution factors s and t are less than 1, normally s is greater than t, and kV_M/kV_H is less than 1. As a result, the ratio $t(kV_M/kV_H)$ will always be less than s, so that I_N' is always positive and the current is up the neutral.

With the high probability of neutral current flowing down for high-side faults and up for low (M)-side faults, the netural is not a reliable polarizing reference. In a given case where current does flow up the neutral for the high-side fault, care should be taken that this is correct for all possible variations of the low-voltage system impedance, Z_{0M}. Switching or changes during operation could result in a neutral current reversal.

Usually, the autotertiary current is in the same direction for high- or low-voltage system faults, and so can be utilized as a polarizing current. As shown in Fig. 12.14, one CT measures the I_0 current circulating. If load is connected or a generator tied into this tertiary, a

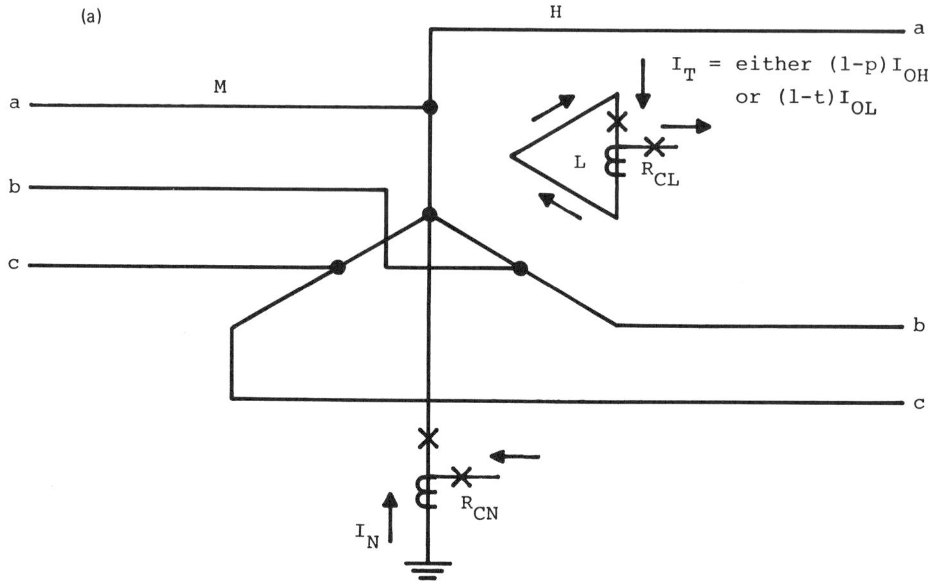

Fig. 12.14 Ground relaying directional polarizing from an autotransformer bank with tertiary (grounded neutral CT seldom usable for polarizing): (a) three-phase diagram showing possible polarizing CT's; (b) zero-sequence network and current distribution for high-side (H) fault; (c) zero-sequence network and current distribution for medium (low) fault.

CT in each of the three winding is required, with the secondaries in parallel to cancel out positive and negative sequence and supply only $3I_0$ for relaying.

However, as documented in Chapter 4, the tertiary I_0 current can reverse for faults on one side of the bank, thus making the tertiary unsuitable for polarizing. The reversal results when a relatively small MVA autotransformer with a negative branch in the equivalent circuit is connected to a very solidly grounded system. In other words, in Fig. 12.14b, if Z_M is negative and larger than Z_{0M}, the current $(1 - p)I_{0H}$ will be negative — a tertiary current reversal. Similarly, in Fig. 12.14c, if Z_H is negative and larger than Z_{0H}, current in the tertiary $(1 - t)I_{0L}$ is negative, to reverse the tertiary current. The equivalent-circuit negative for either Z_M or Z_H is small on the transformer base but can become large when transferred to the larger system base, where either Z_{0M} or Z_{0H} of the interconnected system is small because of the multigrounds.

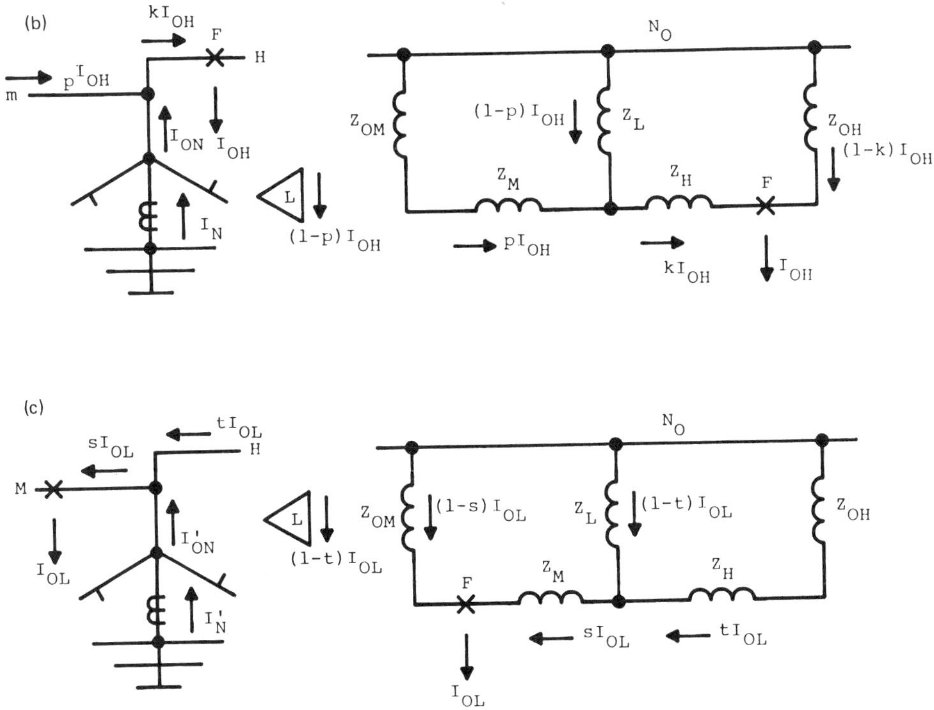

Fig. 12.14 (continued)

Zero-sequence voltage can be utilized for polarizing, but it can be quite small. At least it does not reverse. The system equivalents Z_{0M} and Z_{0H} are always positive in all cases and the zero-sequence voltage at the bank terminals consists of the drops across these impedances. Perhaps a large series capacitor might cause a problem in this area, but this has not been experienced.

12.19 VOLTAGE POLARIZATION LIMITATIONS

Current polarization is preferred when it is available and can be used. Voltage polarization with $3V_0$ can be used at terminals with or without a grounding bank. At stations with large, solidly grounded transformer banks, the $3V_0$ levels should be checked for the remote faults to assure adequate magnitude. This concern with zero-sequence voltage

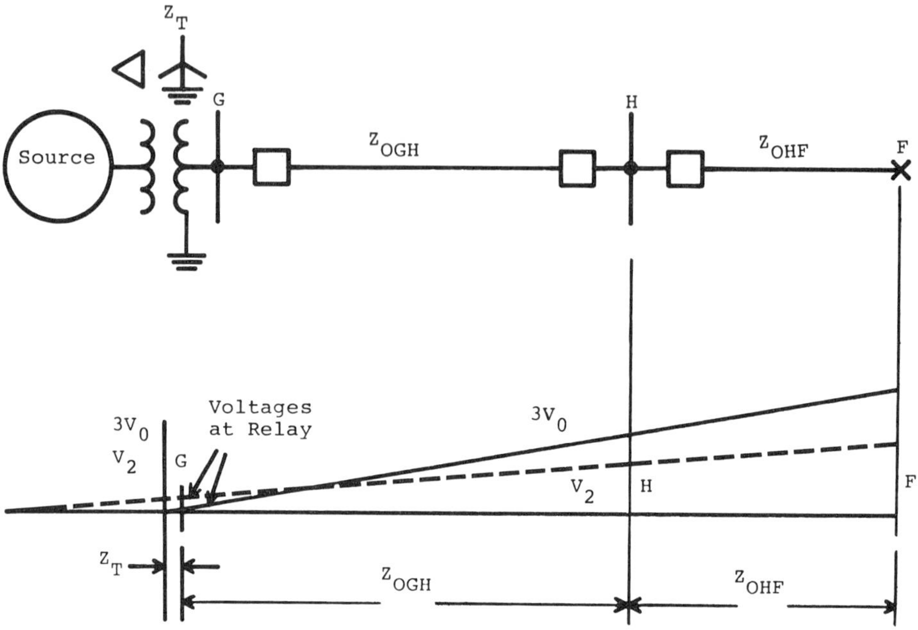

Fig. 12.15 Typical voltage profiles of V_2 and $3V_0$ for ground faults.

is illustrated in Fig. 12.15. With $3V_0$ maximum at the fault, the value at a grounding bank may be quite small. Z_T of the transformer is small because of the large bank, and Z_0 of the lines is large because of distance and the $\pm 3Z_1$ factor. Modern ground relay directional units are quite sensitive, so this may not pose a problem unless long lines and/or settings are made well into the remote adjacent lines.

12.20 DUAL POLARIZATION FOR GROUND RELAYING

A common practice is to use current and voltage polarization jointly. Many ground relays are "dual polarized." Some utilize two separate directional units, one voltage polarized and the other current polarized. They operate in parallel, so that either one can release the overcurrent units. Other designs use a hybrid circuit with one directional unit that can be energized by current or voltage or both. These types offer flexibility of application to various parts of the system.

12.21 GROUND DIRECTIONAL SENSING WITH NEGATIVE SEQUENCE

An excellent alternative is the use of negative sequence to operate the directional unit. Zero-sequence $3I_0$ is still used for the fault-detecting overcurrent elements, with V_2 and I_2 for the directional unit. It is applicable generally but particularly useful at autotransformer stations, with problems as discussed above and where mutual induction is involved, as discussed in Section 12.22.

It is also applicable at stations where only open-delta VTs exist or where VTs are available only on the opposite side of wye-delta power transformers. In these cases, neither current nor voltage polarization is available.

Negative-sequence units are easy to field check for correct connections and operation, a problem with the relays connected in neutral circuits. Interchanging two phases to the negative-sequence current or voltage filter or inputs produces positive operation on balanced voltages or load currents.

The V_2, I_2 quantities for ground faults generally are lower than the $3V_0$, $3I_0$ values and should be checked. V_2 can be smaller or larger than $3V_0$, as indicated in Fig. 12.15. General experience indicates that in most instances the V_2 and I_2 values are sufficient to operate available sensitive relays. Computer fault studies should be programmed to include these negative-sequence values, and this method of polarizing has considerable merit in present-day power systems.

12.22 MUTUAL COUPLING AND GROUND RELAYING

Zero-sequence coupling between lines that are parallel for part or all of their length can induce false information in the unfaulted circuit and cause protection problems in both. The mutual impedance Z_{0M} can be as high as 50 to 70% of the self-impedance Z_0. To review this effect, Fig. 12.16 shows the general case of paralleled circuits. The voltage drops are

$$V_{GH} = Z_{0GH}I_{0GH} + Z_{0M}I_{0RS} \tag{12.6}$$

$$V_{RS} = Z_{0RS}I_{0RS} + Z_{0M}I_{0GH} \tag{12.7}$$

For network analysis these equations or the equivalent network shown in Fig. 12.16 b can be used. The ideal or perfect transformer with a 1:1 ratio is used, so that the mutual Z_{0M} is in both circuits but

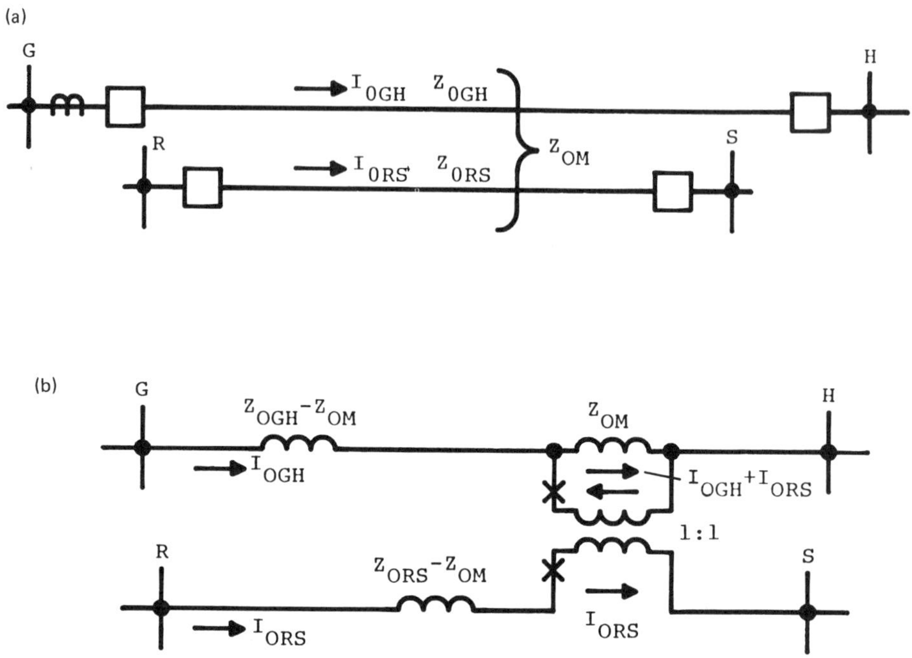

Fig. 12.16 General case of paralleled lines with mutual coupling: (a) the mutual coupled lines; (b) equivalent network for a.

isolated from direct electrical connection. The drops across this equivalent are the same as those of Eqs. (12.6) and (12.7).

If three or more lines are paralleled, there will be a Z_{0M} value between each pair. That value times the respective current in the coupled line is added to Eqs. (12.6) and (12.7). Additional ideal transformers with the respective Z_{0M} values would be added to the equivalent network shown in Fig. 12.16b.

Lines GH and RS can be at the same or different voltages. With Z_{0M} determined in ohms from Carson's formula as modified for zero sequence, the per unit value from Eq. (2.15) is

$$Z_{0M} = \frac{MVA_{Base} \times Z_{0M(ohms)}}{kV_G kV_R} \quad pu \tag{12.8}$$

when line GH operates at kV_G and line RS at kV_R. If both lines are at the same voltage, the bottom term is kV^2, as in Eq. (2.15).

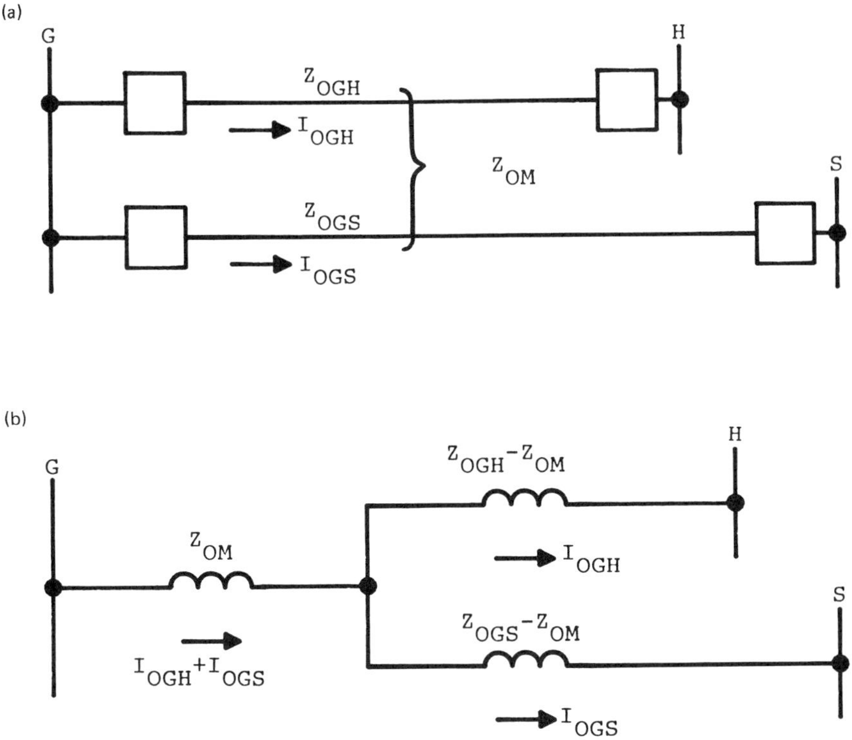

Fig. 12.17 Parallel lines bused at one terminal: (a) the mutual coupled lines; (b) equivalent network for a.

Quite often the paralleled lines are bused at either one or both terminals. These circuits and their equivalent networks are shown in Figs. 12.17 and 12.18.

Where the two bused and coupled lines (Fig. 12.18a) have the same impedance Z_{0GH}, the equivalent impedance between bus G and H from Fig. 12.18b will be

$$Z_{eqGH} = Z_{0M} + \frac{1}{2}(Z_{0GH} - Z_{0M})$$

$$= \frac{1}{2}(Z_{0GH} + Z_{0M}) \tag{12.9}$$

If $Z_{0M} = 0.7Z_{0GH}$, then,

$$Z_{eqGH} = 0.85Z_{0GH}$$

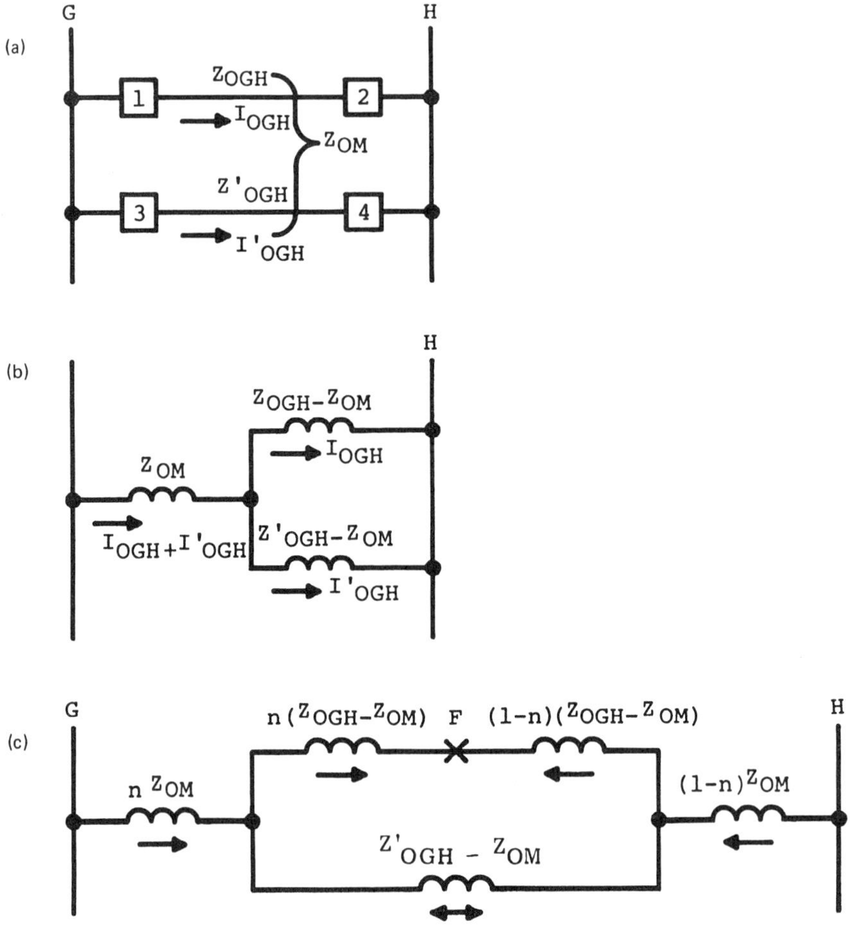

Fig. 12.18 Paralleled lines bused at both terminals: (a) the mutual coupled lines; (b) the equivalent network for faults at the terminals; (c) the equivalent network for faults or taps in one-line sections.

Thus with currents flowing in the same direction, the mutual increases the impedance between the two buses, which would be $0.50Z_{0GH}$ without the mutual.

The discussion above has emphasized mutual coupling by zero sequence only. There is a coupling by the positive- and negative-sequence currents, but this induction effect is usually less than 5 to 7% and hence has negligible effect on protection. As a result, negative-sequence directional units can be applied in most cases for correct directional sensing.

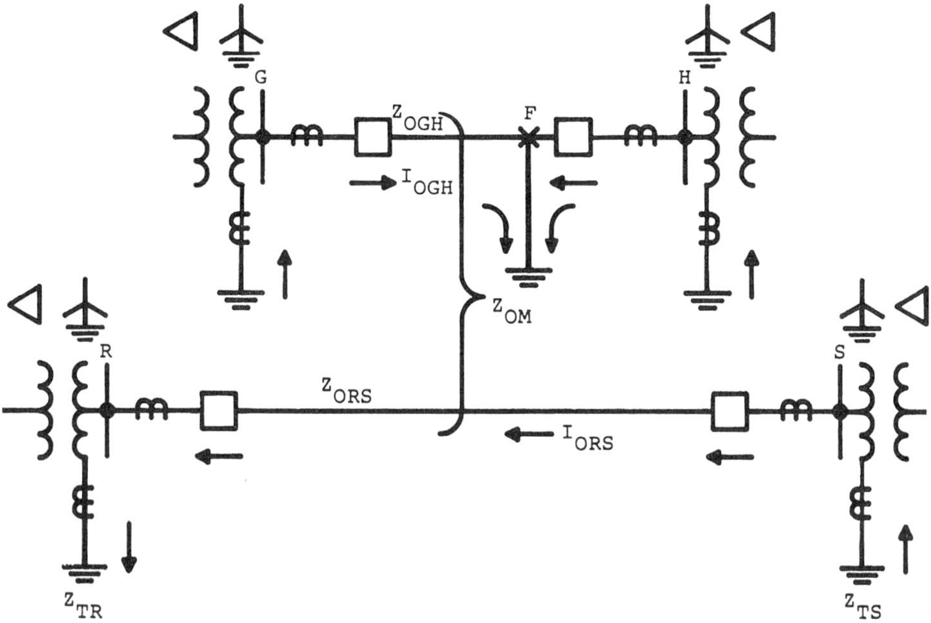

Fig. 12.19 Fault current flow in lines with mutual coupling.

In the systems shown in Fig. 12.19, the zero-sequence network for the line GH system is isolated electrically from that for the line RS system. The electromagnetic coupling acts as a transformer, so that current in one line tends to circulate current in the coupled line. A ground fault at or near one end will cause currents to flow as shown. In the faulted line GH, current up the two neutrals and out into the line is in the operating direction for the directional ground relays at both G and H. They should operate to open up the breakers at G and H. Before that happens, current I_{0RS} is induced in line RS. The directional unit at S will operate as the current is up the neutral and out into the line. The directional unit at R also operates, as current in and down the neutral is equivalent to up and out the line. The magnitude can be large enough to operate the overcurrent units, with the result that either breaker at R or S or both may be tripped incorrectly. The magnitude of the current in line RS is

$$I_{0RS} = \frac{Z_{0M} I_{0GH}}{Z_{TR} + Z_{0RS} + Z_{TS}} \tag{12.10}$$

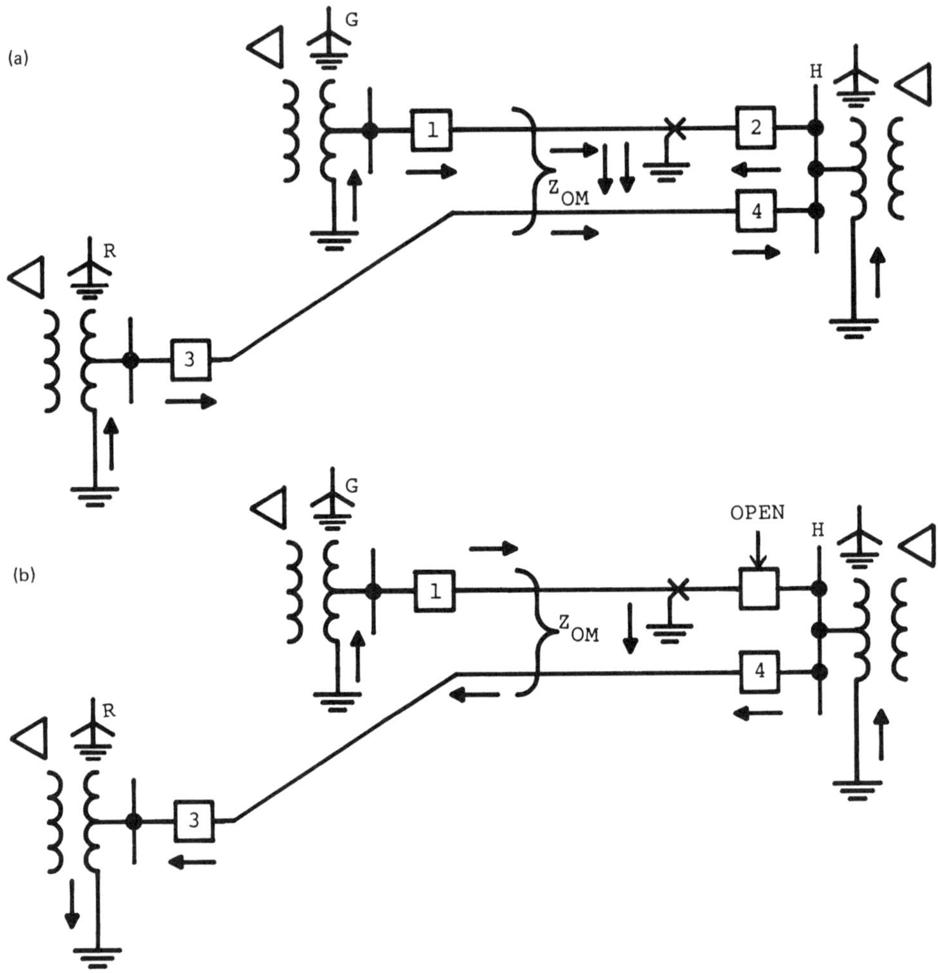

Fig. 12.20 Zero-sequence electrical circuit isolation by switching, causing mutual to reverse current: (a) initial fault near breaker 2; (b) breaker 2 trips by instantaneous relays.

If the G and R buses and/or the H and S buses are in close proximity, so that the grounded neutral CT in the two banks can be paralleled, correct polarizing can be obtained. The faulted line current up the neutral will be greater than the induced current down the neutral.

The $3V_0$ voltage for the case of Fig. 12.19 also will not provide correct polarizing. Correct directional sensing can be obtained by the use of negative sequence, but the induced $3I_0$ may be sufficient to operate the overcurrent units. For Fig. 12.19 this would possibly cause the S terminal ground relays to misoperate.

Circuit switching in a zero-sequence electrically interconnected system can result in zero-sequence isolation and induced circulating currents. An example is shown in Fig. 12.20. Lines GH and RH are mutually coupled and are connected together at bus H. For the ground fault close to breaker 2, the zero-sequence current flows are shown in Fig. 12.20a. This is defined by the equivalent circuit of Fig. 12.17. If the fault is in the zone of the instantaneous units at 2, they will operate fast to open breaker 2. This now isolates the two circuits electrically, as shown in Fig. 12.20b, and until breaker 1 opens, an induced current circulating in the line RH system reverses the line current and causes current to flow down instead of up in the transformer neutral at station R. Zero-sequence directional units, either current or voltage polarized, would operate at both ends to indicate an internal fault on line RH. Thus incorrect tripping of either breaker 3 or 4 or both can occur. This is a "race" with a possibility that breaker 1 will open to clear the fault before relays 3 and/or 4 can operate. This situation has explained intermittent incorrect tripping in several systems, which was solved by applying negative-sequence directional sensing.

Multiple mutual coupling can cause current reversals in systems that are not zero-sequence electrically isolated. An example is shown in Fig. 12.21. The two lines G to H are both coupled with the line R to S. Station H is a very large and solidly grounded station. Thus very high ground-fault currents flow from H to the fault near station G. With the system interconnected as shown, the normal zero-sequence current flow to the fault should be up the neutral at station S and over the line from S to R to join the current at R and continue via line RG to the fault. However, the very large currents coupled to line RS can result in the reversal of line RS current from its normal anticipated direction. This reversal is shown in the figure. Again, zero-sequence current or voltage directional units would operate at both R and S, to suggest a fault on line RS with possible incorrect tripping before the faulted line relays can operate and the breakers clear the fault.

There is a very high probability of this condition occurring more frequently as environmental and economics force the use of existing rights-of-way for more lines. As a result, mutual coupling is an increasing problem. Hence it is very important that all mutuals be included in the fault program and carefully considered in protection.

As has been suggested, negative-sequence directional sensing can provide correct indications in many cases, but it does not solve all problems. In Fig. 12.21 negative-sequence current would flow from S to R, so a negative-sequence directional unit at S senses a fault in line RS and with overcurrent sufficient to operate the zero-sequence

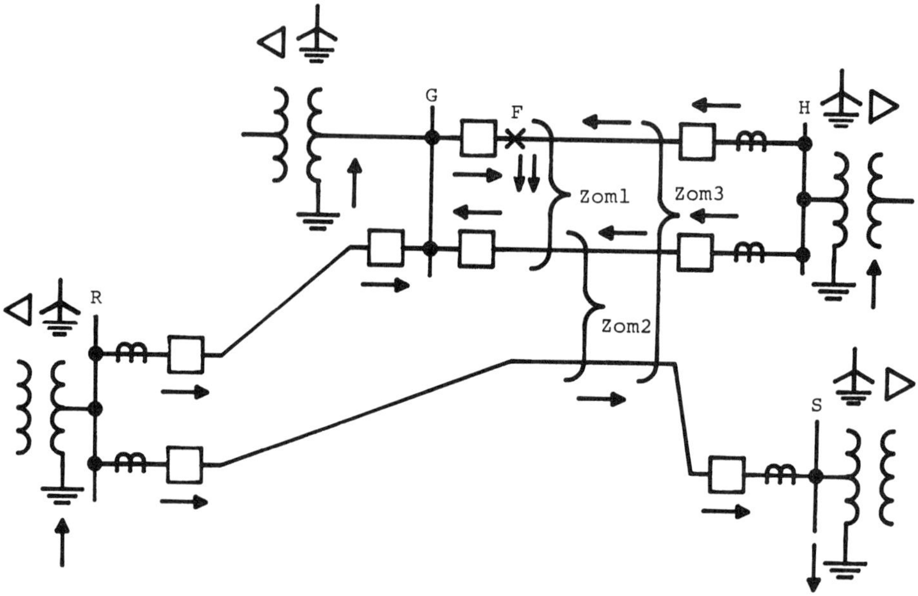

Fig. 12.21 Current reversal by mutual without zero-sequence electrical circuit isolation.

fault detectors, relay S tends to operate. At R the negative-sequence directional unit blocks this relay operation.

An incorrect operation with negative-sequence ground directional relays occurred to open breakers 3 and 5 for a ground fault out on line from breaker 1 (Fig. 12.22). The several lines were coupled as shown. The ground fault induced sufficient current to circultate in the secondary loop to operate the overcurrent zero-sequence units. Negative sequence from the industrial plant flows through both paralleled lines, so both directional units at the industrial plant substation closed. The result was incorrect tripping of both breakers 3 and 5, and unfortunately, dumping of an important load. This particular problem was corrected by raising the overcurrent settings at breakers 2, 3, 4, and 5 above the induced current, as the exposure to line 1 was short, so the higher setting still provided good protection for faults on lines 2-3 and 4-5.

12.23 GROUND DISTANCE RELAYING WITH MUTUAL INDUCTION

The mutual coupling will produce either a higher apparent impedance and relay underreach or a lower apparent impedance and relay

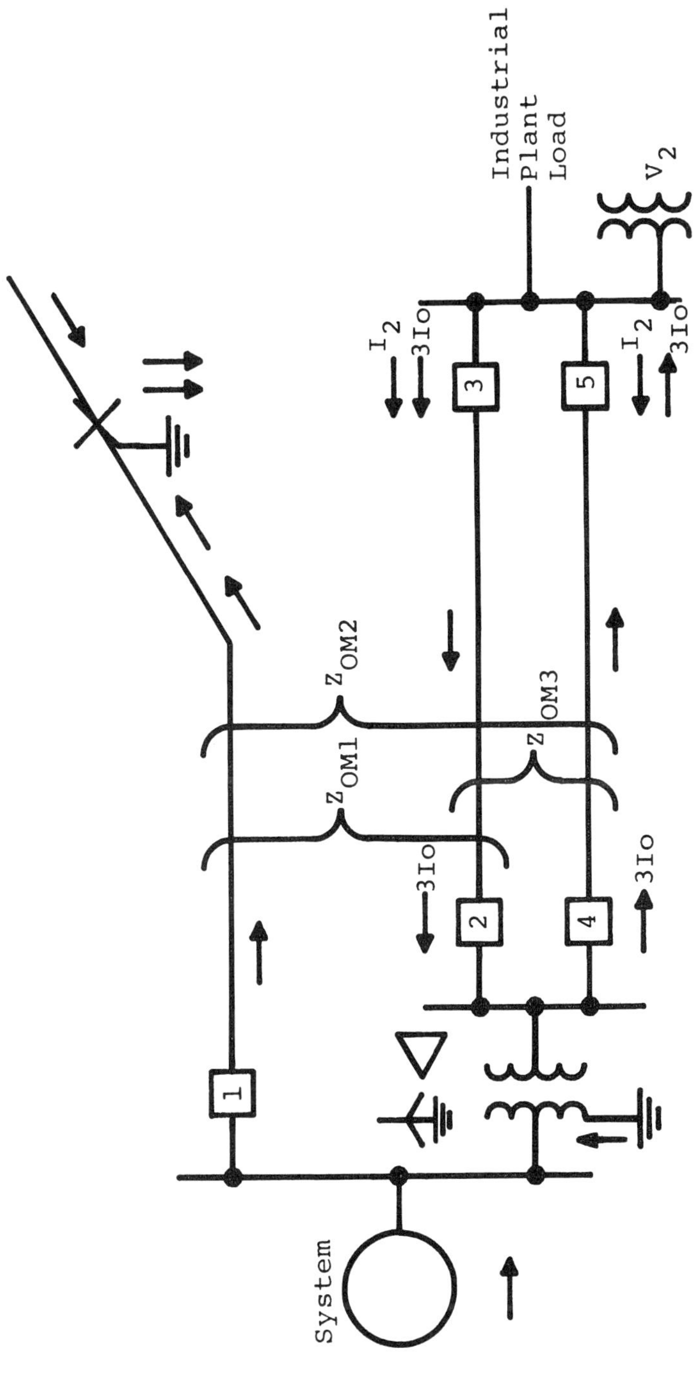

Fig. 12.22 Case of misoperation with mutual coupling and negative-sequence directional sensing.

overreach. Current flowing in the paralleled line in the same direction as the fault current in the protected line will cause a higher apparent impedance [Eq. (12.6)] and underreach. Current flowing in the opposite direction produces a lower apparent impedance and overreach.

Consider ground distance relays at breaker 3 (Fig. 12.18a) with zone 1 set for 85% of the line 3-4 impedance, without any mutual compensation. Zone 1 will reach approximately 70%, with current in line 1-2 flowing from 1 to 2 as shown. If breaker 4 opens first, the current in line 1-2 reverses to cause zone 1 at 3 to reach out and cover approximately 100% of the opened line.

Zone 2 should be set for approximately 150% of the line section to assure that it will cover all on the line for primary protection with current in the parallel flowing in the same direction. With this setting, zone 2 of breaker 3 should coordinate correctly with breaker 2, zone 2.

As a general rule, mutual compensation is not recommended for ground distance relays. If it is used, care must be taken to assure proper operation when the paralleled line current changes direction from that for which the compensation was set.

12.24 FAULT RESISTANCE AND RELAYING

Ground faults on lines usually result as flashover of the insulators caused by lightning induction or failure of the insulators. The current path for ground faults include the arc, tower impedance, and the impedance between the tower foundation and earth (tower footing resistance). When used, ground wires provide a parallel path to the earth return. Phase faults on lines are often the result of high wind swinging the conductors close enough to arc over.

The possibility of significant fault resistance thus exists, with the potential of affecting or inhibiting relay performance. Arcs are resistance, tower and ground wires are complex impedances, and tower footing impedance essentially resistance. The resistance of arc between 70 and 20,000 A has been expressed as

$$R_{arc} = \frac{440 \times l}{I} \quad \Omega \tag{12.11}$$

where l is the arc length in feet and I is the current in the arc. This can be used as R_F in Figs. 4.13 through 4.16. Actual arcs are variable, tending to start at a low value, build up exponentially to a high value, and then the arc breaks over, returning to a lower value of resistance. A typical value of 1 or 2 Ω exists for around 0.5 s, with peaks of 25 to 50 Ω later. Tower footing resistance at various towers can run from less than 1 Ω to several hundred ohms. With so many

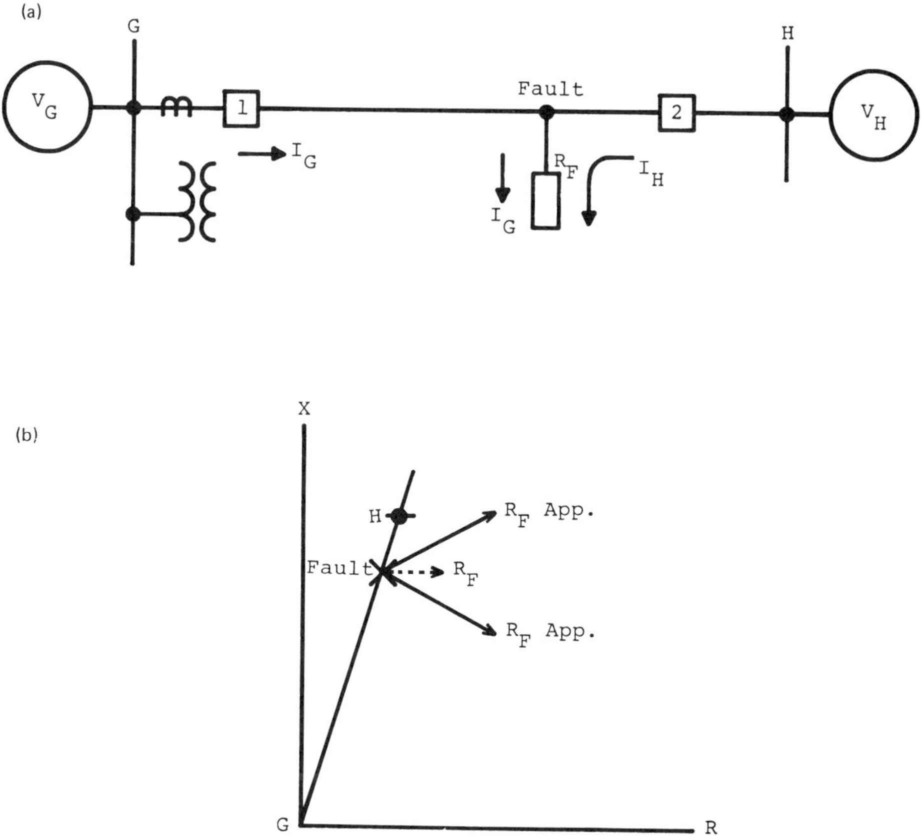

Fig. 12.23 Typical line and representation of fault impedance: (a) single-line diagram; (b) representation on the R-X diagram.

variables, it is quite difficult to represent fault resistance with any degree of certainty or realistically.

It has been a common assumption that the fault impedance is all resistance and so can be represented on an R-X diagram as a horizontal value, such as that shown dotted in Fig. 12.23b. With this concept the reactance characteristic as shown in Fig. 6.11f has been promoted as excellent coverage for large values of R_F and to provide more protection than a circular characteristic such as the mho type. If the arc drop is really a major factor, this document on the R-X diagram is correct only for a radial feeder where distance relays are seldom used, or for an unloaded loop line. However, loop lines generally are carrying load, so that V_G and V_H are not in phase. Real power across a circuit such as GH is transmitted by the angular difference between the source voltages. Thus when a fault occurs, actual currents supplied from the two ends are out of phase. Thus for relays at breaker 1, station G, the voltage drop measured includes the $I_H Z_F$ drop, which is not compensated for since only I_G flows through the relay. Thus relay 1 "sees" (1) a higher impedance than the actual impedance of the line and Z_F, and (2) an apparent reactive component even if Z_F is pure resistance. As a result, R_F will appear as shown either tilted up when V_H leads V_G, or tilted down when V_G leads V_H. Looking from H toward G, relay H will "see" a higher impedance when V_G leads V_H, and a lower value when V_H leads V_G.

Thus if V_G leads V_H in Fig. 12.23, so that the higher apparent value of R_F tilts downward, it is quite possible that an external fault in the system to the right of bus H will have this value of R_F fall into the operating area of the zone 1 reactance unit at G. Unless the reach of the G zone 1 reactance unit is reduced, this unit will overtrip incorrectly. The mho unit will not so overreach for all practical operating conditions that have been studied. Although the mho type characteristics may provide less coverage, it is much more secure.

The tilt of the apparent R_F upward will prevent operation for both types of characteristics for faults near the set point. At worst the underreach by the tilt and larger magnitude could delay operation of zone 1 until after breaker 2 at bus H opens for faults near the set point. Then R_F becomes more nearly horizontal and lower in value. Zone 2 and 3 units provide additional backup. With pilot relaying, fault resistance seems to be of less importance, as the overreaching units can operate on most values of impedance practically encountered. The one exception are tree faults on EHV lines.

The modern practice of minimum right-of-way trimming has resulted in arcs of 100 or more ohms on EHV lines as the trees grow into the voltage gradient breakdown zone. This usually results during heavy load periods when the midspan line sag is maximum. These are difficult to detect. The best possibility is detection by sensitive ground overcurrent relays. Distance relays are generally insensitive to these faults.

The general conclusion from the performance reported is that fault resistance (except the EHV tree type) presents very few problems to protection. The tree problem is minimized by proper maintenance of the rights-of-way.

12.25 LONG EHV SERIES-COMPENSATED LINE PROTECTION

Series capacitor banks are frequently inserted in long power lines to reduce their total impedance, thereby permitting the transmission of

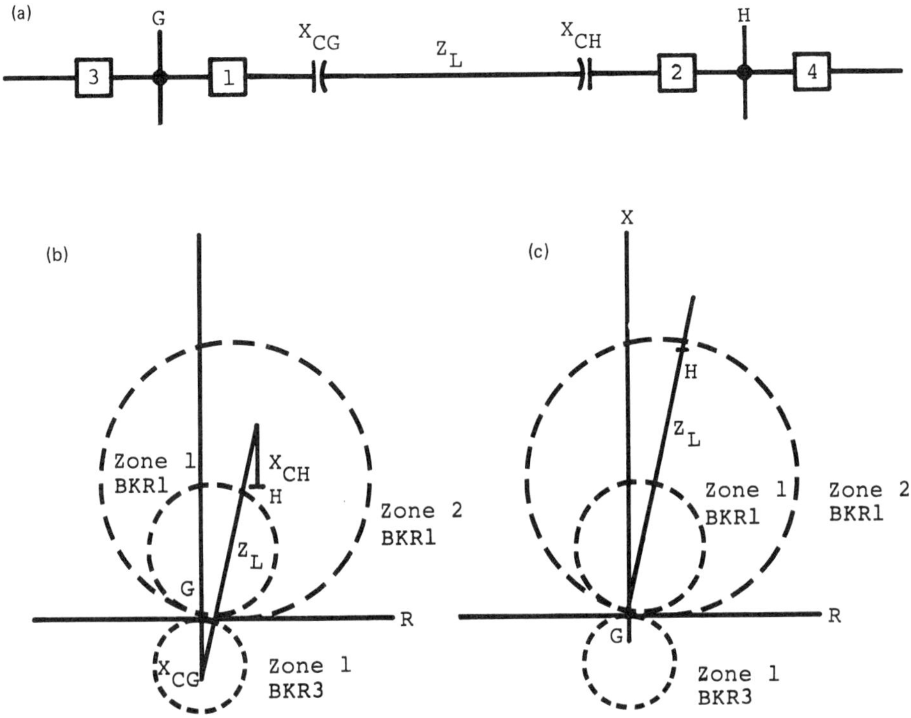

Fig. 12.24 Protection problems encountered with series-compensated transmission lines: (a) a typical series-compensated line with series capacitor banks at the station terminals; (b) R-X and distance relay diagrams with capacitor banks in service; (c) R-X and distance relay diagrams with capacitor banks out of service.

more power with less loss and for higher system stability limits. The capacitors may be inserted in the line at any point, but for economic reasons they are often installed in the line at the terminal stations, as illustrated in Fig. 12.24. In this case directional mho distance relays can operate undesirably, as shown in part b. The distance relays at 1 set to protect line GH cannot "see" faults in capacitor X_{CG} and part of the line, as this impedance area falls outside the circular operating characteristics. However, the distance relays at 3 for the protection of the line to the left of bus G do undesirably operate for faults in the capacitor and part of line GH. As a result, distance relays are not recommended for these lines unless one is willing to assume that the capacitor protective gaps will always short out the capacitor for faults in this problem area. This is a reasonable gamble, as the gaps are fast and reasonably reliable. Zone 1 solid-state relays may require a short-time delay; electromechanical relays probably will not.

When the gaps flash, the line becomes as shown in part c. To cover the long line without the capacitors, a long reach or large ohm setting is required for zone 2 (and zone 3 if used). This setting may be a problem with load and/or recovery stability limits, so that one of the restricted distance characteristics of Fig. 6.7 will be required. Also, with the capacitors in service, zone 2 or zone 3 reach may cause coordination problems with relays at breaker 4 at station H.

Phase comparison pilot systems are preferred for these systems. Current-only systems will not have the problems of correctly sensing internal and external faults. Distance fault detectors do not have to be directional, so can be applied and set to operate for all faults in the line with or without the capacitor banks in service.

The segregated phase comparison system is especially applicable to these types of lines. This is outlined in Chapter 13.

12.26 BACKUP: REMOTE, LOCAL, AND BREAKER FAILURE

Backup protection has been included throughout the protection chapters in two basic forms: redundancy and remote. Redundancy is the additional protection provided in the primary protection zone and sometimes extending into the adjacent system. Examples are: three separate phase relays instead of two or a single unit to serve all three phases, phase relays backing up ground relays, timed overcurrent and distance timed zones backing up the instantaneous or pilot relays, and in EHV and UHV, two separate pilot systems. The degree of independence of the various protection schemes is a measure of the redundancy available. Very high redundancy is obtained for the EHV and UHV systems by operating the two pilot systems from separate CTs and VTs or CCVTs, separate trip coils in the circuit breaker, and separate station batteries. If separate batteries are not available, separate fused supplies from the single battery are used. A similar arrangement is sometimes used for the voltage supply. This provides the maximum redundancy that is economically practical.

Remote backup is the overlapping of the primary relays in one protection area into the adjacent areas. Thus in Fig. 12.25a, relays 1 at station S, relays 5 at station T, and relays 8 at station R should provide backup to the relays and breaker 3 at station G for faults on line GH. That is, if breaker 3 does not open for these faults, the remote breakers 1, 5, and 8 must open to clear line GH faults.

As has been indicated, this becomes quite difficult or impossible, especially for faults near station H because of the infeed effect from the currents of the other lines. This tends to reduce the current or

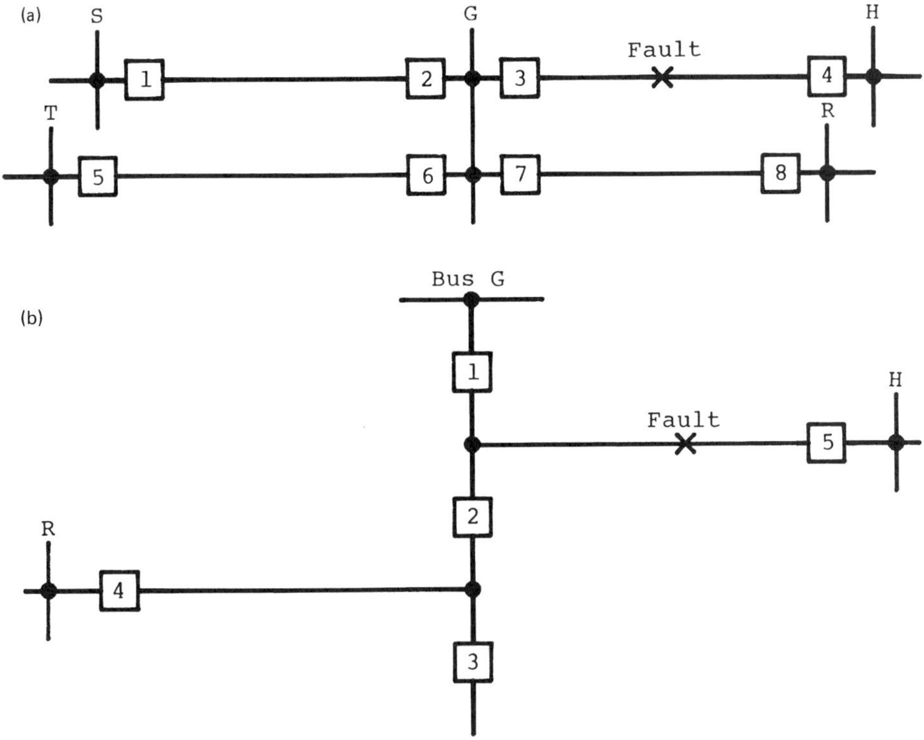

Fig. 12.25 Power system configurations to illustrate backup protection; (a) backup on a single bus system; (b) backup on a ring or breaker-and-half bus system.

increase the impedance seen by the remote relays. If the remote relays can "see" line GH faults, the operating time may be relatively slow because of coordination requirements necessitated by the other lines out of service. Sometimes these difficulties can be solved by sequential remote tripping. If one remote terminal can operate on backup, its removal may result in a redistribution of fault current sufficient to operate the other remote backup relays.

With the advent of EHV and UHV systems in the recent past, two problems developed. One was that initially the higher-voltage circuit breakers were subject to a higher failure incidence than previously encountered. The other was that system stability required much faster backup. Thus breaker failure—local backup systems were applied.

Instead of opening breakers 1, 5, and 8 in Fig. 12.25a for a failure of either relays or breaker 3 to open on line GH faults, local backup

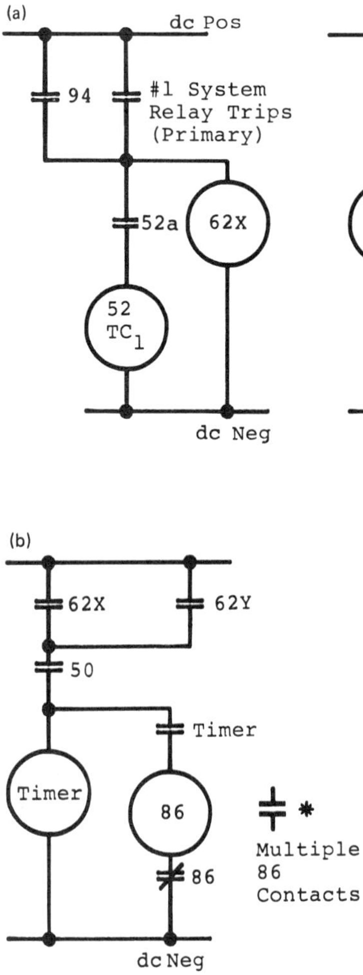

Fig. 12.26 Typical breaker failure — local backup dc (schematic): (a) typical circuit breaker trip systems and auxiliaries for breaker failure — local backup (where breakers have double trip coils, 94 relay coil and contact is omitted and the second trip coil 52 TC_2 connected in place of the 94 relay coil); (b) typical contact logic for breaker failure — local backup; (c) typical solid-state logic for breaker failure — local backup.

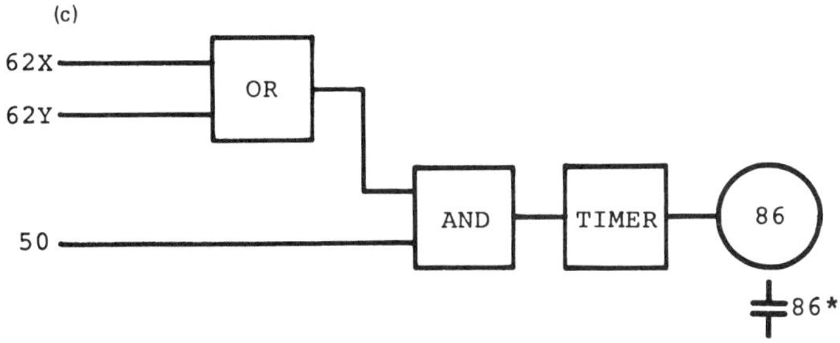

*86 multi-contact relay trips all local breakers necessary to clear fault and optionally to transfer trip the remote end terminal

Fig. 12.26 (continued)

would open the local breakers 2, 6, and 7. This can be done with minimum delay; typical times as low as 150 to 250 ms are in use.

Breaker failure is initiated when the primary protection operated but the breaker does not open. Local backup is identical except that a second and independent primary relay system should be provided to cover the failure of the relays to operate. As indicated, very high redundancy in relay systems is used for higher-voltage systems.

When local backup is applied to lower-voltage systems, careful attention should be given to adequate relay redundancy to cover all possible relay failures. Remote backup by its separate location provides 100% redundancy for the faults within their operating range.

Remote backup is still important as an additional and "last resort" protection, and is necessary with ring or breaker-and-a-half buses. This is illustrated in Fig. 12.25b. For faults on ling GH, breakers 1 and 2 at bus G are tripped. If breaker 1 fails to open, local backup would trip the necessary breakers off bus G (not shown). If breaker 2 fails to open, local backup would trip breaker 3, but the fault is still supplied through breaker 4 at station R. Thus breaker 4 must be opened. This can be accomplished by remote backup operation of the relays at 4. With the infeed through breakers 1 and 3 removed, the possibility of relays 4 "seeing" faults on line GH is increased. Transfer trip of breaker 4 by the locak backup at G is another possibility. This requires information that breaker 2 is in trouble and not breakers 1 or 3.

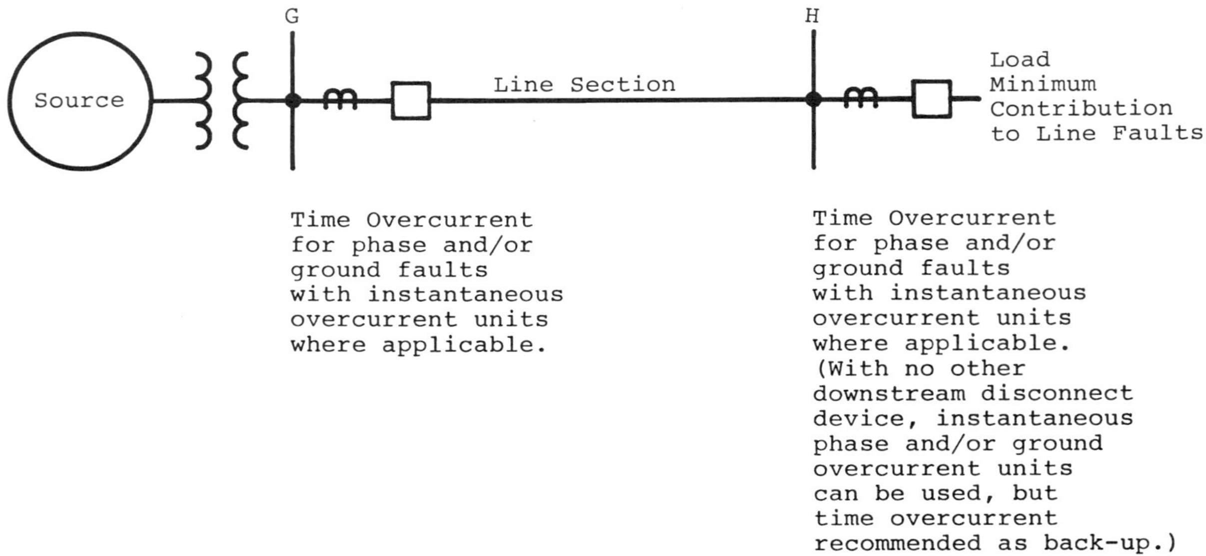

Fig. 12.27 General summary and suggested protection for radial lines and feeder circuits. Fuses and/or reclosers, etc., may be used either downstream or upstream in which case the time and instantaneous overcurrent relays must be selected and set to coordinate with these other devices. Complete coordination may not always be possible under all possible operating conditions. Where minimum fault current is less than the maximum possible load current, apply a distance controlled overcurrent relay.

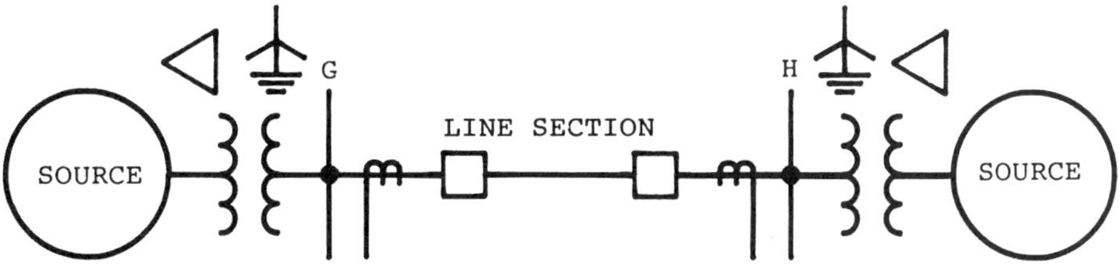

Fig. 12.28 General summary and suggested protection for loop lines and circuits. Protection at both terminals:

1. Up to 34.5 to 69 kV – directional time overcurrent for phase and/or ground with nondirectional or directional instantaneous overcurrent where applicable.
2. 34.5 kV to 115 kV — directional distance (two or three zones) for phase, same as (1) above for ground, alternate-ground distance.
3. 69 kV to 230 kV — pilot relaying (Chapter 13) for phase and ground as primary protection, backup as (2) above.
4. 230 kV and above — two pilot systems (primary and secondary) (Chapter 13) for phase and ground, additional backup as (2) above.
5. For very short lines at any voltage level — pilot wire or pilot type (Chapter 13) for phase and ground, backup per (1) or (2) above but coordination with other protection may not be completely possible.
6. Multiterminal and tapped lines — pilot-type relays generally required unless load type impedance and/or transformer connections will permit proper discrimination for external faults.

Typical local backup-breaker failure schemes are shown in Fig. 12.26. The trip circuits of the two redundant (independent) relay systems are indicated as 1 and 2, primary and secondary. Where the secondary system is a pilot type, its operating times can be equal to or faster than the primary system. With a single trip coil on the breaker, the primary directly trips the breaker by energizing the trip coil $52TC_1$ and the secondary energized relay 94, which in turn operates $52TC_1$. With double trip coils, 94 is omitted and the secondary system directly energizes the second trip coil $52TC_2$.

At the same time, auxiliary relays 62X and 62Y are energized. Either relay operates a timer supervised by a 50 relay. This is a low-pickup nondirectional instantaneous overcurrent relay with a high dropout ratio. It is connected generally in two phases and ground. The phase units should be able to carry the maximum load at their low settings. This 50 relay monitors the current through the breaker and provides a final check on breaker current flow, with fast opening to stop the timer should the breaker open late. The operation of the timer energizes a multicontact auxiliary (86) relay, which is the only one that is manually resettable. The 86 relay contacts initiate tripping of all the breakers locally required to clear the fault, and may also initiate a transfer trip signal to the remote terminal(s) to assure that the relays have operated correctly for the internal fault. There are several variations of the schemes and connections for the various bus configurations.

12.27 SUMMARY: TYPICAL PROTECTION FOR LINES

Typical protection recommended and generally applied for the protection of lines is outlined in Fig. 12.27 for radial lines and feeders, and in Fig. 12.28 for loop lines. Once again there are many variables, circumstances, operating conditions, and local practices that can modify these general suggestions and practices. The pilot systems indicated are discussed in Chapter 13.

BIBLIOGRAPHY

See the Bibliography at the end of Chapter 1 for additional information.

Application and Coordination of Reclosers, Sectionalizers and Fuses, IEEE Tutorial Course, 80 EHO157-8-PWR, IEEE Service Center.

Blackburn, J. L., Voltage Induction in Paralleled Transmission Circuits, *AIEE Trans.*, 81, Pt. III, 1962, pp. 921–929.

Bozoki, B., J. C. Benney, and W. V. Usas, Protective Relaying for Tapped High Voltage Transmission Lines, *IEEE Trans. Power Appar. Syst.*, PAS 104, 1985, pp. 865–872.

Curd, E., and L. E. Curtis, *Procedure for an Overcurrent Protective Device Time-Current Coordination Study*, Vol. 1, No. 1 Section 110, Square D Power Systems Engineering Data, Square D. Co., Middletown, OH, 1979.

Distribution Line Protection Practices, Industry Survey Analysis, Power System Relaying Committee, *IEEE Trans. Power Appar. Syst.*, PAS 102, 1983, pp. 3279–3287.

Distribution System Protection Manual, Bulletin No. 71022, McGraw-Edison Co., Canonsburg, PA.

EHV Protection Problems, Power System Relaying Committee *IEEE Trans. Power Appar. Syst.*, PAS 100, 1981, pp. 2399–2406.

Elmore, W. A., and J. L. Blackburn, *Negative Sequence Directional Ground Relaying, AIEE Trans.*, 81, Pt. III, 1962, pp. 913–921.

Griffin, C. H., Principles of Ground Relaying for High Voltage and Extra High Voltage Transmission Lines, *IEEE Trans. Power Appar. Syst.*, PAS 102, 1983, pp. 420–432.

Guide for Protective Relaying of Consumer-Utility Interconnections, ANSI/IEEE Standard C37.95, IEEE Service Center.

Lewis, W. A., and L. S. Tippett, Fundamental Basis for Distance Relaying on a Three-Phase System, *AIEE Trans.*, 66, 1947, pp. 694–708.

Local Backup Relaying Protection, Power System Relaying Committee, *IEEE Trans. Power Appar. Syst.*, PAS 89, 1970, pp. 1061–1068.

Protection Aspects of Multi-terminal Lines, IEEE-PES Special Publication 79TH0056-2-PWR, IEEE Service Center.

Rook, M. J., L. E. Goff, G. J. Potochnry, and L. J. Powell, Application of Protective Relays on a Large Industrial-Utility Tie with Industrial co-generation, *IEEE Trans. Power Appar. Syst.*, PAS 100, 1981, pp 2804–2812.

Wang, G. N., W. M. Moffatt, L. J. Vegh, and F. J. Veicht, High-Resistance Grounding and Selective Ground Fault Protection for a Major Industrial Facility, *IEEE Trans. Ind. Appl.*, IA20, July-August 1984, pp. 978–985.

13

Pilot Protection

13.1 INTRODUCTION

Pilot protection used for lines provides the possibilities of high-speed simultaneous detection of phase- and ground-fault protection for 100% of the protected section from all terminals: the ideal primary protection goal. It is a type of differential protection where the quantities at the terminals are compared via a communication channel rather than by a direct wire interconnection of the relay input devices. The latter is impractical because of the distances between the several terminals. Like differential schemes, pilot schemes provide primary zone protection and no backup. Thus they do not require coordination with the protection in the adjacent system unless additional backup is included as a part of the pilot scheme.

This protection is applicable at all voltages. In actual practice it is usually applied to short lines at all voltages and to most lines around 69 to 115 kV and above. The application key is the importance of the circuit in the power system and the necessity for rapid clearing of faults for stability and service continuity.

This chapter outlines the fundamentals and basic operation of the several systems in common use in the United States. Today, many pilot systems are solid state, desirable in these schemes since several relay functions must be coordinated where logic circuitry is preferable and faster than relay contact logic. As a result, many variations are possible and exist, each with promoted sophistications and features beyond the scope of this book.

13.2 PILOT SYSTEM CLASSIFICATIONS

These systems can be classified in two categories:

1. *By channel use:*
   a. Channel not required for trip operations; known as *blocking systems*
   b. Channel required for trip operations; known as *transfer trip systems*.
2. *By fault detector principle:* the comparison at the several terminals of:
   a. Power flow, known as *directional comparison*
   b. The relative phase position of the currents, known as *phase comparison*

A particular scheme is usually described or identified by a combination of these two categories. The major schemes in use can be identified more specifically:

A. Directional comparison systems
   1. Directional comparison blocking
   2. Directional comparison unblocking
   3. Overreaching transfer trip
   4. Underreaching transfer trip
      a. Nonpermissive
      b. Permissive
B. Phase comparison systems
   1. "Pilot wire"
   2. Single phase comparison — blocking
   3. Dual phase comparison — unblock
   4. Dual phase comparison — transfer trip
   5. Segregated phase comparison

13.3 COMMUNICATION CHANNELS FOR PROTECTION

A variety of channels are used for protection. This is a field in itself and is outlined only as necessary for a discussion of the protective systems. Historically and today, channels generally have been the weakest link in the protection chain. Tremendous progress has been made from the early applications in the 1930s to the quite sophisticated and highly reliable equipment available today. This is a specialized field, and applications for protection should be made by specialists in this area who are familiar with the protection requirements. Good engineering for channels is mandatory for secure and dependable protection, as it is for the relay protection part.

The channels available and used in protection are:

1. *Pilot wires:* a wire circuit for the transmission of dc, 50 to 60 Hz, or audio-frequency tones. Preferably these should be owned by the user for protection, but leased lines from telephone companies have been and are in use.
2. *Power-line carrier:* radio frequencies between 30 and 300 kHz transmitted principally over the high-voltage transmission line.
3. *Microwave:* a radio signal generally between 2 and 12 GHz, transmitted by line-of-sight paths between terminals. Multiple channels exist with protection utilizing a subcarrier or audio tone modulated on a carrier.
4. *Fiber optic cable:* signals transmitted by light modulation through an electrical nonconducting cable. This technique is now in use, with promise for increased applications. The cable eliminates induction, noise and electrical insulation problems.

13.3.1 Pilot Wires: 50 to 60 Hz Transmission

One of the early and still widely used channels is a twisted pair of "telephone" wires to provide a low-voltage, low-power continuous circuit between the protected zone terminals. Preferably wires on the order of AWG 19 are desired both for mechanical strength and to provide not more than about a 2000-Ω loop for two terminals or not more than 500 Ω per leg for three-terminal applications. It is mandatory that they be twisted pairs, to minimize extraneous voltage differences between the pair from signals on other pairs in the cable and from external voltages outside the cable.

The problems that are experienced with pilot wires result from induction from lightning or a paralleled power circuit, insulation stress from a rise in station ground-mat voltage during faults, direct physical contact by lightning or with the power circuit, physical damage by insulation failure, or gunfire directed at overhead circuits.

Surge arrestors can be used for equipment surge protection as appropriate. Insulation problems generally can be detected by supervisory equipment together with good maintenance and inspection procedures. Mutual induction and station ground rise voltages can be controlled by proper design.

The best pilot-wire channel is one with adequate insulation and shielding to withstand induced voltages and/or ground station rise if possible. Grounded metallic shields with pilot cables or supporting messenger wires can provide about 50% reduction in induced voltage.

Mutual induction occurs when fault current, primarily the $3I_0$ component, flows through the transmission line that is in parallel with the pilot pair. An example is shown in Fig. 13.1a. A voltage equal to $I_F Z_M$ ($3I_0 Z_{0M}$) is induced in the pilot pair between the wires and ground. If the pilot pair is not grounded, this voltage exists across the pair as illustrated in Fig. 13.1b. This assumes uniform exposure along the line and that I_F is for a fault near or at either bus G or bus H. If the gaps flash or the insulation breaks down at one terminal, such as bus H, the full induced voltage will appear at the other end, as shown in Fig. 13.1c.

If protective gaps at both ends break down, the induced voltage causes current to circulate over the two pilot wires in parallel and ground, and thus reduce the voltage to a low level. Telephone gaps, carbon blocks, and gas discharge tubes are used. With 3-mil gaps and gas tubes, typical breakdown voltages are 400 to 500 V; 6-mil gaps have breakdown of about 850 V.

For protective relay applications it is mandatory that a mutual drainage reactor be used with the gaps or tubes. Typical connections are illustrated in Fig. 13.1d. These reactors force simultaneous breakdown of both gaps. Otherwise, the breakdown of one gap before the other will result in a momentary voltage between the pair and is sufficient to cause undesired and incorrect operation of high-speed relays. As indicated, twisted pairs are necessary to keep this voltage difference between pairs to a small value.

Station ground rise is illustrated in Fig. 13.2. The return path for a ground fault is illustrated approximately in Fig. 13.2a. If the fault is within the area of the ground mat, most or all of the return current will be through the mat to the transformer bank neutral. For faults outside the ground-mat area, much or all of the current will return through the resistance represented by R_G between the ground mat and remote ground. R_G is the summation of many parallel paths of contact between the station grounding network and a nonuniform earth. Station-ground-mat design and its grounding is a complex subject with many variables and assumptions. The objective is to keep both the impedance across the mat and to remote earth as low as possible. Physical measurement of R_G is quite difficult.

During a ground fault there will be a voltage difference between the ground mat and remote earth of kI_GR_G. This can be quite large, with high values of current and/or ground resistance.

Equipment that enters the station from outside, such as communication cables, can be subject to this voltage, as illustrated in Fig. 13.2b. All grounded shield circuits should be terminated outside the station if there is the possibility of a significant voltage difference during faults. The pilot pairs that must enter the station will be subject to this voltage rise. Although the pair may not be grounded to remote earth physically, it can be essentially at remote earth potential via the distributed capacitance of the pair to earth. Generally, the rise is expressed as the maximum $I_G(3I_0)$ current for a fault just outside the station times the measured or calculated ground resistance (R_G) This represents a maximum value probably not realized in actual practice. The insulation of the cable and associated terminating

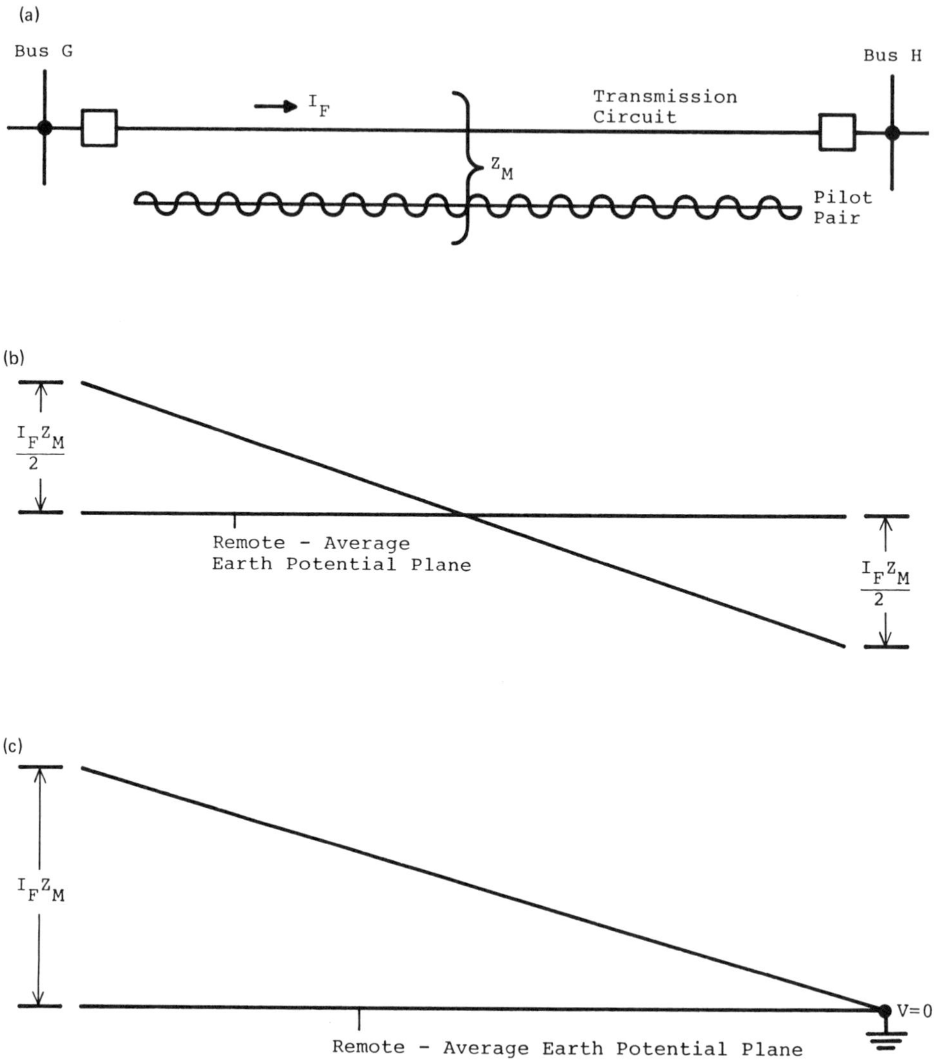

Fig. 13.1 Typical voltage induction in a paralleled pilot circuit: (a) a pilot pair paralleling a transmission circuit; (b) mutual induction voltage profile for uniform induction and pilot pairs ungrounded at both terminals; (c) mutual induction voltage profile for uniform induction and pilot pairs grounded at bus H terminal; (d) mutual induction voltage reduced by circulation over the pilot pair and ground after protective gaps flash over.

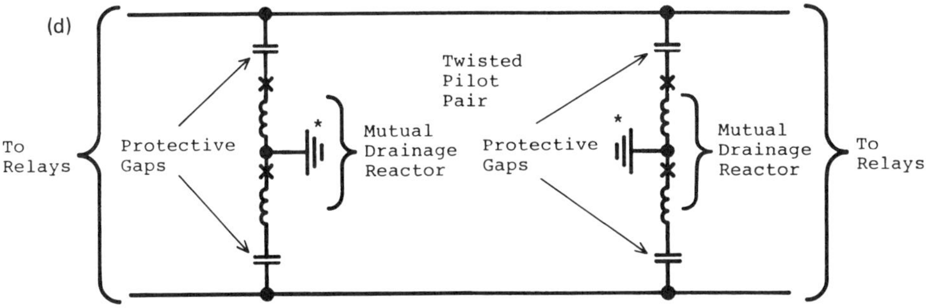

*Ground to Station Ground Mat if Station Ground Rise is Negligible. Otherwise Ground to Remote Ground.

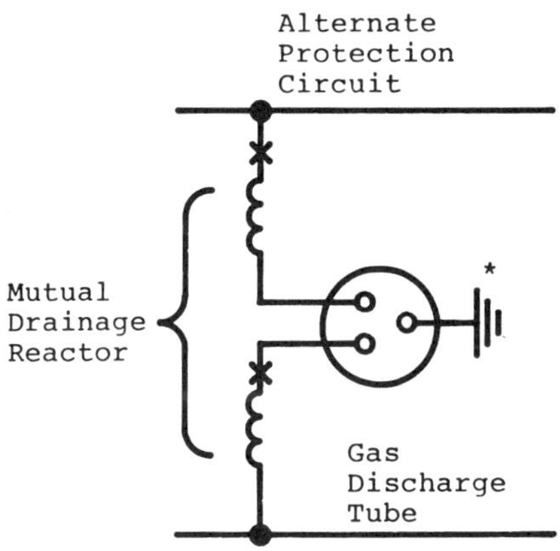

Fig. 13.1 (continued)

terminating equipment must be able to withstand this rise, or neutralizing devices must be applied.

The two types of neutralizing transformers generally used are shown in Fig. 13.2c and d. In c the primary winding is connected between the station and remote earth to sense the station ground rise. This provides a neutralizing voltage across the secondaries to that the station terminals are at the station-ground-mat voltage, while the other end is at remote ground. For multiple pilot pairs, transformers with many multiple secondaries are available.

A two-winding neutralizing reactor applicable to a single pilot pair utilized for protection is shown in Fig. 13.2d. Capacitor C1 with

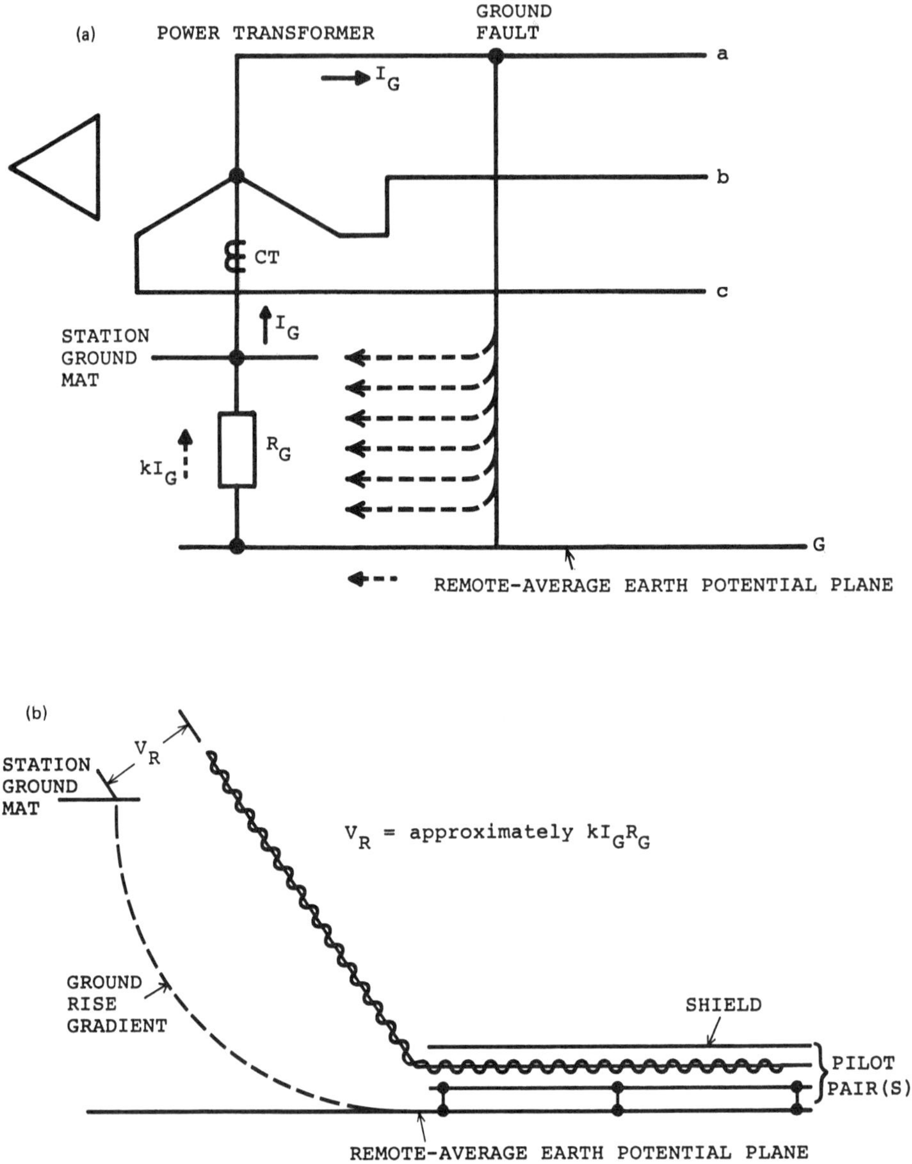

Fig. 13.2 Station-ground-mat voltage rise during system ground faults: (a) phase-to-ground fault and the current return paths; (b) pilot circuit entering station subject to the ground rise voltage; (c) three-winding neutralizing transformer application; (d) two-winding neutralizing transformer application.

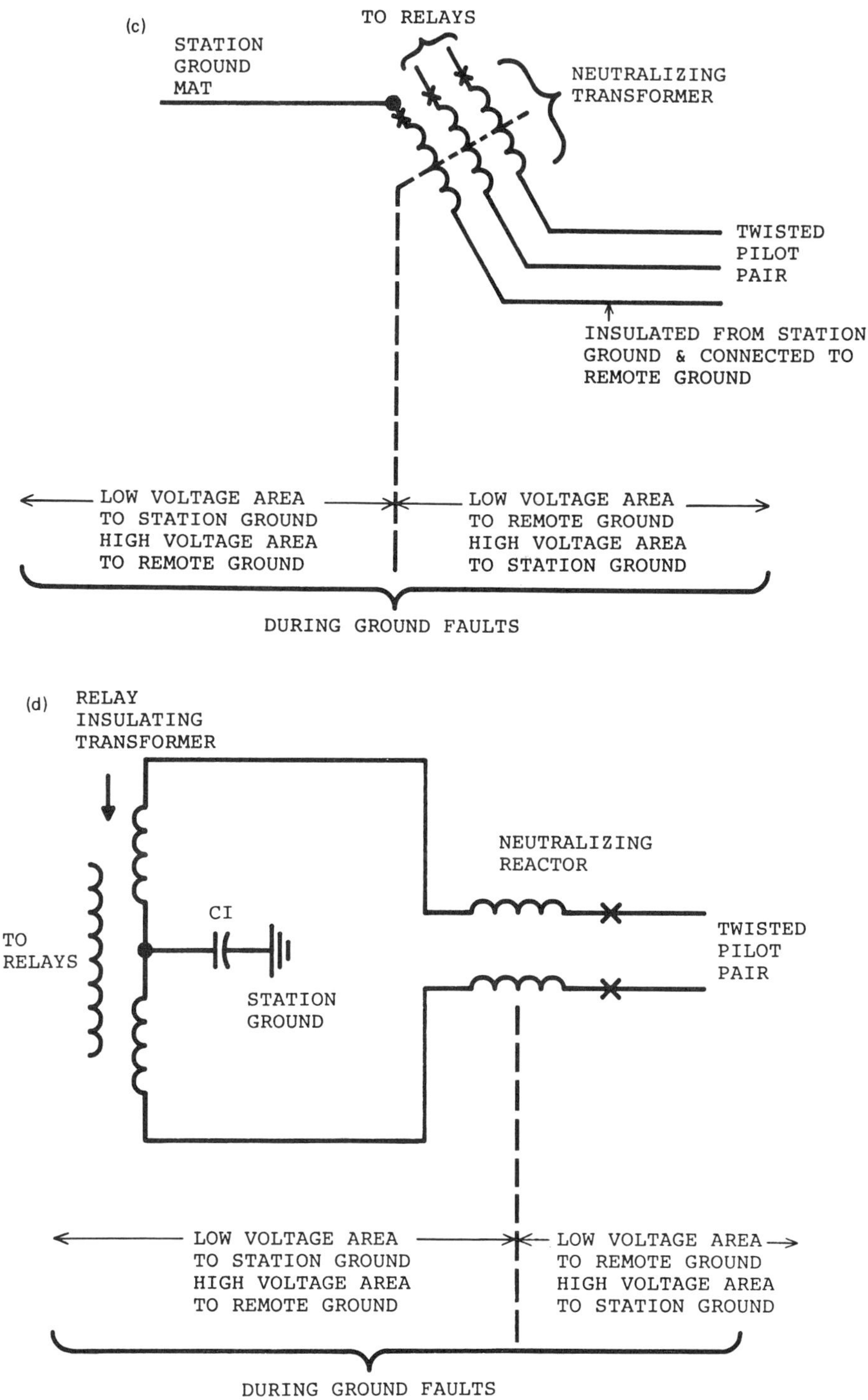

Fig. 13.2 (continued)

equivalent capacitance along the pilot pair or at the remote end provides the exciting current and hence voltage across the windings equivalent to the rise in station ground.

It is important to recognize that with these devices normally mounted in the station environment, one end will be at remote ground and so should be insulated and isolated from personnel contact. This would also be true with high-voltage cable entries without neutralizing devices. Pilot wires are usually limited to short distances on the order of 15 to 20 mi or less.

13.3.2 Pilot Wires: Audio-Tone Transmission

Audio tones in the range of 1000 to 3000 Hz are utilized for protection. They are more compatible for use over leased telephone facilities and so are frequently applied over these channels for protection. The protection hazards and solutions outlined above are applicable. When used, neutralizing transformers should be able to pass the audio frequencies with low losses. Frequency-shift, on-off, and pulse-code equipment is available.

13.3.3 Power-Line Carrier

For longer lines, beginning in the early 1930s radio frequencies between 50 and 150 kHz have been superimposed on power lines. These are known as power-line carrier channels. They are in wide use today, with frequencies between 30 and 300 kHz. Figure 13.3 illustrates a typical channel of this type. In the United States, phase-to-ground coupling is most commonly used. Other types used are phase to phase, two-phase to ground, and so on. The transmitters generate about 1 to 10 W of radio-frequency (RF) power. In the past, 100-W transmitters were available, but they are not in common use today.

The RF signal is connected to the high-voltage line through a line tuner and coupling capacitor, as shown. The tuner, normally mounted at the base of the coupling capacitor unit, cancels the capacitance of the coupling capacitor unit. This provides essentially a low-impedance resistive path for efficient transfer of RF signals to and from the line section.

At the remote end the RF energy passes through similar equipment to the receiver. The transmitter-receiver may be tuned to the same frequency or to different frequencies, which may require a double-frequency tuner.

The inductance between the coupling capacitor and ground presents a high impedance to the Rf signal but low impedance to the system 50 or 60 Hz. This unit may also supply secondary voltage as CCVTs, in which case three are used connected to individual phases, and only one used for coupling, as shown.

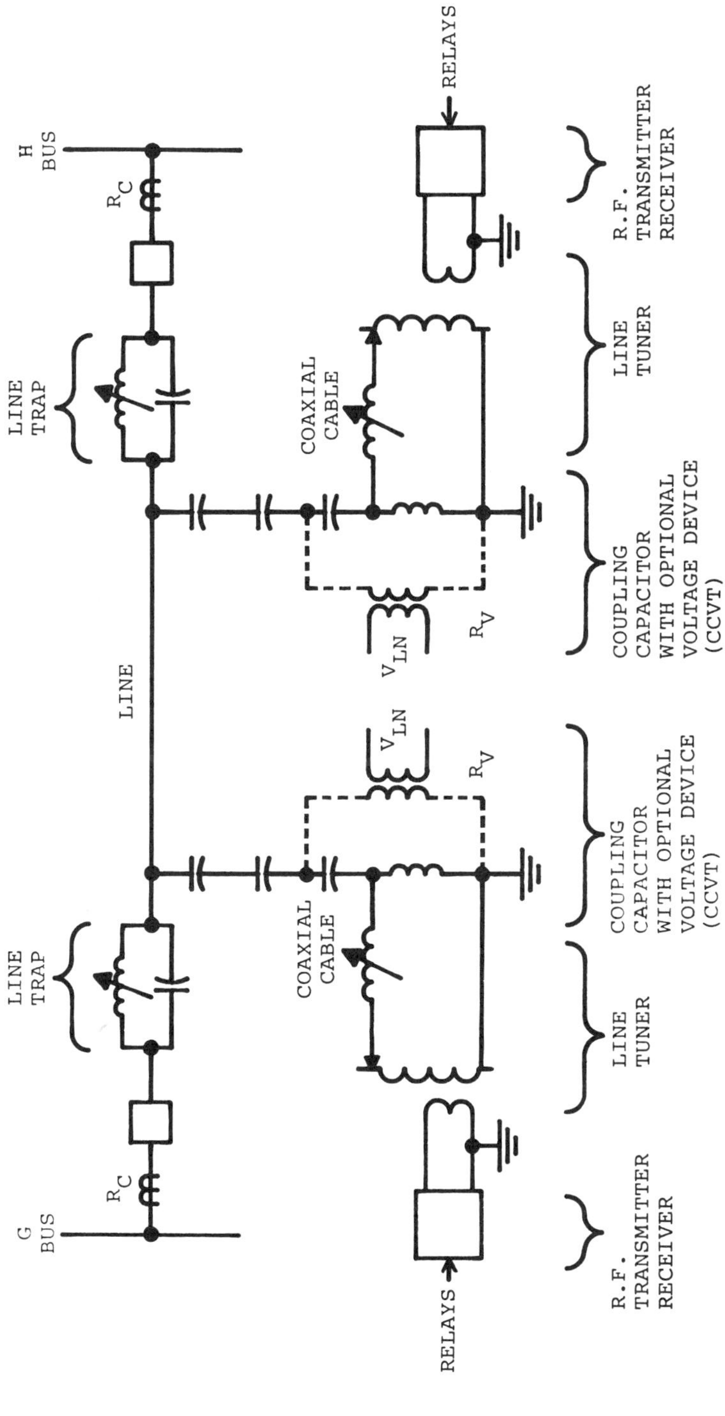

Fig. 13.3 Typical single-line diagram for a phase-to-ground power-line carrier channel.

A line trap is connected in the high-voltage line at each terminal just external to the RF path. It is tuned to provide a high RF impedance, to minimize signal loss into the buses and associated systems, and to prevent external ground faults from potentially shorting out the signal. Traps are available for tuning to a single frequency, double (two) frequencies, or to a broad band of RF frequencies. They are designed to carry 50 to 60 Hz load current continuously at low loss, and to withstand the maximum fault current that can pass through the line.

While the RF signal is introduced to one of the phases, it is propagated via all three phase conductors. There have been several cases where the accidental coupling to different phases at the two terminals has gone unnoticed for a number of years because an adequate signal was received. In fact, the signal may be greater when coupled to different phases. Modal analysis has provided an important modern tool for predicting carrier performance and the best method of coupling and transmission. This is important, especially for long lines. The references provide more information on this.

Overhead power lines tend to have a characteristic impedance (Z_0) between 200 and 500 Ω phase to ground and 400 to 488 Ω between phases. The carrier equipment and coupling essentially match these values for maximum power transfer of the RF. Taps and discontinuities, particularly if they are at quarter wavelengths, can result in high signal losses. The RF signals should be selected to avoid these problems.

Power-line carrier applications to power cables may be impossible or very difficult, as their characteristic impedance is low and the losses much higher than for overhead lines. Carrier transmitter-receivers are operated "on-off," "frequency shift" with mark-space signals, or single sideband, depending on the design and application.

Fault arc noise has not been a problem or a significant factor in the use of power-line carrier for protection. Disconnect switches with currents less than 200 A (amperes) can cause operation of carrier on-off receivers, but this does not impair the protection used with this type of equipment.

13.4 TRANSFER TRIP SYSTEMS

These are used to transfer a tripping signal to circuit breakers or other circuit interrupters that are located at an area remote from the protective relays. This is necessary either (1) because a breaker for fault isolation does not exist at the local station, or (2) for backup to assure that the remote terminal is opened for system faults.

The term "transfer trip" is used in two contexts, which can generally be described as "equipment transfer trip" or "line transfer trip." A common example of equipment transfer trip is the installation of a transformer bank without a local breaker associated with one winding and where fault current can be supplied through that circuit to an internal transformer fault. This was discussed in Chapter 9. Operation of the transformer protection can directly trip the local breaker(s), but a transfer trip system is required to open the remote breaker(s), especially when the current through the remote terminal is too low to operate fault detectors there. An alternative to transfer trip in this application is a local fault switch actuated by the local relays. Such fault is well above the pickup values of the remote protection.

Transfer trip signals can be sent via any of the channels indicated above. Since high-speed operation of the remote terminal is desired, it is important that the channel and equipment used have very high security against transient operation from any spurious signal. Fault detectors at the remote terminal generally cannot be used to supervise the transfer trip signal because of the possibility of a low fault contribution.

Audio-tone transfer trip systems are in wide use, generally with two separate systems for increased security. Signals from both systems must be received in order to trip the remote breaker(s).

In the line transfer trip systems, fault detectors are used at the terminals. Thus tripping can be initiated only after the local fault detector has operated and a transfer trip signal has been received. Thus channel security, while important, is not quite as critical as for the equipment transfer trip systems. Line transfer trip is described further later.

13.5 PILOT-WIRE RELAYING

A widely used scheme for short-line protection is shown in Fig. 13.4. It operates on the three-phase and neutral currents from the line CTs. Voltage is not required. These currents are fed to a sequence network at each terminal to produce a single-phase voltage output V_F:

$$V_F = k_1I_1 + k_2I_2 + k_0I_0 \tag{13.1}$$

I_1, I_2, and I_0 are the positive-, negative-, and zero-sequence current components derived from the line currents I_a, I_b, and I_c, and k_1, k_2, and k_0 are network factors adjustable by the taps and design of the sequence network. For many years the design provided a V_F output proportional to positive and zero sequence only ($k_2 = 0$). All types and combinations of faults have positive sequence, with ground faults adding zero sequence. Thus all faults above the relay sensitivity could be

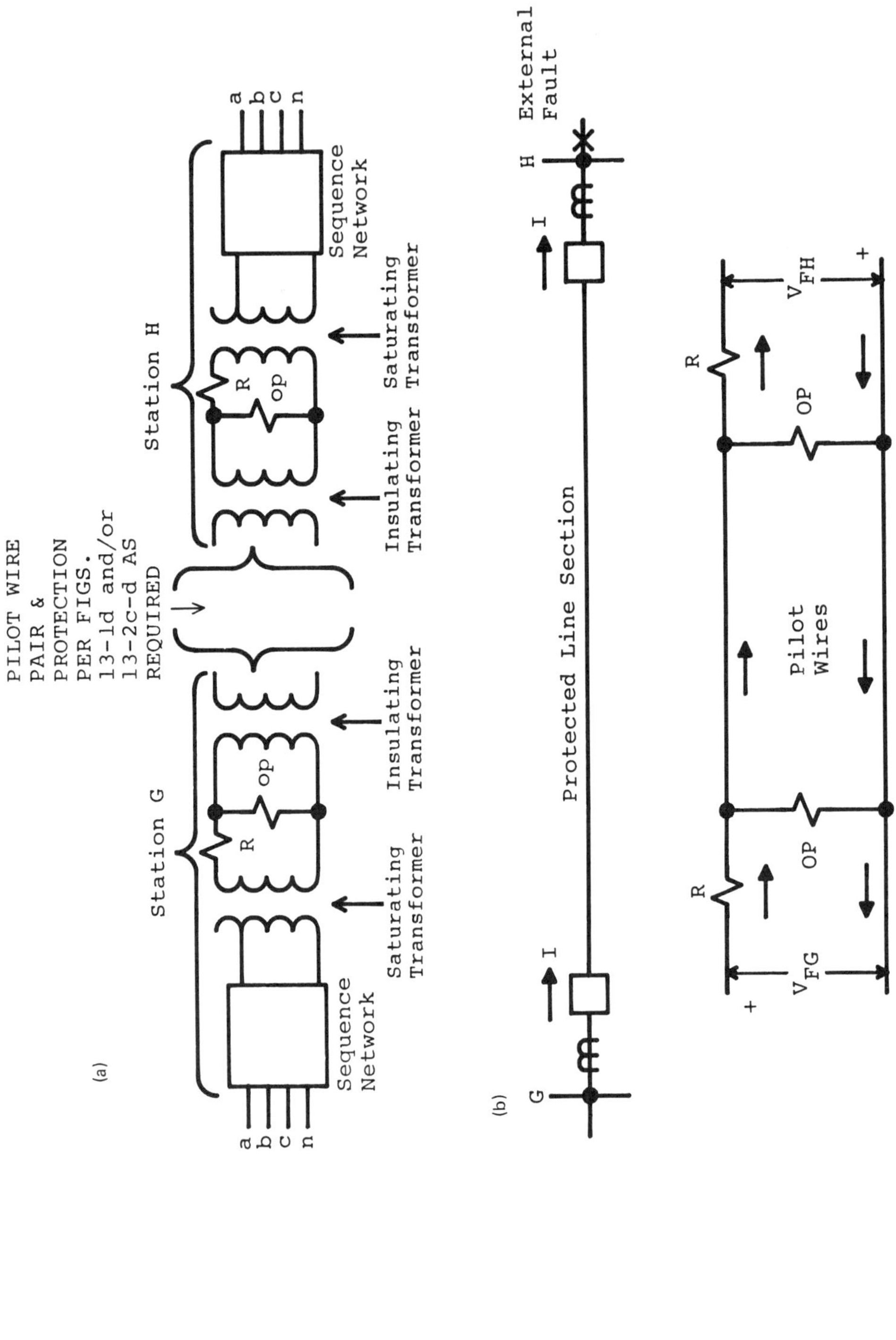
(a)
PILOT WIRE
PAIR &
PROTECTION
PER FIGS.
13-1d and/or
13-2c-d AS
REQUIRED
Station G
Station H
R
op
R
op
a
b
c
n
a
b
c
n
Sequence
Network
Sequence
Network
Saturating
Transformer
Insulating
Transformer
Insulating
Transformer
Saturating
Transformer
(b)
G
H
I
I
External
Fault
Protected Line Section
Pilot
Wires
R
OP
R
OP
+
+
V_{FG}
V_{FH}

(c)

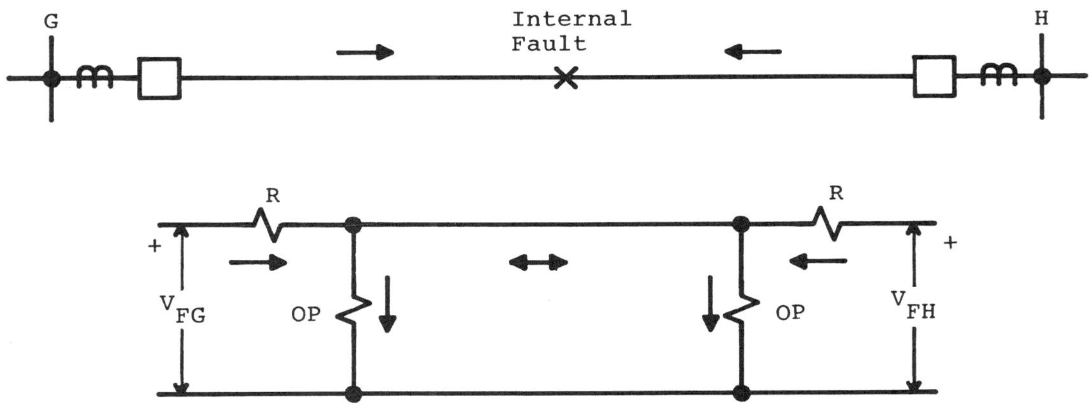

Fig. 13.4 Typical pilot-wire relaying system in common use for line protection: (a) basic schematic of the system; (b) simplified circuit illustrating operation for external faults; (c) simplified circuit illustrating operation for internal faults.

detected. For ground faults k_0 can be made large for high ground-fault sensitivity. The phase-to-phase fault pickup with $Z_1 = Z_2$ is higher by $\sqrt{3}$ than the three-phase fault pickup, with the phase-to-phase actual fault current being 0.866 less than three-phase current. For short lines this lower sensitivity to the phase-to-phase faults is seldom a problem. Higher phase-to-phase fault sensitivity (lower relay pickup) is available by networks that include a response to I_2 ($k_2 \neq 0$).

The output voltage from the sequence network or similar networks is limited by the saturating transformer. This limits the voltage to essentially a constant magnitude of around 15V, so that it is relatively independent of the wide variations that can occur in the fault currents. This voltage is connected to the pilot wires through a restraint coil (R) with an operating coil (OP) across the pair and the insulating transformer. Typically, a 4:1 or 6:1 ratio transformer is used, providing a maximum pilot wire voltage of about 60 or 90 V. The pilot wire side insulation of this transformer is about 10 to 15 kV to ground and between the windings.

Current in the operating (R) coil tends to prevent relay trip operation, while current in the operating (OP) coil produces relay operation. With equal current in (R) and (OP) and above the pickup value, this relay operates to energize the breaker trip coil.

The operation of the system for external faults is illustrated in Fig. 13.4b. Through current, either load or to external faults, results in a circulating current through the pilot wire pair, as shown, with only a small portion passing through the operating (OP) coil. The higher current through the restraint (R) coil prevents relay operation at both terminals.

For internal faults (Fig. 13.4c), current flowing to the fault essentially circulates through the restraint (R) and operating (OP) coils in series with a small current over the pilot pair. If the contribution to the fault from stations G and H were equal, no pilot wire current would flow. The other extreme possibility is, say, no contribution to the fault from bus H; then the current through G would divide between the local operating coil and, via the pilot wires, the remote operating coil at H. If the fault current through G in this case is high enough, both terminal relays can operate to provide simultaneous high-speed tripping of both ends. This is quite desirable in some cases.

Thus over a wide range of current magnitude and distribution, this system provides high-speed simultaneous protection for both phase and ground faults. As can be seen, this is a differential type of protection operating on the total fault current, and a 50- to 60-Hz phase comparison type.

Three terminal applications are possible with this system. The added terminal relay is connected across the pilot pair with its pilot pair forming a wye connection with the other two pilot wire branches. The pilot wire wye connection should have legs of equal impedance. Since different distances will exist between the terminals, balancing resistors are utilized to provide this balance.

Supervision of the pilot pair for opens, shorts, and/or grounds is obtained by circulating about 1 mil of dc current over the pair, with sensing relays at each terminal. These are connected at the midpoint of the insulating transformer on the pilot-wire side. This winding is split at the midpoint and a small capacitor connected between the two halfs to permit the 50- to 60-Hz signals to pass. The supervisory dc voltage is applied across this capacitor. At the other end(s) sensing relays are connected across similar capacitors at those terminals. Interruption and/or increase of the circulating current provides alarm indications. This equipment can also be used for a transfer trip channel in either or both directions. For this the dc current in the supervisory equipment is reversed and increased to operate the transfer trip relays.

With this supervision equipment connected to the pilot wires, special attention must be given to possible high voltage from induction and/or station ground rise. For personnel safety, neutralizing and mutual draingage reactors and gaps should be used in most applications.

13.6 DIRECTIONAL COMPARISON BLOCKING PILOT SYSTEMS

This is the oldest type of system, being first used in the 1930s. It is still in very wide use and is the most versatile and flexible system, especially applicable for multiterminal lines. The direction of the power flow at the terminals is compared. For internal faults the power (current) flows into the line at the terminals, so simultaneous high-speed tripping at the terminals is permitted. For external faults the information that current flows out at one of the terminals is used to block tripping of all terminals.

This system is usually applied with an "on-off" power-line carrier channel and the transmitter-receivers at the two terminals tuned to a common RF frequency. Separate frequencies can be used, as well as other types of channels. A basic schematic of this system is shown in Fig. 13.5. In general, distance relays for phase faults and directional instantaneous overcurrent relays for ground faults are used as fault detectors (FD). The channel signal is initiated by distance phase and instantaneous overcurrent units known as carrier start (S) relays. The fault detectors (FD) must be directional and set to overreach all remote terminals under all operating conditions. As this is also the

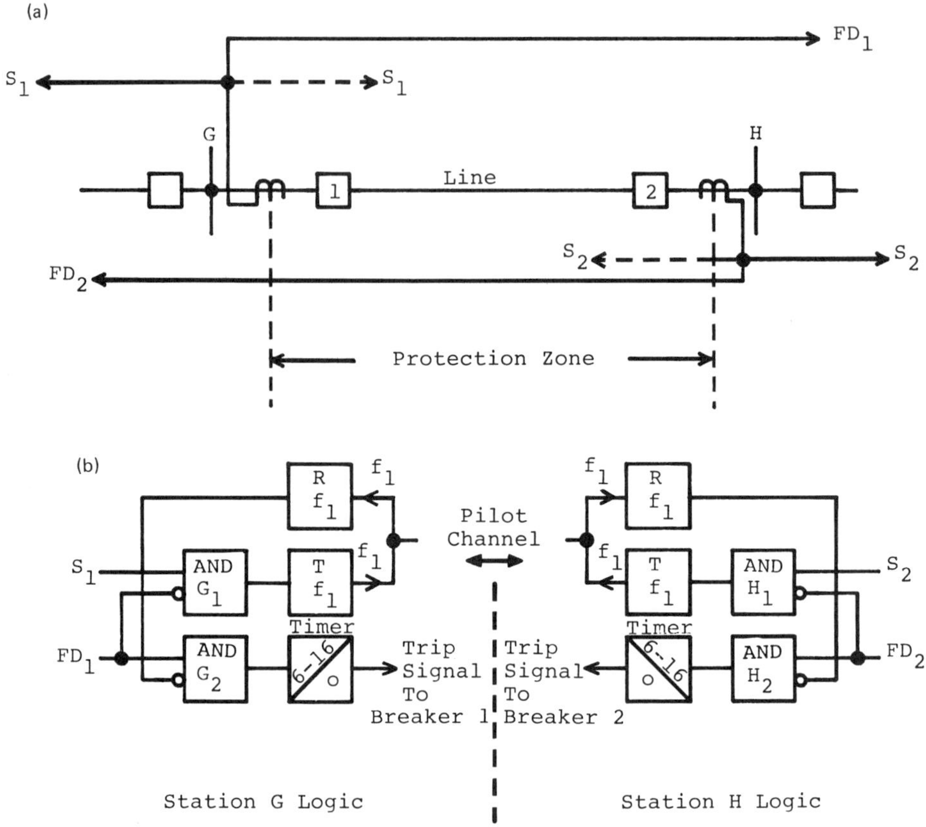

Fig. 13.5 Basic operating principles of the directional comparison blocking pilot system: (a) power system and relay setting diagram; (b) system logic diagrams.

requirement for zone 2 distance relays, these are used in the pilot system. The pilot trip for internal faults is in parallel with the time-delay T_2 trip.

The carrier (channel) start units must be set more sensitively or to reach farther out of the line than the remote fault detector. In other words, in Fig. 13.5a, S_1 units at breaker 1 bus G must operate for all phase or ground faults to the left of 1 that can operate FD_2 at breaker 2, bus H. Similarly S_2 must operate for all phase and ground faults at bus H and to the right than can operate FD_1 at G. Normally, a zone 3 distance relay is used, but connected to "see" out of the protected line section instead of into the line, as discussed in Chapter 12. The conventional mho characteristic which passes through the origin cannot respond adequately to provide blocking for the possibility of a solid zero-volt fault just external to the relay. Accordingly, the zone 3 distance relay has a small current-only torque or bias to provide positive operation for these faults.

Both electromechanical and solid-state systems are in service and the basic operation is the same. A typical operating logic diagram is shown in Fig. 13.5b. For the logic boxes, "1" is used to indicate a logic input or output signal, and "0" for no or an insufficient input or output signal. The "AND" logic box requires all inputs to be "1" in order to have a "1" output. The small circle (o) at the box indicates that a "1" appears to the box as a "0" and vice versa. The top numbers in the timer logic box represent the time in milliseconds for an output to occur after an input (pickup time). The lower number is the reset (dropout) time.

The operation for an external fault, for example, on bus H or on the line(s) to the right of the bus, is as follows:

Relays 2 at bus H: FD_2 does not operate, S_2 does operate. "AND H_1" thereby has both inputs "1" to key the carrier transmitter on. The RF signal f_1 is received locally and transmitted to station G. "AND H_2" has two "0" inputs; neither will provide an output. Thus breaker 2 is not tripped.

Relays 1 at bus G: The RF signal f_1 received provides a "0" on "AND G_2" so no output to trip breaker 1 even though FD_1 operates. FD_1 operation provides an "0" on "AND G_1," so the transmitter is not keyed even though S_1 may operate on this external fault.

Thus the carrier (channel) signal at H is used to block the overreaching trip relays at G from operating on this external fault. Correspondingly, an external fault on or to the left of bus G will provide a blocking signal from G to H to block the overreaching H relays from tripping.

For internal faults in the protected zone:

Relays 1 at bus G: FD_1 operates for a "1" input to "AND G_2" and a "0" input on "AND G_1." Thus the station G transmitter will not be turned on or will be stopped should S_1 operate before FD_1. No signal is transmitted to H.

Relays 2 at bus H: FD_2 operates for a "1" input to "AND H_2" and a "0" input to "AND H_1." Thus the station H transmitter will not be turned on or will be stopped should S_2 operate before FD_2. No signal is transmitted to G.

With no channel signal from either terminal, both "AND G_2" at G and "AND H_2" at H energize their respective timers and both breakers are tripped at high speed. the typical 6- to 16-ms delay is for coordination between the various operating relays. The channel is not required for tripping, so that should an internal fault interrupt the channel signal, tripping can still occur.

13.7 DIRECTIONAL COMPARISON UNBLOCKING PILOT SYSTEM

This system developed around frequency-shift (FSK) channels, such as FSK power-line carrier. This equipment provides excellent narrowband transmission at low power with receivers that are highly insensitive to noise. An RF signal is transmitted continuously in one of two modes, known as "block" or "unblock." Typical shifts for these are ±100 Hz from a center RF frequency. One watt is used for the block mode, and either 1 or 10 W for the unblock mode.

With a continuous signal transmitted, it can be used to block relay operation and thereby eliminate the channel start relays (s) that are required in the preceding system. For internal faults the relay blocking signal is shifted to unblock to permit the relays to trip instantaneously. Since a signal must be transmitted, the unblock signal can be used to augment tripping. Also, this type of channel can be monitored continuously, which is not possible with "on-off" channels.

A typical unblock system is shown in Fig. 13.6. As illustrated in part a, only fault detectors are required. They are of the same types and setting as described in Section 13.6. Both the phase and ground relays must always overreach all remote terminals for all operating conditions to provide 100% internal line-fault protection.

As documented by the logic diagrams of Fig. 13.6b, the operation is as described below. This is a typical system; other arrangements are possible.

Normal Operating Condition (No Faults)

Both frequency shift transmitters at station G (FSK T_G) and station H (FSK T_H) are transmitting in the block mode, so that from their respective receivers (FSK R_G) and (FSK R_H), the block signal is "1" and the unblock signal "0." Thus the outputs of "0R G_1" and "0R G_2" at station G and of "0R H_1" and "0R H_2" at station H are all "0." FD_1 at G and FD_2 at H are not operated, so the "AND G_3" and "AND H_3" inputs are all "0" and no trip is indicated.

As shown, separate frequency channels are required between the stations indicated by f_1 and f_2. Typical spacings between these are about 1 kHz for narrowband equipment. Transmission of these frequencies is continuous during their operation.

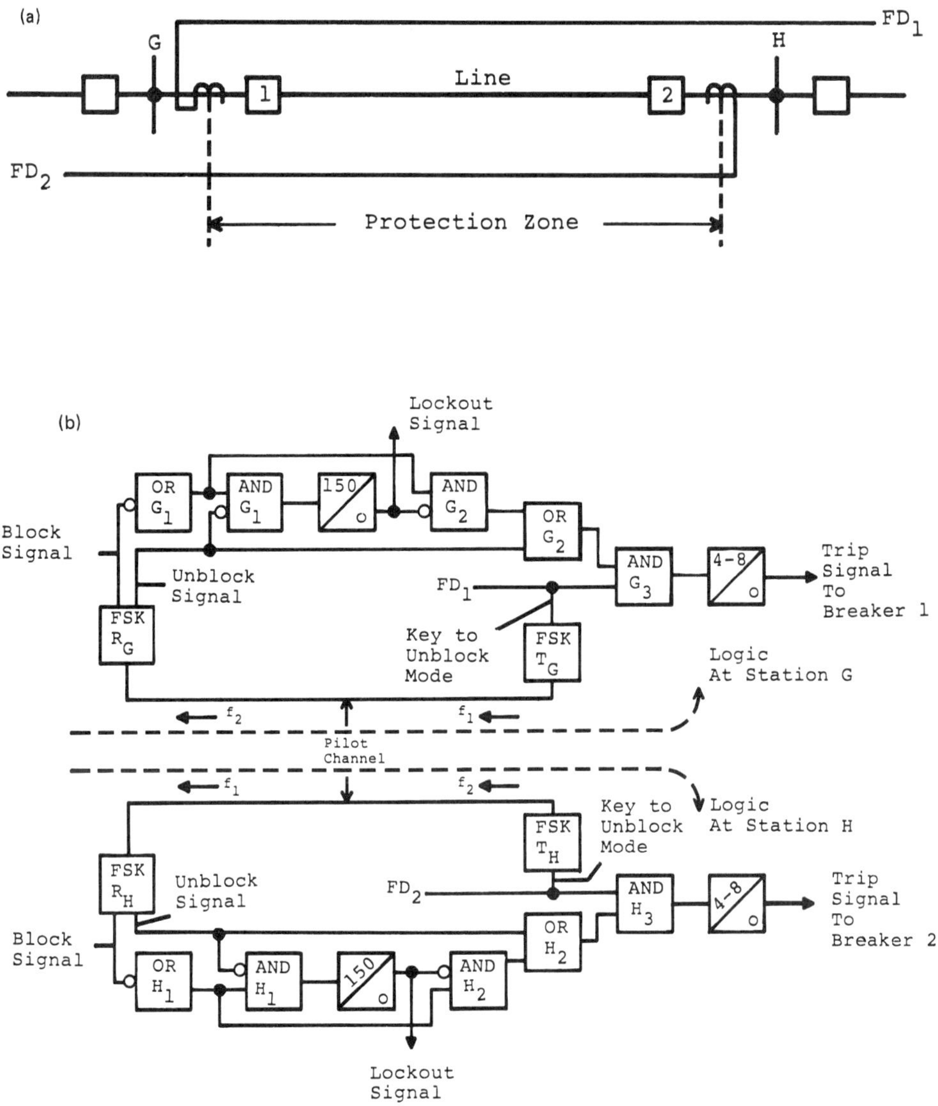

Fig. 13.6 Basic operation principles of the directional comparison unblocking pilot system using a frequency shift channel: (a) power system and relay setting diagram; (b) system logic diagram.

Channel Failure

If one channel is lost for any reason, so that there is not block signal output, the protection is locked out of service and an alarm is sounded. In Fig. 13.6b, if the block signal from "FSK R_G" is lost (goes to "0") without a shift to unblock mode, so that it remains at "0" at station G, "OR G_1" has a "1" input and output. Now "AND G_1" has both inputs "1" to energize the timer. After 150 ms the relay system is locked out with an alarm.

External Fault on Bus G or in the System to the Left

Relays 1 at station G: FD_1 relays do not operate. This directly prevents tripping of breaker 1 and permits the "FSK T_G" transmitter to continue in the blocking mode.

Relays 2 at station H: Block signal "1" continues, so "OR H_1" has no input, hence no output. Thus the lower input to "AND H_3" is "0," so it can have no output even though FD2 has operated with a "1" on "AND H_3." Thus tripping is blocked by the blocking signal from station G. The operation of FD_2 keys the transmitter "FSK T_H" to unblock.

Back at station G: Reception of the unblock signal from station H by "FSK R_G" operates "OR G_2," but "AND G_3" cannot operate since FD_1 has not operated.

Internal Faults in the Protected Zone

At station G: FD_1 operates to key "FSK T_G" to unblock and to input "1" to "AND G_3."

At station H: FD_2 operates to key "FSK T_H" to unblock and to input "1" to "AND H_3."

At both stations: The operation is the same. Considering station G, the unblock signal "1" with the FD_1 signal "1" operates "AND G_3" to initiate breaker 1 tripping after about a 4- to 8-ms delay for operating coordination. This unblock signal is not necessary for tripping. Removal of the block signal operates "OR G_1" and in turn "AND G_1" if the unblock signal is delayed or not received because of the fault. The bypass from "OR G_1" to "AND G_2" causes it to operate with "0" unblock signal on "AND G_1." This "AND G_2" output lasts until the timer operates, which is sufficient to permit a trip signal via "OR G_2" and "AND G_3."

Thus this system operates as a "blocking type," where no channel signal is required for the timer inverval (150 ms as shown), then becomes a "transfer trip" type. This provides a combination of the advantages of both types, as reviewed in a later section. It is applicable to power-line carrier channels, which should not be used for the transfer trip systems.

13.8 DIRECTIONAL COMPARISON OVERREACHING TRANSFER TRIP PILOT SYSTEMS

Power-line carrier channels are not used or recommended for these systems. A signal must be received from the remote terminal(s) in order to trip, and with power-line carrier the signal could be interrupted or shorted out by a fault. Therefore, these systems are used normally with audio tones over telephone circuits or modulated on microwave channels. "Trip-guard" tones are used. The guard is monitored. The arrangement is similar to that described above. Loss of channel (guard) for about 150 ms locks out the protective relays from tripping and alarms. After the guard signal returns, the relays are restored to operation after about 150 ms.

A typical system is shown in Fig. 13.7. The same types and settings of directional phase distance and ground directional instantaneous overcurrent relays are used as for the previous systems. Again it is important that these be set to overreach the remote terminal(s) under all operating conditions.

External Fault on Bus G or in the System to the Left

Relays 1 at station G: FD_1 relays do not operate. Tripping cannot take place. Transmitter T_G continuous to operate in the guard mode.

Relays 2 at station H: FD_2 relays operate but cannot trip, as receiver R_H does not input to "AND H" in the guard mode. FD_2 does shift transmitter T_H to the trip mode, so at station G this receiver energizes the "AND G," but as above, FD_1 does not.

Thus both terminals are blocked from tripping on the external faults.

Internal Faults in the Protected Zone

The operation is the same at both terminals. FD_1 and FD_2 both operate to shift their respective transmitters to the trip mode. This is received at the remote receivers and provides an input into the "AND G" and "AND H," which with the FD_1 and FD_2 inputs provide a trip output and both breakers are operated simultaneously at high speed. The 4- to 8-ms delay provides coordination time between the various components.

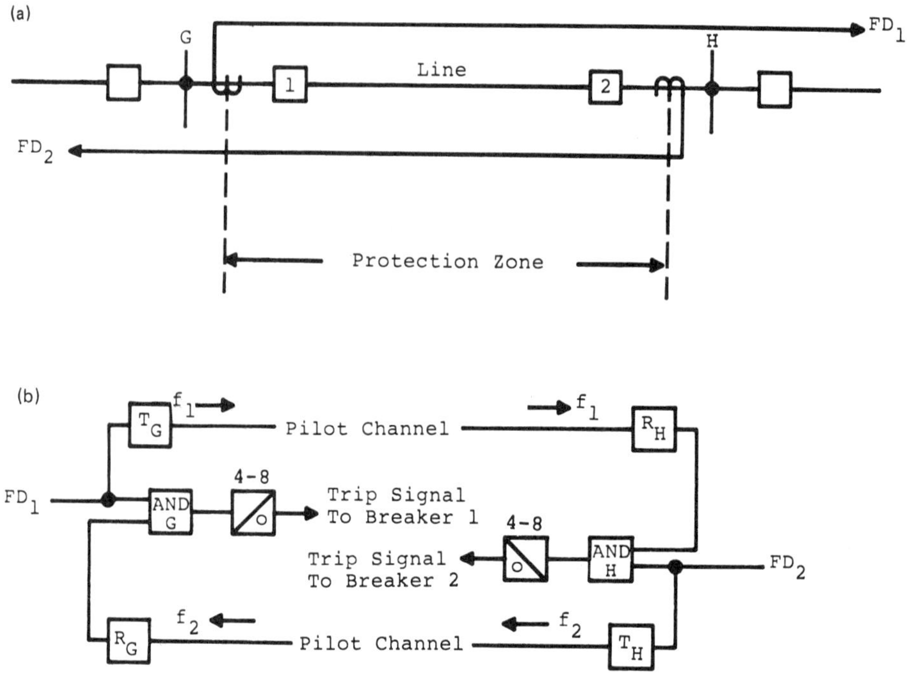

Fig. 13.7 Basic operating principles of the directional comparison overreaching transfer trip pilot system: (a) power system and relay setting diagram; (b) system logic diagram.

13.9 DIRECTIONAL COMPARISON UNDERREACHING TRANSFER TRIP PILOT SYSTEMS

This system requires that the fault detectors be set so as always to overlap but not to overreach any remote terminal under all operating conditions. Phase directional distance zone 1 units meet this requirement and so are used in this system. Instantaneous overcurrent relay "reaches" vary with current magnitude, and it becomes difficult, and in some cases impossible, to assure that these relays for ground faults always overlap but not overreach. Thus distance ground relays are recommended for this system and set as above for zone 1.

The channels required are the same as for the overreaching transfer trip systems. Two types exist: (a) nonpermissive, shown in Fig. 13.8b, and (2) permissive, Fig. 13.8c. With the setting indicated, all external faults do not operate any of the fault detectors. On the internal faults FD_1 at bus G and FD_2 at bus H both trip directly for

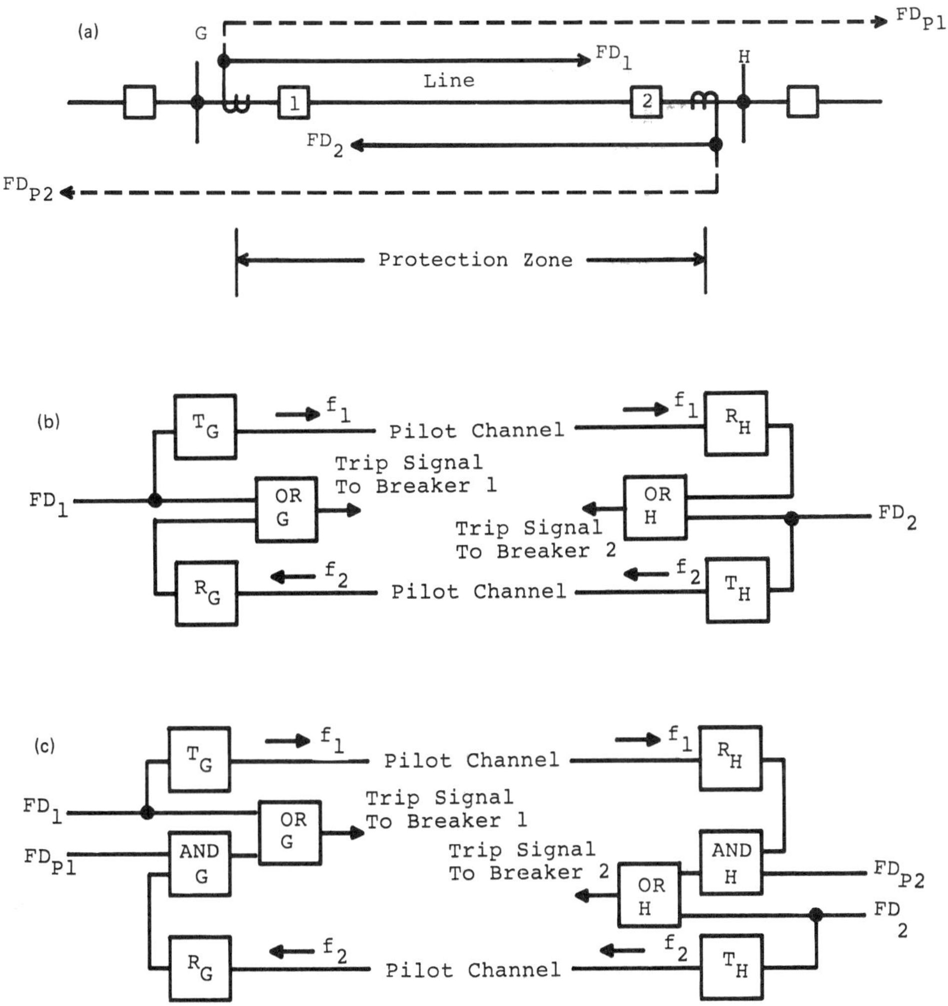

Fig. 13.8 Basic operating principles of the directional comparison underreaching transfer trip pilot system: (a) power system and relay setting diagram; (b) system logic diagrams — nonpermissive underreaching transfer trip; (c) system logic diagrams — permissive underreaching transfer trip.

faults in the area of the overlap. These detectors both key their respective channel transmitters to the trip mode. This provides a trip output at the receivers to directly trip the terminal breakers. No time delay is required.

This system is not in general use because of the very high security requirements of the channels. Any transient or spurious operation of the receiver will result in incorrect operation. Consequently, overreaching fault detectors are added of the type and setting described above. These are used to supervise the channel. As seen in Fig. 13.8c, a trip signal via the channel requires that the overreaching fault detectors operate so that "AND G" and/or "AND H" can have an output to trip the breakers.

13.10 PHASE COMPARISON SYSTEMS

One phase comparison system, the pilot-wire type, was described in Section 13.5. As indicated, it requires a 50- to 60-Hz wire line channel, which limits its application to short lines. This principle of a current-only scheme not requiring system voltages was extended to long lines from the mid-1940s. The voltage derived from the three-phase currents to detect all types of phase and ground faults is passed through a squaring amplifier. This produces square waves for transmission and comparison at the remote terminal.

As for the pilot-wire system, various networks are used to produce the current-derived voltage. One system uses the same sequence network as described, with the output of Eq. (13.1). Negative-sequence response is important to provide increased sensitivity to phase-to-phase faults for the longer lines.

For security, two-level fault detectors are set, both overreaching so as to operate for all internal phase and ground faults. Normally, these are current-operated units. Where the minimum internal fault currents are on the order of or less than maximum load current, distance units are applied for the phase-fault detectors. This, of course, means that the system now requires VTs and is not a current-only system. In either case the high-level detector is set at about 125 to 250% of the low-level unit.

A typical system and its operation are illustrated in Fig. 13.9. At breaker 1, station G the fault detectors are FD_{1S} and ED_{1T} and at breaker 2, station H, FD_{2S} and FD_{2T}. The (S) units are set more sensitively (lower current pickup or longer distance reach) than the (T) units. As shown in the figure, the squaring amplifier produces square waves which are maximum on one half of the 50- to 60-Hz wave and zero on the other half of the wave.

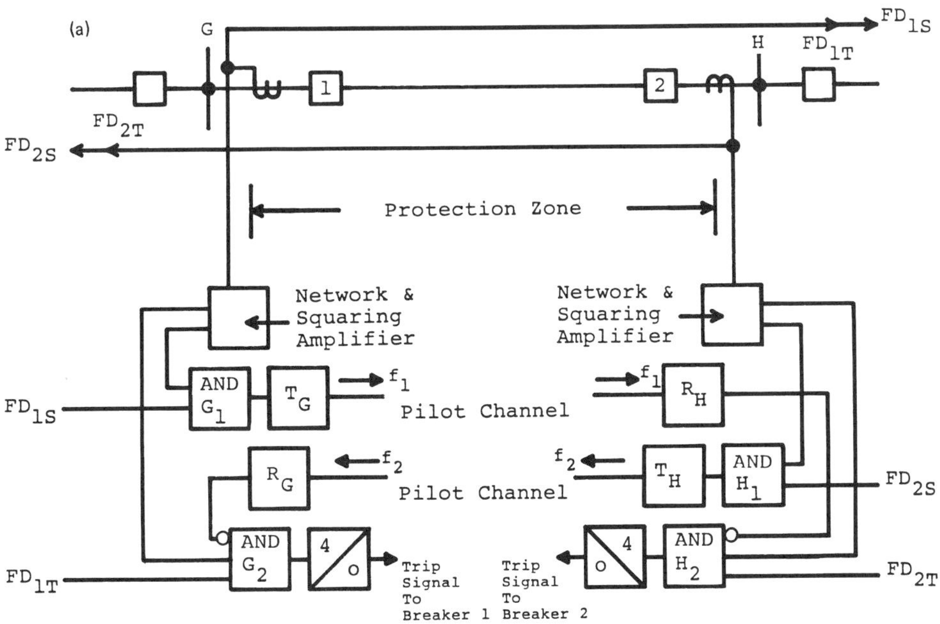

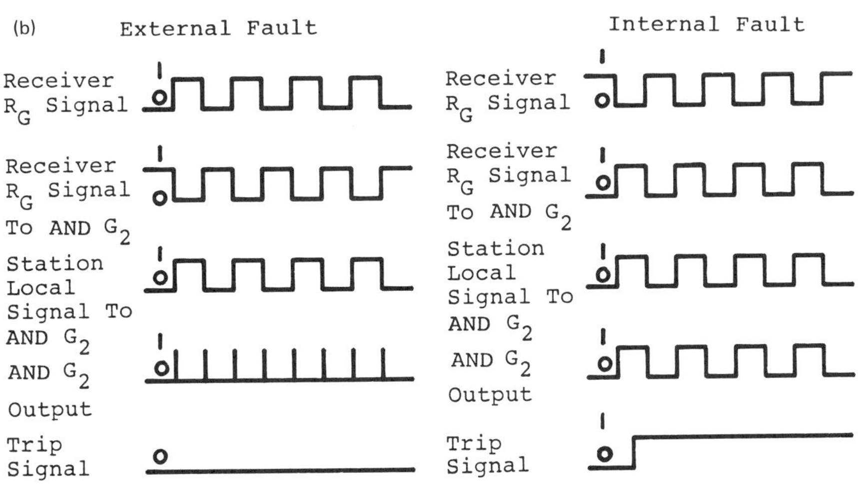

Fig. 13.9 Basic operating principles of phase comparison pilot systems. Single half-wave comparisons over audio-tone channels: (a) system logic diagrams; (b) typical system operation for relays at 1 bus G for external faults at bus H and to the right, and for internal faults.

With similar equipment and settings at the two terminals, the operation at bus G for relays 1 is as follows:

External Fault on Bus H or in the System to the Right

FD_{1S} and FD_{1T} operate to input "1" to "AND G_1" and "AND G_2," respectively. At station H, FD_{2S} and FD_{2T} (both non-directional) operate to input "1" on "AND H_1" and "AND H_2." Thus FD_{2S} and the squaring amplifier output start transmission of square waves from transmitter T_H over pilot channel f_2 to receiver R_G. The polarities are such that this train is as shown, and is negated as input to "AND G_2." Thus "AND G_2" is energized only for a very short interval, if at all, for short spikes of output considerably less than the 4 ms required for operation. For the external faults assumed, the currents into the line at 1 should be in phase with the currents out of the line at 2 for all practical purposes (line losses and phase shifts are normally negligible except for very long lines). Time delay is provided for the local signal to compensate for the differences and the channel delays, thus tripping does not occur for the external faults. The system at breaker 2, station H operates in a similar manner, and both terminals operate similarly for faults on bus G and in the system to the left.

Internal Faults in the Protected Zone

All fault detectors at both terminals operate. At station H the relays 2 current has reversed for the internal faults, so the received signal and the input to "AND G_2" is essentially in phase with the local square waves from relays at 1. These provide positive square output from "AND G_2" and after 4 ms a trip output.

The full half-wave output shown assumes that the currents at stations G and H are 180° out of phase or flowing into the line in phase. with load flowing across the line, this is not true, but these systems can trip with the currents at the two stations out of phase up to 90°. In 60-Hz systems a phase angle difference of 90° corresponds to 4 ms. Similar operation for the internal faults occurs at relays 2, station H.

This system can be used with "on-off" power-line carrier channels. For the internal faults, if a receiver signal is not received, "AND G_2" will both have a "1" input continuously. Trip operation can occur at both ends by the local signal input plus the local FD_T operation.

The systems described above provide comparison on only one half cycle. If the fault should occur on the zero-output half cycle, tripping for internal faults is delayed until the next half cycle.

If frequency-shift (FSK) power-line carrier or "mark-space" audio-tone channels are available, comparison can be made on either half cycle of the 50- to 60-Hz wave. This is known as "dual phase comparison." Equipment is required similar to that of Fig. 13.9 for comparing each half cycle. The square waves of one half cycle are transmitted by "1" on one frequency and "0" by the shift frequency. The other half cycle is transmitted similarly by separate channel. Similar to Figure 13.6, the unblock logic permits operation on internal faults if the channel signal is not received. The transfer trip mode is used in general for the audio tone type channels.

13.11 SEGREGATED PHASE COMPARISON PILOT SYSTEMS

As indicated in Chapter 12, phase comparison pilot systems are preferred for the protection of series-compensated transmission lines. Dual phase comparison systems are in service with good operating records. However, these types of lines can have very severe harmonics, so it is possible that networks to provide the current-derived operating voltages may provide incorrect information on the potential highly distorted current waves from faults on the system. These networks basically are "tuned" to the 50- to 60-Hz current waveform.

Accordingly, a system of individually comparing the phases rather than a single voltage for the three phases is used. The voltage derived from the single current magnitude is independent of power system frequency and waveform. However, multiple channels are required in these systems. The two common systems are (1) comparing $I_a - I_b$ at the terminal ends over one channel, and $3I_0$ at the terminals over another channel; or (2) comparing each phase current I_a, I_b, and I_c at the terminals via three channels. The current square waves are compared basically as described for other type of phase comparison systems.

13.12 SINGLE POLE—SELECTIVE POLE PILOT SYSTEMS

The large majority of faults on the higher-voltage overhead transmission lines result from overvoltage transients induced by lightning. These cause a flashover of the phase(s) generally to ground, with most faults being single-phase to ground. In many cases these can be cleared by rapidly opening the circuit or faulted phase and, since the transient has passed , the arc will be extinguished by deenergizing. Then rapid automatic reclosing will restore service. This is discussed further in Chapter 14.

With the large percentage of single-phase-to-ground faults, it would be necessary to only open the faulted phase, leaving the other phases closed to exchange synchronizing power between the terminals. This technique is known as single-pole trip reclose. Upon reclosing the opened phase, if the fault still exists, all three phases are opened and further reclosing is blocked.

A more sophisticated scheme opens one phase for single-line-to-ground faults, two phases for line-to-line and double-line-to-ground faults, and three phases for three-phase faults. High-speed reclosing occurs for one or two opened phase faults and is optional for three-phase faults.

The advantages of opening one or two phases is a higher stability limit capability and less "shock" to the power system. During the open phase(s) period, zero-sequence and negative-sequence currents flow, so ground backup relays must be set to avoid operation. Typical open times are on the order of 0.5 to 1.0 s.

Pilot relaying is necessary when high-speed reclosing is used to provide reasonable assurance that all terminals are opened to deionize the fault. The schemes for single-pole and selective-pole relaying are more complex. Several techniques are utilized for selecting the faulted phase and a discussion of these is beyond the scope of this book. Most of the schemes have difficulties with some types of faults for which they are not to operate, so additional relays or auxiliaries are required to avoid incorrect operations. The three-subsystem segregated phase comparison scheme of the preceding section provides an excellent means for single-pole or segregated-pole relaying. This type opens only the faulted phase or phases involved in the fault, and with its independent comparison of the three phases, the scheme avoids the difficulty of correctly identifying the phase(s) that is faulted, which plagues many other schemes.

Single-pole trip-reclose systems are common in Europe but are used infrequently in the United States. The basic reason that it is not used was that until the advent of EHV, the U.S. circuit breakers were three-pole type, with only one trip coil. Thus to apply single-pole reclose relaying, special or more expensive breakers were required. With separate pole mechanisms now common, there is increased interest in these relaying systems.

The most important application for these systems is for the single line connecting two major power sources with no other or weak ties between them. Three-phase trip reclose cannot be used for these, as the two systems are separated by opening the line and will be too far out of synchronism to reclose and maintain stability.

13.13 SUMMARY AND GENERAL EVALUATION OF PILOT SYSTEMS

Field experience over many years has indicated that all the pilot system types provide very high reliability. Thus the choice if there are no application restraints or limitations is largely one of personal preference and of economics. Within this high reliability the following general tendencies and comments can be observed.

The blocking systems tend toward higher dependability than security. Failure to establish a blocking signal from a remote terminal can result in overtripping for external faults.

On the other hand, transfer trip systems tend toward higher security than dependability. A failure to receive the channel signal results in a failure to trip for internal faults. The transfer trip systems require extra logic for internal fault operation at a local terminal when the remote terminal breaker is open, or for a "weak feed," where the fault contribution is too low to send a trip signal. This is not a problem with blocking and unblocking systems.

Unblocking systems offer a good compromise of both high dependability (channel not required to trip) and high security (blocking is continuous).

The directional comparison blocking system is most adaptable to multiterminals or taps on the line. A weak feed, with its very low fault contribution to internal faults, does not prevent the other terminals from tripping. However, tripping the weak-feed terminal may require a separate transfer trip system and channel. For nonsource taps where the relays at the other terminals cannot be set selectively with the tap, a blocking-only terminal at the tap can be used to block remote terminal tripping for external faults at the tap.

Historically, directional comparison systems use the same distance relays for both pilot primary and for zone 1 and backup protection. Phase comparison offers complete independence between the pilot and backup and also the possibility of not being dependent on system voltage measurement.

In EHV and UHV protection, where two pilot systems are commonly used, one directional comparison and one phase comparison system provide two completely different types, which can appear attractive. However, to some this can be a burden—two different systems to be familiar with and to maintain.

Pilot-wire relays are recommended and widely applied for short lines such as exist in utility distribution and industrial plant complexes. Experience over many years indicates that nearly all problems are pilot-wire related, often the result of inadequate design, protection, and/or maintenance. When these are resolved, the performance is very high. Fiber optic pilot circuits are being introduced which offer the advantage of having no electrical-related problems.

All types and combinations of pilot channels are in use. The choice is based primarily on availability and economics moderated by personal preference. Power-line carrier has been in use for many years and has the advantage that the channel is under the control of the protection group or within the user company. However, available frequencies can be a problem, as the spectrum is crowded. Frequencies can be repeated within the system where interference does not occur. These RF frequencies are not licensed and so may be used by outside users in the area.

Microwave channels are used where they are available or justified for other uses. They are normally owned by the channel user.

Audio tones on leased or private-line channels have very good operating records. It is very important that the leasing company be aware of the critical nature and requirements of protection channels.

It is interesting to observe again that all systems are in use. Where two pilot systems are applied, no one combination appears to predominate. All of which again emphasizes that relaying is largely a matter of "personality."

BIBLIOGRAPHY

See the Bibliography at the end of Chapter 1 for additional information.

Bayless, R. S., Single Phase Switching Scheme Protects 500 kV Line, *Transmission Distribution*, 35, No. 1, 1983, pp. 24–29, 47.

Bratton, R. E., Transfer Trip Relaying over a Digitally Multiplexed Fiber Optic Link, *IEEE Trans. Power Appar. Syst.*, PAS 103, 1984, pp. 403–406.

Edwards, L., J. W. Chadwick, H. A. Riesch, and L. E. Smith, Single Pole Switching on TVA's Paradise-Davidson 500 kV, Line Design concepts and Staged Fault Test Results, *IEEE Trans. Power Appar. Syst.*, PAS 90, 1971, pp. 2436–2450.

Guide for Protective Relay Applications of Audio Tones Over Telephone Channels, ANSI/IEEE Standard C37.93-1976, IEEE Service Center.

Guide for the Maximum Electric Power Station Ground Potential Rise and Induced Voltage from a Power Fault, IEEE Standard 367–1979, IEEE Service Center.

Guide for the Protection of Wire Line Communication Facilities Serving Electric Power Stations, IEEE Standard 487–1980, IEEE Service Center.

Service Conditions for Power System Communications Apparatus, IEEE Standard 281–1968, IEEE Service Center.

Sun, S. C., and R. E. Ray, A Current Differential Relay System Using Fiber Optics Communication, *IEEE Trans. Power Appar. Syst.*, PAS 102, 1983, pp 410–419.

W. H. Van Zee, and R. J. Felton, 500-kV System Relaying, Design and Operating Experience, CIGRE, 1978, Paper 34-7.

14

Stability, Reclosing, and Load Shedding

14.1 INTRODUCTION

The emphasis in the preceding chapters has been directed toward the protection of the various components of the power system. Now we examine the operation of the power system as a whole or unit, how the protection is effected, and how relaying can be applied or modified to minimize system disturbances.

The normal operation of generating, transmitting, distributing, and utilizing ac electric power is of little concern to the protection systems as long as the load current does not exceed the specified maximums. Regulating relays and equipment are involved, but these are beyond the scope of this book. However, faults, system abnormalities, sudden increases or losses of large blocks of load or generation, and loss of equipment cause power system disturbances that can affect the protection. Also, incorrect application, setting, or operation of the protection can result in system disturbances and loss of service.

14.2 ELECTRIC POWER AND POWER TRANSMISSION

Power in an electrical system is

$$P + jQ = \dot{V}\hat{I} = \bar{V}\bar{I}e^{j\theta} = \bar{V}\bar{I}\ \angle\theta^{o} \qquad (14.1)$$

where P is the real power (W, kW, MW), Q the quadrature or reactive power (var, kvar, Mvar), $\dot{V}$ the phasor voltage (V, kV, MV), and $\hat{I}$ the conjugate phasor current (A, kA, MA). $\bar{V}$ and $\bar{I}$ are the scaler magnitudes, with θ the angle the current lags the voltage. Thus Q is positive for lagging reactive power, commonly designated simply as reactive power. Negative Q indicates leading reactive power. Power transmission across a system can be expressed in terms of sending-end power ($P_S + jQ_S$) and receiving-end power ($P_R + jQ_R$).

For simplification it is practical to neglect the resistance and losses over the system, so that

$$P_S = P_R = P = \frac{V_S V_R}{X} \sin \phi \tag{14.2}$$

where V_S and V_R are the sending- and receiving-end voltages, ϕ the angle by which V_S leads V_R, and X the total reactance between V_S and V_R.

Equation (14.2) can be plotted as shown in Fig. 14.1. At no load across the system, $V_S = V_R$, $\phi = 0°$, and $P = 0$. The maximum power transfer occurs when $\phi = 90°$ and is

$$P_{Max} = \frac{V_S V_R}{X} \tag{14.3}$$

Thus more power can be transmitted between parts of the system by (1) raising the voltage V_S, V_R, and/or (2) reducing the reactance X. The first is the logic for the higher-voltage systems (HV, EHV, UHV). The second is accomplished by more interconnecting lines and/or connecting series capacitors in the line. Both are in use in a variety of combinations.

14.3 STEADY-STATE OPERATION AND STABILITY

The normal operation of the power system involves an exchange of power between the generators and the various loads. In the typical simplified system of Fig. 14.1, V_S and V_R may represent the voltages developed by generators supplying loads connected in the system, such as at buses G and H, or V_S can represent the voltage of generators and V_R the voltage of motor loads. Although the operations are similar in either case, the latter case is easier to visualize. Assume an equilibrium state at a power level P'. With no losses the mechanical input P'_{MS} to the generators equals their output, P'_{ES}. This is transferred across the system to equal the input P'_{ER} to the motors, which in turn equals the mechanical output of the motors, P'_{MR}. Thus before any fault and with all lines in service, the electrical transmission requirements of Fig. 14.1b show that V_S, the voltage behind direct synchronous reactance, leads V_R, the motor internal voltage, by ϕ'.

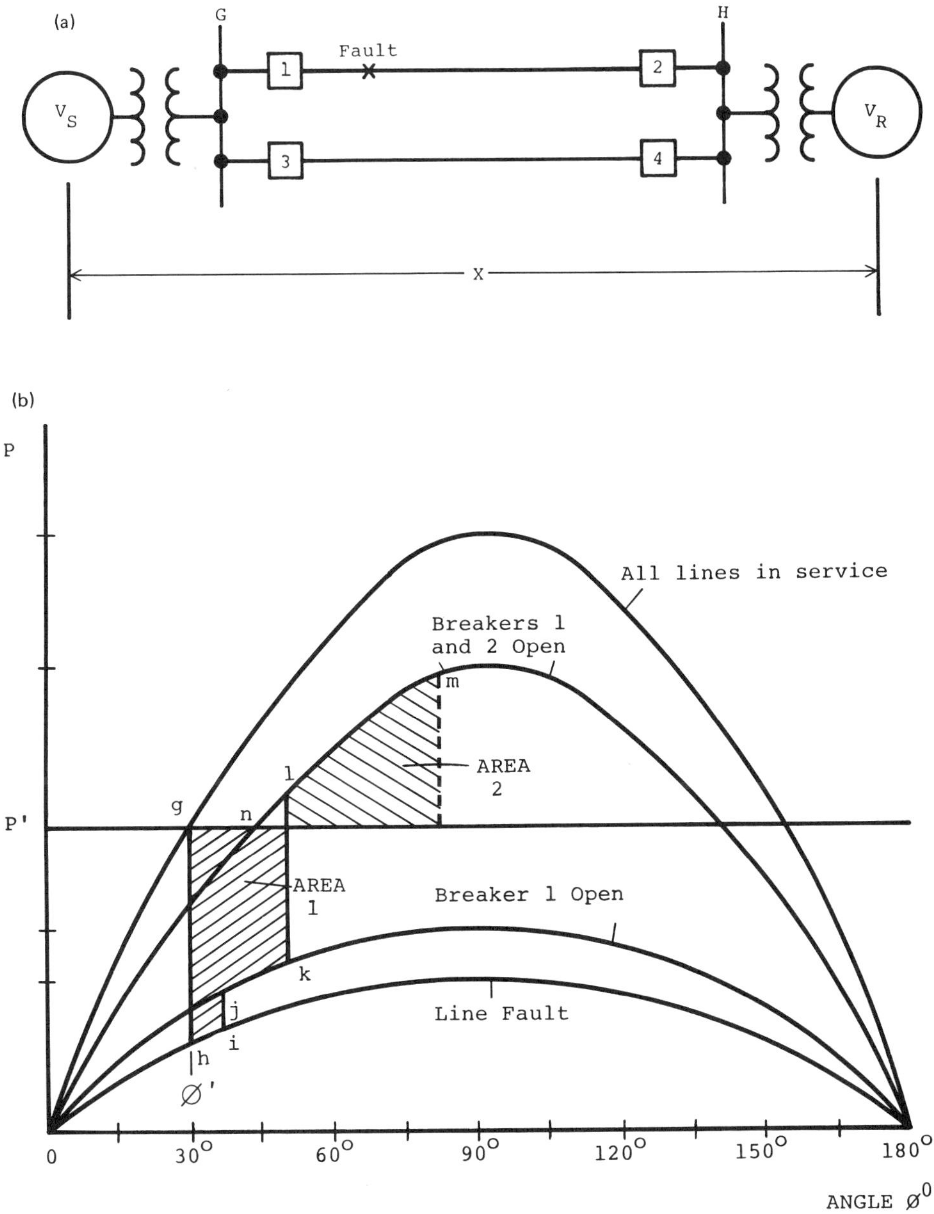

Fig. 14.1 Typical power-angle curves for a simplified two-machine power system; (a) system diagram; (b) power angle curves for various system conditions.

As the load requirements change P' will increase or decrease and the system will adjust to a new operating angle Ø. Steady-state stability is the ability of the system to adjust to gradual changes, and the limit is 90° for the two-machine no-loss system. Equation (14.3) defines the maximum power that the electrical system can transfer.

14.4 TRANSIENT OPERATION AND STABILITY

System faults, line switching, and the loss or application of large blocks of load can result in sudden changes in the electrical characteristics while the mechanical inputs and outputs remain relatively constant. It is important that the system readjust to these changes in order to continue operation and service. If the changes are too severe, instability results and the various parts of the system no longer operate together in synchronism. The resulting loss of synchronism or out-of-step operation requires that the parts be separated, stabilized, and resynchronized in order to continue complete service. Transient stability defines the ability of the power system to adjust to large and sudden changes.

The simplified two-machine system provides a good and adequate basis for understanding transient system operation and the interactions with the protection. A generator plant with a single tie to the power system or an industrial tie with rotating equipment and possibly generation to a utility essentially are two-machine systems. Most utilities, especially the large ones, consist of multiple generators at different locations, connected together by a network of lines and intermediate stations. although their transient performance is much more complex, the effect on the protection is similar to that of a two-machine system. Modern computer programs provide the necessary data for specific systems and operating conditions.

The most common sudden disturbance results from shunt faults. When a fault occurs, the transfer impedance X (Fig. 14.1), between V_S and V_R suddenly increases and instantly changes the power transmission capability of the system. For a fault out on line 1-2, the power angle curve instantly moves from "all lines in service" to the "line fault" curve. With prefault operation at power level P' and angle Ø', the change resulting from the fault reduces the electrical power transmission capability from g to h (Fig. 14.1b). The mechanical input and outputs momentarily are the same, so the generator accelerates and the motor deaccelerates, both increasing the angle Ø. The relays at 1 sense the fault and operate to open breaker 1, which changes the transmission capability to the "breaker 1 open" power curve. During this time to the opening of breaker 1, angle Ø moves from h to i. Acceleration continues to increase angle Ø from j to k when breaker 2 is opened by the relays at 2. This again changes the transmission capability, now to the power curve "breakers 1 and 2 open." At 1 the electrical transmission capability is now greater than the mechanical requirements, so the generators decelerate and the motors accelerate. Although the accelerating power goes through zero as kl crosses P', the swing continues (angle Ø increases) since the velocity of the rotating masses cannot be changed instantly. The shaded area 1 represents the kinetic energy added to the rotating masses; area 2, kinetic energy returned to the circuit. Thus the swing will continue to m, where area 2 equals area 1. At m the swing reverses, oscillating below and above the mechanical requirement, line P', with the final operating at n. This assumes no change in the power P' requirements. The oscillation is damped by the resistance in the system, the voltage regulator, and governor action.

The above represents a stable system. If the area above the P' operating power line were less than area 1 below, the system cannot maintain synchronism and will go out of step. This represents an unstable system.

Thus it is desirable to keep area 1 as low as possible. This can be aided by fast relaying and breaker operation, and as discussed in Chapter 13, pilot protection provides this possibility. With high-speed simultaneous tripping and high-speed breakers, area 1 can be kept small, so that more power (P') can be transferred across the system without it exceeding area 2 possibility.

The type and location of the disturbance is an important factor. In Fig. 14.1b system, the worse fault is a solid three-phase fault at bus G or bus H. In these cases no power can be transmitted across the system during the fault. Usually in a highly interconnected network there will be multiple ties between V_S and bus G, and between bus H and V_R, so some power can be transferred for these three-phase solid faults. This is similar to a solid three-phase fault out on line 1-2, where the power-angle curve typically is as shown as "line fault." As three-phase faults are not too common, stability studies often use the two-phase-to-ground fault as the criterion.

The time for Ø to change during system swings involves the solution of a second-order differential equation which includes the inertia constants or characteristics of the rotating masses. A common method is a step-by-step solution where during small time intervals the accelerating power and angular acceleration are assumed constant. This provides a time-angle curve or data where if the system is stable the angle eventually decreases after the initial disturbance increase.

The voltages and currents during a system swing, particularly during out-of-step operation, are shown in Fig. 14.2. Considering the simplified two-machine system, V_S is considered fixed, with V_R lagging relative to V_S. For normal-load operation, V_R will lag, as defined by the system power-angle curve and the loading.

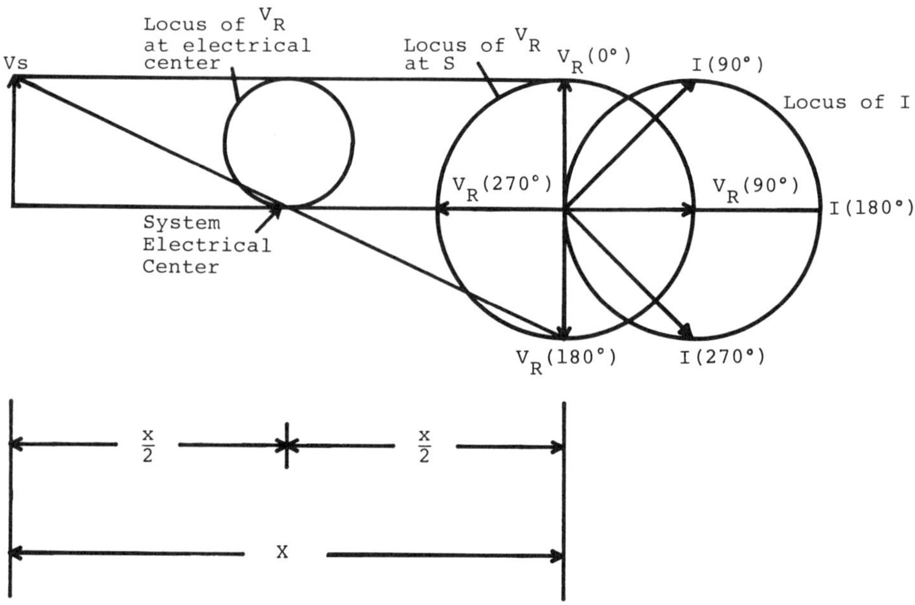

Fig. 14.2 Typical two-machine system voltages and currents as V_R rotates relative to V_S.

As the result of a sudden large disturbance, V_R can swing to as much as 120° approximately in a stable system. If the system is unstable, V_R continues to lag and at 180° a "pole is slipped," synchronism between V_S and V_R is lost, and the two parts operate out of phase. As a result, the voltage and current fluctuates. The electrical center is the point in the system where the impedance to the two sources is equal. At this center the voltage varies from maximum to zero, as shown by the locus circle. At other points the voltage varies from maximum to the diagonal line between V_S and V_R at 180°. In the simple reactance system the current changes from normal to around twice normal when V_S and V_R are 180° out of phase. It should be recognized that during transient operation, the transient reactance of the machine is used rather than the synchronous impedance for normal-load operation. The difference in these values is large, but the effect on the total impedance between the rotating masses may or may not be appreciable. In general, for low- to high-voltage systems, most of the impedance is in the system, so the total X is not affected appreciably. However, in EHV and UHV systems with very large generating units, the effect is quite significant. In fact, in these systems the electrical center is often within the unit generator transformer or within the generator.

After a system disturbance the rate of the swing and the voltage-current fluctuations is quite variable. At first the swing angle moves relatively fast, slowing down as the maximum angle is reached, and if the system is stable, stopping, then slowly changing to decrease the angle, thus oscillating with smaller angular changes until equilibrium is reached.

If the system cannot maintain stability, the angle changes slowly to around 180°, when a pole is slipped and a sudden large angular change occurs. This process is repeated with increasing speed each time a pole is slipped, until the system is separated.

14.5 SYSTEM SWINGS AND PROTECTION

As a rule the phase overcurrent relays should not operate on system swings and out-of-step operation. The reasons are that this type of protection normally is applied in the lower-voltage parts of the system, where the stability limits are seldom as critical, and that these relays are set well above load current. With the out-of-step currents not being too high, the relay operating times at these low multiples would be long. If the out-of-step condition were permitted to continue, there would be a possibility of the time-overcurrent relays "notching up" and operating after repeated current peaks. Instantaneous overcurrent relays normally set well above the maximum load may or may not operate on an out-of-step system swing.

Stability tends to be more critical in the higher-voltage systems. The reduced voltage and increased current during system swings appear as a variable impedance to the distance relays commonly applied for protection. This effect can be illustrated with the simplified two-machine system and the R-X diagram. In Fig. 14.3b the total system equivalent impedance after the fault is cleared by opening line 1-2 (Fig. 14.1a) is plotted between V_S and V_R on the R-X diagram. Arbitrarily, the origin is at a relay location such as 3 at bus G.

The locus of the system swing assuming $|V_S| = |V_R|$ is along a perpendicular line bisecting the line drawn between the V_S and V_R points as shown. On this swing locus the impedance points at various angles between voltages V_S and V_R are the points of the corresponding angle between lines from points V_S and V_R and the swing locus. Thus on the $|V_S| = |V_R|$ swing locus, impedance points are plotted for 60°, 90°, 120°, 180°, 240°, and 270° angles that V_R lags V_S. The impedances "seen" by distance relays 3 at bus G and relays 4 at bus H are the phasors from points G and H, respectively, to the locus line. The change in this impedance is relatively slow and variable, as discussed earlier.

If V_S lags V_R, the normal load is in the second quadrant rather than the first quadrant, and the swings move from left to right.

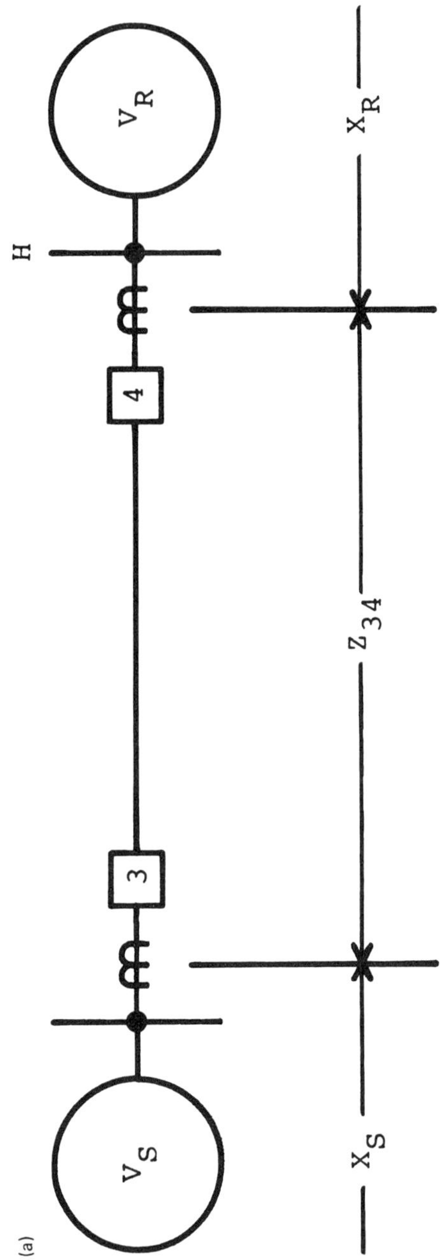

Fig. 14.3 Typical impedance swing loci for the simplified two-machine system after a fault is cleared: (a) system of Fig. 14.1a after line 1.2 is opened to clear fault; (b) typical distance relay settings at 3 bus G and the system-swing loci after faulted line 1–2 opened.

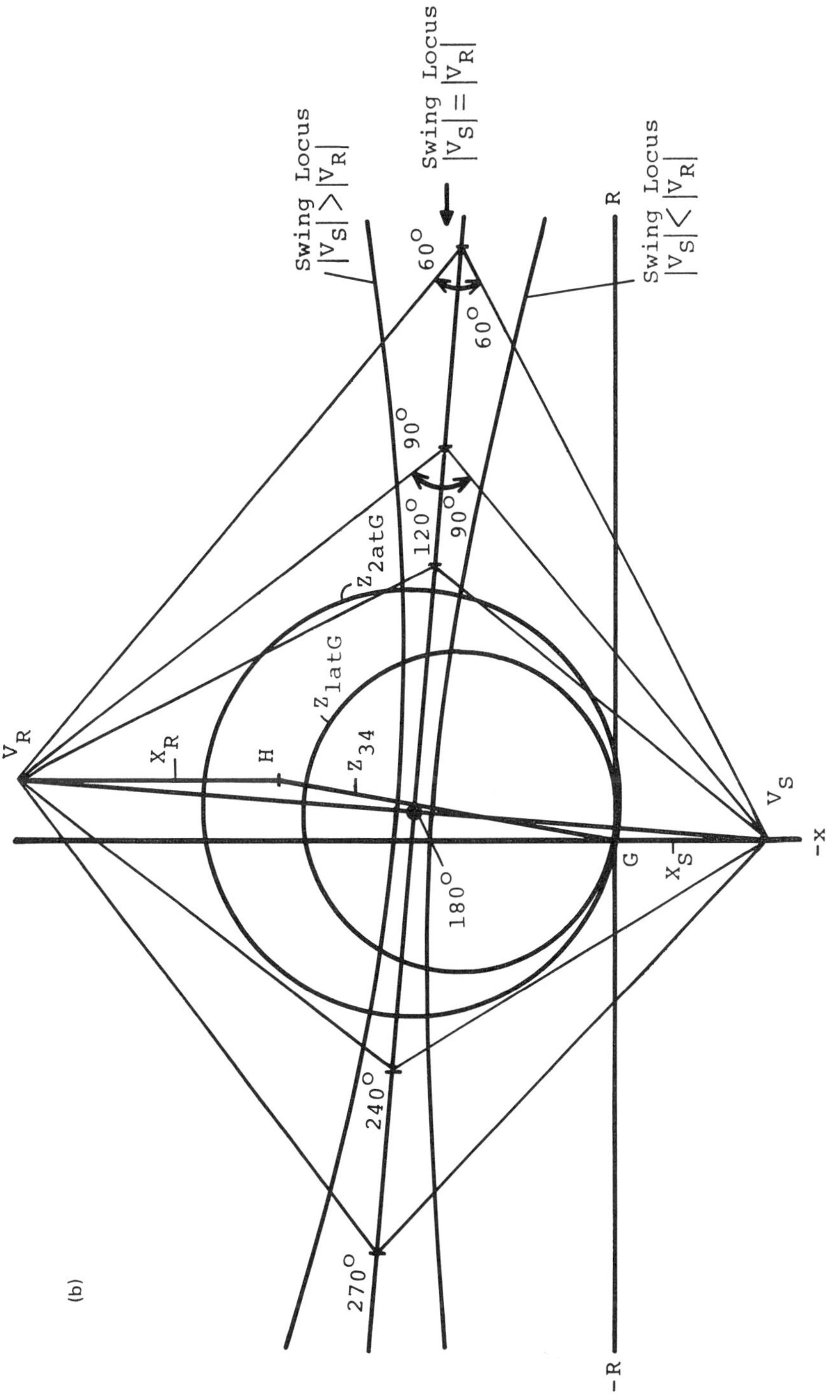
Swing Locus |V_S| > |V_R|
Swing Locus |V_S| = |V_R|
Swing Locus |V_S| < |V_R|
R
-R
-X
V_R
V_S
X_R
X_S
H
G
Z_34
Z_1atG
Z_2atG
60°
60°
90°
90°
120°
180°
240°
270°

(b)

Actual swing curves are more complex and are arcs of curves which are functions of the difference in V_S and V_R magnitudes, system impedance and voltage variables, governor and regulator actions, and so on. Typical swing locuses for the differences in the two source voltage magnitudes are illustrated in fig. 14.3b. In actual practice, the swing impedance as "seen" from any given relay may be anywhere in the four quadrants and move in any direction during a swing segment. With the disturbance usually resulting from a fault, there can be a major shift of the impedance seen by the relays as the fault is cleared to change the transfer impedance (as illustrated in Fig. 14.1b) across the system. Computer studies provide more accurate information for specific systems and conditions and should be utilized for determining relay operation, settings, and applications.

However, the R-X diagram scenario is similar for all cases. Plotting only a specific line impedance under consideration on the R-X diagram (since the impedance behind each line bus will vary with switching), normal operating conditions (load) should appear as impedances outside the relay operating characteristics. Normally for the relay at the origin, load into the line is in the first quadrant, load out of the line is in the second quadrant.

When a system disturbance occurs, the impedance changes instantly. If the fault is the disturbance and is within the line under consideration, the impedance seen by the distance relays moves into their operating zone and the line is tripped. If the disturbance is outside the line under study the impedance seen instantly shifts to a new location somewhere in the four quadrants. During a fault there is a small swing period, then another shift as the two line breakers clear the fault. If the breakers do not open simultaneously, there will be the swing during the fault, an instant shift to a new location when one breaker opens, another small swing change, then an instant shift to still another location as the second breaker opens. With the fault now cleared, the swing continues, typically as illustrated in Fig. 14.3b. For the example of Fig. 14.1 the swing is assumed to start, as shown in Fig. 14.3b, at about an angle Ø of 50°, which corresponds to the point "l" angle in Fig. 14.1b.

If the system is stable, the impedance locus will move with decreasing speed to about 80°, which corresponds to point m in Fig. 14.1b, and where area 1 = area 2. Then the impedance phasor momentarily stops, reverses, and moves out to the right. Eventually, after short oscillations, a new equilibrium is reached at which the system will operate until the next load change or disturbance occurs.

If stability is lost, the impedance phasor will move beyond the 80° point in the example to 180°, where a pole is slipped and the two parts now operate out of synchronism. The impedance change generally slows down to the 180° point, then moves rapidly to the left and back to the right to pass through the 180° point again unless the two parts are separated somewhere. As indicated earlier, generally systems can swing on a transient basis to around 120° and remain stable.

Thus distance relays that respond to balanced three-phase conditions and faults will operate when the swing or out-of-step impedance phasor moves into and remains long enough within their operating characteristic. Directional comparison pilot systems utilizing phase distance fault detectors are also subject to operation as a swing phasor impedance moves through their operating zone. Phase comparison pilot systems operating on current magnitudes are not subject to operation on swings and out-of-step conditions.

14.6 OUT-OF-STEP DETECTION BY DISTANCE RELAYS

From the previous discussion, the "handle" for relay detection of out-of-step conditions is the rate of impedance change "seen" by distance relays. Detection can be accomplished with two distance units, one set for a longer reach and essentially surrounding a unit set with a shorter reach. With this combination a fault falling within the operating zone will operate both units essentially simultaneously. However, a swing that occurs within their operating zones will first operate the outer or longer reach set unit, and as it progresses, later operate the inner or shorter-reach set unit. Logic is arranged so that output will occur only for a sequential operation of the two distance units. Typically, an out-of-step condition is indicated when the time difference between the two distance units' operation is approximately 60 ms or more, with a typical distance unit setting about 20 to 30% apart. The two types most commonly applied are illustrated in Fig. 14.4. These distance relay characteristics were described in Section 6.7 (Fig. 6.11).

The concentric circle application is generally applicable to the shorter lines up to around 100 mi, and where the outer circle 21_{OS} can be set not to operate on the maximum load impedance phasor. On long and heavily loaded lines, the blinder type generally will be required for line protection as well as swing detection, to provide extended reach for the line with restricted tripping for loads.

Application of out-of-step relays is complex, depending on system factors, operating practices, and individual protection philosophies. As a result, there is little uniformity among the utilities. However, the basic objectives are the same.

1. Avoid tripping for any disturbance for which the system can recover. Thus relay operation should not occur for all stable swings. Stability studies can provide the minimum impedance the relays will "see" for stable operation, so that the phase distance relays can be set to avoid operation on these impedance values.

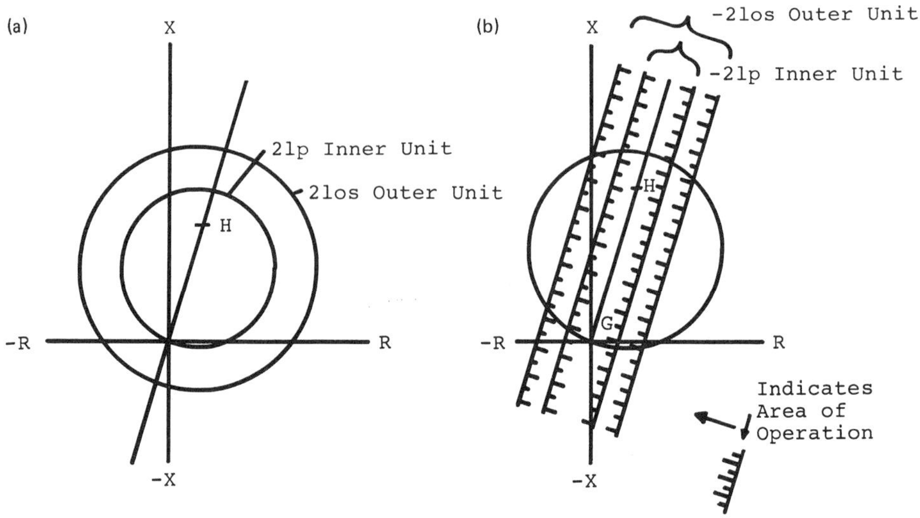

Fig. 14.4 Typical distance units applied for out-of-step detection: (a) concentric circle type; (b) double blinder type.

2. For unstable system operation, it is mandatory to separate the system, as after it goes out of step it cannot recover without separation. This separation preferably should be done at a point where (a) the generation-load balance will permit the parts to continue service temporarily, and (b) at a convenient location where the parts can be resynchronized and reconnected. In most cases these requirements will be in conflict with (a), varying depending on the system operation at the time of the disturbance. As a result, considerable compromise and judgment is required; there are no easy answers.

In the past the general practice was to apply out-of-step blocking in the transmission system. The aim was to keep the system intact to avoid possible long delays after a system disturbance in having to reconnect the various lines that may have opened. In practice it is doubtful that many line circuits would be involved and trip during a system disturbance. Thus the trend seems to be toward more out-of-step blocking of automatic reclosing after the line relays operate, as this only adds more "shock" to an already disturbed system.

Out-of-step tripping has not been used widely over the years but is receiving considerably more attention, particularly with very large generator units connecting into EHV and UHV networks. The higher generator reactances and lower inertia constants have reduced critical clearing times. Thus stability limits tend to be more critical and early separation more important should the system become unstable.

On the other hand, opening a circuit breaker when the two parts of the system are essentially 180° apart may result in breaker damage. This is because of the high recovery voltage across the breaker contacts when interrrupting at around 180° between V_S and V_R. An ANSI standard specifies that out-of-phase switching breakers must be able to interrupt a current 25% of the rated interrupting current with a recovery voltage of 3.53 pu rated line-to-neutral voltage. Various out-of-step tripping schemes are available to detect out-of-step conditions as the impedance phasor moves into the operating areas, such as shown in Fig. 14.4, but to inhibit breaker tripping until the out-of-step swing impedance phasor passes out of the outer unit 21_{OS}. At that time the angle of V_S and V_R is less than 180° and decreasing to a more favorable angle for breaker interruption.

Specific and additional information on the application and details of available schemes for line and generator out-of-step indication and protection should be obtained from the various manufacturers.

14.7 AUTOMATIC LINE RECLOSING

Somewhere around 80 to as high as 90% of faults on most overhead lines are transient. These result principally from flashover of the insulators by high transient voltages induced by lightning, by wind causing the conductors to move together to flashover, or from temporary tree contact, usually by wind. Therefore, by deenergizing the line long enough for the fault source to pass and the fault arc to deionize, service can be restored more expeditiously by automatically reclosing the breaker. This can significantly reduce the outage time, to provide a much higher service continuity to customers.

Reclosing can be one single attempt (one-shot) or several attempts at various time intervals (multiple-shot). The first attempt may be either "instantaneous" or with time delay. "Instantaneous" here signifies no intentional time delay, where the circuit breaker closing coil is energized very shortly after the trip coil has been energized by the protective relays. This is accomplished by a high-speed (52bb) auxiliary switch on the circuit breaker which closes as soon as the breaker mechanism starts moving to open the main interrupting contacts. Thus both the trip and close coils are energized essentially together, so that the breaker contacts are open for only a very short time. The time from energizing the trip coil to the reclosing of the breaker contacts on an instantaneous reclose cycle is a function of the breaker design. typical values are 20 to 30 cycles. Generally, this is sufficient to permit the fault arc to deionize during the period when the circuit is open. Long experience has indicated that in a three-phase circuit, on the average the deenergized (open or dead) time for the fault arc to deionize and not restrike is

$$t = \frac{kV}{34.5} + 10.5 \text{ cycles} \tag{14.4}$$

This formula applies where all three phases are opened. For single-pole trip-reclose operation, longer deionizing times are required, as energy coupled from the unopened phases can keep the arc active.

Many varieties of reclosing relays (79) are available with varying degrees of sophistication for single and multiple shots. After their cycle of operation, if reclosures are successful at any point, they reset after a preset time interval. If reclosure is not successful, they move to a lockout position so that no further automatic operation is possible. After service is restored by manual operation of the circuit breaker, the reclosing relays reset after a time delay. This is to prevent an automatic reclose operation should the breaker inadvertently be closed manually with a fault still existing on the line.

As many stations and substations are unattended and are operated by various remote control systems, the reclosing requirements can be programmed into the automation and control from a central dispatch office. A few operations of this type are now in service.

Reclosing practices are quite variable depending on the type of circuit and individual company philosophies. A brief review will outline the more common practices.

14.8 DISTRIBUTION FEEDER RECLOSING

For an overhead radial feeder without synchronous machines or with minimum induction motor load, fast reclosing at the source offers a relative high probability of rapid restoration of service. The almost universal practice for these types of circuits is to use three and occasionally four attempts to restore service before locking out. Subsequent energizing is by manual closing. Generally, the first reclose is "instantaneous," the second around 10 to 20 seconds after retrip, the third attempt about 20 to as high as 145 s after retrip. Specific times vary widely. When applying four reclosure attempts, the times between them are smaller.

Reclosing is generally not used for underground feeders, as cable faults are more inclined to be permanent. Reclosing is used in practice for combination overhead-underground circuits, with more inclination to use it where the percentage of overhead to underground is high.

For industrial ties with large motor loads and/or synchronous machines and for ties connecting "cogeneration" sources, care is required that reclosing at the utility substation does not cause damage to the rotating equipment. Generally, the connections are by single-circuit ties, so that during the open period the two parts will be operating independently of each other. Reclosing then will connect the two parts together out out of phase, resulting in power surges and large mechanical shaft torques that can cause severe damage.

With synchronous machines, reclosing should be delayed until the machines are removed from the circuit; then reclosing can restore voltage and pick up the static and nonsynchronous loads at the station. One effective way to drop the machines is by using underfrequency relays, operable as long as the load at the plant exceeds the available local generation.

For induction motors, reclosing should not take place until their residual voltage has decreased to 33% of rated value. In general, the induction motor voltage decreases very rapidly, as there is no field to supply excitation.

Joint cooperation is very important with these types of interconnections.

Where there are mutliple fused taps off a distribution feeder, a common application is to apply an instantaneous trip unit set to overreach these fuses out to the first main feeder sectionalizing point. Fast reclosing is then used and if unsuccessful, further operation of the low-set instantaneous relays is blocked. With the high percentage of transient faults, this prevents the fuse from operating should the fault be on the down side of the fuse, thereby minimizing long delays in fuse replacement. If the fault is permanent, the fuse will clear faults within their operating zone and range. This is known as "fuse saving."

14.9 SUBTRANSMISSION AND TRANSMISSION-LINE RECLOSING

Circuits used to transfer power between stations require reclosing of both terminals to restore service. This can be done at high speed only if there are sufficient parallel ties in the network to provide an exchange of power between the two parts of the system. This is necessary to hold the source voltages in synchronism and essentially in phase during the line-open period. This reclosing requires high-speed simultaneous tripping of the line and so is limited to circuits protected by pilot protection. If this protection is not used or is out of service, fast reclosing should not be used or should be blocked.

Where the effectiveness of the parallel ties is in doubt, several reclosing possibilities exist:

1. Single-pole pilot trip and reclose. This is described in Chapter 13.

2. With pilot relay protection for simultaneous line tripping, reclose one end instantaneously and the other end after a check to assure that the line and bus voltages are synchronous and within a preset angle difference. Synchroverifier (25) relays operate when the "beat" frequency across an open breaker is within prescribed limits, the two voltages on either side are of sufficient magnitude, and the angle between them within a preset value. Typical angle adjustments are about 20 to 60°.
3. Without pilot protection, or when preferred, reclose with (a) live line, dead bus; (b) dead line, live bus, and/or (c) live line, live bus with synchro-check. It is desirable to operate these through reclosing relays to avoid potential breaker "pumping."

Different combinations for reclosing are applicable and depend on the operating and the system requirements.

The successful reclosing of both terminals of a line increases area 2 availability (Fig. 14.1b) by moving from the "breaker 1 and 2 open" curve to the higher "all lines in service" curve shortly after point "1." This aids stability. On the other hand, if reclosing is not successful, such as for a permanent fault, area 1 is increased by the additional fault and retrip time operation along the low "line fault" curve. Area 2 is also reduced, so that the stability limit is reduced, since area 2 = area 1 at a much lower angle ϕ.

However, since most line faults are transient and in many cases the stability limit is never reached, simultaneous instantaneous reclosing is widely used for lines of 115 kV and up. Generally, only one attempt is made, but in some cases subsequent attempt(s) are made with the voltage check and/or synchro-check equipment outlined above. Again, this is only for overhead lines. Transmission cable circuits are not reclosed. For overhead-cable combinations, separate protection is sometimes used for the cable and overhead section, with reclosing permitted for the overhead line faults only.

Questions have been raised as to the application of instantaneous reclosing at or near large generating stations. This is because of the potential damage to the long turbine shafts of large modern turbine generator units. An example is a solid or near-solid three-phase fault near or at a generator bus. This reduces the voltage of the three phases essentially to zero, so that no power can be transferred until the fault is cleared. Even with very fast relaying and circuit breakers, the voltages on the two sides of the breaker are at a different angle, so reclosing results in a sudden shock and movement of the rotor and to transient oscillations and stresses. These are cumulative, so actual known damage may occur some years in the future. Past experience has not documented any large amount of damage but remains as a potential worry. Some utilities are removing reclosing near large generator stations; others are not. The issue is a very difficult one and not easily resolved. It is further complicated by the fact that the most severe types of faults (three-phase and two-phase to ground) occur rather infrequently. Thus perhaps the cumulation of the stresses will not exceed a critical damage point during the equipment lifetime. In the meantime, more benefits of reclosing are obtained.

This raises another difficult question: How far apart can the voltages be before damage and problems with stable operation occur? Out in the system, voltage differences on the order of 60° or more have existed before reclosing, and reclosing has been successful. Here the system impedance network can absorb this large difference. This would not be tolerated near a generating unit. It has not been possible to provide general limits, as each application depends on the specific system.

Reclosing on multiterminal lines is more complex, especially if more than two terminals have synchronous voltage sources. It may be practical to reclose one or two terminals instantaneously and the others after voltage or synchro-check. Where the transmission line has load taps, so that it serves the dual function of transmission and distribution, circuit reclosing of the source terminals can facilitate service to the tap loads. In these cases the requirements outlined above under distribution feeder reclosing also apply.

14.10 RECLOSING ON LINES WITH TRANSFORMERS OR REACTORS

Lines that terminate in a transformer bank without a breaker between them, or lines with shunt reactors, should not be reclosed automatically unless it can be assured that the fault is in the line section. Differential or transformer reactor protection can be used to block reclosing at the local station with transfer trip to block the remote station. This generally means some delay in reclosing to assure that the trouble is not in the transformer or reactor. If the line protection excludes the transformer bank, this protection can only initiate reclosing. Breakers are not generally used with shunt reactors, so the protection zone for the line would include the reactors.

Transformers are sometimes connected to a line through a motor-operated air break switch. For faults in the transformer the protection trips the low-side transformer breaker, initiates a remote trip via a grounding switch or transfer trip channel, and opens the air switch. Reclosing of the remote terminal(s) may be initiated after a coordinating time delay for the air switch to operate. This may be supervised by receipt of the remote trip signal.

14.11 AUTOMATIC SYNCHRONIZING

This equipment can be applied for automatic synchronizing at attended or unattended stations, or to assist manual synchronizing. It does not adjust the two separated systems, but will program closing as the voltages arrive in phase when the frequency difference is small. The synchro-verifier type is not recommended for this service but has been used for small machines on the order of 500 kVA.

14.12. FREQUENCY RELAYING FOR LOAD SHEDDING—LOAD SAVING

The primary application of underfrequency relays is to detect overload. Load shedding or load saving is the attempt to match load to the available generation after a disturbance that has left a deficiency in the generation relative to the connected loads. Thus the overloaded system or part of a system (island) begins a frequency decay which if not halted can result in a total system shutdown such as was experienced in the 1965 Northeast blackout. Generating plants generally cannot operate below 56 to 58 Hz (60-Hz base).

Normal load changes and moderate overloads can be absorbed by the spinning reserve in the system, as all the generators are usually not operating at full capacity. Thus these overloads result in small increments of reduced speed and frequency which activate the governors to increase the prime-mover input. As has been outlined above, transient changes such as those that result from faults involve the exchange of kinetic energy of the rotating masses to the system until the system can readjust to a new equilibrium.

When the load requirements significantly exceed the generation capabilities, the frequency of the system decreases. The power system can survive only if enough load is dropped until the generator outputs equal or are greater than that of the connected loads. This imbalance most often results from the loss of a key or major transmission line(s) or transformer(s) which are involved in a major transfer of power either within the system or between two interconnected systems. The causes can be faults cleared without high-speed reclosing, incorrect or accidental relay or manual operations, or other situations which interrupt large power flows.

The rate at which the frequency drops is a variable function with time, depending on the amount of overload, the system inertia constant, and the load and generator variations as the frequency changes. This is covered in detail in the articles by Dalziel and Steinbach and Berry et al. in the Bibliography.

The "handle" to detect these problems is underfrequency or the rate of frequency decline in the power system. In a large system there can be an almost infinite number of possibilities that can result in load-generator imbalance, so it becomes difficult to determine quickly and accurately where and what action should be taken. The present practices are to apply underfrequency relays at various load points set to progressively remove load blocks until the frequency decay is stopped and returns to normal. Generally, underfrequency is used, but rate of change of frequency has some use.

Any measurement at one point in the system, as is the case with underfrequency or rate-of-change relays, is an approximation of the problem. As a result, generally more load than may actually be necessary is probably shed, but it is far more important to avoid a massive shutdown.

It is desirable to locate the relays throughout the system to minimize possible "islanding" and heavy power flows. They are set at different frequency levels to trip varying amounts of load. General practice has been to use three frequency steps between about 59.8 to 58 or 57 Hz (60 Hz normal), although as many as five steps have been applied. As far as possible, nonessential loads are removed first, sometimes with rotation among different loads.

The early frequency relays (81) were electromechanical, with present relays of the digital type. The latter operate by counting the zero crossings of the filtered ac system voltage, and are extremely accurate.

Frequency relays can be used to restore or supervise the restoration of load after the system is stabilized and generation capability is available to meet the dropped load. If load restoration is done automatically, the loads should be added in small increments, with sufficient time intervals for the system to adjust, to avoid reversion to reduced frequency.

The application and setting of underfrequency relays is not standardized and is based for the large systems on a study of the most probable and worst-case possibilities seasoned with general experience and judgment. For interconnected systems such as exist in U.S. power pools, it is important that a common pool program be developed and implemented through the interconnected systems. This can involve separating the system when a heavy power interchange occurs that is unfavorable to an undisturbed system.

The present wide use of load shedding has both prevented many blackouts and limited some to small areas with minimum outage times. However, not all possibilities can be anticipated, so blackouts will continue to occur, hopefully quite limited and for very short outage times.

Ultimately perhaps, multiple measurements across the system can be sent to a central location for a programmed analysis of the problem, with solutions quickly dispatched to the proper locations for rapid matching of load and available generation. It appears that this will involve considerable data and communication facilities operating at high speeds.

14.13 FREQUENCY RELAYING FOR INDUSTRIAL SYSTEMS

In general, one underfrequency step is sufficient for the industrial ties that have local generation. The relays would trip loads on the loss of the utility supply, so that the remaining loads match the local generation capability. In this manner the most essential load for the plant can be maintained. The very rapid decline of frequency when the loads are large relative to the local generation probably makes more than one step of underfrequency relaying impossible.

It may be desirable to supervise plant load shedding with an undercurrent relay in the utility tie. This would prevent a utility frequency disturbance from necessarily shedding plant load as long as the tie is closed and the utility is able to restore frequency by shedding other, low-priority loads. The undercurrent relay operates to permit local load shedding only when the tie current (or power) is below a set value. A reverse power relay to measure power flowing from the plant to the utility may be necessary to prevent the local generator from supplying power to other loads connected on the same feeder when the utility substation breaker is open.

BIBLIOGRAPHY

See the Bibliography at the end of Chapter 1 for additional information.

A Status Report on Methods Used for System Preservation During Underfrequency Conditions. IEEE Power System Relaying Committee Working Group. *IEEE Trans. Power Appar. Sys.* PAS 94, 1975, pp. 360–366.

Automatic Reclosing of Transmission Lines, IEEE Power System Relaying Committee, *IEEE Trans. Power Appar. Syst.*, PAS 103, 1984, pp. 234–245.

Berry, Brown, Redmond, and Watson, Underfrequency Protection of the Ontario Hydro System, CIGRE, 1970, Paper 32-14.

Dalziel, C. E., and E. W. Steinbach, Underfrequency Protection of Power Systems for System Relief, Load Shedding System Splitting, *IEEE Trans. Power Appar. Syst.*, PAS 78, 1959, pp. 1227–1238.

Lokay, H. E., and V. Burtnyk, Application of Underfrequency Relay, for Automatic Load Shedding, *IEEE Trans. Power Appar. Syst.*, PAS 87, 1968, pp. 776–783.

Proposed Terms and Definitions for Power System Stability, Power Systems Engineering Committee, *IEEE Trans. Power Appar Syst.*, PAS 101, 1982, pp. 1894–1898.

Problems

These practical problems, developed over many years from actual experience, provide the opportunity to apply the above material by practicing the basic techniques. Problems involving application choices are generally so subjective that I have avoided them whenever possible. Each problem has a "message" hopefully "unlocked" with a reasonable minimum of "labor"—that is, with a good RH factor, where R equals relative minimum labor and H equals high educational value.

CHAPTER 2

2.1 A wye-connected generator has a nameplate rating of 200 MVA, 20 kV, and its subtransient reactance (X''_d) is 1.2 pu. Determine its reactance in ohms.

2.2 The generator of Problem 2.1 is connected in a power system where the base is specified as 100 MVA, 13.8 kV. What is the generator reactance (X''_d) in per unit on this system base?

2.3 Convert the per unit answer calculated in Problem 2.2 to ohms. Does this check the value determined in Problem 2.1?

2.4 Three 5 MVA single-phase transformers, each rated 8:1.39 kV, have a leakage impedance of 6%. These can be connected in a number of different ways to supply three identical 5 ohm resistive loads. Various transformer and load connections are outlined in the following table. Complete the table columns. Use a three-phase base of 15 MVA.

Table P 2.4

Case No.	Transformer connection		Load connection to secondary	Line-to-line base kV		Load R in per unit	Total Z as viewed from the high side	
	Primary	Secondary		HV	LV		Per unit	Ohms
1	Wye	Wye	Wye					
2	Wye	Wye	Delta					
3	Wye	Delta	Wye					
4	Wye	Delta	Delta					
5	Delta	Wye	Wye					
6	Delta	Wye	Delta					
7	Delta	Delta	Wye					
8	Delta	Delta	Delta					

2.5 A three-phase generator feeds three large synchronous motors over a 16 km, 115 kV transmission line, through a transformer bank, as shown. Draw an equivalent single-line reactance diagram with all reactances indicated in per unit on a 100 MVA, 13.8 kV or 115 kV base.

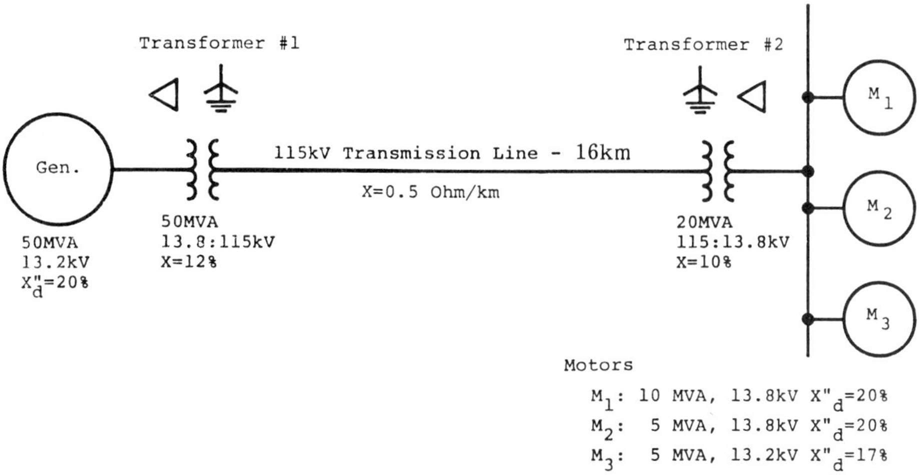

Fig. P2.5

2.6 In the system of Problem 2.5 it is desired to maintain the voltage at the motor bus of $1\angle 0°$ per unit. The three motors are operating at full rating and 90% PF.

a. Determine the voltage required at the generator terminals assuming that there is not voltage regulating taps or similar equipment in this system.

b. What is the voltage required behind the subtransient reactance?

CHAPTER 3

3.1 Four boxes represent an ac generator, reactor, resistor, and capacitor and are connected to a source bus XY as shown. From the circuit and phasor diagrams identify each box.

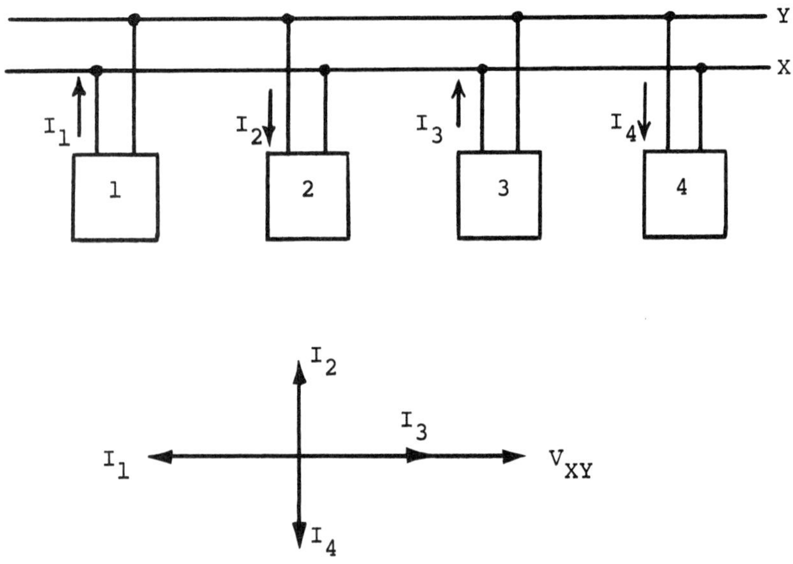

Fig. P3.1

3.2 Two transformer banks are connected to a common bus as shown on p. 503. What are the phase relations between the voltages V_{AN} and $V_{A'N'}$; V_{BN} and $V_{B'N'}$, V_{CN} and $V_{C'N'}$?

3.3 Reconnect transformer bank Number 2 of Problem 3.2 with the left windings in wye instead of delta, and the right windings in delta instead of wye so that V_{AN} and $V_{A'N'}$ are in phase, V_{BN} and $V_{B'N'}$ are in phase, and V_{CN} and $V_{C'N'}$ are in phase.

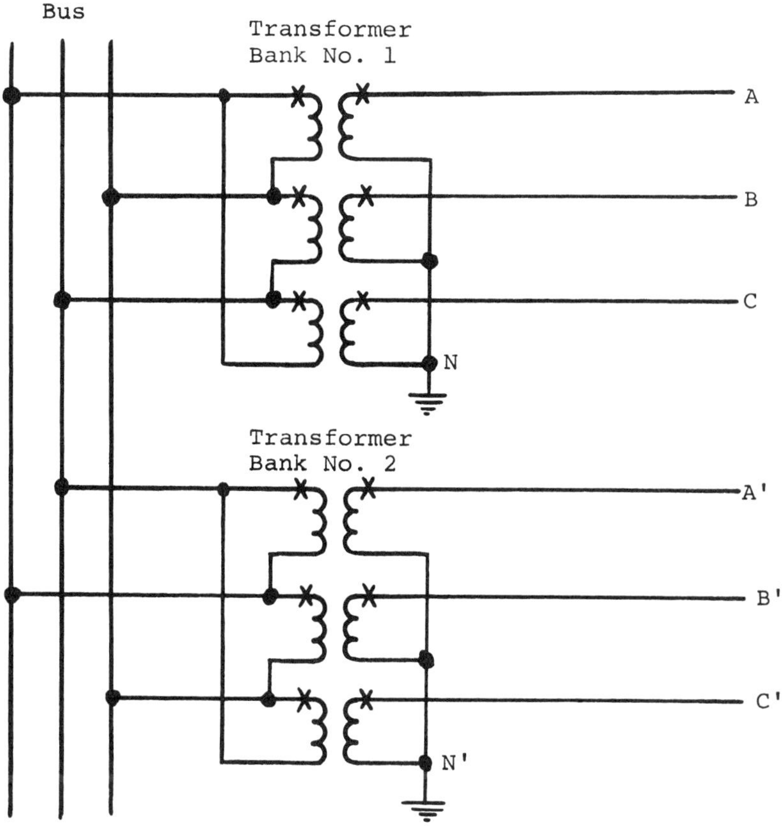

Fig. P3.2

3.4 The power transformer connections shown are nonstandard and quite unusual with today's standardization. However, this connection provides an excellent exercise in understanding phasors, polarity, and directional sensing relay connections.

Connect the three directional phase relays A, B, C to line-side CTs and bus-side VTs for proper operation for phase faults out on the line. Use the 90°–60° connection. Each directional relay has maximum torque when the applied current leads the applied voltage by 30°. The auxiliary VTs should be connected to provide the relays with equivalent line side voltages.

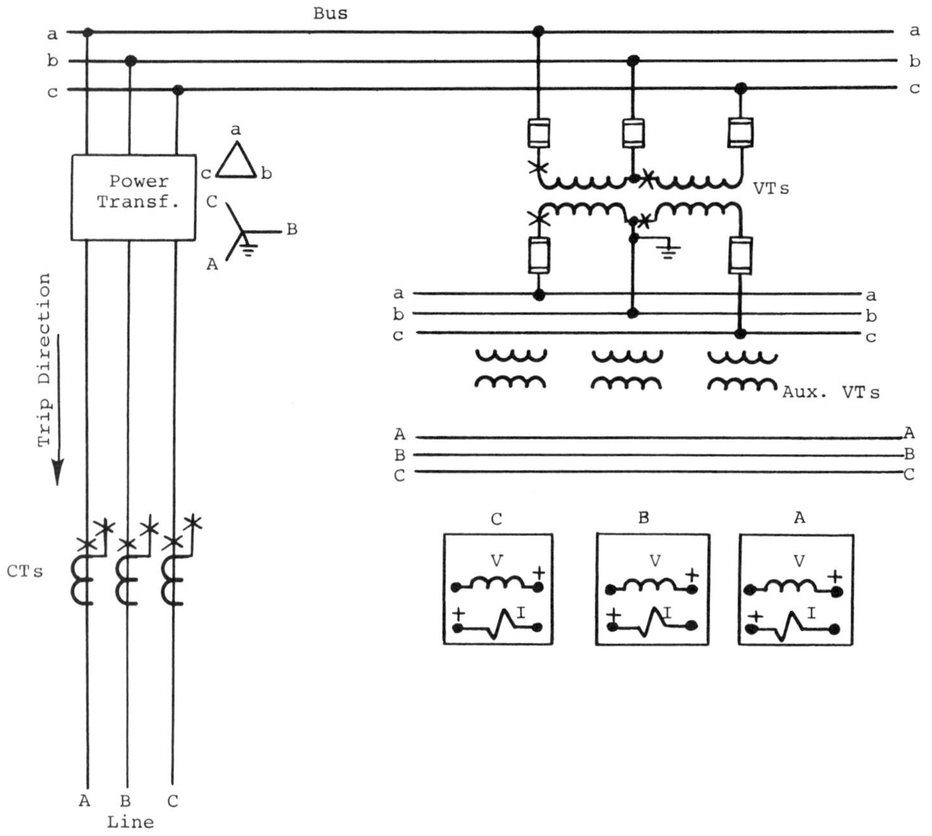

Fig. P3.4

CHAPTER 4

4.1 The per-unit currents for a phase-a-to-ground fault are shown in the diagram. Assume that the system is reactive with all resistances neglected, and that the generator(s) are operating at j 1.0 per unit voltage.

Draw the positive, negative, and zero sequence diagrams and describe the system that must exist to produce the current flow as shown.

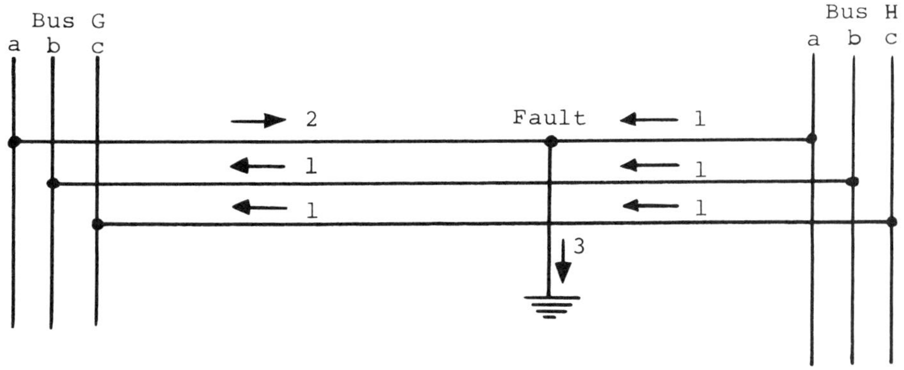

Fig. P4.1

4.2 For the system shown on p. 506:

a. Determine the source and equivalent star reactances of the transformer on a 30 MVA base.

b. Set up the positive, negative, and zero sequence networks. There are no fault sources in the 13.8 and 6.9 kV systems. Reduce these networks to single-sequence reactances for faults on the 13.8 kV side.

c. Calculate a three-phase fault at the 13.8 kV terminals of the transformer.

d. Calculate a single-phase-to-ground fault at the 13.8 kV transformer terminals.

e. For the fault of part d, determine the phase-to-neutral voltages at the fault.

f. For the fault of part d, determine the phase currents and the phase-to-neutral voltages on the 115 kV side.

g. For the fault of part d, determine the current flowing in the delta winding of the transformer in per unit and amperes.

h. Make an ampere-turn check for the fault currents flowing in the 115 kV, 13.8 kV, and 6.9 kV windings of the transformer.

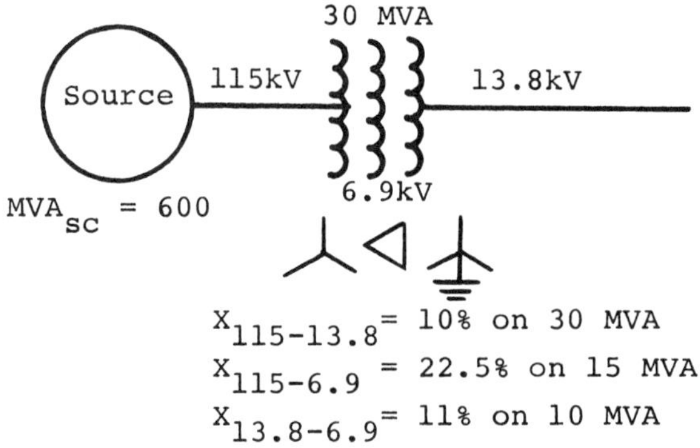

Fig. P4.2

4.3 For the system shown in Fig. P4.3:
   a. Determine the current flowing to the load. Assume that the generators of the equivalent source behind the 13.8 kV bus are operating at 1 per unit voltage at 0°.
   b. Calculate the voltage at the load bus.
   c. Calculate the fault current for a solid phase-a-to-ground fault at the load bus. Neglect load for this part.
   d. Calculate the currents flowing in the line when an open circuit occurs in the line just at the load bus. Assume phase a opens while carrying the load as indicated.

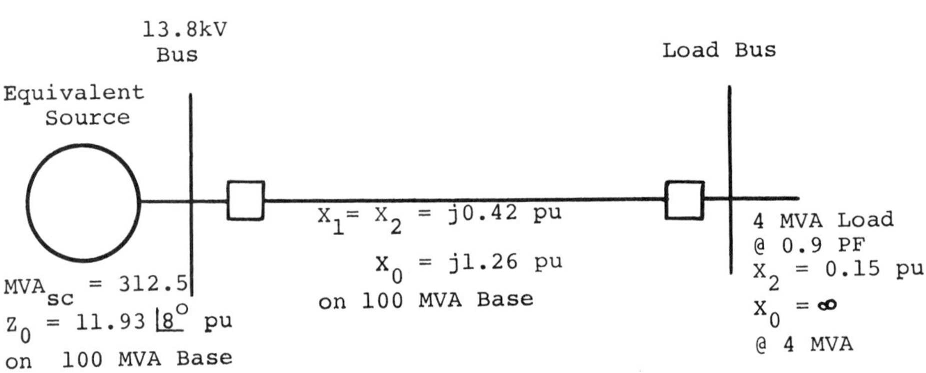

Fig. P4.3

e. Calculate the currents flowing when the open phase of a conductor of part d falls to ground on the source side of the open. Assume a solid fault.
f. Repeat part e but for the case when the opened conductor falls to ground on the load side.

4.4 Repeat the example of Section 4.14 for a solid phase-to-ground fault on the 161 kV terminals of the autotransformer. Compare the directions of the currents in the bank neutral and the tertiary with those for the fault on the 345 kV side.

CHAPTER 5

5.1 A 13.8 kV feeder circuit breaker has a 600:5 multiratio current transformer with characteristics as shown in Figure 5.10. The maximum load on the feeder is 80 amp primary. Phase time inverse overcurrent relays are connected to the CT secondaries. The relay burden is 3.2 VA at the tap values selected, and the lead burden is 0.38 ohm.
a. If the 100:5 CT ratio is used, then a relay tap of 5 amp is required in order for the relay pickup to be 125% above the maximum load. With these, determine the minimum primary current to just operate the relays.
b. For the selection of part a what is the approximate maximum symmetrical fault current for which the CTs will not saturate (use the ANSI/IEEE knee point)?
c. If the 200:5 CT ratio is used so that the 2.5 amp relay tap can be used, determine the minimum primary current to just operate the relays.
d. Repeat part b for the selection of part c.
e. Which of these two CT and relay selections would you recommend?

5.2 Determine the minimum CT ratio that might be used with a 0.5–2.5-amp ground relay with an instantaneous trip unit set at 10 amp. The total ground relay burden is 285 VA at 10 amp. See Figure 5.10 for CT characteristics.

5.3 A circuit has 800:5 wound-type CTs with characteristics as shown in Fig. 5.7. The maximum symmetrical fault for which the associated relays are to operate is 15,200 amp. Approximately what will be the error in percent if the total connected burden is 2.0 ohms? If 4.0 ohms?

5.4 The feeder of Problem 5.1 has a ground relay connected in the CT circuit which has a burden of 4.0 VA at tap value. The taps available are 0.5, 0.6, 0.8, 1.0, 1.5, 2.0, and 2.5, which represent the minimum pickup current. What is the maximum sensitivity that can be obtained in primary amperes for a phase-a-to-ground fault? Assume that $I_b = I_c = 0$ for the fault and that the phase relay burden is 0.032 ohm if 50:5 CT ratio is used, 0.128 ohm with 100:5 tap, 0.261 ohm with 150:5 tap, and 0.512 ohm with 200:5 tap.

5.5 Phase and ground relays are connected to a set of voltage transformers (VTs) as shown in Fig. P5.5. The secondary winding voltages are 69.5 for the phase relays and 120 volts for the ground relays. The equivalent line-to-neutral burden of the phase relays is 25 VA resistive each phase at 69.5 V. The burden of the ground relays is 15 VA, 120 V at 25° leading PF angle.

a. Calculate the total burden on each of three voltage transformers during a phase-a-to-ground fault which reduces the phase a voltage to 0.15 per unit.

b. What is the minimum-capacity voltage transformer that can be used for this application?

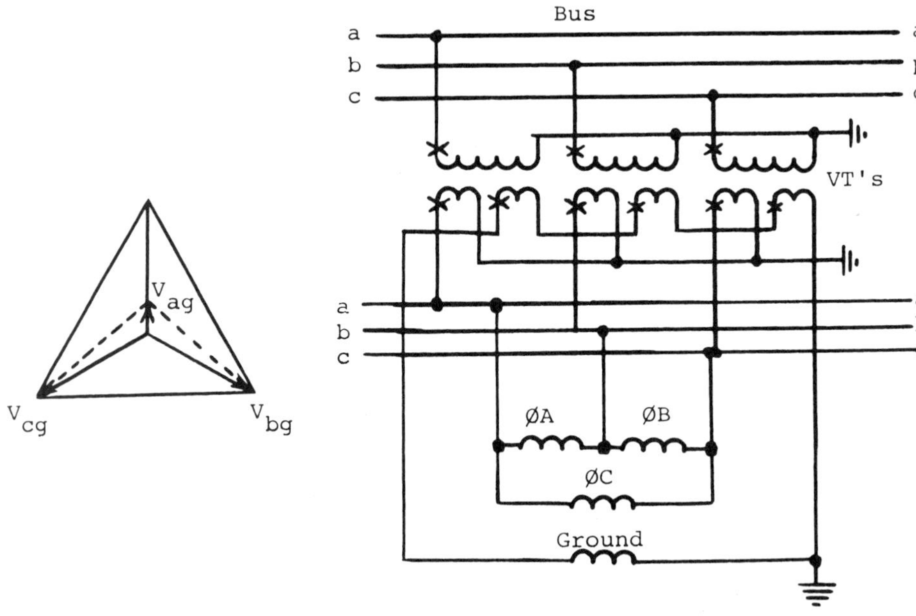

Fig. P5.5

CHAPTER 7

7.1 Phase a of a three-phase 4.16 kV ungrounded system is solidly grounded. For this fault, calculate the magnitude of the positive, negative, and zero sequence voltages at the fault. Explain your answers with reference to the sequence networks and interconnections used to calculate line-to-ground faults on three-phase systems.

7.2 An ungrounded 4.16 kV system has a capacitance to ground of 0.4 microfarads per phase. In this system:

a. Calculate the normal charging current in amperes per phase.
b. Calculate the fault current for a phase-a-to-ground fault.
c. Will this fault current operate a ground overcurrent relay set at 0.5 ampere pickup and connected to 100:5 current transformers, or a ground sensor connected to a toroidal type CT with a primary pickup of 5 amp?
d. It has been decided to ground this system with a zig-zag transformer and neutral resistor. The source to the 4.16 kV bus has $X_1 = 10\%$ on 5000 kVA. If this system X_1 is 2.4% on the zig-zag bank rating, what is the kVA of the zig-zag bank?
e. In order to limit the overvoltage on the unfaulted phases to a maximum of 250% for possible restriking ground faults, it is necessary that

$$\frac{X_0}{X_1} \leqq 20 \quad \text{and} \quad \frac{R_0}{X_0} \geqq 2.0$$

This requires a zig-zag transformer reactance of 6.67% and a ground resistor of 0.292 + j 0.124 per unit, all on the zig-zag transformer rating. Verify that these ratios requirements have been met.

f. Calculate the solid phase-a-to-ground fault current in the 4.16 kV system with the zig-zag transformer and resistor grounding.
g. Provide specifications for purchasing the zig-zag transformer and resistor.
h. Will the relays in part c operate for ground faults of part f?

7.3 Verify that the unusual connection of voltage transformers shown on p. 510 provides zero sequence voltage during a ground fault on an ungrounded system for operation of the overvoltage relay 59G. (This connection is in use by a large power system for ground detection on ungrounded systems supplied from delta tertiaries.)

Open-delta connected VTs are used for three-phase voltage, and the desire was to utilize these with minimum additions. The Fig. 7.5a connections using three auxiliary VTs in broken delta had a tendency in their system to go into ferro-resonance, probably because the resistor was not effective with the high leakage impedance of the auxiliaries. The scheme shown was evolved using the two existing VTs and adding a VT connected phase to ground. This added VT is either a 60 Hz rated twice line-to-line voltage unit or a 25 Hz unit rated line-to-line so that they operate low on the saturation curve. No ferro-resonance has been encountered with this scheme.

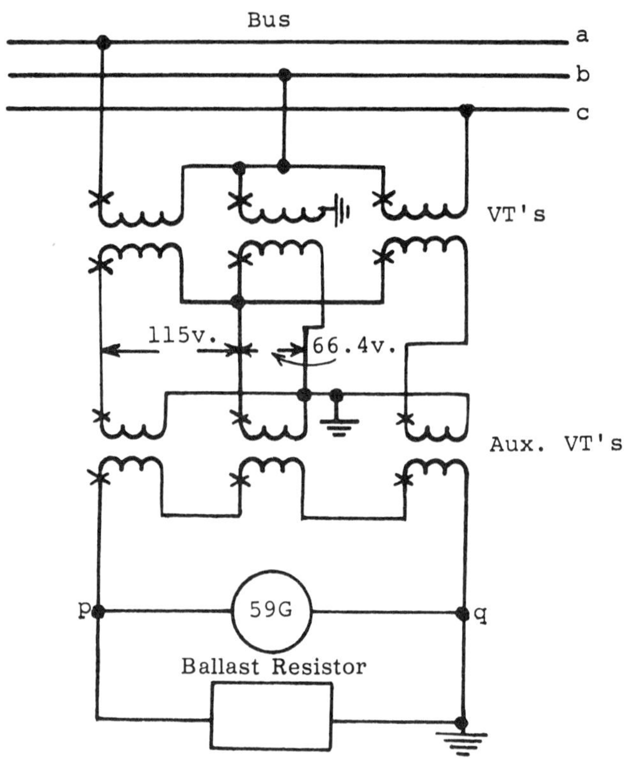

Fig. P7.3

7.4 a. In order to limit ground faults, a reactor is to be connected in the grounded neutral of the 13.8 kV winding of the transformer (Fig. P7.4). Calculate the value of the reactor in ohms required to limit the solid single-phase-to-ground current on the 13.8 kV side to 4000 amp.

b. What percentage reduction does this represent if the wye winding were solidly grounded instead of through the reactor?

c. Repeat part a above except use a resistor instead of a reactor. Determine the resistor value in ohms.

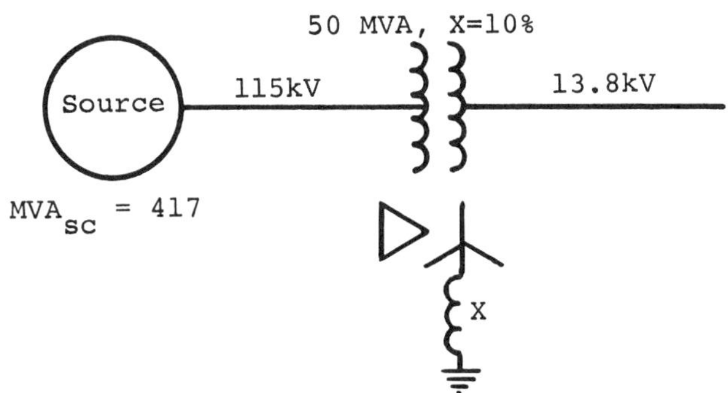

Fig. P7.4

7.5 The directional ground relay has been connected (p. 512) for operation on ground faults out on the line. The relay has maximum torque when the current lags the voltage by 60°, with the relative instantaneous polarities as shown.

a. Are the connections correct? Check by assuming a line-to-ground fault in the tripping direction. Make any corrections as required.

b. With the correct connections of a, field checks are to be made to verify the connections. Assuming 100% PF load, determine whether these checks provide relay-directional unit operation or not. Support your answer with a phasor diagram: Test A — Short phase c current transformer and open the secondary lead. Open phase a voltage transformer lead and short the secondary winding of that transformer. Restore connections after test; Test B — Short phase b current transformer and open the secondary lead. Open phase c voltage transformer lead and short the secondary winding of that transformer. Restore connections after test.

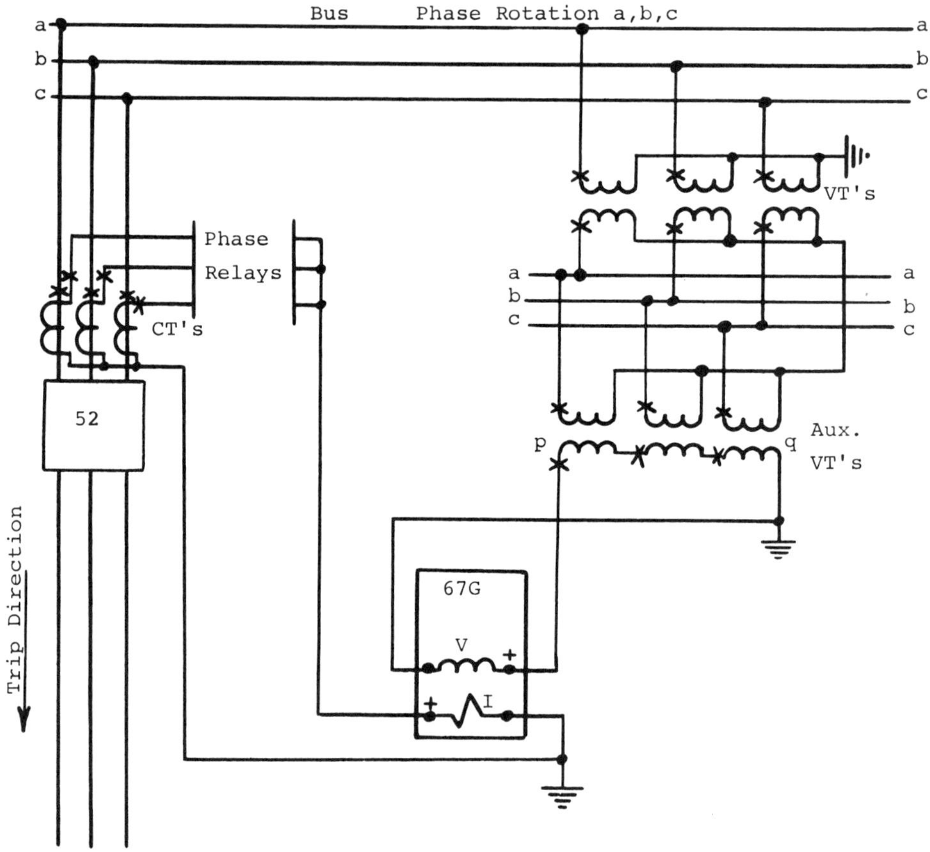

Bus
Phase Rotation a,b,c
a
b
c
a
b
c
VT's
Phase
Relays
CT's
a
b
c
a
b
c
52
p
q
Aux.
VT's
Trip Direction
67G
V
I
+
+

Fig. P7.5

CHAPTER 8

8.1 Three 21,875 kVA, 13.8 kV generators with $X_d'' = 13.9\%$ are connected to individual buses, from which various loads are supplied. These buses are connected to another bus through 0.25 ohm reactors as shown in Fig. P8.1. The generators are all ungrounded. In this system:

a. Calculate a three-phase fault at the terminals of one of the generators.

b. Choose a current transformer ratio for differential relays to protect the generators. If the generator differential relays have a minimum pickup of 0.14 ampere, how many times pickup does the three-phase fault provide?

c. Calculate a single-phase-to-ground fault at the terminals of one of the generators.

d. Will this ground fault operate the generator differential relays? If so, how many times pickup will the ground fault provide?

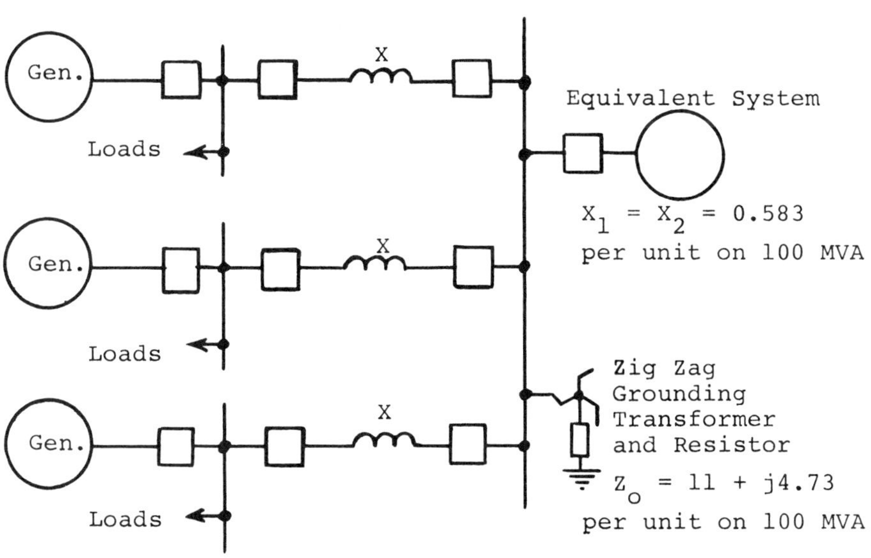

Fig. P8.1

8.2 The unit generator shown in Fig. P8.2 has the following capacitance-to-ground values in microfarads per phase:

Generator windings	0.24
Generator-surge capacitors	0.25
Generator-to-transformer leads	0.004
Power transformer low-voltage windings	0.03
Station-service-transformer high-voltage windings	0.004
Voltage-transformer windings	0.0005

The ground resistor R has a 64.14 kW rating at 138 volts.

a. Determine the fault current magnitude for a single line-to-ground fault between the generator and the power transformer.
b. Determine the three-phase fault current magnitude for a fault between the generator and the power transformer.
c. Choose a CT ratio for the generator differential protection. Compare the fault currents of parts a and b with the generator relay pick-up value of 0.14 amp.
d. How much voltage is available to operate an overvoltage relay 59G when connected across the grounding resistor? What is the multiple of pick-up if 59G minimum operating value is 5.4 volts.
e. How much current flows through the resistor? Select a CT and suggested overcurrent pickup values for the 50/51 relay.

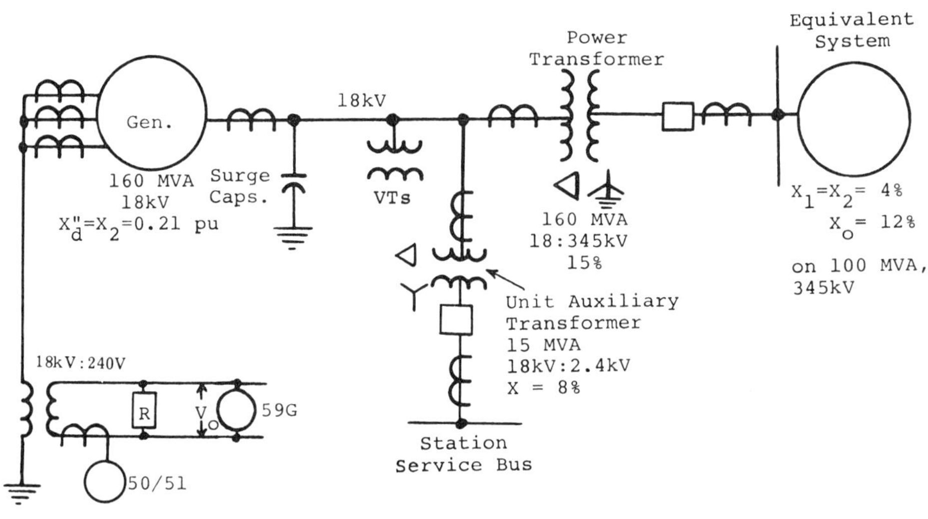

Fig. P8.2

8.3 The per unit kVA capability and steady state stability curves at rated terminal voltage for a 50 MVA, 13.2 kV, 60 Hz generator are shown in the diagram in Fig. P8.3. The current transformers used are 3000:5. For loss-of-excitation protection:

a. Translate the steady state stability limit to a per unit R-X diagram for a terminal voltage of 1.0 per unit.

b. Translate the 15 psi capability curve to a per unit R-X diagram for a terminal voltage of 1.0 per unit.

c. With these limits plotted on an R-X diagram, draw a distance relay-offset mho circle to provide protection for low or loss of excitation on this machine.

d. For the relay mho circle selected in part c, determine the per unit offset (distance of the circle center from the R-X origin) and the per-unit circle radius. Translate these values to relay ohms for setting a loss-of-excitation relay. R_c = 3000:5. R_V = 120.

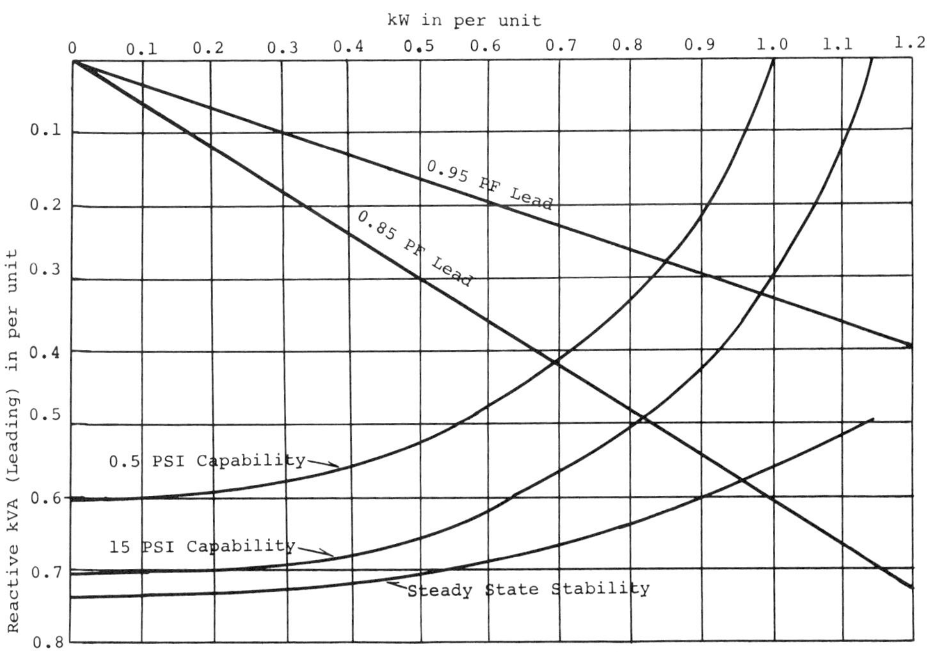

Fig. P8.3

CHAPTER 9

9.1 Assume for this problem that the 69 kV system (Fig. P9.1) is open and make the following calculation:

a. Calculate the fault current in the three phases for a solid phase-a-to-ground fault on the 69 kV terminals.

b. Calculate the three phase voltages existing at the fault.

c. For this 69 kV ground fault, determine the currents flowing in the 13.8 kV system.

d. What are the voltages for the three phases at the 13.8 kV transformer terminals for the 69 kV fault?

e. Compare the current and voltage phasors on the two sides of the bank for the 69 kV ground fault.

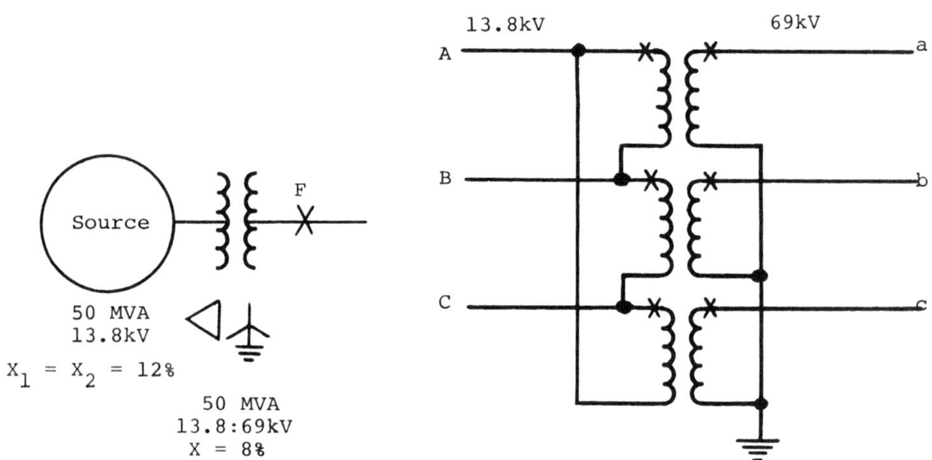

Fig. P9.1

9.2 For the transformer bank of Problem 9.1, assume that phases A, B, C on the 13.8 kV side have 3000:5 CTs with taps at 1500, 2000, 2200, and 2500 amp, and that the 69 kV circuits a,b,c have 600:5 multiratio CTs with taps, as indicated in Fig. 5.10.

a. Show the three-phase connections for transformer differential relays to protect this bank.

b. Select suitable 69 kV and 13.8 kV CTs ratios for this transformer differential application.

c. If the differential relay has taps of 4, 5, 6, and 8, select two taps to be used with the CT ratios selected in part b so that the percent mismatch is less than 10%.

d. With this application and setting, how much current can flow to operate the differential relay(s) if the phase-a-to-ground fault of Problem 9.1 part a is within the differential zone? How many of the three relays operate for this ground fault?

9.3 The transformer bank (Fig. P9.3) shown connected between the 13.8 kV and 2.4 kV buses, consists of three single-phase units, each rated 1000 kVA, 13.8:2.4 – 1.39 kV.

a. Connect a two-restraint type differential relay for protection of the transformer bank.
b. Select proper current transformer ratios and relay taps. Assume the differential relay has ratio adjusting taps of 5:5 to 5:10 with ratios of 1, 1.1, 1.3, 1.5, 1.6, 1.8 and 2.0. The CTs on the 13.8 kV breaker are 200:5 with 150, 100, and 50:5 taps, and on the 2.4 kV breaker; 2000/1500/1000/500:5 CTs.
c. If one of the single-phase transformers is damaged, can service be continued with the remaining two banks? If so, show the connections, including any modifications required for the differential relaying.
d. What is the maximum three-phase load that can be carried with any temporary connections?

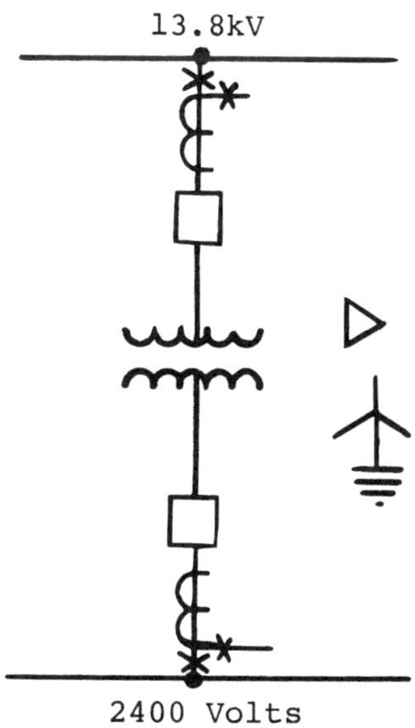

Fig. P9.3

9.4 A 50 MVA transformer bank(Fig. P9.4), wye-grounded to a 115 kV bus, and delta to a 13.8 kV bus, supplies power to the 13.8 kV system. Transformer breakers are available on both sides of the bank with 300:5 (115 kV side) and 2200:5 (13.8 kV side) current transformers. In order to ground the 13.8 kV system, a 1200 kVA zig-zag transformer has been connected between the power transformer and the 13.8 kV bus and within the differential zone. For this arrangement as shown:

a. Connect three two-restraint type transformer differential relays to protect the 50 MVA bank utilizing the two sets of CTs on the breakers. Only these are available.

b. The system $X_1 = X_2$ reactance to the 13.8 kV bus is 13% on 50 MVA, and the zig-zag bank reactance is 6% on its rating base. Calculate the current for a solid single phase-to-ground fault on the 13.8 kV system. If the transformer differential relays have a pickup of 1.8 amperes, will they operate for a ground fault within the differential zone? What would you recommend for protection of the zig-zag bank?

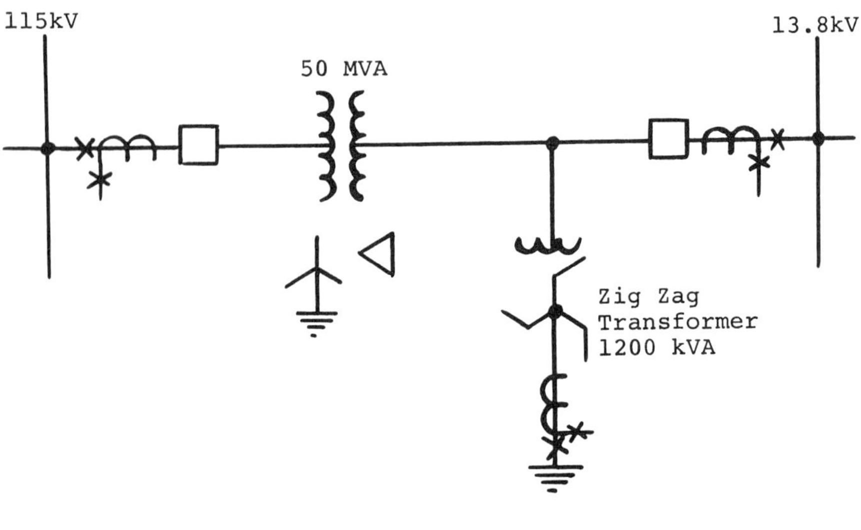

Fig. P9.4

9.5 Two separate transformer banks are connected as shown (Fig. P9.5) without high-side breakers for economy. High-side transformer CTs are not available. The banks are connected per ANSI Standards. For this arrangement:

a. Show complete three-phase connections for protecting these two transformer banks using three three-winding type transformer differential relays and the three sets of CTs shown.
b. Discuss the advantages and disadvantages of this protection compared with separate transformer differentials if separate 115 kV transformer CTs had been available.

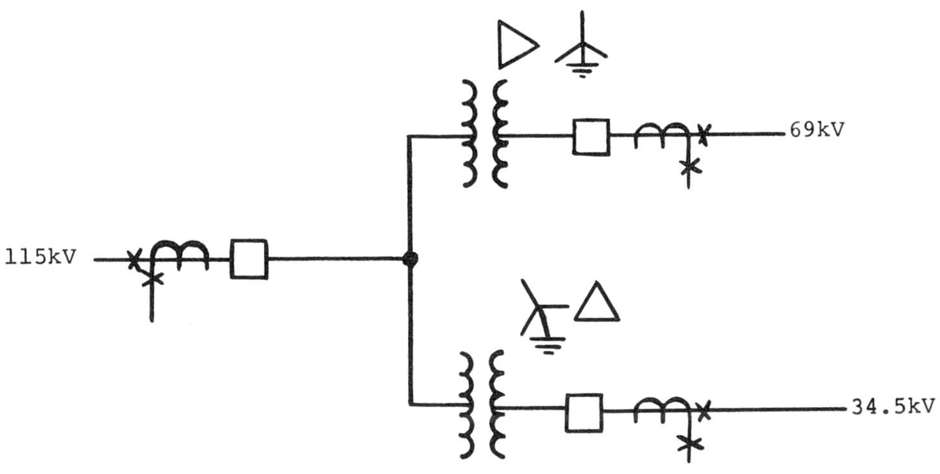

Fig. P9.5

9.6 For the application shown in Fig. 9.12 and 9.13, determine the currents that will flow in the relays for an 800-ampere ground fault. The neutral CT ratio is 250:5 and the line CT ratios are 1600:5. In the following determinations, choose a value of "n" to provide a good level of current in the 87G relay windings:

a. For the ground fault external to the differential protection zone.
b. For the ground fault internal and within the differential protection zone. Assume the low voltage feeders supply zero current to the internal fault.

9.7 A 1 MVA transformer bank, 13.8 kV delta, 480 V wye, solidly grounded with X = 5.75%, supplies a group of induction motors. The source $X_1 = X_2$ is 0.0355 per unit on 5 MVA, 13.8 kV. 13.8 kV, 65 ampere fuses are used to protect the transformer bank and the 480 V bus. In order to determine the clearing time of the fuses for 480 V arcing faults, determine the following:

a. What is the maximum possible ground-fault current at the 480 V bus?
b. With a typical arc voltage of 150 V essentially independent of current magnitude, determine the magnitude of the arcing fault at the 480 V bus.
c. What is the magnitude of this arcing fault on the 13.8 kV primary?
d. Estimate the total clearing time for the 13.8 kV, 65 ampere fuses used in the primary supply to the bank for the 480 V arcing fault. The total clearing time for these fuses are as follows:

Amperes	Total clearing time (sec)
150	500
175	175
200	115
250	40
300	20
350	9
400	6

CHAPTER 10

10.1 High-impedance voltage-differential relays are to be applied to protect a three-breaker bus, as shown in Fig. 10.9. The CTs are all 600:5 multiratio type with characteristics per Fig. 5.10. For this application, determine the relay-pickup setting voltage and the minimum primary-fault current for which the relays will operate. The maximum external fault is 8000 amp rms. Assume the lead resistance $R_L = 0.510$ ohms for the maximum resistance from any CT to the junction point.

For the particular relays applied, the pickup setting voltage is:

$$V_R = 1.6k(R_S + pR_L)\ \frac{I_F}{N}\ V \tag{10-3}$$

where 1.6 is a margin factor, k a CT performance factor (assume k = 0.7 for this problem), p = 1 for three-phase faults and p = 2 for single-phase-to-ground faults (Figure 5.9 reference), I_F the primary rms external maximum fault current, and N the CT ratio. R_S is the CT resistance. p = 2 should be used to determine the value of V_R setting.

The maximum setting of the relay voltage element should not exceed 0.67 times the secondary exciting voltage of the poorest CT in the differential circuit at 10 amp exciting current.

The minimum internal fault primary current to operate the relays is

$$I_{Min} = (nI_e + I_R + I_T)N \text{ primary amperes} \tag{10-4}$$

where n is the number of circuits, I_e the exciting current of the individual CT at the pickup voltage, I_R the relay current at the pickup setting voltage, and I_T the current required by a high voltage protective device across the relay coil (not shown in Fig. 10.9). For this problem, assume I_T = 0.02 ampere. The relay impedance and generally negligible resistance of the leads from the junction to the relay is 1700 ohms. nI_e is applicable in this problem since all three breaker CTs are the same, otherwise this is a summation of the different CT exciting currents at the V_R pickup voltage.

10.2 A feeder circuit is added to the bus of Problem 10.1, making a four-circuit bus. The new breaker has the same type 600:5 multiratio CTs. With this addition, the maximum external fault increases to 10,000 amperes rms. All other circuit values remain the same. For this change, calculate the relay-pickup setting voltage and the minimum primary-fault current for which the relays will operate.

CHAPTER 11

11.1 A 2850 hp, 4 kV induction motor is connected to the supply system through a 2.5 MVA transformer, 13.8:4 kV with a reactance of 5.6%. The motor full-load current is 362 amp and its locked rotor current is 1970 amp. The supply system short circuit MVA at the 13.8 kV terminals of the transformer is 431 maximum, 113 minimum, on 100 MVA base. Determine if a phase instantaneous overcurrent relay can be applied if it is set at half the minimum fault current and twice the locked rotor current.

11.2 Review the application of Problem 11.1 if a time delayed instantaneous unit is applied and set at 1.1 times locked-rotor current.

11.3 Another feeder is supplied by the same source as in problem 11.1 through a 2.5 MVA, 13.8:2.4 kV transformer with 5.88% reactance. The largest motor connected to this bank is rated at 1500 HP, 2.3 kV, with a full load current of 330 amp, locked rotor current of 2213.5 amp. Can an instantaneous phase overcurrent relay be applied set at half the minimum fault current and twice the locked-rotor current?

11.4 The same source supplies a 460 V feeder through a 2 MVA transformer, 13.8 kV:480 V transformer with 5.75% reactance. The largest motor on this feeder is 125 hp, 460 V with 90.6 amp full-load, 961 amp locked-rotor current. Can a phase-instantaneous overcurrent be applied if set at half the minimum-fault current and twice the locked-rotor current?

11.5 In the system shown on p. 523:
   a. Calculate the fault currents flowing for a solid three-phase fault on the 4160 volt bus. For this problem consider the 500 hp induction motor as one of the sources.
   b. What percent of the fault current does this induction and each of the two synchronous motors supply?
   c. Calculate the current flowing for a solid single-line-to-ground fault on the 4.16 kV bus.
   d. Select CT ratios and instantaneous overcurrent relay settings for protecting the motors for both phase and ground faults.

11.6 A fully loaded motor is connected to a supply source through a transformer as shown in Fig. P11.6. The phase sequence is different on the two sides. Assume that the positive sequence current into the motor does not change after the fuse operations.
   a. For phase b fuse open on the source side, plot the sequence and total currents existing on both sides of the transformer. With one per unit positive sequence current, determine the magnitudes of the phase currents on both sides.
   b. Repeat part 1 with all source side fuses in service but with phase A fuse on the motor side open.
   c. What effect does grounding the transformer neutral have?

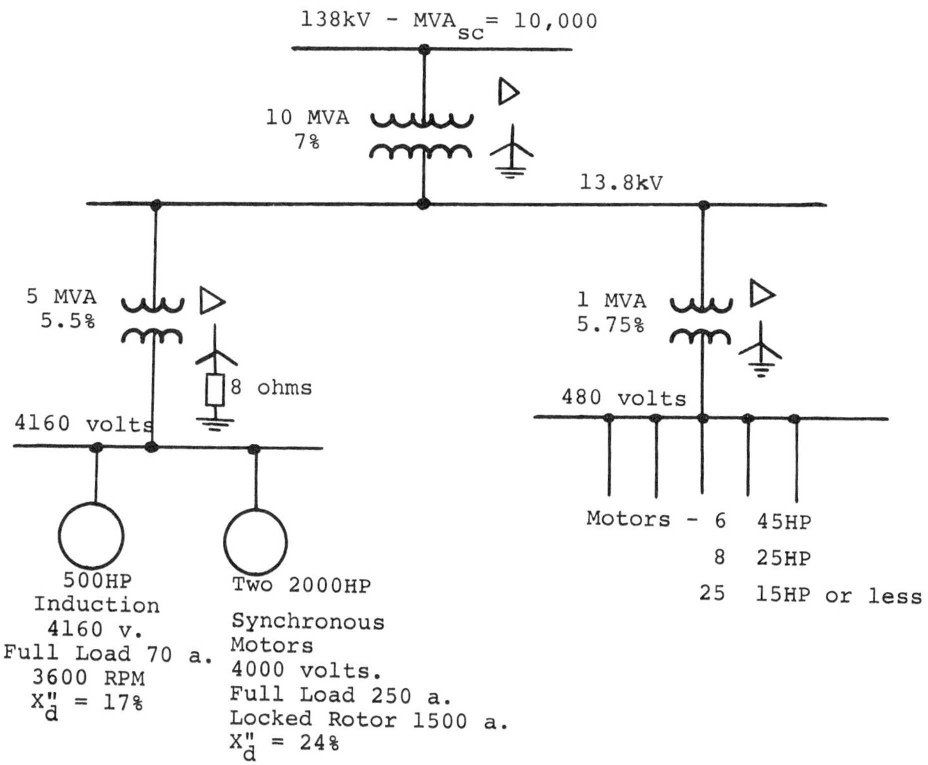

Fig. P11.5

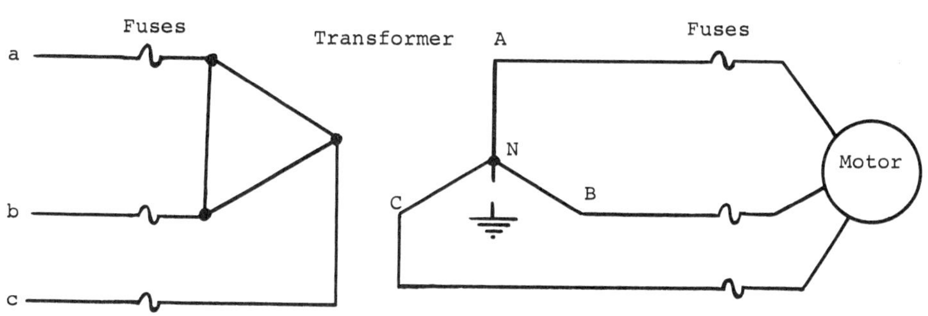

Fig. P11.6

CHAPTER 12

12.1 The 12.5 kV distribution feeder (Fig. P12.1) has two taps. One is protected by three oil circuit reclosers with 70/140 amp coils set as in Table 1. The other tap is a single phase circuit protected by one 30 amp fuse operating as shown in Table 2. The data for the 46 kV fuse is in Table 3. The phase and ground relays are very inverse time overcurrent with instantaneous units. Their time-overcurrent characteristics are shown in the typical curves of P12.11, p. 531. Fault currents are in amperes at 12.5 kV.

a. Determine the 46 kV fuse time-current characteristic in terms of 12.5 kV amperes for 12.5 kV three-phase, phase-to-phase and phase-to-ground faults. Draw these high side fuse curves along with the recloser and 30 amp fuse curves on time-current log paper, such as K & E 48 5257, with 12.5 kV amperes as the abscissa and time in seconds as the ordinate.

Table 1. Circuit reclosers

Current (amp)	Time (sec)
140	20
185	10
200	7.5
275	5
320	4
400	3
480	2
600	1
650	0.8
720	0.7
800	0.6
900	0.5
1200	0.4
1600	0.3
2200	0.25

Table 2. 30-Ampere fuse

Approximate by a 120° line passing through 1000 amp at 0.06 sec for the minimum melt curve, and through 0.11 sec for the maximum clearing curve.

Table 3. 65-Ampere fuse, minimum melt

46 kV (amp)	Time (sec)
130	300
260	10
500	1
1500	0.1

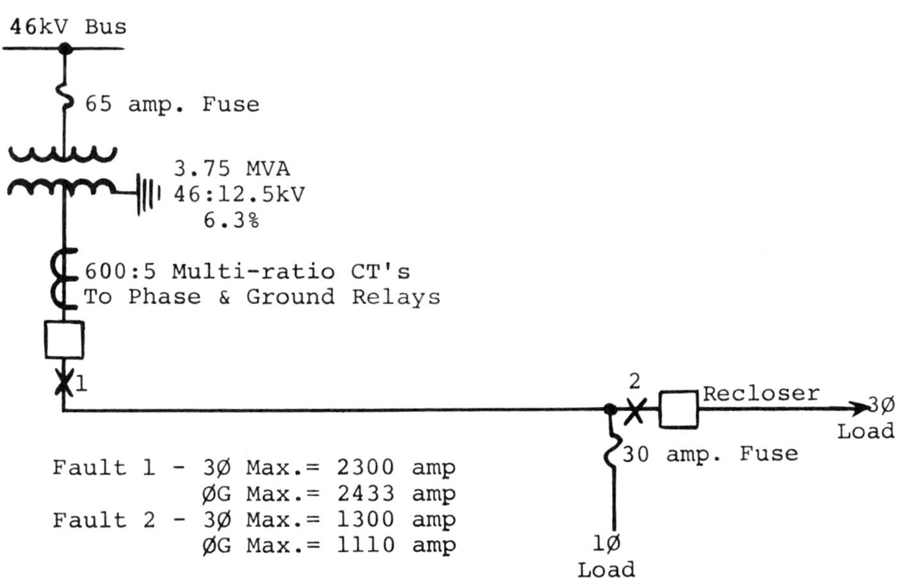

Fig. P12.1

b. Select a suitable ratio for the current transformers to the phase and ground relays.

c. Set and coordinate the phase and ground relays. Provide a minimum of 0.2 sec coordination interval between the recloser and the relays, and a minimum of 0.5 sec between the 46 kV fuse and the relays. Specify the time-overcurrent relay tap selected (available taps are 1-1.2-1.5-2-2.5-3-3.5-4-5-6-7-8-10), the time dial, and the instantaneous current pick-up for both phase and ground relays. Plot the coordination on the curve of part 1.

12.2 In the loop system of Fig. 12.4, set and coordinate the phase overcurrent type relays around the loop in the counterclockwise direction for breakers 4, 6, and 9. Use criteria and the settings for the other relays involved from the example in the text.

12.3 Apply and set phase instantaneous relays where they are applicable for the breakers 4, 6, and 9 in the system of Fig. 12.4.

12.4 A 12/16/20 MVA transformer is connected to a 115 kV source through a high side 125E fuse and through a low side recloser to supply a 12.5 kV feeder. The transformer is delta-connected on the high side and solidly wye-grounded on the low side.

The total reactance to the 12.5 kV bus is $X_1 = X_2 = 0.63$ per unit, $X_0 = 0.60$ per unit on 100 MVA. The lines from the 12.5 kV bus have a positive sequence impedance of 0.82 ohm per mile, and a zero sequence impedance of 2.51 ohms per mile. Ignore the line angle in this problem.

As a result of a problem it is necessary to operate temporarily with the low side recloser bypassed. Determine how many miles out on the line that can be protected by the high side fuse for solid line-to-ground faults. The minimum current to open the fuse is 300 amp.

12.5 a. Apply and set distance-type relays at Stations H and R for the protection of line HR in the system on p. 527. Set zone 1 units for 90% of the protected line, zone 2 to reach 50% into the next line section beyond the protected line, and zone 3 for 120% of the next line section.

b. Plot this system on an R-X diagram with the origin at bus H. Plot the relay settings of part 1 using mho type characteristics. The mathematical formula for a circle through the origin or relay location is where Z_S is the relay setting at 75°,

$$Z = \frac{1}{2}(Z_s - Z_s \angle \phi)$$

The first term is the offset from the origin at 75° and the second term the radius. Thus when ϕ is 75°, $Z = 0$, the relay location; when ϕ is 255°, $Z = Z_S$ the forward reach.

c. What is the maximum load in MVA at 87% P.F. that can be carried over line HR without the distance relays operating? Assume the voltage transformer ratio $R_V = 1000$ and the current transformer ratio is $R_C = 80$.

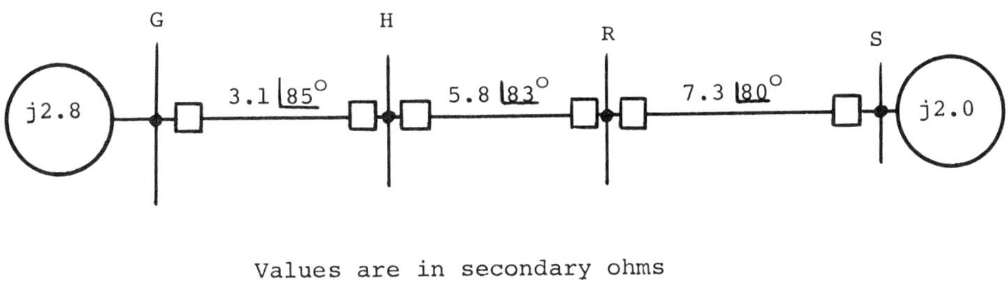

Fig. P12.5

12.6 a. Apply and set distance relays for line HR as in Problem 12.5 except set the zone 3 unit in the reverse direction to reach 150% of the line section behind the relay.

b. Plot these settings (zone 1 and 2 as in Problem 12.5) and zone 3 as above on the R-X diagram with the origin at bus H.

c. For this application, what is the maximum load in MVA at 87% PF that can be carried over line HR without the distance relays operating? $R_V = 1000$ and $R_C = 80$.

12.7 The line impedance values for the system on p. 528 are in percent on a 100 MVA, 161 kV base. The fault values are in MVA at 161 kV for three-phase faults at the buses as indicated. The first value is for maximum conditions, and the second for minimum conditions.

a. The zone 2 distance relay at station M is set for 70% impedance reach for the protection of line MS and into the lines SL and SP. The zone 3 distance relay is set for 100% impedance also into the line MS and the lines SL and SP. Determine the apparent impedance seen by these units at M under the maximum and minimum operation.

b. What percentage of the lines SL and SP are protected during these two operating conditions?

c. Determine the maximum load in MVA at 87% P.F. that can be transmitted over line MS without operating the distance relays set as in part 2. Assume that the voltage transformer ratio R_V = 1400 and the current transfromer ratio R_C = 100. Assume that the distance relay mho characteristic has a circle angle of 75°.

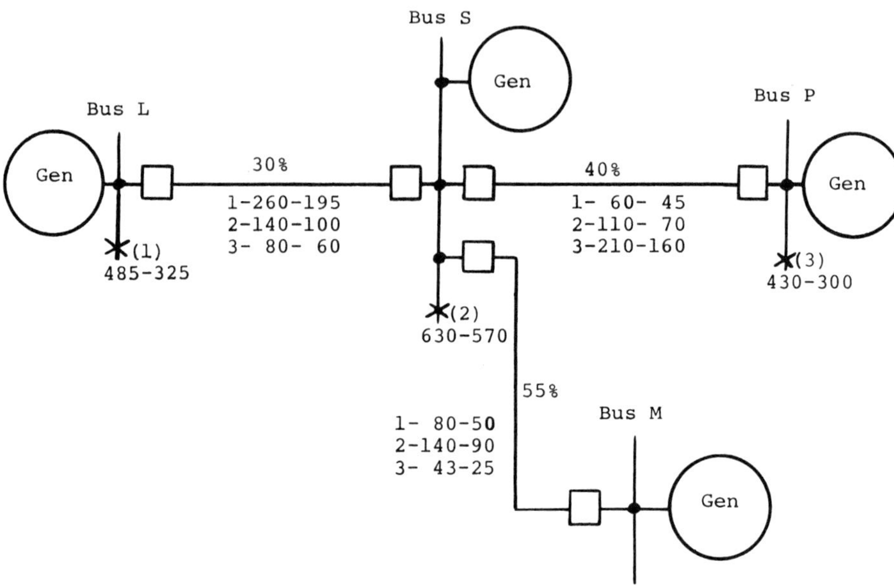

Fig. P12.7

12.8 The 60 mile 115 kV line GH (p. 529) is operating with the voltages at each end 30° out of phase when a three-phase fault occurs at 80% of the distance from bus G. This fault has 12 ohms arc resistance. The currents flowing to the fault are as shown and are in per unit at 100 MVA, 115 kV.

a. Determine the apparent impedance seen by the distance relays at G for this fault.

b. Determine if the zone 1 mho unit at G set for 90% of the line GH can operate on this fault. Assume the angle of the mho characteristic (Fig. 6.11b) is 75°.

c. Determine the apparent impedance seen by the distance relays at H for this fault.

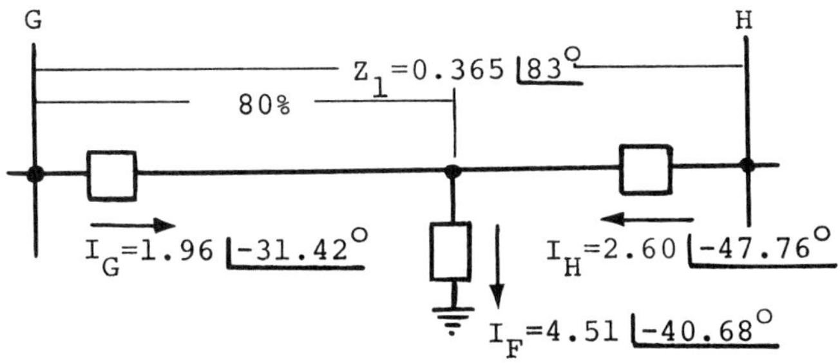

All values in per unit on 100 MVA, 115kV.

Fig. P12.8

d. Determine if the zone 1 mho unit at H set for 90% of the line GH can operate for this fault. Assume the angle of the mho characteristic is 75°.

e. Describe how this three-phase fault can be cleared by the line distance relays.

12.9 The 40 MVA transformer bank (Fig. P12.9) has tap changing under load (TCUL) with low voltage ±10% taps. The reactances at the high, mid, and low voltage taps are 7.6% at 38 kV, 8% at 34.5 kV and 8.5% at 31 kV respectively. This bank is connected directly to a 115 kV transmission line without a high side breaker. There are no 115 kV voltage or current transformers available at G. In order to provide phase distance line protection, the relays must be set to look through the transformer into the line. Assume that any phase shift through

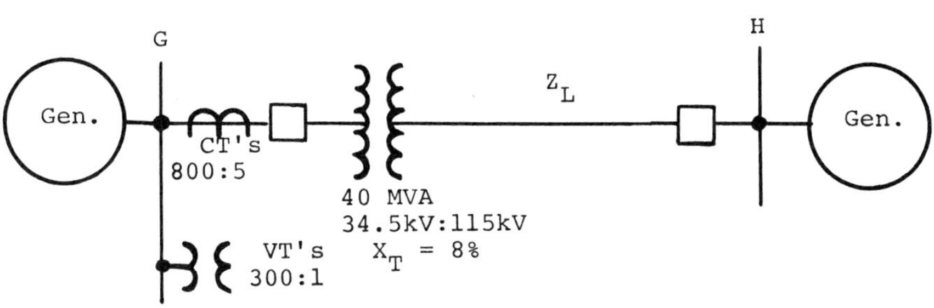

Fig. P12.9

the transformer does not change the relay reach either by the connections or by relay design.

a. Set Zone 1 phase distance relays at G for a 12 mile line GH where $Z_L = 10 \angle 80°$ ohms. Note that it is necessary to determine which transformer bank tap gives the lowest value of ohms to Bus H as viewed from Bus G to prevent the relays from overreaching Bus H as taps are changed. Set Zone 1 for 99% X_T + 90% Z_L.
b. With this setting of part 1, what percent of the line is protected by Zone 1?
c. What percent of the line will be protected when the other taps are in service with the setting of part 1?
d. In view of the above analysis, what recommendations would you make for line protection?

12.10 Repeat Problem 12.9 but with a 50 mile 115 kV line where Z_L = 40 $\angle 80°$ ohms. Compare the protection for the 12 mile line of Problem 12.9 with the protection for the 50 mile line.

12.11 Ground directional overcurrent relays are to be applied to the 69 kV and 138 kV breakers for the protection of the 138 kV line which includes the auto transformer as shown. To determine the

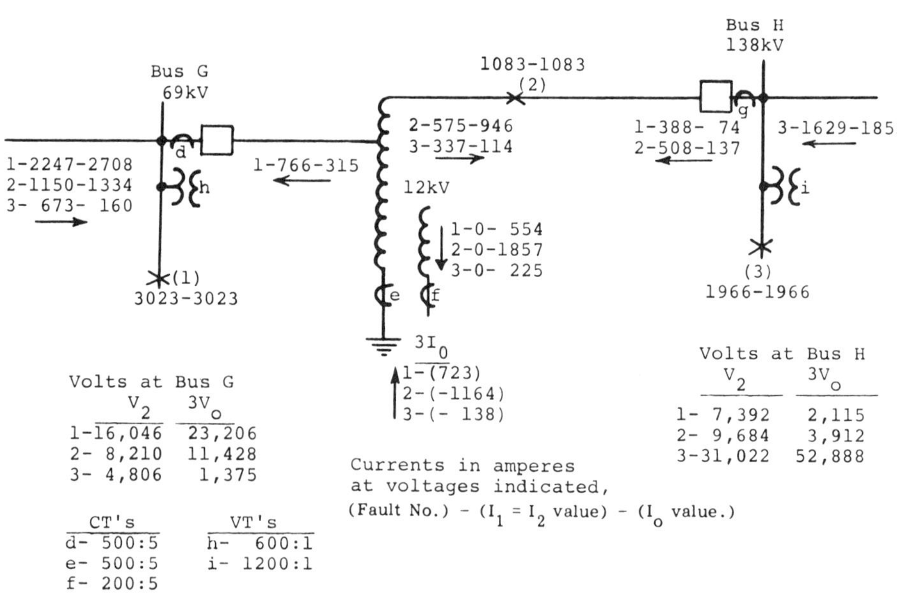

Fig. P12.11

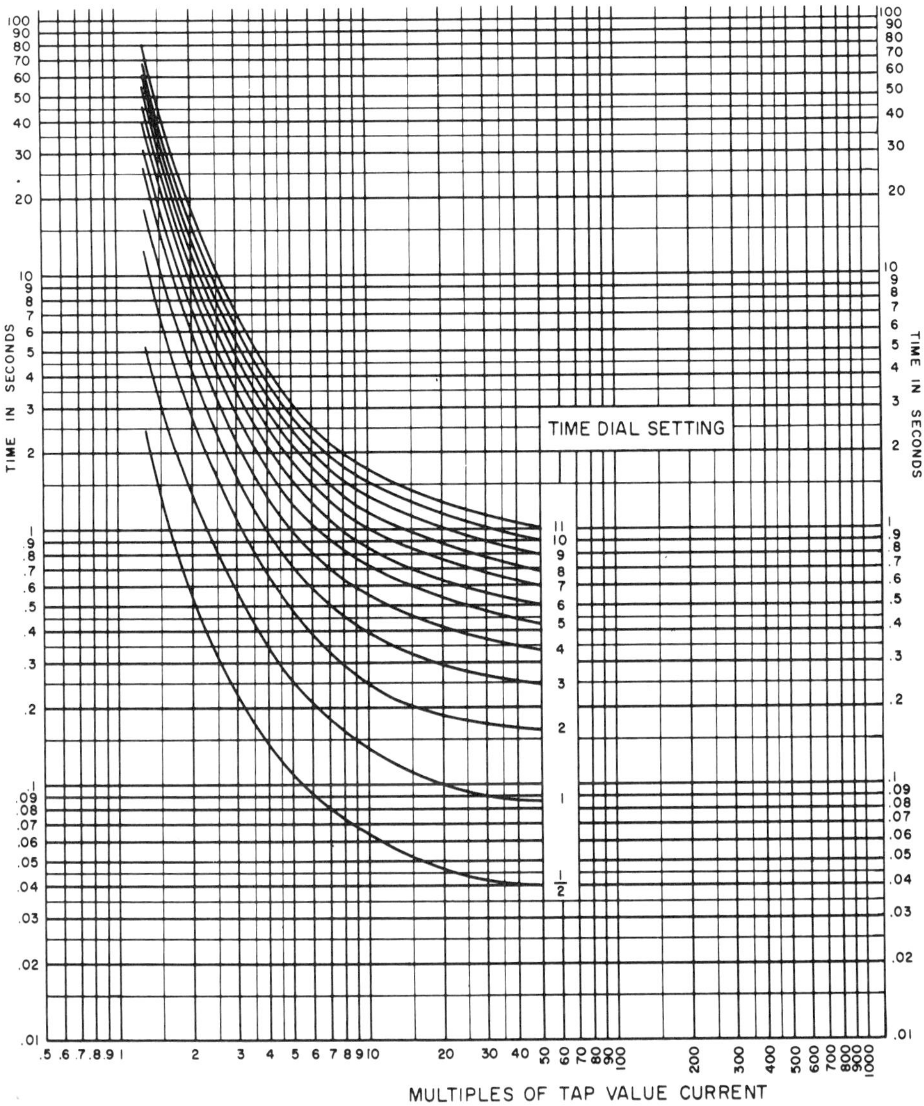

Fig. P 12.11 (Courtesy of Westinghouse Electric Corp.)

best method of directional sensing fault $I_1 = I_2$ and I_0 currents and V_2 and V_0 voltages are indicated for the three different line-to-ground faults.

a. Determine the secondary (relay) quantities that could be used to polarize and operate ground relays at both the G and H terminals.
b. Make recommendations for the preferred method to polarize and operate the ground relays at G and H.

CHAPTER 14

14.1 For the system of Problem P12.5:

a. Draw the locus of the surge ohms seen by the relays at H and R as the generators at the two ends of the system slip a pole. Assume that the two generator voltages remain equal in magnitude throughout the swing. Locate the 60°, 90°, 120°, 180°, 240°, 270°, and 300° points.
b. What is the magnitude of impedance as seen from bus H and from bus R for a 120° swing?
c. With the distance settings applied in Problem 12.5, determine which distance relays will operate on the swing and at what swing angle this will occur.
d. Repeat part 3 but with the distance settings as applied in Problem 12.6.

Index

Number(s) in parentheses are relay device numbers.